SCHOLAR Study Guide
Advanced Higher Mathe
Course materials
Part 2: Topics 5 to 8

Authored by:
Fiona Withey (Stirling High School)
Karen Withey (Stirling High School)

Reviewed by:
Margaret Ferguson

Previously authored by:
Jane S Paterson
Dorothy A Watson

Heriot-Watt University
Edinburgh EH14 4AS, United Kingdom.

First published 2019 by Heriot-Watt University.

This edition published in 2019 by Heriot-Watt University SCHOLAR.

Copyright © 2019 SCHOLAR Forum.

Members of the SCHOLAR Forum may reproduce this publication in whole or in part for educational purposes within their establishment providing that no profit accrues at any stage, Any other use of the materials is governed by the general copyright statement that follows.

All rights reserved. No part of this publication may be reproduced, stored in a retrieval system or transmitted in any form or by any means, without written permission from the publisher.

Heriot-Watt University accepts no responsibility or liability whatsoever with regard to the information contained in this study guide.

Distributed by the SCHOLAR Forum.

SCHOLAR Study Guide Advanced Higher Mathematics: Course Materials: Topics 5 to 8

Advanced Higher Mathematics Course Code: C847 77

    ISBN 978-1-911057-77-2

Print Production and Fulfilment in UK by Print Trail www.printtrail.com

Acknowledgements

Thanks are due to the members of Heriot-Watt University's SCHOLAR team who planned and created these materials, and to the many colleagues who reviewed the content.

We would like to acknowledge the assistance of the education authorities, colleges, teachers and students who contributed to the SCHOLAR programme and who evaluated these materials.

Grateful acknowledgement is made for permission to use the following material in the SCHOLAR programme:

The Scottish Qualifications Authority for permission to use Past Papers assessments.

The Scottish Government for financial support.

The content of this Study Guide is aligned to the Scottish Qualifications Authority (SQA) curriculum.

All brand names, product names, logos and related devices are used for identification purposes only and are trademarks, registered trademarks or service marks of their respective holders.

Contents

5 Binomial theorem — 1
   5.1 Looking back — 3
   5.2 Factorials — 8
   5.3 Binomial coefficients — 11
   5.4 Pascal's triangle — 17
   5.5 Binomial theorem — 20
   5.6 Finding coefficients — 27
   5.7 Sigma notation and binomial theorem applications — 29
   5.8 Learning points — 32
   5.9 Proofs — 34
   5.10 Extended information — 36
   5.11 End of topic test — 37

6 Complex numbers — 39
   6.1 Looking back — 43
   6.2 Introduction to Imaginary numbers — 53
   6.3 The arithmetic of complex numbers — 59
   6.4 The modulus, argument and polar form of a complex number — 76
   6.5 Geometric interpretations — 95
   6.6 Fundamental theorem of algebra and solving complex equations — 110
   6.7 De Moivre's theorem — 119
   6.8 Conjugate properties — 147
   6.9 Learning points — 153
   6.10 Proofs — 158
   6.11 Extended information — 161
   6.12 End of topic test — 163

7 Sequences and series — 167
   7.1 Looking back — 171
   7.2 Sequences and recurrence relations — 188
   7.3 Arithmetic and geometric sequences — 193
   7.4 Fibonacci and other sequences — 203
   7.5 Convergence and limits — 207
   7.6 Definitions of e and π as limits of sequences — 212
   7.7 Series and sums — 216
   7.8 Arithmetic and geometric series — 219
   7.9 Sums to infinity — 228
   7.10 Power series — 236
   7.11 Maclaurin series for simple functions — 237
   7.12 Maclaurin's theorem — 239

	7.13	The Maclaurin series for $\tan^{-1}(x)$	244
	7.14	Maclaurin's series expansion to a given number of terms	245
	7.15	Composite Maclaurin's series expansion	247
	7.16	Learning points	256
	7.17	Extended information	259
	7.18	End of topic test	262
8	**Curve sketching**		**267**
	8.1	Looking back	270
	8.2	Functions	296
	8.3	Inverse functions	314
	8.4	Odd, even or neither functions	326
	8.5	Critical and stationary points	333
	8.6	Derivative tests	337
	8.7	Concavity	343
	8.8	Continuity and asymptotic behaviour	349
	8.9	Sketching and rational functions	363
	8.10	Type 1 rational function: Constant over linear	368
	8.11	Type 2 rational function: Linear over linear	371
	8.12	Type 3 rational function: Constant or linear over quadratic	375
	8.13	Type 4 rational function: Quadratic over quadratic	383
	8.14	Type 5 rational function: Quadratic over linear	387
	8.15	Summary of shortcuts to sketching rational functions	391
	8.16	Graphical relationships between functions	392
	8.17	Learning points	398
	8.18	Extended Information	402
	8.19	End of topic test	403

Glossary	**410**
Hints for activities	**417**
Answers to questions and activities	**418**

Topic 5

Binomial theorem

Contents

- 5.1 Looking back 3
  - 5.1.1 Revision 4
- 5.2 Factorials 8
- 5.3 Binomial coefficients 11
- 5.4 Pascal's triangle 17
- 5.5 Binomial theorem 20
- 5.6 Finding coefficients 27
- 5.7 Sigma notation and binomial theorem applications 29
- 5.8 Learning points 32
- 5.9 Proofs 34
- 5.10 Extended information 36
- 5.11 End of topic test 37

TOPIC 5. BINOMIAL THEOREM

Learning objective

By the end of this topic, you should be able to:

- understand the notation $n!$;
- evaluate $n!$ for a value of $n \in \mathbb{N}$;
- rewrite $n!$ in terms of factorials smaller than n;
- identify what the binomial coefficient is and evaluate it;
- use the different notations for the binomial coefficient;
- determine relationships between binomial coefficients;
- construct Pascal's triangle;
- understand the connection between Pascal's triangle and the binomial coefficients;
- use the binomial theorem to expand an expression of the form $(x + y)^n$ where $x, y \in \mathbb{Q}$ and $n \in \mathbb{N}$;
- write down an expression for the general term in $(x + y)^n$;
- find the coefficient of a particular term in a given expression of the form $(ax + by)^n$ or $\left(ax + \frac{b}{y}\right)^n$;
- use the binomial theorem to evaluate a decimal to a power e.g. $1 \cdot 07^3$.

5.1 Looking back

Summary of prior knowledge
Expanding double bracket expressions: National 5

- When expanding double brackets, remember that each term in the second bracket is multiplied by each term in the first bracket by turning the expression into two single bracket expressions or using the *rainbow method* or *FOIL*.

When adding and subtracting fractions: National 5

- Determine a common denominator by multiplying the denominators together or finding the lowest common multiple of the denominators.
- Change each fraction to an equivalent fraction using the common denominator, then add/subtract the numerators - lastly, simplify the resulting fraction if possible.

Simplifying fractions: National 5

- To simplify fractions, factorise the numerator and denominator where possible, then cancel common factors of the numerator and denominator.

Multiplying fractions: National 5

- To multiply fractions, multiply the numerators together and multiply the denominators together.
- Simplify the resulting fraction if possible.

Dividing fractions: National 5

- When dividing by a fraction, take the divisor (the fraction that you are dividing by) and turn it upside down. At the same time, change the division to a multiplication.
- Now multiply the fractions and simplify the resulting fraction if possible.

5.1.1 Revision

Expanding brackets

> **Example**
>
> **Problem:**
>
> Expand and simplify $(3y + 2)(2y - 5)^2$
>
> **Solution:**
>
> To multiply out this set of brackets we multiply two brackets out and get an answer. We then take this answer and multiply it by the remaining bracket.
>
> Remember that to multiply out brackets we take each term in one bracket and multiply it by each term in the second bracket.
>
>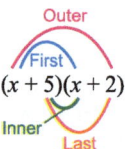
>
> When there are only two terms in each bracket this method is known as the *FOIL* method (*First Outer Inner Last*).
>
> Written out in full: $(3y + 2)(2y - 5)^2 = (3y + 2)(2y - 5)(2y - 5)$
>
> Multiply out $(2y - 5)(2y - 5)$ first: $(3y + 2)(2y - 5)^2 = (3y + 2)(4y^2 - 20y + 25)$
>
> Multiply each term in the first bracket by each term in the second bracket, then simplify:
>
> $$(3y + 2)(2y - 5)^2 = 3y\left(4y^2 - 20y + 25\right) + 2\left(4y^2 - 20y + 25\right)$$
> $$= \left(12y^3 - 60y^2 + 75y\right) + \left(8y^2 - 40y + 50\right)$$
> $$= 12y^3 - 52y^2 + 35y + 50$$

Factorising into brackets

> **Examples**
>
> **1. Problem:**
>
> Factorise $2y^2 + 5y + 2$
>
> **Solution:**
>
> We want two terms which multiply to make $2y^2$.
>
> $2y \times y = 2y^2$ so our brackets start $(2y\ \ )(y\ \ )$.
>
> Now we want two numbers which multiply to make $+2$, i.e. 2×1 or 1×2.
>
> In this example we need to check that the sum of the products of the inner and outer terms gives us the middle term $5y$.

TOPIC 5. BINOMIAL THEOREM

Multiplying the inner terms gives $2y$ and multiplying the outer terms gives $2y$ (sum $4y$), but the middle term we want is $5y$.

Multiplying the inner terms gives $1y$ and multiplying the outer terms $4y$ (sum $5y$).

This gives us the middle term that we want. So we have:

$2y^2 + 5y + 2 = (2y + 1)(y + 2)$

....

2. Problem:

Factorise $14g^2 - 20g + 6$

Solution:

We should always check for a simple common factor first. This question has a common factor of 2 giving $2(7g^2 - 10g + 3)$

Next we want two terms to make $7g^2$, $7g \times g = 7g^2$, so our brackets start $(7g\quad)(g\quad)$

Now we want two numbers which multiply to make $+3$, i.e. 3×1 or 1×3.

In this example we need to check that the sum of the products of the inner and outer terms gives the middle term $-10g$.

$(7g + 3)(g + 1)$ makes the product of the inner terms $3g$ and the outer terms $7g$ (sum $10g$); we nearly have it, but we wanted $-10g$.

The sign is wrong, but we know that -3×-1 also makes 3, so we have:

$(7g - 3)(g - 1) = 7g^2 - 10g + 3$

Hence:

$$14g^2 - 20g + 6 = 2\left(7g^2 - 10g + 3\right)$$
$$= 2\left(7g - 3\right)\left(g - 1\right)$$

Adding fractions

Example

Problem:

Add $\frac{2}{5}$ and $\frac{3}{4}$ together.

Solution:

$\frac{2}{5} + \frac{3}{4} = ?$ we need a common denominator

$= \frac{2}{5} \times \frac{4}{4} + \frac{3}{4} \times \frac{5}{5}$ multiply for a common denominator

$= \frac{8}{20} + \frac{15}{20}$ now we can add the fractions together

$= \frac{23}{20}$ change the improper fraction to a mixed number

$= 1\frac{3}{20}$

Multiplying fractions

Example
Problem:

$\frac{3}{4} \times \frac{5}{6} \times \frac{2}{7}$

Solution:

To multiply fractions:

- multiply the numerators together;
- multiply the denominators together;
- simplify the resulting fraction.

$$\frac{3}{4} \times \frac{5}{6} \times \frac{2}{7} = \frac{30}{168}$$
$$= \frac{5}{28}$$

Division

Example
Problem:

Without a calculator divide 1260 by 15 using long division.

Solution:

$15 \overline{) \begin{matrix} 8 \\ 1260 \\ 120 \end{matrix}}$ 15 does not go into 1 or 12, but it goes into 126 eight times.

$15 \overline{) \begin{matrix} 8 \\ 1260 \\ 120 \\ \hline 6 \end{matrix}}$ Multiply 15 by 8 to give 120, write that underneath 126 and then subtract.

$15 \overline{) \begin{matrix} 8 \\ 1260 \\ 120\downarrow \\ \hline 60 \end{matrix}}$ Bring the 0 of 1260 down beside the 6 to make 60.

$15 \overline{) \begin{matrix} 84 \\ 1260 \\ 120 \\ \hline 60 \end{matrix}}$ 15 goes into 60 four times.

TOPIC 5. BINOMIAL THEOREM

```
      84
15 ) 1260
     120     Multiply 15 by 4 to give 60, write that underneath the 60 already there and then
      60    subtract. The answer is 0 which means there is no remainder.
      60
       0
```

Therefore, $1260 \div 15 = 84$

The following revision practice questions should help to identify any areas of weakness in techniques which are required for the study of this topic. Some revision may be necessary if any of the questions seem difficult.

Revision exercise   Go online

Q1: Expand and simplify $(2y - 1)(5y + 3)^2$

Q2: Expand and simplify $(3y - 4)^3$

Q3: Factorise $(3g^2 + 4g + 1)$

Q4: Factorise $5h^2 + 3h - 2$

Q5: Factorise $2j^2 + 2j - 12$

Q6: Factorise $4k^2 - 2k - 2$

Q7: $\frac{4}{5} + \frac{3}{7}$

Q8: $\frac{7}{8} - \frac{2}{3}$

Q9: $\frac{1}{5} \times \frac{4}{7} \times \frac{2}{3}$

Q10: $\frac{5}{6} \times \frac{3}{8} \times \frac{2}{5}$

Q11: Without a calculator, divide 1568 by 28 using long division.

Q12: Without a calculator, divide 2720 by 32 using long division.

© HERIOT-WATT UNIVERSITY

5.2 Factorials

The Binomial Theorem is a useful tool in many branches of mathematics. One importance is that is gives an efficient way of counting combinations. For instance, with the Binomial Theorem you can quickly calculate the number of distinct ways of choosing six numbers for your lottery ticket.

As you progress through the topics you will see the Binomial Theorem appear in other seemingly unrelated topics.

In order to use the Binomial Theorem we need to understand what a factorial is, how to evaluate and manipulate it. The following section will cover this.

The definition of n!

n! (called n factorial) is the product of the integers $n, n-1, n-2, ..., 2, 1$, i.e.
$n! = n \times (n-1) \times (n-2) \times ... \times 2 \times 1 \; for \; n \in \mathbb{N}$

Examples

1.
$n = 5$
$n! = n \times (n-1) \times (n-2) \times ... \times 2 \times 1$
$5! = 5 \times (5-1) \times (5-2) \times (5-3) \times (5-4)$
$5! = 5 \times 4 \times 3 \times 2 \times 1$
$5! = 120$

....

Factorial value

Problem:

What is the value of $6!$?

Solution:

$6! = 6 \times 5 \times 4 \times 3 \times 2 \times 1 = 720$

Q13: What is $3!$?

....

Q14: What is $4!$?

Note that as n increases, $n!$ rapidly increases.

Key point

$$n! = n \times (n-1) \times (n-2) \times ... \times 3 \times 2 \times 1$$

TOPIC 5. BINOMIAL THEOREM

Factorials calculator activity

Find the largest value of n that can be entered in a calculator as $n!$ without giving an error message.

What is the value of this factorial?
Is this an accurate answer?

Factorial formula 1

Given that $n!$ is the product of the integers $n, n-1, n-2, \ldots, 2, 1$ for $n \in \mathbb{N}$ and $(n-1)!$ is the product of the integers $(n-1), (n-2), \ldots, 2, 1$, it follows that:

$$\frac{n!}{(n-1)!} = \frac{n \times (n-1) \times (n-2) \times \ldots \times 2 \times 1}{(n-1) \times (n-2) \times \ldots \times 2 \times 1}$$

$$= \frac{n!}{(n-1)!}$$

$$= n$$

so $n! = n \times (n-1)!$

Examples

1. Problem:
What is 7! in terms of 6!?
Solution:
Using $n! = n \times (n-1)!$:
$7! = 7 \times 6 \times 5 \times 4 \times 3 \times 2 \times 1$
$7! = 7 \times (6 \times 5 \times 4 \times 3 \times 2 \times 1)$
$7! = 7 \times 6!$

....

2. Problem:
What other factorials could 7! be written in terms of ?
Solution:
Using $n! = n \times (n-1)!$:
$7! = 7 \times 6 \times 5 \times 4 \times 3 \times 2 \times 1$
$7! = 7 \times 6!$
or
$7! = 7 \times 6 \times 5! = 42 \times 5!$
or
$7! = 7 \times 6 \times 5 \times 4! = 210 \times 4!$
and so on …

....

© HERIOT-WATT UNIVERSITY

3. Problem:
What is $9!$ in terms of $8!$?
Solution:
Using $n! = n \times (n-1)!$:
$9! = 9 \times 8 \times 7 \times 6 \times 5 \times 4 \times 3 \times 2 \times 1$
$9! = 9 \times (8 \times 7 \times 6 \times 5 \times 4 \times 3 \times 2 \times 1)$
$9! = 9 \times 8!$

Factorials exercise Go online

Q15: What is $8!$ in terms of $7!$?
....

Q16: What is $12!$ in terms of $11!$?
....

Q17: Write $8!$ in terms of $5!$?
....

Q18: Write $10!$ in terms of $8!$?
....

Q19: Write $11!$ in terms of $9!$?
....

Q20: Write $6!$ in terms of $3!$?

$n!$ has been defined for $n \in \mathbb{N}$ as $1! = 1 \times 0! = 1$

By convention, $0!$ is given the value 1.

Key point

Zero factorial has a value of 1.

That is, $0! = 1$

Note that this still fits the rule $n! = n \times (n-1)!$ because $1! = 1$ and $1 \times 0! = 1$

TOPIC 5. BINOMIAL THEOREM

> **A video explaining why** $0! = 1$ **Go online**
>
> Please watch the following YouTube video which explains why $0! = 1$:
>
> https://www.youtube.com/watch?v=Mfk_L4Nx2Zl
>
> Regarding the equation at the end: James says it should be $e^{-t}\, dt$ NOT $e^{-n}\, dn$ ... sorry for the mix-up!

5.3 Binomial coefficients

For integers $n \in \mathbb{N}$ and $0 \leq r \leq n$, the number given by $\frac{n!}{r!(n-r)!}$ is called a *Binomial Coefficient*.

It is denoted by $\binom{n}{r}$

> **Key point**
>
> The binomial coefficient formula is:
>
> $$\binom{n}{r} = \frac{n!}{r!(n-r)!}$$

Example : Binomial coefficient

Problem:

Evaluate $\binom{7}{2}$ using the binomial coefficient $\binom{n}{r} = \frac{n!}{r!(n-r)!}$

Solution:

$$\binom{7}{2} = \frac{7!}{2!\,(7-2)!}$$
$$= \frac{7!}{2! \times 5!}$$
$$= \frac{7 \times 6 \times 5 \times 4 \times 3 \times 2 \times 1}{2 \times 1 \times 5 \times 4 \times 3 \times 2 \times 1}$$
$$= \frac{7 \times 6}{2 \times 1}$$
$$= 21$$

The binomial coefficient $\binom{n}{r}$ is also denoted $^{n}C_{r}$ and is used in another related maths topic called combinatorics.

© HERIOT-WATT UNIVERSITY

TOPIC 5. BINOMIAL THEOREM

Combinatorics is not covered in this course, but the term indicates the number of ways of choosing r elements from a set of n elements.

Example : ^nC_r

Problem:

How many ways can two chocolates be chosen from a box containing twenty chocolates?

Solution:

$$^{20}C_2 = \binom{20}{2}$$
$$= \frac{20!}{2!18!}$$
$$= \frac{20 \times 19 \times 18!}{2 \times 1 \times 18!}$$
$$= \frac{20 \times 19}{2 \times 1}$$
$$= 190$$

That is, two chocolates be chosen from a box containing twenty chocolates 190 ways.

Binomial coefficients: ^nC_r **exercise**   Go online

Q21: What is $\binom{5}{2}$?

....

Q22: What is $\binom{6}{4}$?

....

Q23: What is $\binom{5}{3}$?

....

Q24: How many ways can four pupils be chosen from a group of seven?

First rule of binomial coefficients

From the binomial coefficient $\binom{n}{r} = \frac{n!}{r!(n-r)!}$

TOPIC 5. BINOMIAL THEOREM

$$\binom{5}{2} = \frac{5!}{2!\,(5-2)!}$$
$$= \frac{5!}{2!\,3!}$$
$$= \frac{5\times 4\times 3\times 2\times 1}{2\times 1\times 3\times 2\times 1}$$
$$= \frac{5\times 4}{2\times 1}$$
$$= 10$$

notice the symmetry

$$\binom{5}{3} = \frac{5!}{3!\,(5-3)!}$$
$$= \frac{5!}{3!\,2!}$$
$$= \frac{5\times 4\times 3\times 2\times 1}{3\times 2\times 1\times 2\times 1}$$
$$= \frac{5\times 4}{2\times 1}$$
$$= 10$$

Notice that $\binom{5}{2} = 10$ and $\binom{5}{3} = 10$ and so $\binom{5}{2} = \binom{5}{3}$

This illustrates the first rule for binomial coefficients.

Key point

The first binomial coefficient rule is:

$$\binom{n}{r} = \binom{n}{n-r}$$

The proof is not required for exam purposes, but is available simply for your interest.

Proof 1: First binomial coefficient rule

Proof 1

Prove that $\binom{n}{r} = \binom{n}{n-r}$, the same as proving that $\binom{n}{n-r} = \binom{n}{r}$

$$\binom{n}{r} = \binom{n}{n-r}$$
$$= \frac{n!}{(n-r)!\,(n-(n-r))!}$$
$$= \frac{n!}{(n-r)!\,(n-n+r)!}$$
$$= \frac{n!}{(n-r)!\,(r)!}$$
$$= \frac{n!}{r!\,(n-r)!}$$
$$= \binom{n}{r}$$

© HERIOT-WATT UNIVERSITY

First binomial coefficient rule exercise   Go online

Q25: Find another binomial coefficient equal to $\binom{7}{4}$

....

Q26: Find another binomial coefficient equal to $\binom{21}{17}$

Second rule of binomial coefficients

From the earlier questions $\binom{5}{3} = 10$, $\binom{5}{4} = 5$ and $\binom{6}{4} = 15$

Therefore $\binom{5}{3} + \binom{5}{4} = \binom{6}{4}$

This illustrates the second rule for binomial coefficients.

> **Key point**
>
> The second binomial coefficient rule is:
>
> $$\binom{n}{r-1} + \binom{n}{r} = \binom{n+1}{r}$$

The proof is not required for exam purposes, but is available simply for your interest.

Proof 2: Second binomial coefficient rule

Proof 2

Prove that $\binom{n}{r-1} + \binom{n}{r} = \binom{n+1}{r}$

In this proof, the following facts are required:

1. $r! = r \times (r-1)!$
2. $(n-r+1)! = (n-r+1) \times (n-r)!$
3. Two fractions can be combined over a common denominator.

$$\binom{n+1}{r} = \binom{n}{r-1} + \binom{n}{r}$$
$$= \frac{n!}{(r-1)!\,(n-(n-r))!} + \frac{n!}{r!\,(n-r)!}$$
$$= \frac{r \times n!}{r\,(r-1)!\,(n-r+1)!} + \frac{(n-r+1) \times n!}{r!\,(n-r+1)\,(n-r)!}$$
$$= \frac{r \times n! + (n-r+1) \times n!}{r! \times (n-r+1)!}$$
$$= \frac{n! \times (r+n-r+1)}{r!\,(n-r+1)!}$$
$$= \frac{n! \times (n+1)}{r!\,(n+1-r)!}$$
$$= \frac{(n+1)!}{r!\,(n+1-r)!}$$
$$= \binom{n+1}{r}$$

Second binomial coefficient rule exercise

Go online

Q27: Write down $\binom{8}{6} + \binom{8}{7}$ as a binomial coefficient.

....

Q28: Write down $\binom{14}{11} + \binom{14}{12}$ as a binomial coefficient.

16 TOPIC 5. BINOMIAL THEOREM

Example : Finding n given the value of a binomial coefficient and r

Problem:

If n is a positive integer such that $\binom{n}{2} = 15$, find n.

Solution:

From the binomial coefficient:

$$15 = \binom{n}{2}$$
$$= \frac{n!}{2!\,(n-2)!}$$

Now simplify by cancelling factors: $15 = \frac{n \times (n-1)}{2 \times 1}$

Multiplying out and simplifying: $30 = n^2 - n$

To solve a quadratic, we need to factorise:

$$n^2 - n - 30 = 0$$
$$(n-6)(n+5) = 0$$
$$n = 6 \text{ or } n = -5$$

Since n is a member of the natural numbers ($n \in \mathbb{N}$), then $n = 6$

Finding n given the value of a binomial coefficient and r exercise   Go online

Q29: Find the positive integer n such that $\binom{n}{2} = 10$

Q30: Find the positive integer n such that $\binom{n}{2} = 21$

Q31: Find the positive integer n such that $\binom{n}{2} = 36$

Binomial coefficients exercise   Go online

Q32: Evaluate $0!$

Q33: Evaluate $7!$

© HERIOT-WATT UNIVERSITY

TOPIC 5. BINOMIAL THEOREM

Q34: What is $100!$ as a product of an integer and the factorial of 99?

Q35: What is $1!$ as a product of an integer and the factorial of 0?

Q36: Evaluate $^6C_2$

Q37: Evaluate $\binom{9}{5}$

Q38: Evaluate $\binom{7}{4}$

Q39: Find another binomial coefficient equal to $\binom{13}{12}$

Q40: Find another binomial coefficient equal to $\binom{6}{4}$

Q41: Write down $\binom{7}{4} + \binom{7}{5}$ as a binomial coefficient.

Q42: Write down $\binom{12}{9} + \binom{12}{10}$ as a binomial coefficient.

5.4 Pascal's triangle

Blaise Pascal was credited with discovering what is known as 'Pascal's triangle'.

- It is made up of integers set out as a triangle.
- The number 1 appears at the top and at each end of subsequent rows.
- The numbers in the body of the triangle follow the rule:
  To find a number, add the two numbers from the above left and above right of it.
- The rows are numbered from Row 0.

© HERIOT-WATT UNIVERSITY

Pascal's triangle

 Go online

Q43: Complete rows 1 to 7 of Pascal's triangle.

Row 0
Row 1
Row 2
Row 3
Row 4
Row 5
Row 6
Row 7

....

Q44: The following is a table of the same design as Pascal's triangle, only this time the entries are the binomial coefficients as shown.

Row 0 $\binom{0}{0}$

Row 1 $\binom{1}{0}$ $\binom{1}{1}$

Row 2 $\binom{2}{0}$ $\binom{2}{1}$ $\binom{2}{2}$

Row 3 $\binom{3}{0}$ $\binom{3}{1}$ $\binom{3}{2}$ $\binom{3}{3}$

What are the corresponding values of these binomial coefficients?

Remember that $\binom{n}{r} = \frac{n!}{r!(n-r)!}$

TOPIC 5. BINOMIAL THEOREM

Q45: Complete rows 4 to 7 with binomial coefficients and their corresponding values.

Row 4

Row 5

Row 6

Row 7

Q46: What is the connection between Pascal's triangle and the binomial coefficients table?

Q47: What is the binomial coefficient for the fourth term in row 5?

Q48: What is the binomial coefficient for the third term in row 7?

Q49: What would be the binomial coefficient for the seventh term in row 10?

Q50: What would be the binomial coefficient for the fifth term in row 12?

Now if row 9 is required, instead of writing out Pascal's triangle, it is simply a matter of taking the binomial coefficients of row 9, namely:

$$\binom{9}{0}, \binom{9}{1}, \binom{9}{2}, \binom{9}{3}, \binom{9}{4}, \binom{9}{5}, \binom{9}{6}, \binom{9}{7}, \binom{9}{8} \text{ and } \binom{9}{9}$$

Either method can be used.

The entries in Pascal's triangle and the corresponding binomial coefficients are equal.

Recall that the second rule for binomial coefficients is $\binom{n}{r-1} + \binom{n}{r} = \binom{n+1}{r}$

This is the definition of how to construct the entries in Pascal's triangle.

If the numbers in the nth row of Pascal's triangle are the binomial coefficients, then the rth entry in the next row, row $(n + 1)$, is the sum of the $(r - 1)^{th}$ entry and the rth entry in row n.

The rth entry in the $(n + 1)^{th}$ row is $\binom{n+1}{r}$

© HERIOT-WATT UNIVERSITY

Hence, the entries in the $(n+1)^{\text{th}}$ row of Pascal's triangle are also the binomial coefficients.

An example of this computation follows:

Row 6   ...   .

Row 7   ...   ... $\binom{7}{2}$ $\leftarrow sum \rightarrow$ $\binom{7}{3}$ ...

Row 8   ...   . $\binom{8}{3}$ ...   ...

Row 9   ...   .

Key point

$$\binom{n}{r-1} + \binom{n}{r} = \binom{n+1}{r}$$

Pascal's triangle exercise   Go online

Q51: What is the binomial coefficient equivalent to $\binom{3}{1} + \binom{3}{2}$?

Q52: What is the binomial coefficient equivalent to $\binom{5}{3} + \binom{5}{2}$?

Q53: What is the binomial coefficient equivalent to $\binom{6}{2} - \binom{5}{2}$?

Q54: What is the binomial coefficient equivalent to $\binom{7}{4} - \binom{6}{3}$?

5.5 Binomial theorem

Look at the following expansions of $(x+y)^n$ for $n = 0, 1, 2, 3$

Row 0	$(x+y)^0$	1
Row 1	$(x+y)^1$	$x+y$
Row 2	$(x+y)^2$	$x^2 + 2xy + y^2$
Row 3	$(x+y)^3$	$x^3 + 3x^2y + 3xy^2 + y^3$

TOPIC 5. BINOMIAL THEOREM

The coefficients still follow the same pattern as Pascal's triangle and the equivalent table of binomial coefficients.

					Coefficients			
Row 0	1				1			
Row 1	$1x + 1y$				1	1		
Row 2	$1x^2 + 2xy + 1y^2$				1	2	1	
Row 3	$1x^3 + 3x^2y + 3xy^2 + 1y^3$				1	3	3	1

Or equivalently, since Pascal's triangle and the triangle of Binomial coefficients are the same:

Binomial coefficients

Row 0: $\binom{0}{0}$ $\quad\quad\quad\quad\quad\quad\quad\quad\quad\quad$ $\binom{0}{0}$

Row 1: $\binom{1}{0}x \quad \binom{1}{1}y \quad\quad\quad\quad\quad\quad$ $\binom{1}{0} \quad \binom{1}{1}$

Row 2: $\binom{2}{0}x^2 \quad \binom{2}{1}xy \quad \binom{2}{2}y^2 \quad\quad\quad$ $\binom{2}{0} \quad \binom{2}{1} \quad \binom{2}{2}$

Row 3: $\binom{3}{0}x^3 \quad \binom{3}{1}x^2y \quad \binom{3}{2}xy^2 \quad \binom{3}{3}y^3 \quad$ $\binom{3}{0} \quad \binom{3}{1} \quad \binom{3}{2} \quad \binom{3}{3}$

Features of Binomial expansion

Look carefully at the expansion of $(x + y)^4$: $(x + y)^4 = x^4 + 4x^3y + 6x^2y^2 + 4xy^3 + y^4$

Notice that the x terms start with a power of 4 which decreases by 1 in successive terms.

Notice that the y terms start with a power of 0 which increases by 1 in successive terms.

$$(x + y)^4 = x^4 + 4x^3y + 6x^2y^2 + 4xy^3 + y^4$$

with $3+1=4$, $1+3=4$, 4, $2+2=4$, 4

Notice that for each set of terms, the powers of x and y add up to 4.

Using these observations, we can re-write the expansion of $(x + y)^4$ in a different way:

For example:

$4x^3y$ can be written as $\binom{4}{1} x^{4-1}y^1$ and y^4 can be written as $\binom{4}{4} x^{4-4}y^4$

This is an example of the binomial theorem and leads to two definitions:

- one for the expansion, and
- one for any term within the expansion.

© HERIOT-WATT UNIVERSITY

Key point

The binomial theorem states that if $x, y \in \mathbb{R}$ and $n \in \mathbb{N}$ then:

$$(x+y)^n = \binom{n}{0}x^n + \binom{n}{1}x^{n-1}y + \binom{n}{2}x^{n-2}y^2 + \ldots + \binom{n}{r}x^{n-r}y^r + \ldots + \binom{n}{n}y^n$$

Key point

The general term of $(x+y)^n$ is given by:

$$\binom{n}{r}x^{n-r}y^r$$

The proof is not required for exam purposes, but is available simply for your interest.

Proof 3: Binomial theorem

Proof 3

Prove the binomial theorem which states that if $x, y \in \mathbb{R}$ and $r, n \in \mathbb{N}$, then:

$$(x+y)^n = \binom{n}{0}x^n + \binom{n}{1}x^{n-1}y + \binom{n}{2}x^{n-2}y^2 + \ldots + \binom{n}{r}x^{n-r}y^r + \binom{n}{n}y^n$$

$$= \sum_{r=0}^{n} \binom{n}{r} x^{n-r} y^r$$

This proof is by induction, the method of which will be covered later in the course.

Let $n = 1$, then:
LHS $= (x+y)^n$
$= (x+y)^1$
$= x + y$

RHS $= \binom{1}{0}x^1 y^0 + \binom{1}{1}x^0 y^1$
$= x + y$

So LHS = RHS and the theorem holds for $n = 1$

Now, suppose that the result is true for $n = k$ (where $k \geq 1$), then:

$$(x+y)^k = \binom{k}{0}x^k + \binom{k}{1}x^{k-1}y + \binom{k}{2}x^{k-2}y^2 + \ldots + \binom{k}{k-1}xy^{k-1} + \binom{k}{k}y^k$$

TOPIC 5. BINOMIAL THEOREM

Consider:

$$(x+y)^{k+1} = (x+y)(x+y)^k = x(x+y)^k + y(x+y)^k$$

$$= \binom{k}{0} x^{k+1} + \binom{k}{1} x^k y + \binom{k}{2} x^{k-1} y^2 + \ldots + \binom{k}{k-1} x^2 y^{k-1} + \binom{k}{k} x y^k +$$

$$\binom{k}{0} x^k y + \binom{k}{1} x^{k-1} y^2 + \binom{k}{2} x^{k-2} y^3 + \ldots + \binom{k}{k-1} x y^k + \binom{k}{k} y^{k+1}$$

$$= \binom{k}{0} x^{k+1} + \left[\binom{k}{1} + \binom{k}{0}\right] x^k y + \left[\binom{k}{2} + \binom{k}{1}\right] x^{k-1} y^2 + \ldots +$$

$$\left[\binom{k}{k-1} + \binom{k}{k-2}\right] x^2 y^{k-1} + \left[\binom{k}{k} + \binom{k}{k-1}\right] x y^k + \binom{k}{k} y^{k+1}$$

$$= 1 x^{k+1} + \binom{k+1}{1} x^k y + \ldots + \binom{k+1}{k} x y^k + 1 y^{k+1}$$

$$= \sum_{r=0}^{k+1} \binom{k+1}{r} x^{k+1-r} y^r$$

Hence, if it is true for $n = k$, it is also true for $n = k + 1$.

However, it was also true for $n = 1$ so it is true for all values of n.

Remember that $\binom{r}{0} = \binom{r}{r} = 1$ for all r and n, and that $\binom{k}{j+1} + \binom{k}{j} = \binom{k+1}{j}$

(See the second binomial coefficient rule.)

Example

Problem:

Use the binomial theorem to expand $(x + y)^5$

Solution:

In this case $n = 5$

So we have:

$$(x+y)^n = \binom{n}{0} x^n + \binom{n}{1} x^{n-1} y + \binom{n}{2} x^{n-2} y^2 + \ldots + \binom{n}{r} x^{n-r} y^r + \ldots + \binom{n}{n} y^n$$

24 TOPIC 5. BINOMIAL THEOREM

At this stage we can evaluate coefficients using:

- binomial coefficients, e.g.

$$\binom{n}{r} = \frac{n!}{r!(n-r)!}$$

$$\binom{5}{0} = \frac{5!}{0!(5-0)!}$$

$$= \frac{5!}{0!5!}$$

$$= 1 \quad \text{and so on...}$$

- Pascal's triangle

```
            1
          1   1
        1   2   1
      1   3   3   1
    1   4   6   4   1
  1   5  10  10   5   1
```

$$(x+y)^5 = \binom{5}{0}x^5 + \binom{5}{1}x^4y + \binom{5}{2}x^3y^2 + \binom{5}{3}x^2y^3 + \binom{5}{4}xy^4 + \binom{5}{5}y^5$$
$$= x^5 + 5x^4y + 10x^3y^2 + 10x^2y^3 + 5xy^4 + y^5$$

Binomial expansion   Go online

Example

Problem:

Using the binomial theorem, expand $(x+y)^7$

$$(x+y)^n = \binom{n}{0}x^n + \binom{n}{1}x^{n-1}y + \binom{n}{2}x^{n-2}y^2 + \ldots + \binom{n}{r}x^{n-r}y^r + \ldots + \binom{n}{n}y^n$$

Solution:

$$(x+y)^7 = \binom{7}{0}x^7 + \binom{7}{1}x^6y + \binom{7}{2}x^5y^2 + \ldots + \binom{7}{5}x^2y^5 + \binom{7}{6}xy^6 + \binom{7}{7}y^7$$
$$= x^7 + 7x^6y + 21x^5y^2 + 35x^4y^3 + 35x^3y^4 + 21x^2y^5 + 7xy^6 + y^7$$

© HERIOT-WATT UNIVERSITY

TOPIC 5. BINOMIAL THEOREM

Remember that:

- the Binomial coefficients take the form: $\binom{n}{0}, \binom{n}{1}, \binom{n}{2} \ldots \binom{n}{n}$ where n is the power that $(x+y)$ has been raised to;
- for the expansion $(x+y)^n$, since x is the first term in the bracket its power will decrease from n, and since y is the second term in the bracket to be expanded its powers will increase to n, e.g.
$(x+y)^4 = x^4 + 4x^3y + 6x^2y^2 + 4xy^3 + y^4$

Example

$$(x+y)^8 = \binom{8}{0}x^8 + \binom{8}{1}x^7y + \binom{8}{2}x^6y^2 + \binom{8}{3}x^5y^3 + \binom{8}{4}x^4y^4 +$$
$$\binom{8}{5}x^3y^5 + \binom{8}{6}x^2y^6 + \binom{8}{7}xy^7 + \binom{8}{8}y^8$$
$$= x^8 + 8x^7y + 28x^6y^2 + 56x^5y^3 + 70x^4y^4 + 56x^3y^5 + 28x^2y^6 + 8xy^7 + y^8$$

The binomial theorem works also with multiples of x and y, and other symbols (such as a, b or α, β).

Example : Binomial expansion

Problem:

Use the binomial theorem to expand $(2x-3y)^4$

Solution:

From the binomial theorem:

$$(2x-3y)^4 = \binom{4}{0}(2x)^4 + \binom{4}{1}(2x)^3(-3y) + \binom{4}{2}(2x)^2(-3y)^2 +$$
$$\binom{4}{3}(2x)(-3y)^3 + \binom{4}{4}(-3y)^4$$

In this case, x is replaced by $2x$ and y is replaced by $-3y$

Now we can evaluate the binomial coefficients using Pascal's triangle:

$(2x-3y)^4 = (2x)^4 + 4(2x)^3(-3y) + 6(2x)^2(-3y)^2 + 4(2x)(-3y)^3 + (-3y)^4$

Evaluate the powers of $2x$ and $-3y$:

$(2x-3y)^4 = 16x^4 + 4 \cdot 8x^3(-3y) + 6 \cdot 4x^2 9y^2 + 4 \cdot 2x(-27)y^3 + 81y^4$

Simplify each term by multiplying the coefficients together:

$(2x-3y)^4 = 16x^4 - 96x^3y + 216x^2y^2 - 216xy^3 + 81y^4$

© HERIOT-WATT UNIVERSITY

Further binomial expansion

Example

Problem:

Using the binomial theorem, expand $(-x + 2y)^3$

$$(x+y)^n = \binom{n}{0} x^n + \binom{n}{1} x^{n-1}y + \binom{n}{2} x^{n-2}y^2 + \ldots + \binom{n}{r} x^{n-r}y^r + \ldots + \binom{n}{n} y^n$$

Solution:

$$(-x+2y)^3 = \binom{3}{0}(-x)^3 + \binom{3}{1}(-x)^2(2y) + \binom{3}{2}(-x)(2y)^2 + \binom{3}{3}(2y)^3$$
$$= -x^3 + 6x^2y - 12xy^2 + 8y^3$$

Remember that:

- the binomial expansion takes the form:

$$(x+y)^n = \binom{n}{0} x^n + \binom{n}{1} x^{n-1}y + \binom{n}{2} x^{n-2}y^2 + \ldots + \binom{n}{r} x^{n-r}y^r + \ldots + \binom{n}{n} y^n$$

  - replace the first term x with your first term;
  - replace the second term y with your second term (remember negatives);
- the coefficients of x and y are also raised to a power as well $(2y)^3 = 2^3 y^3$
- -1 is usually represented by just a negative sign.

Rational coefficients are sometimes found as is shown in the following activity.

Binomial expansion: Rational coefficients

Example

Problem:

Use the binomial theorem to expand $\left(\frac{x}{2} - y\right)^4$

Solution:

$$\left(\frac{x}{2} - y\right)^4 = \binom{4}{0}\left(\frac{x}{2}\right)^4 + \binom{4}{1}\left(\frac{x}{2}\right)^3(-y) + \binom{4}{2}\left(\frac{x}{2}\right)^2(-y)^2 + \binom{4}{3}\left(\frac{x}{2}\right)(-y)^3 + \binom{4}{4}(-y)^4$$
$$= \frac{x^4}{16} - \frac{x^3}{2}y + \frac{3}{2}x^2y^2 - 2xy^3 + y^4$$

TOPIC 5. BINOMIAL THEOREM

Remember that:

- the binomial expansion takes the form:
$$(x+y)^n = \binom{n}{0}x^n + \binom{n}{1}x^{n-1}y + \binom{n}{2}x^{n-2}y^2 + \ldots + \binom{n}{r}x^{n-r}y^r + \ldots + \binom{n}{n}y^n$$
  - replace the first term *x* with your first term;
  - replace the second term *y* with your second term (remember negatives);
- the coefficients of *x* and *y* are also raised to a power as well $\left(\frac{x}{2}\right)^3 = \frac{x^3}{2^3}$

5.6 Finding coefficients

Sometimes only one power in an expansion is required.

By using the general term formula given earlier, the coefficient of any term in an expansion can be found.

Examples

1. Problem:

Find the coefficient of $x^3 y^4$ in the expansion of $(x+y)^7$

Solution:

Start with the general term of $(x+y)^n$ which is: $\binom{n}{r} x^{n-r} y^r$

In this problem, $n = 7$, so the general term now has the form: $\binom{7}{r} x^{7-r} y^r$

But we want: $x^3 y^4$

We need to find r so we equate the powers: $x^3 = x^{7-r}$

So $3 = 7 - r$

Then $r = 4$

This gives the term: $x^3 y^4$

Replace r with 4 and evaluate the coefficient:

$$\binom{7}{4} = \frac{7!}{4!3!}$$
$$= 35$$

2. Problem:
Find the coefficient of xy^4 in the expansion of $(x - y)^5$
Solution:
The general term in this problem is: $\binom{5}{r} x^{5-r}(-y)^r = \binom{5}{r} x^{5-r}(-1)^r y^r$

The coefficient is then $\binom{5}{4}(-1)^r$

For $5 - r = 1$, we require $r = 4$
Therefore, the coefficient is:
$$\binom{5}{4}(-1)^4 = 5 \times 1$$
$$= 5$$

....

3. Problem:
Find the coefficient of x^2 in the expansion of $\left(x + \frac{2}{x}\right)^4$
Solution:
The general term in this problem is:
$$\binom{4}{r} x^{4-r} \left(\frac{2}{x}\right)^r = \binom{4}{r} x^{4-r}(2)^r \frac{1}{x^r}$$
$$= \binom{4}{r} x^{4-2r}(2)^r$$

We need to use the rules of indices to simplify the power of x.

The coefficient is then $\binom{4}{r}(2)^r$

For $4 - 2r = 2$, we require $r = 1$.
Therefore, the coefficient is:
$$\binom{4}{1}(2)^1 = 4 \times 2$$
$$= 8$$

....

4. Problem:

Find the coefficient of x^6 in the expansion of $(1 + x^2)^8$

Solution:

The general term is given by: $\binom{8}{r}(1)^{8-r}(x^2)^r = \binom{8}{r}x^{2r}$

The coefficient is then $\binom{8}{r}$

For $2r = 6$, we require $r = 3$

Therefore, the coefficient is:

$$\binom{8}{3} = \frac{8!}{3!5!}$$
$$= 56$$

Finding coefficients exercise   Go online

Q55: Find the coefficient of x^5 in the expansion of $\left(x - \frac{3}{x}\right)^7$
....

Q56: Find the coefficient of x^4 in the expansion $(1 + 2x^2)^3$
....

Q57: Find the coefficient of $x^2 y^6$ in the expansion of $(3x - 2y)^8$

5.7 Sigma notation and binomial theorem applications

If $a_0, a_1, \ldots, a_n$ are real numbers, then the sum $a_0 + a_1 + \ldots + a_n$ is sometimes written in shorthand form as $\sum_{r=0}^{n} a_r$

Here, the symbol $\sum$, called sigma, means 'the sum of'.

The expression $r = 0$ at the bottom of the sigma sign means 'starting from $r = 0$'.

The letter n at the top of the sigma sign means 'until $r = n$'.

The term a_r is the sequence of terms to be added together.

Sigma notation of binomial expansion

$$(x+y)^n = \sum_{r=0}^{n} \binom{n}{r} x^{n-r} y^r \quad \text{for } r, n \in \mathbb{N}$$

$$= \binom{n}{0} x^n + \binom{n}{1} x^{n-1} y + \binom{n}{2} x^{n-2} y^2 + \binom{n}{3} x^{n-3} y^3 + \ldots + \binom{n}{n} y^n$$

Notice in the:

- 1st term, r is replaced by 0
- 2nd term, r is replaced by 1
- 3rd term, r is replaced by 2, and so on.

Example

Problem:

Using the binomial theorem expand $(x+y)^5$

Solution:

$$(x+y)^5 = \sum_{r=0}^{5} \binom{5}{r} x^{5-r} y^r$$

$$= \binom{5}{0} x^5 + \binom{5}{1} x^4 y + \binom{5}{2} x^3 y^2 + \binom{5}{3} x^2 y^3 + \binom{5}{4} xy^4 + \binom{5}{5} y^5$$

$$= x^5 + 5x^4 + 10x^3 y^2 + 10x^2 y^3 + 5xy^4 + y^5$$

The binomial theorem can be used to find powers of a real number z.

The technique is to split z into two parts x and y where x is the closest integer to z and y is the remaining part.

For example, if $z = 1 \cdot 9$ then it can be split as $z = (2 - 0 \cdot 1)$

Examples

1. Evaluating a decimal to a power using binomial expansion

Problem:

Using the binomial theorem, find $1 \cdot 04^3$

Solution:

By writing $1 \cdot 04^3$ in the form $(x+y)^n$, we can use the binomial theorem to expand this.
$1 \cdot 04^3 = (1 + 0 \cdot 04)^3$ where $x = 1$ and $y = 0 \cdot 04$

TOPIC 5. BINOMIAL THEOREM

$$(1 \cdot 04)^3 = \sum_{r=0}^{3} \binom{3}{r} 1^{3-r}(0 \cdot 04)^r$$
$$= \binom{3}{0} 1^3 + \binom{3}{1} 1^2 (0 \cdot 04) + \binom{3}{2} 1^1 (0 \cdot 04)^2 + \binom{3}{3} (0 \cdot 04)^3$$
$$= 1 + (3 \times 1 \times 0 \cdot 04) + (3 \times 1 \times 0 \cdot 0016) + (1 \times 0 \cdot 000064)$$
$$= 1 + 0 \cdot 12 + 0 \cdot 0048 + 0 \cdot 000064$$
$$= 1 \cdot 124864$$

....

2. Problem:

Using the binomial theorem find $0 \cdot 7^4$

Solution:

$$0 \cdot 7^4 = (1 + (-0 \cdot 3))^4$$
$$(0 \cdot 7)^4 = \sum_{r=0}^{4} \binom{4}{r} 1^{4-r}(-0 \cdot 3)^r$$
$$= \binom{4}{0} 1^4 (-0 \cdot 3)^0 + \binom{4}{1} 1^3 (-0 \cdot 3) + \binom{4}{2} 1^2 (-0 \cdot 3)^2 + \binom{4}{3} 1^1 (-0 \cdot 3)^3 +$$
$$\binom{4}{4} 1^0 (-0 \cdot 3)^4$$
$$= 1 + (4 \times -0 \cdot 3) + (6 \times 0 \cdot 09) + (4 \times -0 \cdot 027) + 0 \cdot 0081$$
$$= 0 \cdot 2401$$

Remember that:

- this expansion is for $(x + y)^n$, so if you have $(x - y)^n$, you have to remember that − (the negative sign) is part of your number;
- there are many ways in which you can split the number - it is recommended to choose one of these numbers to be a whole number. Why is this?

Binomial applications exercise   Go online

Q58: Using the binomial theorem expand and evaluate 0.5^5

....

Q59: Using the binomial theorem expand and evaluate 2.7^3

© HERIOT-WATT UNIVERSITY

5.8 Learning points

Binomial theorem
Factorials

- $n! = n \times (n-1) \times (n-2) \times \ldots \times 2 \times 1 \; for \; n \in \mathbb{N}$
  e.g. $5! = 5 \times 4 \times 3 \times 2 \times 1$
  $= 120$

- $n! = n \times (n-1)!$
  e.g. $5! = 5 \times 4!$

- $0! = 1$

Binomial coefficients

- The binomial coefficient is $\binom{n}{r} = \frac{n!}{r!(n-r)!}$ for integers $n, r \in \mathbb{N}$ and $n < r = n$

  e.g. $\binom{5}{2} = \frac{5!}{2!(5-2)!}$
  $= 10$

- The binomial coefficient can be represented by ^nC_r or $\binom{n}{r}$

- Binomial coefficients are symmetrical so the following will give you the same answer $\binom{n}{r} = \binom{n}{n-r}$

  e.g. $\binom{5}{2} = \binom{5}{3}$

- $\binom{n}{r-1} + \binom{n}{r} = \binom{n+1}{r}$

  e.g. $\binom{5}{2} + \binom{5}{3} = \binom{6}{3}$

Pascal's triangle

- Pascal's triangle is made up of integers set out as a triangle.
- To find the terms in the triangle you follows these rules:
    - The number one appears at the top and the ends of each row.
    - To find a number in the main body, add the two numbers from the above left and above right of it:

TOPIC 5. BINOMIAL THEOREM

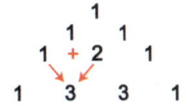

```
        1
      1   1
    1 + 2   1
    1   3   3   1
```

- When the binomial coefficients are set up in a similar way and evaluated, they form Pascal's triangle:

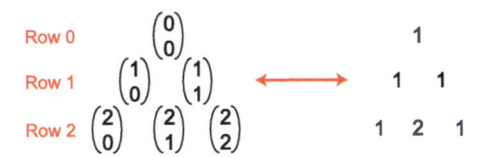

Row 0: $\binom{0}{0}$ $\longleftrightarrow$ 1

Row 1: $\binom{1}{0}$ $\binom{1}{1}$ $\longleftrightarrow$ 1 1

Row 2: $\binom{2}{0}$ $\binom{2}{1}$ $\binom{2}{2}$ $\longleftrightarrow$ 1 2 1

- Pascal's triangle is a quick way to evaluate the binomial coefficients.

Binomial theorem

- The *binomial theorem* states that if $x, y \in \mathbb{R}$ and $n \in \mathbb{N}$ then:

$$(x+y)^n = \binom{n}{0}x^n + \binom{n}{1}x^{n-1}y + \binom{n}{2}x^{n-2}y^2 + \ldots + \binom{n}{r}x^{n-r}y^r + \ldots + \binom{n}{n}y^n$$

- The *general term* of $(x+y)^n$ is given by:

$$\binom{n}{r}x^{n-r}y^r$$

- In both cases, the variables x and y can be replaced by any other given term.

Finding coefficients

- To find the coefficient of a given term in the expression $(x+y)^n$, we use the general term

$$\binom{n}{r}x^{n-r}y^r$$

  - equate the powers of x and y in the given term with that in the general term to find r;
  - once we have found r, we can evaluate the binomial coefficient and any other coefficients in the general term.

Sigma notation and binomial theorem application

- To expand 1.07^3, we write this number in the form $(x+y)^n$, choosing x to be the integer closest to 1.07, then use the binomial expansion:

$$(x+y)^n = \binom{n}{0}x^n + \binom{n}{1}x^{n-1}y + \binom{n}{2}x^{n-2}y^2 + \ldots + \binom{n}{r}x^{n-r}y^r + \ldots + \binom{n}{n}y^n$$

replacing x with 1, y with 0.07 and n with 3 before evaluating.

© HERIOT-WATT UNIVERSITY

5.9 Proofs

Proof 1: First binomial coefficient rule

Prove that $\binom{n}{r} = \binom{n}{n-r}$, the same as proving that $\binom{n}{n-r} = \binom{n}{r}$

$$\binom{n}{r} = \binom{n}{n-r}$$
$$= \frac{n!}{(n-r)!\,(n-(n-r))!}$$
$$= \frac{n!}{(n-r)!\,(n-n+r)!}$$
$$= \frac{n!}{(n-r)!\,(r)!}$$
$$= \frac{n!}{r!\,(n-r)!}$$
$$= \binom{n}{r}$$

Proof 2: Second binomial coefficient rule

Prove that $\binom{n}{r-1} + \binom{n}{r} = \binom{n+1}{r}$

In this proof, the following facts are required:

1. $r! = r \times (r-1)!$
2. $(n-r+1)! = (n-r+1) \times (n-r)!$
3. Two fractions can be combined over a common denominator.

$$\binom{n+1}{r} = \binom{n}{r-1} + \binom{n}{r}$$
$$= \frac{n!}{(r-1)!\,(n-(n-r))!} + \frac{n!}{r!\,(n-r)!}$$
$$= \frac{r \times n!}{r\,(r-1)!\,(n-r+1)!} + \frac{(n-r+1) \times n!}{r!\,(n-r+1)\,(n-r)!}$$
$$= \frac{r \times n! + (n-r+1) \times n!}{r! \times (n-r+1)!}$$
$$= \frac{n! \times (r+n-r+1)}{r!\,(n-r+1)!}$$
$$= \frac{n! \times (n+1)}{r!\,(n+1-r)!}$$
$$= \frac{(n+1)!}{r!\,(n+1-r)!}$$
$$= \binom{n+1}{r}$$

Proof 3: Binomial theorem

Prove the binomial theorem which states that if $x, y \in \mathbb{R}$ and $r, n \in \mathbb{N}$, then:

$$(x+y)^n = \binom{n}{0} x^n + \binom{n}{1} x^{n-1} y + \binom{n}{2} x^{n-2} y^2 + \ldots + \binom{n}{r} x^{n-r} y^r + \binom{n}{n} y^n$$

$$= \sum_{r=0}^{n} \binom{n}{r} x^{n-r} y^r$$

This proof is by induction, the method of which will be covered later in the course.

Let $n = 1$, then:
LHS $= (x+y)^n$
$= (x+y)^1$
$= x+y$

RHS $= \binom{1}{0} x^1 y^0 + \binom{1}{1} x^0 y^1$
$= x+y$

So LHS = RHS and the theorem holds for $n = 1$

Now, suppose that the result is true for $n = k$ (where $k \geq 1$), then:

$$(x+y)^k = \binom{k}{0} x^k + \binom{k}{1} x^{k-1} y + \binom{k}{2} x^{k-2} y^2 + \ldots + \binom{k}{k-1} xy^{k-1} + \binom{k}{k} y^k$$

Consider:

$$(x+y)^{k+1} = (x+y)(x+y)^k = x(x+y)^k + y(x+y)^k$$

$$= \binom{k}{0} x^{k+1} + \binom{k}{1} x^k y + \binom{k}{2} x^{k-1} y^2 + \ldots + \binom{k}{k-1} x^2 y^{k-1} + \binom{k}{k} xy^k +$$

$$\binom{k}{0} x^k y + \binom{k}{1} x^{k-1} y^2 + \binom{k}{2} x^{k-2} y^3 + \ldots + \binom{k}{k-1} xy^k + \binom{k}{k} y^{k+1}$$

$$= \binom{k}{0} x^{k+1} + \left[\binom{k}{1} + \binom{k}{0}\right] x^k y + \left[\binom{k}{2} + \binom{k}{1}\right] x^{k-1} y^2 + \ldots +$$

$$\left[\binom{k}{k-1} + \binom{k}{k-2}\right] x^2 y^{k-1} + \left[\binom{k}{k} + \binom{k}{k-1}\right] xy^k + \binom{k}{k} y^{k+1}$$

$$= 1 x^{k+1} + \binom{k+1}{1} x^k y + \ldots + \binom{k+1}{k} xy^k + 1 y^{k+1}$$

$$= \sum_{r=0}^{k+1} \binom{k+1}{r} x^{k+1-r} y^r$$

Hence, if it is true for $n = k$, it is also true for $n = k + 1$.

However, it was also true for $n = 1$ so it is true for all values of n.

Remember that $\binom{r}{0} = \binom{r}{r} = 1$ for all r and n, and that $\binom{k}{j+1} + \binom{k}{j} = \binom{k+1}{j}$

(See the second binomial coefficient rule.)

5.10 Extended information

Pascal

Blaise Pascal (1632-1662) was a French mathematician, physicist and philosopher. He published the triangle in 'Traite du Triangle Arithmetique' in 1665 but did not claim recognition for it. His interest in the triangle arose from his study of the theory of probabilities linked to his gambling with his friend Fermat. He was the inventor of the first mechanical calculator and a programming language (pascal) is named after him.

Chu Shih - Chieh

This Chinese mathematician (1270-1330) published a version of the triangle in his 'Precious Mirror of the Four Elements' in 1303.

Open-ended challenge

Try to create a 3-d version of Pascal's Triangle using a triangular pyramid.

TOPIC 5. BINOMIAL THEOREM

5.11 End of topic test

End of topic 5 test — Go online

Q60:

a) Simplify the following sigma notation: $\sum_{r=0}^{4} \binom{4}{r} (2x)^{4-r}(-5)^r$

b) What are the binomial coefficients?

c) Write down the binomial expansion of $(2x - 5)^4$ and simplify your answer.

....

Q61:

a) Simplify the following sigma notation: $\sum_{r=0}^{5} \binom{5}{r} (5u)^{5-r}(-3v)^r$

b) What are the binomial coefficients?

c) Write down the binomial expansion of $(5u - 3v)^5$ and simplify your answer.

....

Q62:

a) Simplify the following sigma notation: $\sum_{r=0}^{5} \binom{5}{r} (y^2)^{5-r}(7)^r$

b) What are the binomial coefficients?

c) Write down the binomial expansion of $(y^2 + 7)^5$ and simplify your answer.

....

Q63:

a) Simplify the following sigma notation: $\sum_{r=0}^{4} \binom{4}{r} (k)^{4-r}\left(-\frac{5}{k^2}\right)^r$

b) What are the binomial coefficients?

c) Write down the binomial expansion of $\left(k - \frac{5}{k^2}\right)^4$ and simplify your answer.

....

Q64:

a) Write down and simplify the general term in the expansion of $\left(2x - \frac{3}{x^2}\right)^9$.

b) Hence, or otherwise, obtain the term independent of x; that is, find the term that is a constant where the power of x is zero.

....

© HERIOT-WATT UNIVERSITY

Q65:

a) Write down and simplify the general term in the expansion of $\left(3x^2 - \frac{2}{x}\right)^{10}$

b) Hence, or otherwise, obtain the term x^{14}.

....

Q66:

a) Write down the binomial expansion of $(1 + x)^5$

b) Hence, calculate $0 \cdot 9^5$.

....

Q67:

a) Write down the binomial expansion of $(1 + x)^4$

b) Write down the expression for $1 \cdot 2^4$ in terms of $(1 + x)^4$ and evaluate.

Topic 6

Complex numbers

Contents

- 6.1 Looking back ... 43
  - 6.1.1 Multiplying out brackets ... 44
  - 6.1.2 Factorising a trinomial ... 45
  - 6.1.3 Arithmetic of surds ... 47
  - 6.1.4 Trigonometric identities ... 50
- 6.2 Introduction to Imaginary numbers ... 53
  - 6.2.1 Square roots of a negative number and quadratic equations ... 54
  - 6.2.2 Complex numbers and the complex plane ... 57
- 6.3 The arithmetic of complex numbers ... 59
  - 6.3.1 Addition and subtraction of complex numbers ... 60
  - 6.3.2 Multiplying complex numbers ... 64
  - 6.3.3 Conjugates of complex numbers ... 70
  - 6.3.4 Division of complex numbers ... 73
- 6.4 The modulus, argument and polar form of a complex number ... 76
  - 6.4.1 Cartesian form ... 78
  - 6.4.2 Principal argument ... 81
  - 6.4.3 Strategy to find modulus and argument ... 86
  - 6.4.4 Polar form ... 90
- 6.5 Geometric interpretations ... 95
  - 6.5.1 Geometric interpretations: Circle representations ... 95
  - 6.5.2 Geometric interpretations: Straight line representations ... 104
- 6.6 Fundamental theorem of algebra and solving complex equations ... 110
  - 6.6.1 Fundamental theorem of algebra ... 110
  - 6.6.2 Solving complex equations ... 111
- 6.7 De Moivre's theorem ... 119
  - 6.7.1 Multiplication using polar form ... 119
  - 6.7.2 Division using polar form ... 120
  - 6.7.3 De Moivre's theorem ... 122
  - 6.7.4 De Moivre's theorem and multiple angle formulae ... 130
  - 6.7.5 De Moivre's theorem with fractional powers ... 136
  - 6.7.6 De Moivre's theorem and n^{th} roots ... 138
  - 6.7.7 De Moivre's theorem and roots of unity ... 141
- 6.8 Conjugate properties ... 147

6.9	Learning points	153
6.10	Proofs	158
6.11	Extended information	161
6.12	End of topic test	163

TOPIC 6. COMPLEX NUMBERS

Learning objective

By the end of this topic, you should be able to:

- identify the following numbers and their symbols:
    - Natural numbers ($\mathbb{N}$);
    - Integers ($\mathbb{Z}$);
    - Rational numbers ($\mathbb{Q}$);
    - Real numbers ($\mathbb{R}$);
- use $I^2 = -1$ to find the roots of quadratic equations;
- identify the real and imaginary part of a complex number;
- plot and read points off the complex plane or Argand diagram;
- add/subtract complex numbers by adding/subtracting the real and imaginary parts;
- multiply, divide and find the square root of complex numbers;
- calculate the:
    - modulus, r, of a complex number from the Cartesian form $z = a + ib$;
    - principal argument, π, of a complex number from the Cartesian form $z = a + ib$;
- convert a complex number to Cartesian form when given the modulus, r, and principal argument, π;
- give the locus and a geometrical interpretation of the equation:
    - $|z - a| = b$, where a is a point in the complex plane;
    - $|z - a| = |z - b|$;
- understand the triangle inequality $|z + w| \leq |z| + |w|$;
- understand the fundamental theorem of algebra;
- verify that a given solution is a root of a polynomial;
- find the complex conjugate (which is also a root) given a root to a polynomial which is a complex number;
- use long division to find remaining roots;
- multiply and divide complex numbers in polar form using de Moivre's theorem;
- use de Moivre's theorem to simplify z^n, where z is a complex number and n is an integer or rational number;
- use the binomial theorem and de Moivre's theorem to state $\sin(n\pi)$, $\cos(n\pi)$ and $\tan(n\pi)$ in terms of powers of $\sin \pi$, $\cos \pi$ and $\tan \pi$;

© HERIOT-WATT UNIVERSITY

Learning objective continued

- use $2\cos(n\theta) = \left(z^n + \frac{1}{z^n}\right)$ and $2i\sin(n\theta) = \left(z^n + \frac{1}{z^n}\right)$ to state powers of $\sin \pi$ and $\cos \pi$ in terms of multiple angles of $\sin \pi$ and $\cos \pi$;
- use de Moivre's theorem to find the roots of $z^n = r(\cos\theta + i\sin\theta)$;
- use the properties of the complex conjugate.

TOPIC 6. COMPLEX NUMBERS

6.1 Looking back

Pre-requisites from Advanced Higher

You should have covered the following in the *Binomial theorem* topic. If you need to reinforce your learning go back and study this topic.

Binomial theorem

- The binomial theorem states that if $x, y \in \mathbb{R}$ and $n \in \mathbb{N}$, then:

$$(x+y)^n = \binom{n}{0} x^n + \binom{n}{1} x^{n-1} y + \binom{n}{2} x^{n-2} y^2 + \ldots + \binom{n}{r} x^{n-r} y^r + \ldots + \binom{n}{n} y^n$$

Summary of prior knowledge
Multiplying out brackets: National 5

- When expanding double brackets, remember that each term in the second bracket is multiplied by each term in the first bracket.

Factorising into brackets: National 5

- When factorising, always ask yourself three questions.
    1. Is there a simple common factor?
    2. Is it a difference of two squares?
    3. Is it a trinomial?

    Remember, you could have a simple common factor and a difference of two squares *or* a simple common factor and a trinomial.

Arithmetic of Surds: National 5

- To simplify surds, we find the biggest square number that goes into the surd and simplify.
- We can only add or subtract surds that are the same (like in algebra).
- When multiplying surds, we multiply the numbers under the square root then simplify.
- When dividing by a surd, we instead multiply by its conjugate.

Trigonometric identities: Higher

- $\cos^2 A + \sin^2 A = 1$
- $\tan A = \frac{\sin A}{\cos A}$
- $\sin(A + B) = \sin A \cos B + \cos A \sin B$
- $\sin(A - B) = \sin A \cos B - \cos A \sin B$
- $\cos(A + B) = \cos A \cos B - \sin A \sin B$
- $\cos(A - B) = \cos A \cos B + \sin A \sin B$

© HERIOT-WATT UNIVERSITY

- $\sin(2A) = 2\sin A \cos A$
- $\cos(2A) = \cos^2 A - \sin^2 A$
  $ = 2\cos^2 A - 1$
  $ = 1 - 2\sin^2 A$

6.1.1 Multiplying out brackets

Examples

1. Problem:
Expand and simplify $(3x + 1)(2x - 5)$
Solution:

$(3x + 1)(2x - 5)$

Multiply the first term $3x$ by everything in the second bracket.
$3x(2x - 5) = 6x^2 - 15x$
Multiply the second term $+1$ by everything in the second bracket.
$1(2x - 5) = 2x - 5$
Combining the results:
$(3x + 1)(2x - 5) = 6x^2 - 15x + 2x - 5$
$ = 6x^2 - 13x - 5$

2. Problem:
Expand and simplify $(g - 5)(g^2 + 3g - 1)$
Solution:
$(g - 5)(g^2 + 3g - 1) = g(g^2 + 3g - 1) - 5(g^2 + 3g - 1)$
$ = g^3 + 3g^2 - g - 5g^2 - 15g + 5$
$ = g^3 - 2g^2 - 16g + 5$

Ensure that you multiply negative numbers correctly.

Multiplying out brackets exercise — Go online

Q1: Expand and simplify $(3x + 1)(2x + 5)$

Q2: Expand and simplify $(3x + 2)(4x - 1)$

Q3: Expand and simplify $(4 - 3x)(x + 2)$

TOPIC 6. COMPLEX NUMBERS

Q4: Expand and simplify $(2 - 3x)(1 - x)$
....

Q5: Expand and simplify $(b + 2)(b^2 - 4b + 3)$
....

Q6: Expand and simplify $(g - 3)(2g^2 + 5g - 7)$

6.1.2 Factorising a trinomial

The following example shows you how to factorise a trinomial expression.

Examples

1. Problem:

What is the factorised form of $x^2 + 11x + 30$?

i.e. $(x + a)(x + b)$ where $a \times b = +30$ and $a + b = +11$

Solution:

Step 1: Begin by finding the factors of $+30$

Step 2: Find the pair which add to make $+11$

Step 1	Step 2
-1 and -30	$(-1) + (-30) = -31$
-2 and -15	$(-2) + (-15) = -17$
-3 and -10	$(-3) + (-10) = -13$
-5 and -6	$(-5) + (-6) = -11$
$+1$ and $+30$	$+1 + 30 = +31$
$+2$ and $+15$	$+2 + 15 = +17$
$+3$ and $+10$	$+3 + 10 = +13$
$+5$ and $+6$	$\mathbf{+5 + 6 = +11}$

We have $5 \times 6 = 30$ and $5 + 6 = 11$

That gives us the values $a = 5$ and $b = 6$.

We can state that $x^2 + 11x + 30 = (x + 5)(x + 6)$

Multiply out the brackets to check:

$(x + 5)(x + 6) = x(x + 6) + 5(x + 6)$
$ = x^2 + 6x + 5x + 30$
$ = x^2 + 11x + 30$
....

© HERIOT-WATT UNIVERSITY

2. Problem:

Factorise $14x^2 - 20x + 6$

Solution

We should always check for a simple common factor first. This question has a common factor of 2 giving $2(7x^2 - 10x + 3)$

Next we want two terms to make $7x^2$, e.g. $7x \times x$, so our brackets start $(7x\ \ )(x\ \ )$

Now we want two numbers which multiply to make $+3$, e.g. 3×1 or 1×3.

In this example, we need to check that the sum of the products of the inner and outer terms gives the middle term $-10g$.

$(7x + 3)(x + 1)$ makes the product of the inner terms $3x$ and the outer terms $7x$ so the sum is $10x$. We nearly have it, but we wanted $-10x$.

The sign is wrong, but we know that -3×-1 also makes 3 so:

$(7x - 3)(x - 1) = 7x^2 - 10x + 3$

Hence:
$$14x^2 - 20x + 6 = 2\left(7x^2 - 10x + 3\right)$$
$$= 2(7x - 3)(x - 1)$$

Multiply out the brackets to check:
$$2\left(7x - 3\right)(x - 1) = 2\left(7x\left(x - 1\right) - 3\left(x - 1\right)\right)$$
$$= 2\left(7x^2 - 7x - 3x + 3\right)$$
$$= 2\left(7x^2 - 10x + 3\right)$$
$$= 14x^2 - 20x + 6$$

Factorising a trinomial exercise   Go online

Factorise the following trinomials

Q7: $6x^2 + 17x + 5$

Q8: $7 + 40x - 12x^2$

Q9: $5x^2 - 16x + 3$

Q10: $25 - 4x^2$

Q11: $3 - 14x + 8x^2$

6.1.3 Arithmetic of surds

Examples

1. Simplifying a surd
Problem:
Simplify $\sqrt{40}$
Solution:
Step 1: Find the biggest square number that goes into 40.
$\sqrt{40} = \sqrt{4 \times 10}$
Step 2: Separate using the rule $\sqrt{ab} = \sqrt{a} \times \sqrt{b}$.
$\sqrt{40} = \sqrt{4} \times \sqrt{10}$
Step 3: Evaluate the surd.
$\sqrt{40} = 2\sqrt{10}$

....

2. Adding and subtracting surds
Problem:
Evaluate $\sqrt{40} + 2\sqrt{90}$
Solution:
Step 1: In order to add or subtract surds, you need to simplify them first.
$$\sqrt{40} = \sqrt{4 \times 10}$$
$$= \sqrt{4} \times \sqrt{10}$$
$$= 2\sqrt{10}$$
$$2\sqrt{90} = 2 \times \sqrt{9 \times 10}$$
$$= 2 \times \sqrt{9} \times \sqrt{10}$$
$$= 2 \times 3 \times \sqrt{10}$$
$$= 6\sqrt{10}$$
Step 2: Add or subtract surds that are the same (like with algebra).
$$\sqrt{40} + 2\sqrt{90} = 2\sqrt{10} + 6\sqrt{10}$$
$$= 8\sqrt{10}$$

....

3. Multiplying surds
Problem:
Evaluate $3\sqrt{5} \times \sqrt{8}$
Solution:
Step 1: Multiply the surds together using the rule $\sqrt{a} \times \sqrt{b} = \sqrt{ab}$.
$$3\sqrt{5} \times \sqrt{8} = 3 \times \sqrt{5 \times 8}$$
$$= 3\sqrt{40}$$

© HERIOT-WATT UNIVERSITY

Step 2: Simplify the surd by finding the biggest square number that goes into 40.
$$3\sqrt{5} \times \sqrt{8} = 3\sqrt{4 \times 10}$$
Step 3: Evaluate the surd.
$$3\sqrt{5} \times \sqrt{8} = 3 \times 2\sqrt{10}$$
$$= 6\sqrt{10}$$

4. Expanding brackets with surds

Problem:

Evaluate $(3\sqrt{2} + 3)(5 - \sqrt{6})$

Solution:

Step 1: Expand the brackets by taking the terms from the first bracket and multiplying each by the second bracket.

$$(3\sqrt{2} + 3)(5 - \sqrt{6})$$

$$(3\sqrt{2} + 3)(5 - \sqrt{6}) = 3\sqrt{2}(5 - \sqrt{6}) + 3(5 - \sqrt{6})$$

Step 2: Multiply the surds together using the rule $\sqrt{a} \times \sqrt{b} = \sqrt{ab}$.

$$(3\sqrt{2} + 3)(5 - \sqrt{6}) = 15\sqrt{2} - 3\sqrt{2}\sqrt{6} + 15 - 3\sqrt{6}$$
$$= 15\sqrt{2} - 3\sqrt{12} + 15 - 3\sqrt{6}$$

Step 3: Simplify the surds if possible.

$$(3\sqrt{2} + 3)(5 - \sqrt{6}) = 15\sqrt{2} - 3\sqrt{4 \times 3} + 15 - 3\sqrt{6}$$
$$= 15\sqrt{2} - 3 \times 2\sqrt{3} + 15 - 3\sqrt{6}$$
$$= 15\sqrt{2} - 6\sqrt{3} + 15 - 3\sqrt{6}$$

5. Rationalising the denominator

Problem:

Rationalise $\frac{3\sqrt{2}+\sqrt{5}}{5-\sqrt{3}}$

Solution:

Step 1: It is standard practice to rewrite a fraction so that a surd does not appear on the bottom of the fraction. We do this by multiplying the denominator by its conjugate, i.e. if we have $a - b$ then the conjugate is $a + b$.

$$\frac{3\sqrt{2}+\sqrt{5}}{5-\sqrt{3}} = \frac{(3\sqrt{2}+\sqrt{5})}{(5-\sqrt{3})} \times \frac{(5+\sqrt{3})}{(5+\sqrt{3})}$$

Notice that the conjugate gives a difference of two squares.

Step 2: Multiply out the brackets.

$$\frac{3\sqrt{2}+\sqrt{5}}{5-\sqrt{3}} = \frac{3\sqrt{2} \times 5 + 3\sqrt{2}\sqrt{3} + 5\sqrt{5} + \sqrt{5}\sqrt{3}}{25 - 5\sqrt{3} + 5\sqrt{3} - 3}$$
$$= \frac{15\sqrt{2} + 3\sqrt{6} + 5\sqrt{5} + \sqrt{15}}{22}$$

TOPIC 6. COMPLEX NUMBERS

Arithmetic of surds exercise

Q12: Simplify $\sqrt{50}$

Q13: Evaluate $4\sqrt{20} + 2\sqrt{45}$

Q14: Evaluate $3\sqrt{8} - 2\sqrt{18}$

Q15: Evaluate $5\sqrt{6} \times \sqrt{3}$

Q16: Evaluate $(2\sqrt{2} + 3)(5 - 2\sqrt{3})$

Q17: Rationalise the denominator $\frac{4\sqrt{3}-2}{1+\sqrt{3}}$

Q18: Rationalise the denominator $\frac{2\sqrt{5}+3}{4-\sqrt{2}}$

Q19: Evaluate $2\sqrt{27} + 5\sqrt{12}$

Q20: Evaluate $\sqrt{18} - 3\sqrt{32}$

Q21: Evaluate $2\sqrt{45} - 5\sqrt{80} + \sqrt{2}$

Q22: Evaluate $3\sqrt{28} + 4\sqrt{63}$

Q23: Evaluate $2\sqrt{12} \times \sqrt{3}$

Q24: Evaluate $\sqrt{10} \times 7\sqrt{5}$

Q25: Evaluate $(3\sqrt{2} - 2)(5 + \sqrt{3})$

Q26: Evaluate $(7 - 3\sqrt{5})(4 - 2\sqrt{5})$

Q27: Rationalise the denominator $\frac{5\sqrt{2}+6}{1-\sqrt{5}}$

Q28: Rationalise the denominator $\frac{2\sqrt{2}-3}{4-\sqrt{2}}$

© HERIOT-WATT UNIVERSITY

50 TOPIC 6. COMPLEX NUMBERS

6.1.4 Trigonometric identities

These formulae should be remembered from National 5 and Higher:

- $\sin(A \pm B) = \sin A \cos B \pm \cos A \sin B$
- $\cos(A \pm B) = \cos A \cos B \mp \sin A \sin B$
- $\sin(2A) = 2 \sin A \cos A$
- $\cos(2A) = \cos^2 A - \sin^2 A$
  $ = 2\cos^2 A - 1$
  $ = 1 - 2\sin^2 A$
- $\cos^2 A + \sin^2 A = 1$
- $\tan A = \frac{\sin A}{\cos A}$

Examples

1. Problem:

Show that $\frac{2\tan A}{1+\tan^2 A} = \sin(2A)$

Solution:

$\frac{2\tan A}{1+\tan^2 A} = \frac{\frac{2\sin A}{\cos A}}{\frac{\cos^2 A}{\cos^2 A} + \frac{\sin^2 A}{\cos^2 A}}$ remember $\tan A = \frac{\sin A}{\cos A}, 1 = \frac{\cos^2 A}{\cos^2 A}, \tan^2 A = \frac{\sin^2 A}{\cos^2 A}$

$= \frac{\frac{2\sin A}{\cos A}}{\frac{\cos^2 A + \sin^2 A}{\cos^2 A}}$ add the fractions on the denominator

$= \frac{2\sin A}{\cos A} \times \frac{\cos^2 A}{\cos^2 A + \sin^2 A}$ simplify the fractions

$= \frac{2 \sin A \, \cos A}{\cos^2 A + \sin^2 A}$ cancel out $\cos A$

$= \frac{2 \sin A \, \cos A}{1}$ remember that $\cos^2 A + \sin^2 A = 1$

$= \sin(2A)$ remember that $\sin(2A) = 2 \sin A \cos A$

....

2. Problem:

Prove that $\cos^2 x = \frac{1}{2}(1 + \cos(2x))$

Solution:

We know that $\cos(2x) = 2\cos^2 x - 1$.

$2\cos^2 x - 1 = \cos(2x)$

$2\cos^2 x = \cos(2x) + 1$

$\cos^2 x = \frac{1}{2}(\cos(2x) + 1)$

$\cos^2 x = \frac{1}{2}(1 + \cos(2x))$

....

© HERIOT-WATT UNIVERSITY

TOPIC 6. COMPLEX NUMBERS

3. Problem:
Show that $\cos(4x) = \cos^4 x - 6\sin^2 x \cos^2 x + \sin^4 x$

Solution:

$$\begin{aligned}
\cos(4x) &= \cos 2(2x) & &\text{express in the form } \cos(2A) \\
&= \cos^2(2x) - \sin^2(2x) & &\text{expand as for } \cos(2A) \\
&= (\cos(2x))^2 - (\sin(2x))^2 \\
&= (\cos^2 x - \sin^2 x)^2 - (2\sin x \cos x)^2 & &\text{expand for } \cos(2A) \text{ and } \sin(2A) \\
&= \cos^4 x - 2\sin^2 x \cos^2 x + \sin^4 x - 4\sin^2 x \cos^2 x & &\text{collect like terms} \\
&= \cos^4 x - 6\sin^2 x \cos^2 x + \sin^4 x
\end{aligned}$$

....

4. Problem:
Find the value of $\cos(2x)$ if $2\cos^2 x - 4\sin^2 x + 1 = 3$

Solution:

$$\begin{aligned}
2\cos^2 x - 4\sin^2 x + 1 &= 3 & &\text{remember: } 2\cos^2 x - 1 = \cos(2x) \text{ and } 1 - 2\sin^2 x = \cos(2x) \\
2\cos^2 x - 1 + 2 - 4\sin^2 x &= 3 & &\text{notice that: } -1 + 2 = 1 \\
(2\cos^2 - 1) + 2(1 - 2\sin^2 x) &= 3 & &\text{remember: } 1 - 2\sin^2 x = \cos(2x) \\
\cos(2x) + 2\cos(2x) &= 3 \\
3\cos(2x) &= 3 \\
\cos(2x) &= 1
\end{aligned}$$

....

5. Problem:
Show that $\sin(3\theta) = 3\sin\theta - 4\sin^3\theta$

Solution:

This is a tricky question, but let's start by expressing 3θ as $2\theta + \theta$!

$$\begin{aligned}
\sin(3\theta) &= \sin(2\theta + \theta) & &\text{expand } \sin(A+B) \\
&= \sin(2\theta)\cos\theta + \cos(2\theta)\sin\theta & &\text{expand } \sin(2A) \text{ and } \cos(2A) \\
&= (2\sin\theta \cos\theta)\cos\theta + (1 - 2\sin^2\theta)\sin\theta & &\text{multiply out the brackets} \\
&= 2\sin\theta \cos^2\theta + \sin\theta - 2\sin^3\theta & &\text{common factor } \sin\theta \\
&= \sin\theta(2\cos^2\theta + 1 - 2\sin^2\theta) & &\text{rearrange the bracket} \\
&= \sin\theta(2\cos^2\theta - 2\sin^2\theta + 1) & &\text{common factor 2 in first two bracketed terms} \\
&= \sin\theta(2(\cos^2\theta - \sin^2\theta) + 1) & &\text{replace } \cos^2\theta - \sin^2\theta \\
&= \sin\theta(2(1 - 2\sin^2\theta) + 1) & &\text{expand the inner bracket} \\
&= \sin\theta(2 - 4\sin^2\theta + 1) & &\text{collect like terms} \\
&= \sin\theta(3 - 4\sin^2\theta) & &\text{expand the bracket} \\
&= 3\sin\theta - 4\sin^3\theta
\end{aligned}$$

© HERIOT-WATT UNIVERSITY

Trigonometric identities exercise

Q29: Show that $\sin(x - \pi) = -\sin x$

Q30: Show that $\cos\left(\frac{\pi}{2} + x\right) = -\sin x$

Q31: Give $\cos\theta \sin^2\theta$ in terms of cosine only.

Q32: Give $\cos^4\theta$ in terms of sine only.

Q33: Show that $\cos(\theta - \pi) = -\cos x$

Q34: Show that $\sin\left(x + \frac{3\pi}{2}\right) = -\cos x$

Q35: Give $\cos^2\theta \sin^3\theta$ in terms of $\sin\theta$ only.

Q36: Give $\sin^4 x \cos^2 x$ in terms of cosine only.

This exercise should to help identify any areas of weakness in techniques which are required for the study of this section. Some revision may be necessary if any of the questions seem difficult.

Revision exercise

Q37: Solve the quadratic $3x^2 - 13x + 4 = 0$

Q38: Expand $(2x + 3)(x - 2)$

Q39: Expand $(2x - y)^4$ using the binomial theorem.

Q40: Show that $\cos(a + \pi) = -\cos a$

Q41: Simplify the expression $\frac{(3\sqrt{48} - \sqrt{27})}{(2\sqrt{12} + \sqrt{75})}$

6.2 Introduction to Imaginary numbers

Usually, in mathematics, the first number system that is encountered is the set of natural numbers ($\mathbb{N}$), i.e. 1, 2, 3, ...

This set is subsequently enlarged to cope with various difficulties.

- The set of integers ($\mathbb{Z}$) enables subtractions such as $3 - 5 = -2$
  $-2 \notin \mathbb{N}$, but $-2 \in \mathbb{Z}$
- The set of rationals ($\mathbb{Q}$) enables divisions such as $4 \div 6 = \frac{2}{3}$
  $\left(\frac{2}{3} \notin \mathbb{Z}\right)$, but $\left(\frac{2}{3} \in \mathbb{Q}\right)$
- The set of reals ($\mathbb{R}$) provides solutions to certain quadratic equations such as $x^2 - 2 = 0$, which has the solutions $x = \pm\sqrt{2}$
  $\pm\sqrt{2} \notin \mathbb{Q}$, but $\pm\sqrt{2} \in \mathbb{R}$

However, the set $\mathbb{R}$ will not allow solutions to all quadratic equations, e.g. $x^2 + 1 = 0$; this implies that $x^2 = -1$ which has no real solution since x^2 is always positive for real numbers.

Key point

This difficulty can be overcome by enlarging the set $\mathbb{R}$ to include a new 'number', denoted i, with the property that:

$$i^2 = -1$$

This 'number' i leads to a new and extensive branch of mathematics.

Standard numbers

Here is how the complex numbers fit in with the standard number sets already known.

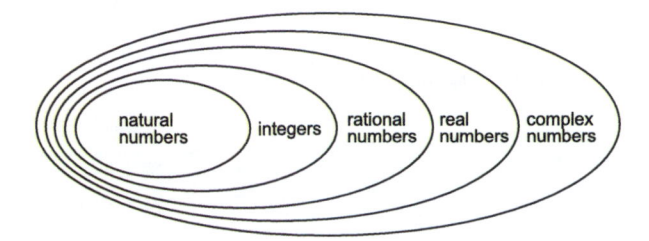

6.2.1 Square roots of a negative number and quadratic equations

The introduction of this new number i overcomes the problem of finding the square roots of a negative number.

Example

Problem:

Find the square roots of -25

Solution:

Separate -25:

$$-25 = ? \times -1$$
$$= 25 \times -1$$

We know that $i^2 = -1$ so: $-25 = 25 \times i^2$

As a squared number:

$$-25 = 5^2 \times i^2$$
$$= (5i)^2$$

So this gives:

$$\sqrt{-25} = \sqrt{(5i)^2}$$
$$= \pm 5i$$

With the existence of the square roots of a negative number, it is possible to find the solutions of any quadratic equation of the form $ax^2 + bx + c = 0$ using the quadratic formula.

Example

Problem:

Solve the quadratic equation $x^2 - 2x + 3 = 0$

We can use the quadratic formula to solve this quadratic equation:

Quadratic formula: $x = \frac{-b \pm \sqrt{b^2 - 4ac}}{2a}$

Where $ax^2 + bx + c = 0$ and $a \neq 0$

Solution:

$x^2 - 2x + 3 = 0$

$a = 1, b = -2, c = 3$

$x = \dfrac{2 \pm \sqrt{(-2)^2 - 4(1)(3)}}{2(1)}$

$= \dfrac{2 \pm \sqrt{-8}}{2}$

$= \dfrac{2 \pm \sqrt{8i^2}}{2}$ Note that $i^2 = -1$

$= \dfrac{2 \pm 2i\sqrt{2}}{2}$ Always simplify surds, remembering here that $\sqrt{8} = 2\sqrt{2}$

$= 1 \pm i\sqrt{2}$

> **Top tip**
>
> The solution can be written as $x = 1 \pm i\sqrt{2}$ or $x = 1 \pm \sqrt{2}\,i$.
> By using the form $i\sqrt{2}$, we avoid confusing it with $\sqrt{2i}$.

Example

Problem:

Give three quadratic equations which have no solution in the real numbers.

Solution:

For there to be no solution to the quadratic equation, we must check that $b^2 - 4ac < 0$

There are an infinite number of solutions to this.

Initially, we guess values for a, b and c, then check whether $b^2 - 4ac < 0$

- Let $a = 1, b = 3$ and $c = 4$
  Now calculate $b^2 - 4ac < 0$:
  $b^2 - 4ac = 3^2 - 4(1)(4)$
  $ = 9 - 16$
  $ = -7$
  $ < 0$

  There are no real roots so the quadratic equation $x^2 + 3x + 4 = 0$ has no solution in the real numbers.

- Let $a = 3$, $b = -4$ and $c = 3$
  Now calculate $b^2 - 4ac < 0$:
  $$b^2 - 4ac = (-4)^2 - 4\,(3)\,(3)$$
  $$= 16 - 36$$
  $$= -20$$
  $$< 0$$
  There are no real roots so the quadratic equation $3x^2 - 4x + 3 = 0$ has no solution in the real numbers.

- Let $a = -3$, $b = -2$ and $c = -1$
  Now calculate $b^2 - 4ac < 0$:
  $$b^2 - 4ac = (-2)^2 - 4\,(-3)\,(-1)$$
  $$= 4 + 12$$
  $$= 16$$
  $$> 0$$
  There are real roots so this is not the quadratic we are looking for.

- Let $a = -3$, $b = -2$ and $c = 1$
  Now calculate $b^2 - 4ac < 0$:
  $$b^2 - 4ac = (-2)^2 - 4\,(-3)\,(1)$$
  $$= 4 - 12$$
  $$= -8$$
  $$< 0$$
  There are no real roots so the quadratic equation $-3x^2 - 2x + 1 = 0$ has no solution in the real numbers.

Therefore, three quadratic equations which have no solution in the real numbers are:

1. $x^2 + 3x + 4 = 0$
2. $3x^2 - 4x + 3 = 0$
3. $-3x^2 - 2x + 1 = 0$

Square roots of a negative number and quadratic equations exercise Go online

Q42: Find the square roots of -16
....

Q43: Find the square roots of -64
....

Q44: Find the square roots of -8
....

TOPIC 6. COMPLEX NUMBERS

Q45: Solve the quadratic equation $3x^2 - 2x + 4 = 0$

....

Q46: Solve the quadratic equation $x^2 - 2x + 5 = 0$ using the quadratic formula. (Your answer will involve the new number i)

....

Q47: Solve the quadratic equation $x^2 + 2x + 6 = 0$ using the quadratic formula.

....

Q48: Solve the quadratic equation $x^2 + x + 1 = 0$ using the quadratic formula.

6.2.2 Complex numbers and the complex plane

The solutions to these problems are in the form $a + bi$ (or $a + ib$). Such expressions are called complex numbers.

Note that it is customary to denote a complex number by the letter z.

If $z = a + ib$ then:

- the number a is called the real part of z;
- the number b is called the imaginary part of z.

These are sometimes denoted Re(z) for a and Im(z) for b

That is, Re($a + ib$) = a and Im($a + ib$) = b

Identifying the real and imaginary parts of complex numbers exercise   Go online

Q49: Identify the real and imaginary part of the complex number $-5 + 3i$

....

Q50: Identify the real and imaginary part of the complex number $-6i - 2$

If the real part of a complex number is zero, the number is said to be *purely imaginary*.

If the imaginary part of a complex number is zero, the number is just a real number. Therefore, the real numbers are a subset of the complex numbers.

The real numbers can be represented on the number line.

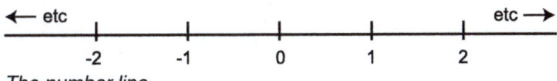

← etc   etc →
-2 -1 0 1 2

The number line

© HERIOT-WATT UNIVERSITY

TOPIC 6. COMPLEX NUMBERS

Is there a similar representation for the complex numbers?

The definition of a complex number involves two real numbers. Two real numbers can be used to define a point on a plane.

Therefore, complex numbers can be plotted on a plane by using the x-axis for the real part and the y-axis for the imaginary part.

This plane is called the *complex plane* or *Argand diagram*.

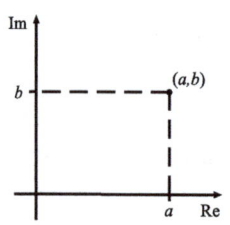

Argand diagram

The point (a, b) on the Argand diagram represents the complex number $z = a + ib$

Example

Problem:

Show the complex numbers $z = -2 + 3i$ and $w = 1 - 2i$ on an Argand diagram.

Solution:

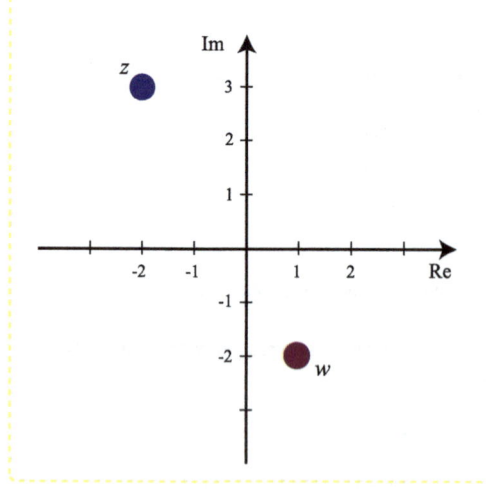

TOPIC 6. COMPLEX NUMBERS

Complex numbers and the complex plane exercise   Go online

The points on the following Argand diagram represent three complex numbers.

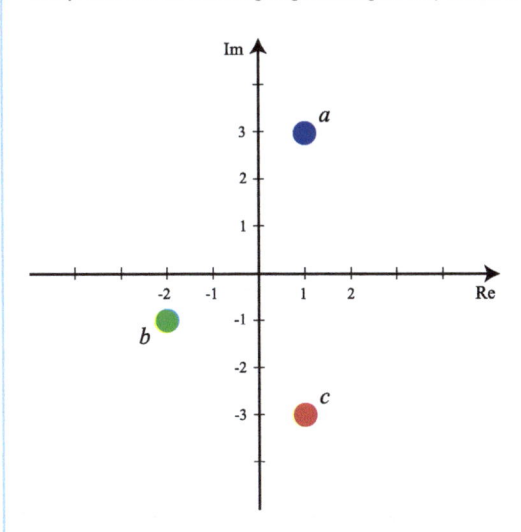

Q51: Identify the point a on the diagram as the complex number which it represents.
....

Q52: Identify the point b on the diagram as the complex number which it represents.
....

Q53: Identify the point c on the diagram as the complex number which it represents.

6.3 The arithmetic of complex numbers

Since complex numbers have real and imaginary components, the rules for adding, subtracting, multiplying and dividing have to be clarified. These rules are not new and have been seen before. They are similar to the rules for algebra and rationalising surds.

© HERIOT-WATT UNIVERSITY

6.3.1 Addition and subtraction of complex numbers

Adding complex numbers

Take two complex numbers: $3 + 4i$ and $2 - i$

To add these numbers together:

1. Add the *real* parts together: $3 + 2 = 5$
2. Add the *imaginary* parts together: $4 + (-1) = 3$

So,
$$(3 + 4i) + (2 - i) = (3 + 2) + (4i - i)$$
$$= 5 + 3i$$

> **Key point**
>
> To add two complex numbers:
>
> 1. Add the *real* parts;
> 2. Add the *imaginary* parts.
>
> $(a + ib) + (c + id) = (a + c) + i(b + d)$

Example : Adding complex numbers

Problem:

Add $\frac{1}{2} - i\frac{\sqrt{3}}{4}$ and $3 + i\frac{\sqrt{3}}{2}$

Solution:

$$\left(\frac{1}{2} - i\frac{\sqrt{3}}{4}\right) + \left(3 + i\frac{\sqrt{3}}{2}\right) = \left(\frac{1}{2} + 3\right) + i\left(\frac{-\sqrt{3}}{4} + \frac{\sqrt{3}}{2}\right)$$
$$= \frac{7}{2} + i\frac{\sqrt{3}}{4}$$

Or, alternatively:

- Adding the *real* parts gives: $\left(\frac{1}{2} + 3\right) = \frac{7}{2}$
- Adding the *imaginary* parts gives: $\left(\frac{-\sqrt{3}}{4} + \frac{\sqrt{3}}{2}\right) = \frac{\sqrt{3}}{4}$

So the sum is $\frac{7}{2} + i\frac{\sqrt{3}}{4}$

TOPIC 6. COMPLEX NUMBERS

Graphic illustration of adding of two complex numbers

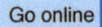

Go online

In the complex plane, the addition of $(1 + 3i) + (5 + 2i) = 6 + 5i$ can be interpreted geometrically as follows.

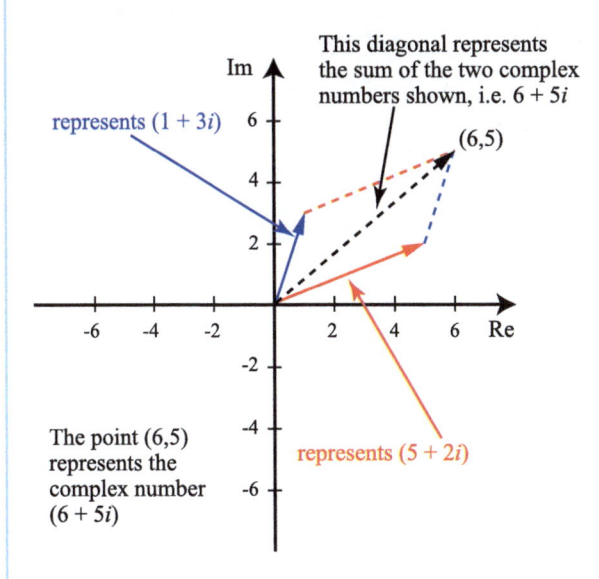

represents $(1 + 3i)$

This diagonal represents the sum of the two complex numbers shown, i.e. $6 + 5i$

The point (6,5) represents the complex number $(6 + 5i)$

represents $(5 + 2i)$

Note the construction of the diagram:

- the points representing the complex numbers are plotted;
- lines are drawn joining each of them to the origin;
- two further lines are drawn to complete a parallelogram - this is like adding vectors and is the geometrical representation of this;
- the diagonal of this parallelogram is drawn, starting from the origin.

This diagonal corresponds to the sum of the complex numbers.

Subtracting complex numbers

Subtract $3 - i$ from $7 + i$

To subtract these numbers:

1. Subtract the *real parts* from one another: $7 - 3 = 4$
2. Subtract the *imaginary parts* from one another: $1 - (-1) = 2$

© HERIOT-WATT UNIVERSITY

So

$(7+i) - (3-i) = (7-3) + i(1-(-1))$
$ = 4 + 2i$

> **Key point**
>
> To subtract two complex numbers:
>
> 1. Subtract the *real* parts;
> 2. Subtract the *imaginary* parts.
>
> $(a + ib) - (c + id) = (a - c) + i(b - d)$

Example : Subtracting complex numbers

Problem:

Subtract $\frac{1}{2} - i\frac{\sqrt{3}}{4}$ from $3 + i\frac{\sqrt{3}}{2}$

Solution:

$$\left(3 + i\frac{\sqrt{3}}{2}\right) - \left(\frac{1}{2} - i\frac{\sqrt{3}}{4}\right) = \left(3 - \frac{1}{2}\right) + i\left(\frac{\sqrt{3}}{2} - \left(-\frac{\sqrt{3}}{4}\right)\right)$$
$$= \frac{5}{2} + i\left(\frac{\sqrt{3}}{2} + \frac{\sqrt{3}}{4}\right)$$
$$= \frac{5}{2} + i\frac{3\sqrt{3}}{4}$$

Or, alternatively:

- Subtracting the *real* parts gives: $\left(3 - \frac{1}{2}\right) = \frac{5}{2}$
- Subtracting the *imaginary* parts gives: $\left(\frac{\sqrt{3}}{2} - \left(-\frac{\sqrt{3}}{4}\right)\right) = \frac{3\sqrt{3}}{4}$

So the sum is $\frac{5}{2} + i\frac{3\sqrt{3}}{4}$

TOPIC 6. COMPLEX NUMBERS

Graphical illustration of subtracting complex numbers Go online

In the complex plane, the subtraction of $(-3 + (-2i)) - (-6 + 4i) = 3 + (-6i)$ can be interpreted geometrically as follows.

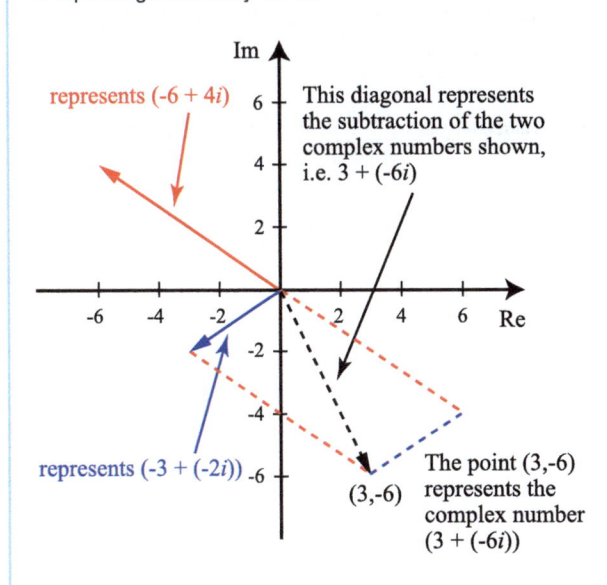

represents $(-6 + 4i)$

This diagonal represents the subtraction of the two complex numbers shown, i.e. $3 + (-6i)$

represents $(-3 + (-2i))$

The point $(3,-6)$ represents the complex number $(3 + (-6i))$

Note the construction of the diagram:

- the points representing the complex numbers are plotted;
- lines are drawn joining each of them to the origin;
- remember, to subtract a vector:
    - first change the direction of the vector to the opposite direction;
    - then add onto the previous vector.

This diagonal corresponds to the direct path from the origin to $3 - 6i$ after subtraction.

© HERIOT-WATT UNIVERSITY

6.3.2 Multiplying complex numbers

Before exploring the multiplication of complex number, we will first remind ourselves about multiplying imaginary numbers.

> **Key point**
>
> $$i^2 = -1$$

Multiplying imaginary numbers exercise   Go online

Q54: $i^2 = ?$

....

Q55: $i^3 = ?$

....

Q56: $i^4 = ?$

....

Q57: $i^5 = ?$

....

Q58: $i^6 = ?$

....

Q59: $-i^2 = ?$

....

Q60: $-i^3 = ?$

....

Q61: $-i^4 = ?$

....

Q62: $-i^5 = ?$

....

Q63: $-i^6 = ?$

The technique for multiplying two complex numbers is similar to that used when multiplying out two brackets.

Example

Problem:

Multiply the complex numbers $3 + 2i$ and $4 + 5i$

TOPIC 6. COMPLEX NUMBERS

Solution:

$(3 + 2i)(4 + 5i)$

- Multiply each term in the second bracket by 3:
$$3 \times (4 + 5i) = 3 \times 4 + 3 \times 5i$$
$$= 12 + 15i$$

- Multiply each term in the second bracket by $2i$, remembering that $i^2 = -1$:
$$2i \times (4 + 5i) = 2i \times 4 + 2i \times 5i$$
$$= 8i + 10i^2$$
$$= -10 + 8i$$

- Therefore:
$$(3 + 2i)(4 + 5i) = 12 + 15i - 10 + 8i$$
$$= 2 + 23i$$

Key point

To multiply complex numbers, use the technique for multiplying out two brackets:

$$(a + ib)(c + id) = (ac - bd) + i(bc + ad)$$

Examples

1. Multiplying two complex numbers

Problem:

Multiply $(1 + 2i)$ and $(2 - 3i)$

Solution:
$$(1 + 2i)(2 - 3i) = (1 \times 2) + (1 \times (-3i)) + (2i \times 2) + (2i \times (-3i))$$
$$= 2 - 3i + 4i + 6$$
$$= 8 + i$$

....

2. Problem:

Multiply $(4 - 2i)$ and $(2 + 3i)$

Solution:
$$(4 - 2i)(2 + 3i) = (4 \times 2) + (4 \times 3i) + (-2i \times 2) + (-2i \times 3i)$$
$$= 8 + 12i - 4i + 6$$
$$= 14 + 8i$$

© HERIOT-WATT UNIVERSITY

6.3.2.1 Finding the square roots of a complex number

Earlier examples showed that the square roots of a negative real number could be found in terms of i in the set of complex numbers. It is also possible to find the square roots of any complex number.

Example

Problem:

Find the square root of the complex number $5 + 12i$

Solution:

Let the square roots of $5 + 12i$ be $a + ib$:

$$5 + 12i = (a + ib)^2$$
$$= a^2 + 2iab + i^2b^2$$
$$= a^2 - b^2 + 2iab$$

Equate the real and imaginary parts:

$$a^2 - b^2 = 5 \quad (1)$$
$$2ab = 12 \quad (2)$$

Rearranging (2): $\quad b = \dfrac{6}{a} \quad (3)$

Substituting (3) into (1): $\quad a^2 - \left(\dfrac{6}{a}\right)^2 = 5$

Simplifying: $\quad a^2 - \dfrac{36}{a^2} = 5$

Multiplying through by a^2: $\quad a^4 - 5a^2 - 36 = 0$

Factorising and solving: $\quad (a^2 + 4)(a^2 - 9) = 0$

$$a^2 = -4 \text{ or } a^2 = 9$$

$a \in \mathbb{R}$ so $a^2 = -4$ is impossible

so $a = \pm 3$

Substituting into (2):

When $a = 3$, then $b = \dfrac{6}{3} \Rightarrow b = 2$ and when $a = -3$, then $b = \dfrac{6}{-3} \Rightarrow b = -2$.

The square roots of $5 + 12i$ are $3 + 2i$ and $-3 - 2i$.

TOPIC 6. COMPLEX NUMBERS

Finding the square roots of a complex number exercise   Go online

Q64: Find the square roots of the complex number $8 + 6i$
....

Q65: Find the square roots of the complex number $6 - 8i$
....

Q66: Find the square roots of the complex number $-8 + 15i$
....

Q67: Find the square roots of the complex number $-20 - 21i$

6.3.2.2 Multiplication by i

Multiplication by i has an interesting geometric interpretation. These examples should demonstrate what happens.

Examples

1. Problem:
Take the complex number $1 + 7i$ and multiply it by i.

Solution:
$$(1 + 7i)\,i = i + 7i^2$$
$$= -7 + i$$

....

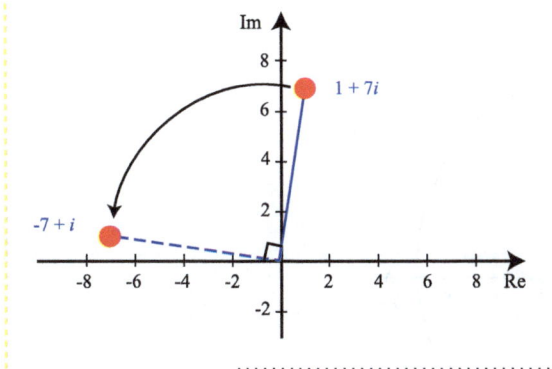

2. Problem:

Take the complex number $-6 - 2i$ and multiply it by i.

Solution:
$$(-6 - 2i)\,i = -6i - 2i^2$$
$$= 2 - 6i$$

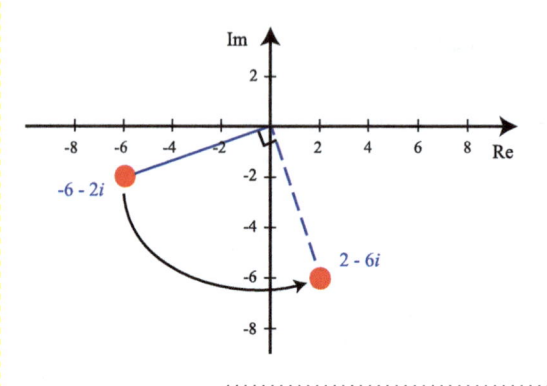

3. Problem:

Take the complex number $8 - 4i$ and multiply it by i.

Solution:
$$(8 - 4i)\,i = 8i - 4i^2$$
$$= 4 + 8i$$

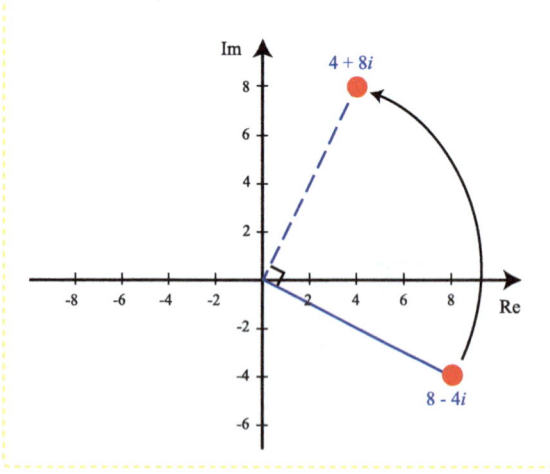

TOPIC 6. COMPLEX NUMBERS

> **Key point**
>
> In each of the examples, it can be seen that the effect of multiplying a complex number by i is an anti-clockwise rotation of the point on the Argand diagram through $90°$ or $\frac{\pi}{2}$.

Repeated multiplication by i can be summed up by the following diagram.

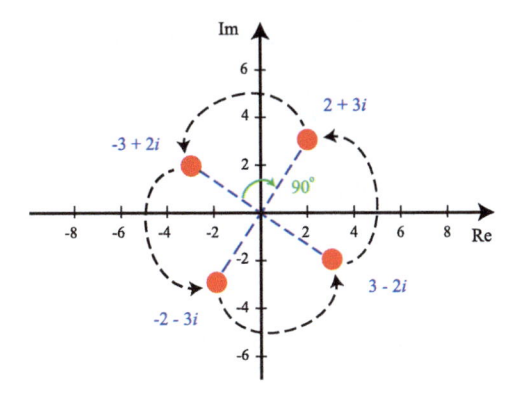

If z is multiplied by i four times, the answer is z; this is not surprising because:

$$i^4 = i^2 \times i^2$$
$$= (-1)(-1)$$
$$= 1$$

A similar effect can be achieved for multiplying by $-i$.

Example Problem:

Take the complex number $2 - 3i$ and multiply by $-i$

Solution:

$$(2 - 3i)(-i) = -2i + 3i^2$$
$$= -3 - 2i$$

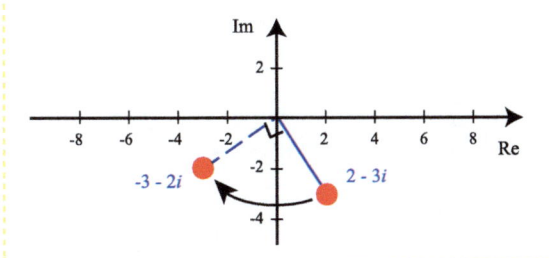

© HERIOT-WATT UNIVERSITY

> **Key point**
>
> Multiplying a complex number by $-i$ is a clockwise rotation of the point on the Argand diagram through $90°$ or $\frac{\pi}{2}$

Multiplication by i exercise

Go online

Q68: Multiply the complex number $3 - 2i$ by i and plot the points on an Argand diagram.

....

Q69: Multiply $-1 - 3i$ by i and plot the points on an Argand diagram.

....

Q70: Multiply $-3 + 2i$ by i and plot the points on an Argand diagram.

....

Q71: Multiply $-3 + 2i$ by i and plot the points on an Argand diagram.

....

Q72: Take the complex number $1 - 4i$. Plot the results on an Argand diagram when this number is repeatedly multiplied by the complex number $-i$.

Hence give a geometric interpretation of multiplication by $-i$ and check this interpretation with other complex numbers.

6.3.3 Conjugates of complex numbers

The basic operations of adding, subtracting and multiplying complex numbers are straightforward. Division is also possible but requires the use of the conjugate of a complex number.

> **Key point**
>
> The conjugate of a complex number $z = a + ib$ is denoted as $\overline{z}$ and defined by:
>
> $$\overline{z} = a - ib$$

The conjugate is sometimes denoted as z^*

TOPIC 6. COMPLEX NUMBERS

Geometrically, the conjugate of a complex number can be shown in an Argand diagram as the reflection of the point in the x-axis.

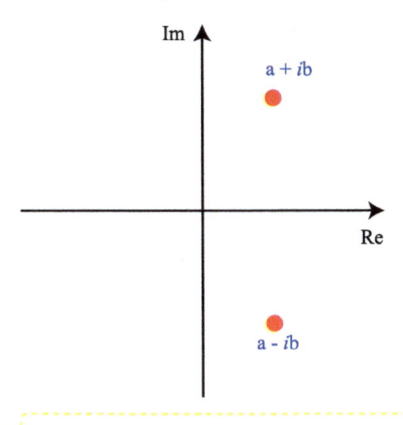

Examples

1. Problem:
Give the conjugate of the complex number $z = -6 + 8i$ and show z and $\bar{z}$ on an Argand diagram.

Solution:
The conjugate is $\bar{z} = -6 - 8i$

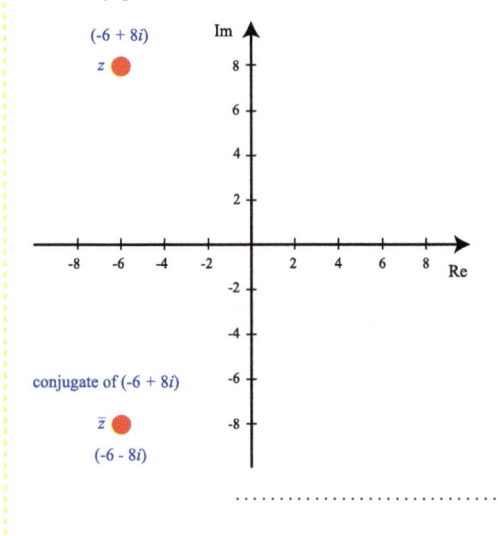

....

© HERIOT-WATT UNIVERSITY

2. Problem:
Given $z = 4 + 3i$, find $z\bar{z}$
Solution:
$\bar{z} = 4 - 3i$
So
$$\begin{aligned}z\bar{z} &= (4+3i)(4-3i) \\ &= 16 - 3i^2 \\ &= 16 + 9 \\ &= 25\end{aligned}$$

3. Problem:
Given $\bar{z} = 2 - 5i$, find $2i(z + 4)$
Solution:
$z = 2 + 5i$
So
$$\begin{aligned}2i(z+4) &= 2i(2+5i+4) \\ &= 2i(6+5i) \\ &= 12i + 10i^2 \\ &= -10 + 12i\end{aligned}$$

Conjugates of complex numbers exercise

Go online

Q73: Give the conjugate of the complex number $z = 4 + 7i$ and show z and $\bar{z}$ on an Argand diagram.

Q74: Give the conjugate of $5 + i$

Q75: Give the conjugate of $-3 - 5i$

Q76: Give the conjugate of $4i$

Q77: Give the conjugate of 5

Q78: Given $z = 3 + 2i$, find $z\bar{z}$

Q79: Given $\bar{z} = -4 - 5i$, find $3iz$

TOPIC 6. COMPLEX NUMBERS

Q80: Given $z = 2 - i$, find $i(\bar{z} + 1)$

6.3.4 Division of complex numbers

The technique used to divide complex numbers is similar to that used to rationalise the denominator of surd expressions such as $\frac{3}{\sqrt{2}}$

The trick is to multiply both top and bottom by a suitable number, in this case $\sqrt{2}$, in order to eliminate the surd from the denominator.

Here, $\frac{3}{\sqrt{2}}$ is multiplied by $\frac{\sqrt{2}}{\sqrt{2}}$ to give $\frac{3\sqrt{2}}{2}$

Examples

1. Problem:

Evaluate $\frac{5+i}{2-3i}$

Solution:

To divide by a complex number, we multiply by the complex conjugate of the denominator in order to change the complex number on the denominator into a real number:

$\frac{(5+i)}{(2-3i)} \times \frac{(2+3i)}{(2+3i)}$

The complex conjugate of $2 - 3i$ is $2 + 3i$.

$\frac{(5+i)}{(2-3i)} \times \frac{(2+3i)}{(2+3i)} = \frac{10+17i-3}{4-9i^2}$

We multiply by the conjugate so that the denominator becomes a real number.

$$\frac{(5+1)}{(2-3i)} \times \frac{(2+3i)}{(2+3i)} = \frac{7+17i}{13}$$

$$= \frac{7}{13} + \frac{17}{13}i$$

2. Dividing one complex number by another

Problem:

If $z_1 = 2 - 3i$ and $z_2 = 3 - 4i$, find $\frac{z_1}{z_2}$

© HERIOT-WATT UNIVERSITY

Solution:
Multiply top and bottom by $\bar{z}_2 = 3 + 4i$
Thus
$$\frac{z^1}{z^2} = \frac{2-3i}{3-4i}$$
$$= \frac{(2-3i)}{(3-4i)} \times \frac{(3+4i)}{(3+4i)}$$
$$= \frac{(6+8i-9i-12i^2)}{9-16i^2}$$
$$= \frac{(18-i)}{25}$$
$$= \frac{18}{25} - \frac{1}{25}i$$

....

3. Division of two complex numbers
Problem:
Given $z_1 = 4 + 3i$ and $z_2 = 2 - 5i$, find $\frac{z_1}{z_2}$
Solution:
$$\frac{4+3i}{2-5i} = \frac{(4+3i)}{(2-5i)} \times \frac{(2+5i)}{(2+5i)}$$
$$= \frac{8+20i+6i+15i^2}{4-25i^2}$$
$$= \frac{-7+26i}{29}$$
$$= -\frac{7}{29} + \frac{26}{29}i$$

....

4. Problem:
Given $z_1 = 5 - i$ and $z_2 = 5i$, find $\frac{z_1}{z_2}$
Solution:
$$\frac{5-i}{5i} = \frac{(5-i)}{(5i)} \times \frac{(-5i)}{(-5i)}$$
$$= \frac{-25i+5i^2}{-25i^2}$$
$$= \frac{-5-25i}{25}$$
$$= -\frac{1}{5} - i$$

....

5. Problem:

Given $z_1 = 7 + 2i$ and $z_2 = -i$, find $\frac{z_1}{z_2}$

Solution:

$$\frac{7+2i}{-i} = \frac{(7+2i)}{(-i)} \times \frac{i}{i}$$

$$= \frac{7i + 2i^2}{-i^2}$$

$$= \frac{-2 + 7i}{1}$$

$$= -2 + 7i$$

The effect of dividing by $-i$ is anticlockwise rotation of $90°$.

....

6. Problem:

Given $z_1 = 1 - 2i$ and $z_2 = 3 + i$, find $\frac{z_1}{z_2}$

Solution:

$$\frac{1-2i}{3-i} = \frac{(1-2i)}{(3+i)} \times \frac{(3-i)}{(3-i)}$$

$$= \frac{3 - i - 6i + 2i^2}{10}$$

$$= \frac{1 - 7i}{10}$$

$$= \frac{1}{10} - \frac{7}{10}i$$

Division by i exercise Go online

Q81: Take the complex number $2 + 3i$.

Plot the results on an Argand diagram when this number is repeatedly divided by the complex number i.

Hence give a geometric interpretation of division by i and check this interpretation with other complex numbers.

6.4 The modulus, argument and polar form of a complex number

Argand diagram: Cartesian form

Remember the Argand diagram in which the point (a, b) corresponds to the complex number $z = a + ib$

When the complex number is stated as $a + ib$, where a and b are real numbers, this is known as the *Cartesian form*.

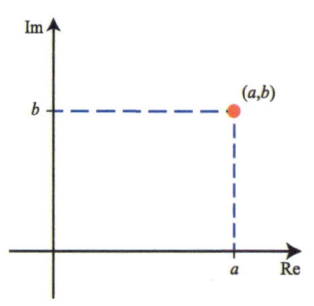

The point (a, b) represents the complex number $z = a + ib$ in Cartesian form

Argand diagram: Polar form

The point (a, b) can also be specified by giving the:

- distance, r, of the point from the origin;
- the angle, θ, between the positive x-axis and the line representing the complex number on an Argand diagram.

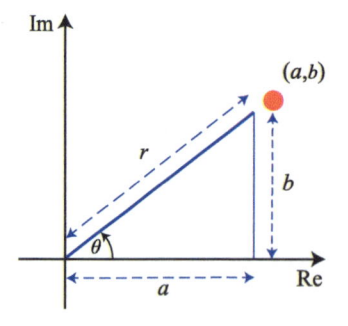

By some simple trigonometry, it follows that $a = r \cos \theta$ and $b = r \sin \theta$

> **Key point**
>
> The complex number $z = a + ib$ can also be written as: $z = r \cos \theta + ir \sin \theta$
> This is known as the *polar form*.

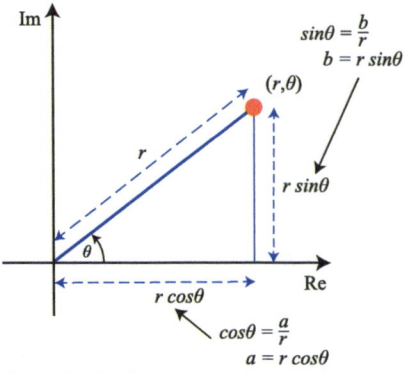

TOPIC 6. COMPLEX NUMBERS

The point (r, θ) represents the complex number $z = r\cos\theta + ir\sin\theta$ in Polar form

From the polar form we have:

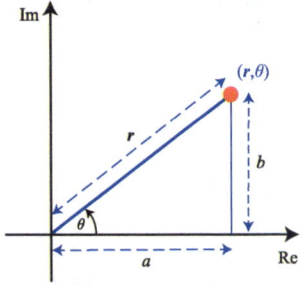

r is the modulus of z

Key point

The modulus r of a complex number $z = a + ib$ is written $|z|$ and is defined by:
$|z| = \sqrt{a^2 + b^2}$

Since we also know that $z = r\cos\theta + ir\sin\theta$ then, from $|z| = \sqrt{a^2 + b^2}$, we have an equivalent statement that:

$$\begin{aligned}|z| &= \sqrt{(r\cos\theta)^2 + (r\sin\theta)^2} \\ &= \sqrt{r^2\cos^2\theta + r^2\sin^2\theta} \\ &= \sqrt{r^2\left(\cos^2\theta + \sin^2\theta\right)} \\ &= \sqrt{r^2 \times 1} \\ &= r\end{aligned}$$

© HERIOT-WATT UNIVERSITY

From the polar form we also have:

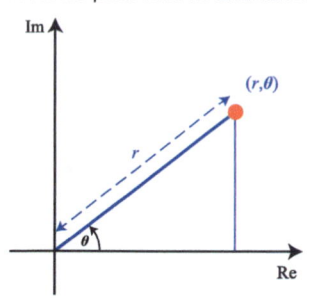

θ is the argument of z

> **Key point**
>
> The argument θ of a complex number is written $\arg(z)$ and is the anticlockwise angle between the positive x-axis and the line representing the complex number on an Argand diagram.

In some cases, calculations in polar form are much simpler so it is important to be able to work with complex numbers in both forms.

There will be times when conversion between these forms is necessary.

6.4.1 Cartesian form

Given the modulus (r) and argument (θ) of a complex number, it is easy to convert the number to Cartesian form.

Use the following steps to do this:

- evaluate $a = r \cos \theta$ and $b = r \sin \theta$
- state the number in the form $a + ib$

Examples

1. Problem:

If a complex number z has modulus of 2 and argument of $-\frac{\pi}{6}$, express z in the form $a + ib$ and plot the point which represents the number on an Argand diagram.

Solution:

$a = 2 \cos\left(-\frac{\pi}{6}\right)$
$= 2 \times \frac{\sqrt{3}}{2}$
$= \sqrt{3}$

$b = 2 \sin\left(-\frac{\pi}{6}\right)$
$= 2 \times -\frac{1}{2}$
$= -1$

So, $z = \sqrt{3} - i$

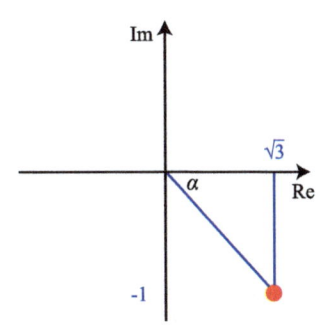

Check that $\sqrt{3} - i$ lies in the fourth quadrant and that $-\frac{\pi}{6}$ is also in the fourth quadrant.

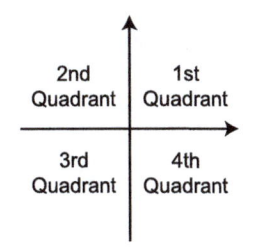

....

2. Problem:

If a complex number z has modulus 2 and argument $\frac{11\pi}{6}$, express z in the form $a + ib$ and plot the point which represents the number on an Argand diagram.

Solution:

$z = r\cos\theta + ir\sin\theta$

$z = 2\cos\left(\dfrac{11\pi}{6}\right) + 2i\sin\left(\dfrac{11\pi}{6}\right)$

$z = 2\cos\left(-\dfrac{\pi}{6}\right) + 2i\sin\left(-\dfrac{\pi}{6}\right)$

$z = 2 \times \dfrac{\sqrt{3}}{2} + 2i \times \dfrac{-1}{2}$

$z = \sqrt{3} - i$

TOPIC 6. COMPLEX NUMBERS

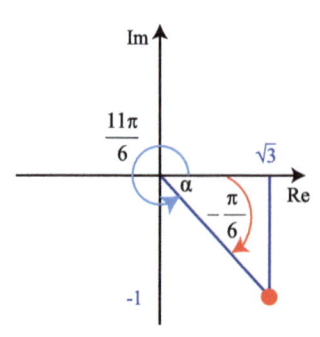

Notice that this is exactly the same point as in the previous example. This duplication demonstrates that different arguments can give the same complex number. To prevent confusion, the argument used is the principal value of the argument.

Cartesian form exercise

Go online

Convert the following complex numbers to Cartesian form.

Q82: A complex number z has modulus 2 and argument $\frac{\pi}{3}$.

....

Q83: A complex number z has modulus 3 and argument $30°$.

....

Q84: A complex number z has modulus 4 and argument $\frac{\pi}{4}$.

....

Q85: A complex number z has modulus 3 and argument $240°$.

....

Q86: A complex number z has modulus 2 and argument $\frac{3\pi}{2}$.

....

Q87: A complex number z has modulus 5 and argument $150°$.

....

Q88: Express $2\left(\cos\left(\frac{7\pi}{6}\right) + i\sin\left(\frac{7\pi}{6}\right)\right)$ in the standard form $(a + ib)$.

....

Q89: Express $3(\cos(25°) + i\sin(25°))$ in the standard form $(a + ib)$.
Give the answer to two decimal places.

6.4.2 Principal argument

The principal value of an argument is the value which lies between $-\pi$ and π

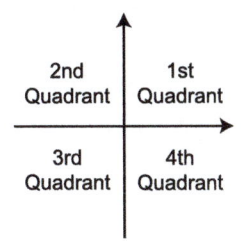

2nd Quadrant	1st Quadrant
3rd Quadrant	4th Quadrant

First quadrant

The principal argument in the first quadrant is shown in the following diagram.

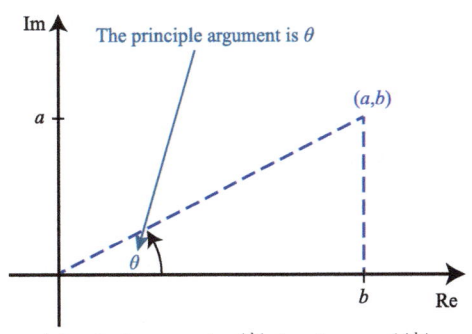

$\arg(z) = \theta$ where $\tan \theta = \left|\frac{b}{a}\right|$, i.e. $\theta = \tan^{-1}\left|\frac{b}{a}\right|$

The principal argument is positive since θ is between 0 and π in the anticlockwise direction from the x-axis.

The modulus of $\frac{b}{a}$ is used because we are not interested in whether the length is negative. All we need to know when calculating the acute angle is the value of $\frac{b}{a}$.

Note that taking

$$\tan^{-1}\left|\frac{b}{a}\right|$$

without the modulus sign on a calculator may give a different value than required.

We can determine whether the principal argument is a negative or positive angle depending on whether it is a clockwise or anticlockwise angle as mentioned above.

Second quadrant

The principal argument in the second quadrant is shown in the following diagram.

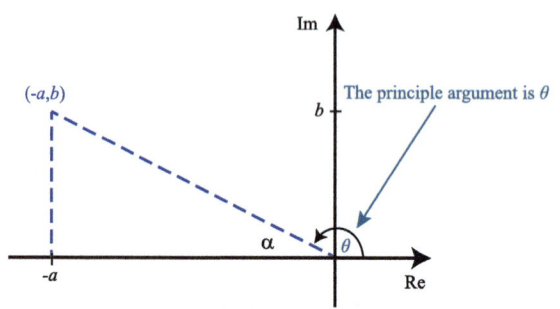

arg(z) = θ where $\theta = \pi - \alpha$ so to find α, calculate $\tan \alpha = \left|\frac{b}{a}\right|$, i.e. $\alpha = \tan^{-1}\left|\frac{b}{a}\right|$

The principal argument is positive since θ is between 0 and π in the anticlockwise direction from the x-axis.

Third quadrant

The principal argument in the 3rd quadrant is shown in the following diagram.

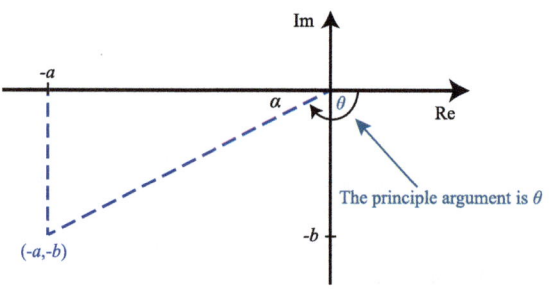

arg(z) = θ where $\theta = -\pi + \alpha$ so to find α, calculate $\tan \alpha = \left|\frac{b}{a}\right|$, i.e. $\alpha = \tan^{-1}\left|\frac{b}{a}\right|$

The principal argument is negative since θ is between 0 and $-\pi$ in the clockwise direction from the x-axis.

Fourth quadrant

The principal argument in the fourth quadrant is shown in the following diagram.

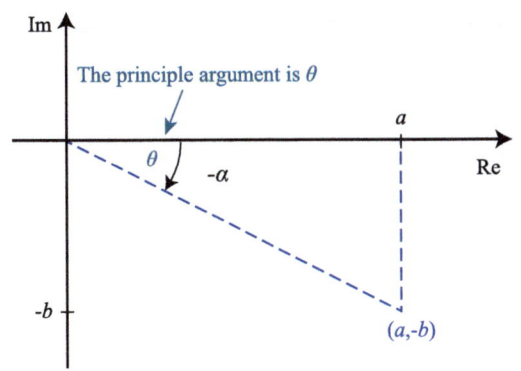

arg(z) = θ where $\theta = -\alpha$ so to find α, calculate $\tan\alpha = \left|\frac{b}{a}\right|$, i.e. $\alpha = \tan^{-1}\left|\frac{b}{a}\right|$

The principal argument is negative since θ is between 0 and $-\pi$ in the clockwise direction from the x-axis.

Summary

In summary, we can use the following diagram (like the CAST diagram from Higher) to find the principal argument θ in radians or degrees.

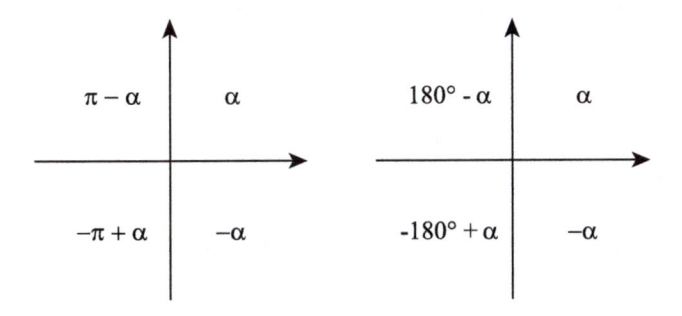

| $\pi - \alpha$ | α | $180° - \alpha$ | α |
| $-\pi + \alpha$ | $-\alpha$ | $-180° + \alpha$ | $-\alpha$ |

Examples

1. Problem:
What is the principal argument for the complex number $z = 3 + 6i$ to the nearest degree?
Solution:

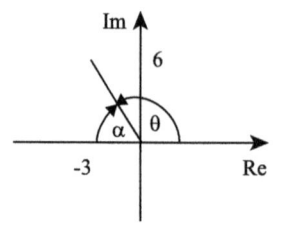

First we need to find α:
$$\alpha = \tan^{-1} \left| \frac{6}{3} \right|$$
$$= 63°$$
z is in the second quadrant so $\theta = 180° - \alpha$ and the argument is positive.
The principal argument is:
$$\theta = 180° - 63°$$
$$= 117°$$

....

2. Problem:
What is the principal argument for the complex number $z = -3 - 6i$ to the nearest degree?
Solution:

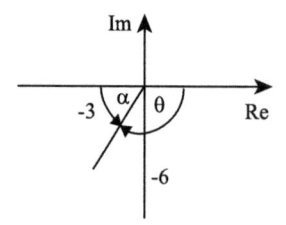

First we need to find α:
$$\alpha = \tan^{-1} \left| \frac{-6}{-3} \right|$$
$$= 63°$$
z is in the third quadrant so $\theta = -180° + \alpha$ and the argument is negative.
The principal argument is:
$$\theta = -180° + 63°$$
$$= -117°$$

....

TOPIC 6. COMPLEX NUMBERS

3. Problem:
What is the principal argument for the complex number $z = 3 - 3i$?
Solution:

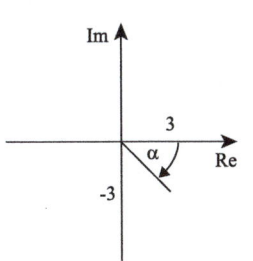

Im ↑

3
α
Re
-3

First we need to find α:
$$\alpha = \tan^{-1}\left|\frac{-3}{3}\right|$$
$$= \frac{\pi}{4}$$
z is in the fourth quadrant so $\theta = -\alpha$ and the argument is negative.
The principal argument is $\theta = -\frac{\pi}{4}$

Principal argument exercise   Go online

Q90: What is the principal argument for the complex number $z = -2 + 5i$?
Give the answer to the nearest degree.
....

Q91: What is the principal argument for the complex number $z = 4 - i$?
Give the answer to the nearest degree.
....

Q92: What is the principal argument for the complex number $z = -4 - 6i$?
Give the answer to the nearest degree.

Q93: What is the principal argument for the complex number $z = 2 - i$?
Give the answer in radians to two decimal places.
....

Q94: What is the principal argument for the complex number $z = -1 + 5i$?
Give the answer in radians to two decimal places.
....

Q95: What is the principal argument for the complex number $z = -5 - 7i$?
Give the answer in radians to two decimal places.

© HERIOT-WATT UNIVERSITY

6.4.3 Strategy to find modulus and argument

For the complex number $z = a + ib$:

- Plot the point (a, b) on an Argand diagram.

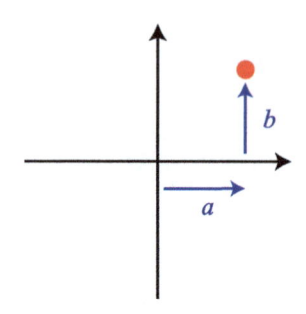

- Find the modulus $r = \sqrt{a^2 + b^2}$

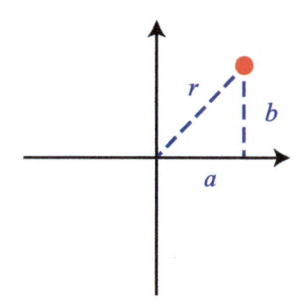

- Evaluate $\alpha = \tan^{-1}\left|\frac{b}{a}\right|$

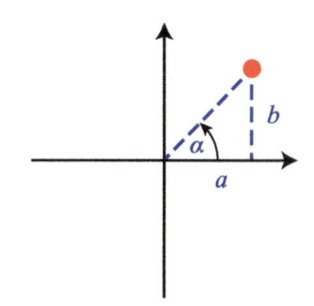

- Use the Argand diagram to find the value of the argument between $-\pi$ and $+\pi$

TOPIC 6. COMPLEX NUMBERS

In carrying out the final step, the following quadrant diagram may be useful:

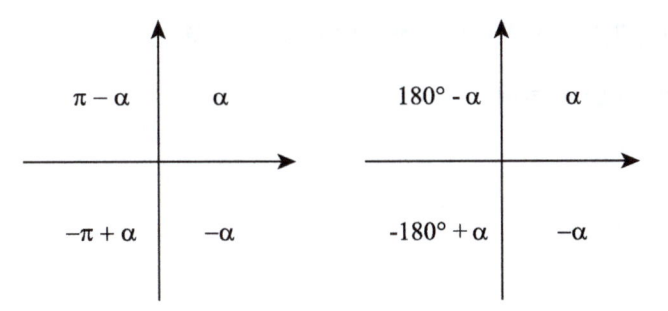

$\pi - \alpha$	α
$-\pi + \alpha$	$-\alpha$

$180° - \alpha$	α
$-180° + \alpha$	$-\alpha$

Once the modulus and argument of z are known, the **polar form** of a complex number can be written down.

Examples

1. Problem:
Find the modulus and principal argument of the complex number $z = 1 + i\sqrt{3}$

Solution:
First, plot the point $1 + i\sqrt{3}$ on an Argand diagram.

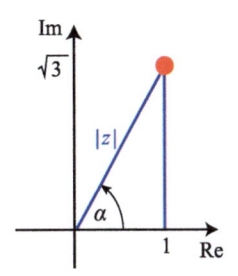

Then:
$$|z| = \sqrt{a^2 + b^2}$$
$$= \sqrt{(1)^2 + \left(\sqrt{3}\right)^2}$$
$$= 2$$

The modulus is 2. Since z is in the first quadrant, $\arg(z) = \alpha$ where $\tan \alpha = \frac{\sqrt{3}}{1}$.

Hence:
$$\alpha = \tan^{-1} \left|\sqrt{3}\right|$$
$$= 60°$$

So $\arg(z) = 60°$

....

2. Problem:

Find the modulus and principal argument of the complex number $z = -3 + 4i$

Solution:

Plot the point $-3 + 4i$ on an Argand diagram.

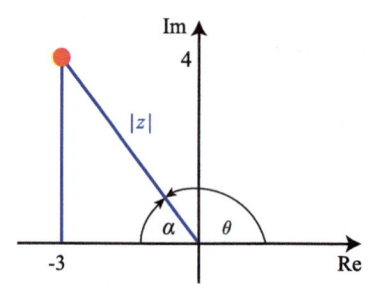

Then:
$$|z| = \sqrt{a^2 + b^2}$$
$$= \sqrt{(-3)^2 + (4)^2}$$
$$= 5$$

The modulus is 5. Since z is in the second quadrant, $\arg(z) = 180° - \alpha$ where $\tan \alpha = \frac{4}{3}$.

Hence:
$$\alpha = \tan^{-1}\left|\frac{4}{3}\right|$$
$$= 53 \cdot 13°$$

So:
$$\arg(z) = 180° - 53.13°$$
$$= 126 \cdot 87°$$

Alternatively, we could work in radians.

Since z is in the second quadrant, $\arg(z) = \pi - \alpha$ where $\tan \alpha = \frac{4}{3}$

Hence:
$$\alpha = \tan^{-1}\left|\frac{4}{3}\right|$$
$$= 0 \cdot 927 \text{ rad}$$

So:
$$\arg(z) = \pi - 0 \cdot 927$$
$$= 2 \cdot 214 \text{ rad to three decimal places}$$

TOPIC 6. COMPLEX NUMBERS

3. Problem:
Find the modulus and principal argument of the complex number $z = -1 - i$

Solution:
Plot the point $z = -1 - i$ on an Argand diagram.

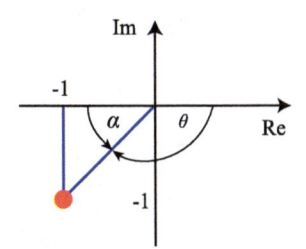

Then:
$$|z| = \sqrt{a^2 + b^2}$$
$$= \sqrt{(-1)^2 + (-1)^2}$$
$$= \sqrt{2}$$

The modulus is $\sqrt{2}$. Since z is in the third quadrant, $\arg(z) = -180° + \alpha$ where $\tan \alpha = \frac{1}{1}$.

Hence:
$$\alpha = \tan^{-1} |1|$$
$$= 45°$$

So:
$$\arg(z) = -180° + 45°$$
$$= -135°$$

....

4. Problem:
Find the modulus and principal argument of the complex number $z = 2 - 3i$

Solution:
Plot the point $z = 2 - 3i$ on an Argand diagram.

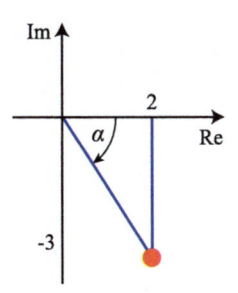

Then:

$|z| = \sqrt{a^2 + b^2}$
$= \sqrt{(2)^2 + (-3)^2}$
$= \sqrt{13}$

The modulus is $\sqrt{13}$. Since z lies in the fourth quadrant, $\arg(z) = -\alpha$ where $\tan \alpha = \frac{3}{2}$.
Hence:

$\alpha = \tan^{-1} \left| \frac{3}{2} \right|$
$= 0 \cdot 983$

So $\arg(z) = -0 \cdot 983$ radians

Strategy to find modulus and argument exercise   Go online

Q96: Find the modulus and the principal argument of the complex number $3 + 5i$
....

Q97: Find the modulus and the principal argument of the complex number $-2 + 4i$
....

Q98: Find the modulus and the principal argument of the complex number $4 - i$
....

Q99: Find the modulus and the principal argument of the complex number $-3 - 6i$
....

Q100: Find the modulus and the principal argument for the complex number $12 + 5i$ and hence state the number in polar form.
....

Q101: Find the modulus and the principal argument of the complex number $-\sqrt{3} - i$

6.4.4 Polar form

Here is a reminder of the quadrant diagram.

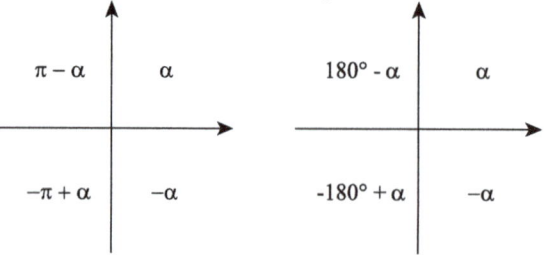

$\pi - \alpha$	α
$-\pi + \alpha$	$-\alpha$

$180° - \alpha$	α
$-180° + \alpha$	$-\alpha$

Examples

1. Problem:
Find the modulus and principal argument of the complex number $4 + i$.

Solution:

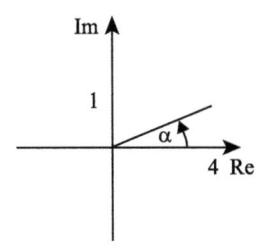

[Diagram: Argand diagram showing point at 4 on Re axis, 1 on Im axis, with angle α from positive real axis]

Modulus:
To find the modulus, we use $r = \sqrt{a^2 + b^2}$:
$$r = \sqrt{4^2 + 1^2}$$
$$= \sqrt{17}$$

Principal argument:
To find the principal argument, we must first find $\alpha = \tan^{-1}\left|\frac{b}{a}\right|$:
$$\alpha = \tan^{-1}\left|\frac{1}{4}\right|$$
$$\alpha = 14°$$

The complex number $4 + i$ is in the first quadrant.
To find the principal argument we therefore use $\theta = \alpha \Rightarrow \theta = 14°$
Since $z = r\cos\theta + ir\sin\theta$ is equivalent to $z = a + ib$, then we can rewrite $z = 4 + i$ as:
$$z = \sqrt{17}\cos(14°) + i\sqrt{17}\sin(14°)$$
$$= \sqrt{17}(\cos(14°) + i\sin(14°))$$

....

2. Problem:
Find the modulus and principal argument of the complex number $-2 + i$.

Solution:

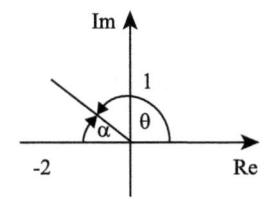

[Diagram: Argand diagram showing point at -2 on Re axis, 1 on Im axis, with angle α in second quadrant and θ measured from positive real axis]

Modulus:
To find the modulus, we use $r = \sqrt{a^2 + b^2}$:
$$r = \sqrt{(-2)^2 + 1^2}$$
$$= \sqrt{5}$$

Principal argument:
To find the principal argument, we must first find $\alpha = \tan^{-1}\left|\frac{b}{a}\right|$:
$$\alpha = \tan^{-1}\left|\frac{1}{-2}\right|$$
$$\alpha = 27°$$
The complex number $-2 + i$ is in the second quadrant.
To find the principal argument we therefore use $\theta = 180° - \alpha$.
$$\theta = 180° - 27°$$
$$= 153°$$
Since $z = r\cos\theta + ir\sin\theta$ is equivalent to $z = a + ib$, then we can rewrite $z = -2 + i$ as:
$$z = \sqrt{5}\cos(153°) + i\sqrt{5}\sin(153°)$$
$$= \sqrt{5}(\cos(153°) + i\sin(153°))$$

....

3. Problem:
Find the modulus and principal argument of the complex number $-1 - 2i$.

Solution:

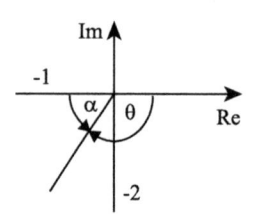

Modulus:
To find the modulus, we use $r = \sqrt{a^2 + b^2}$:
$$r = \sqrt{(-1)^2 + (-2)^2}$$
$$= \sqrt{5}$$

Principal argument:
To find the principal argument, we must first find $\alpha = \tan^{-1}\left|\frac{b}{a}\right|$:
$$\alpha = \tan^{-1}\left|\frac{-2}{-1}\right|$$
$$\alpha = 63°$$
The complex number $-1 - 2i$ is in the third quadrant.

TOPIC 6. COMPLEX NUMBERS

To find the principal argument we therefore use $\theta = -180° + \alpha$.

$\theta = -180° + 63°$
$= -117°$

Since $z = r\cos\theta + ir\sin\theta$ is equivalent to $z = a + ib$, then we can rewrite $z = -1 - 2i$ as: $z = \sqrt{5}\cos(-117°) + i\sqrt{5}\sin(-117°)$

Since $\cos\theta$ is symmetrical about the y-axis, $\cos(-117°) = \cos(117°)$. Check it on a calculator.

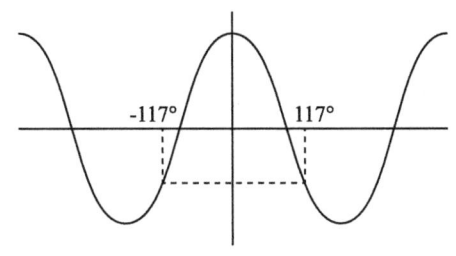

Since $\sin\theta$ has rotational symmetry about the origin, $\sin(-117°) = -sin(117°)$.

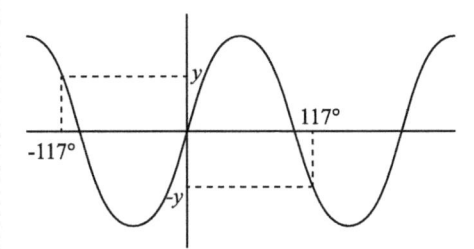

We therefore have: $z = \sqrt{5}(\cos(117°) - i\sin(117°))$

....

4. Problem:

Find the modulus and principal argument of the complex number $4 - 3i$.

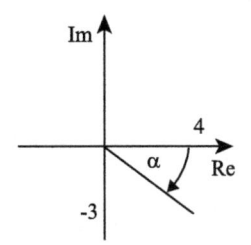

Solution:

© HERIOT-WATT UNIVERSITY

Modulus:

To find the modulus, we use $r = \sqrt{a^2 + b^2}$:

$r = \sqrt{4^2 + (-3)^2}$
$= \sqrt{25}$
$= 5$

Principal argument:

To find the principal argument, we must first find $\alpha = \tan^{-1}\left|\frac{b}{a}\right|$:

$\alpha = \tan^{-1}\left|\frac{-3}{4}\right|$

$\alpha = 37°$

The complex number $4 - 3i$ is in the fourth quadrant.

To find the principal argument we therefore use $\theta = -\alpha \Rightarrow \theta = -37°$.

Since $z = r\cos\theta + ir\sin\theta$ is equivalent to $z = a + ib$, then we can rewrite $z = 4 - 3i$ as: $z = 5\cos(-37°) + 5i\sin(-37°)$

Since $\cos\theta$ is symmetrical about the y-axis, $\cos(-37°) = \cos(37°)$. Check it on a calculator.

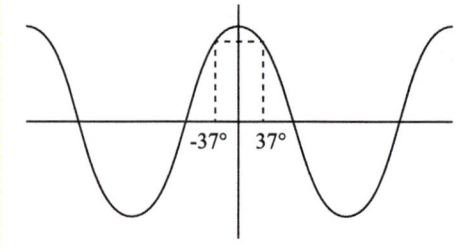

Since $\sin\theta$ has rotational symmetry about the origin, $\sin(-37°) = -sin(37°)$

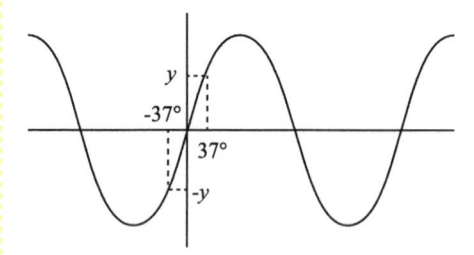

We therefore have: $z = 5(cos(37°) - i\sin(37°))$

6.5 Geometric interpretations

As shown earlier, complex numbers can be represented on an Argand diagram.

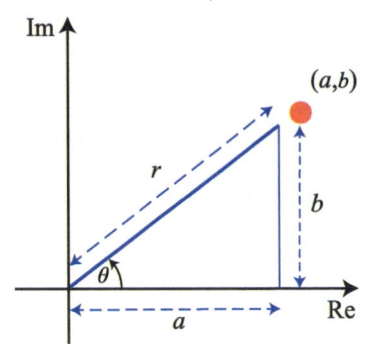

Sometimes equations involving complex numbers have interesting geometric interpretations. For example, the family of all solutions to $|z| = r$ and $|z - a| = |z - b|$ lie on the equation of a circle or straight line.

6.5.1 Geometric interpretations: Circle representations

Consider the equation $|z| = 1$

What are the solutions to this equation? That is, which complex numbers z satisfy $|z| = 1$?

Geometrically, since $|z|$ is the distance between the point representing z and the origin, $|z| = 1$ if and only if the distance between z and the origin is 1.

This can only occur if z lies on the circle centre $(0, 0)$ and radius 1.

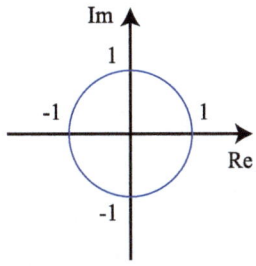

$|z| = 1$

The equation that represents the set of points that lie on the circle is the locus of $|z| = 1$

This can be further explained by understanding the notation $|\ |$. The modulus sign means that you want to know the magnitude or length of something. To do this we use Pythagoras' theorem.

So, given $z = x + iy$, calculate the length of z or, equivalently in mathematical notation, $|z|$.

(Note that the notation for z has changed from $z = a + ib$ to $z = x + iy$ because you will recognise the final formula better using these letters.)

$|z| = \sqrt{x^2 + y^2}$

Now, if we square each side:

$|z|^2 = \left(\sqrt{x^2 + y^2}\right)^2$
$= x^2 + y^2$

Note that in the working the i from $z = x + iy$ is missing. This is because the i represents the direction in the imaginary axis. It does not tell us anything about the length of the line. Therefore, we are only interested in the coefficient of i.

Now remember that for the modulus we used the notation r for the length of the line so we have:
$x^2 + y^2 = r^2$

This is the equation of a circle with centre $(0, 0)$ and radius r.

Usually you will be given what $|z|$ is equal to so you will know what the radius actually is.

Note that if we had the complex number $z = x - iy$, this would become:

$|z| = \sqrt{x^2 + (-y)^2}$
$= \sqrt{x^2 + y^2}$

And the rest would follow as before.

Now consider the inequality |z| > 1:

Clearly |z| > 1 if and only if the distance from z to the origin is greater than 1.

Hence, the solution is the set of all points which lie strictly outside the circle centre $(0, 0)$ and radius 1.

Similarly, |z| < 1 has the solution of the set of all points which lie strictly inside the unit circle.

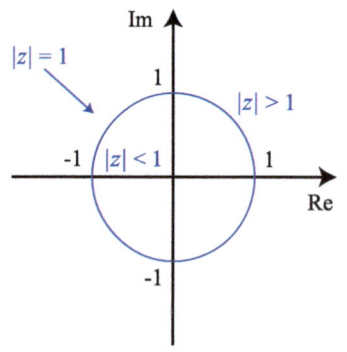

|z| > 1, |z| = 1 and |z| < 1

Examples

1.

Let's take the real number 2 and consider $|z - 2| = 3$

$|z - 2| = 3$ means the distance between the fixed point 2 on the plane and any other complex number z is set at 3.

The easiest way to visualise this is to imagine cutting a piece of string of 3 units in length and fixing one end on the point $z = 2$. Then, taking the other end and keeping the sting taught, scribe a locus of points of distance 3 away from the point $z = 2$. It will be found that this creates a circle with centre $(2, 0)$ and radius 3.

On an Argand diagram, this would look like:

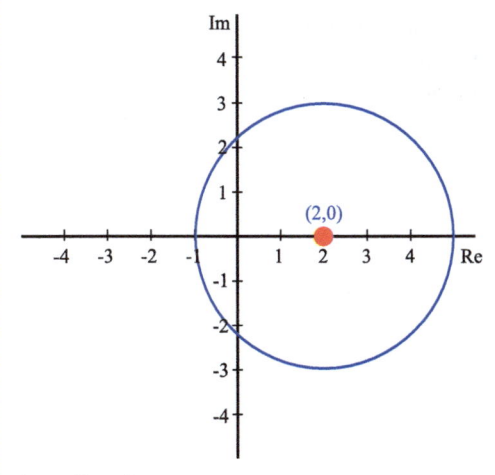

$|z - 2| = 3$

This can also be confirmed algebraically.

Remember that $|z| = \sqrt{x^2 + y^2}$, where $z = x + iy$

Substituting $z = x + iy$ in $|z - 2| = 3$: $|x + iy - 2| = 3$

Grouping the real and imaginary parts together: $|(x - 2) + iy| = 3$

Since we do not want any square roots we will square both sides: $|(x - 2) + iy|^2 = 3^2$

Expanding the left hand side, we use the definition $|z|^2 = x^2 + y^2$: $(x - 2)^2 + y^2 = 3^2$

This is a circle with centre $(2, 0)$ and a radius of 3.

The equals sign means that we are interested in the points on the circle; since the locus of points creating the circle boundary is included, it is drawn as a solid line.

$|z - 2| \geq 3$ would be all the points on and outside the circle centre $(2, 0)$, radius 3.

$|z - 2| \leq 3$ would be all the points on and inside the circle centre $(2, 0)$, radius 3.

Usually, to show the distinction that the locus of points creating the circle is not included, it is drawn as a dashed line.

$|z - 2| > 3$ would be all the points outside the circle centre $(2, 0)$, radius 3.
$|z - 2| < 3$ would be all the points inside the circle centre $(2, 0)$, radius 3.

....

2. Problem:
Identify the locus in the plane given by $|z + 3 - i| = 4$ and interpret this geometrically.
Solution:
Remember that $|z| = \sqrt{x^2 + y^2}$, where $z = x + iy$
Substituting $z = x + iy$ in $|z + 3 - i| = 4$: $|x + iy + 3 - i| = 4$
Grouping the real and imaginary parts together: $|(x + 3) + i(y - 1)| = 4$
Since we do not want any square roots we will square both sides: $|(x + 3) + i(y - 1)|^2 = 4^2$
Expanding the left hand side, we use the definition $|z|^2 = x^2 + y^2$: $(x + 3)^2 + (y - 1)^2 = 4^2$
This is a circle with centre $(-3, 1)$ and a radius of 4.
(Note that when grouping terms, the equation of a circle has to be in the form:
$(x - a)^2 + (y - b)^2 = r^2$ not $(x - a)^2 - (y - b)^2 = r^2$)
Argand diagram: $|z + 3 - i| = 4$

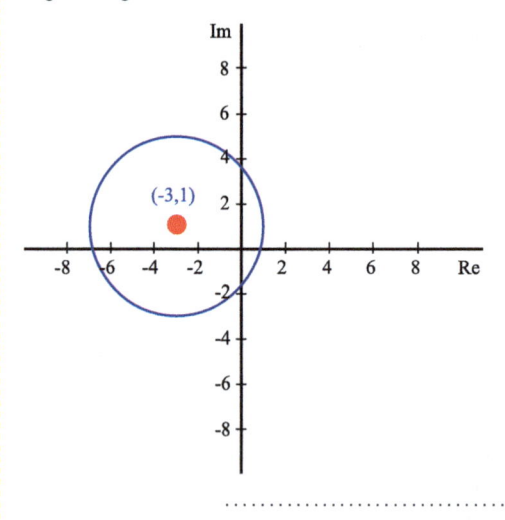

....

3. Problem:

Give the geometrical interpretation of the equation $|z - 4| < 5$

Solution:

Remember that $|z| = \sqrt{x^2 + y^2}$, where $z = x + iy$

Substituting $z = x + iy$ in $|z - 4| < 5$: $|x + iy - 4| < 5$

Grouping the real and imaginary parts together: $|(x - 4) + iy| < 5$

Since we do not want any square roots we will square both sides: $|(x - 4) + iy|^2 < 5^2$

Expanding the left hand side, we use the definition $|z|^2 = x^2 + y^2$: $(x - 4)^2 + y^2 < 5^2$

This is a circle with centre $(4, 0)$ and a radius of 5.

$<$ means that we are interested in the region inside the circle.

Argand diagram: $|z - 4| < 5$

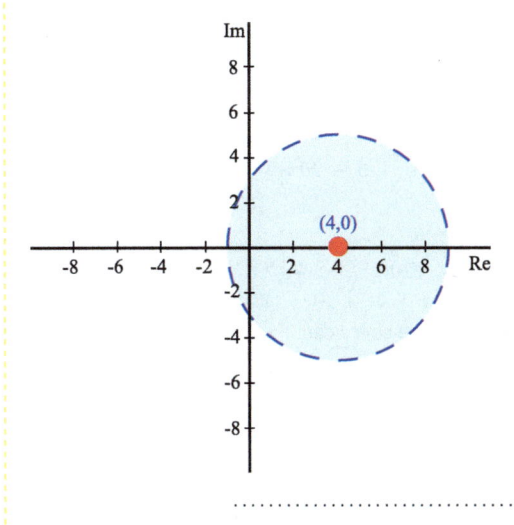

4. Problem:

Give the geometrical interpretation of the equation $|z + 1| = 4$

Solution:

Remember that $|z| = \sqrt{x^2 + y^2}$, where $z = x + iy$

Substituting $z = x + iy$ in $|z + 1| = 4$: $|x + iy + 1| = 4$

Grouping the real and imaginary parts together: $|(x + 1) + iy| = 4$

Since we do not want any square roots we will square both sides: $|(x + 1) + iy|^2 = 4^2$

Expanding the left hand side, we use the definition $|z|^2 = x^2 + y^2$: $(x + 1)^2 + y^2 = 4^2$

This is a circle with centre $(-1, 0)$ and a radius of 4.

$=$ means that we are interested in the points on the circle.

Argand diagram: $|z + 1| = 4$

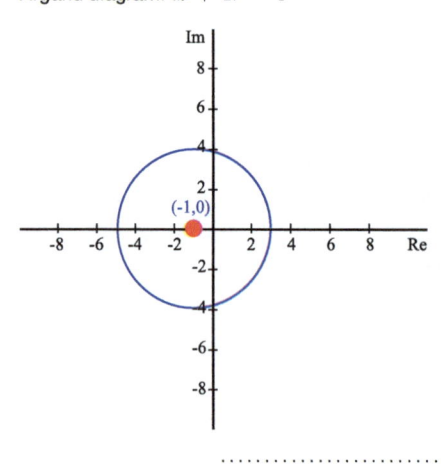

....

5. Problem:

Give the geometrical interpretation of the equation $|z + 3 - 2i| > 2$

Solution:

Remember that $|z| = \sqrt{x^2 + y^2}$, where $z = x + iy$

Substituting $z = x + iy$ in $|z + 3 - 2i| > 2$: $|x + iy + 3 - 2i| > 2$

Grouping the real and imaginary parts together: $|(x + 3) + i(y - 2)| > 2$

Since we do not want any square roots we will square both sides: $|(x + 3) + i(y - 2)|^2 > 2^2$

Expanding the left hand side, we use the definition $|z|^2 = x^2 + y^2$: $(x + 3)^2 + (y - 2)^2 > 2^2$

This is a circle with centre $(-3, 2)$ and a radius of 2.

$>$ means that we are interested in the region outside the circle.

Argand diagram: $|z + 3 - 2i| > 2$

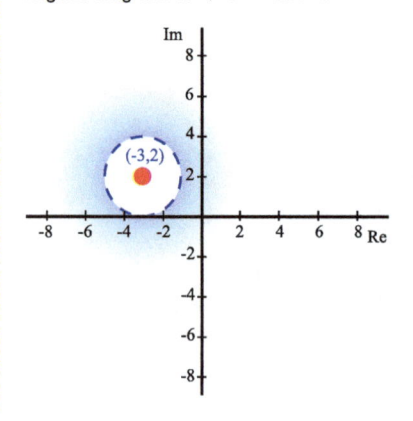

TOPIC 6. COMPLEX NUMBERS

6. Problem:
Give the geometrical interpretation of the equation $|2z - 5i| \geq 6$

Solution:
Remember that $|z| = \sqrt{x^2 + y^2}$, where $|2z - 5i| \geq 6$
Substituting $z = x + iy$ in $|z + 1| = 4$: $|2(x + iy) - 5i| \geq 6$
Grouping the real and imaginary parts together: $|2x + i(2y - 5)| \geq 6$
Since we do not want any square roots we will square both sides: $|2x + i(2y - 5)|^2 \geq 6^2$
Expanding the left hand side, we use the definition $|z|^2 = x^2 + y^2$: $(2x)^2 + (2y - 5)^2 \geq 6^2$
The general equation of a circle is: $(x - a)^2 + (y - b)^2 = r^2$
To find the centre we equate $x - a = 0$ and $y - b = 0$: $2x = 0 \Rightarrow x = 0$
$2y - 5 = 0 \Rightarrow y = \frac{5}{2}$
This is a circle with centre $(0, \frac{5}{2})$ and a radius of 6.
$\geq$ means that we are interested in the region on and outside the circle.
Argand diagram: $|2z - 5i| \geq 6$

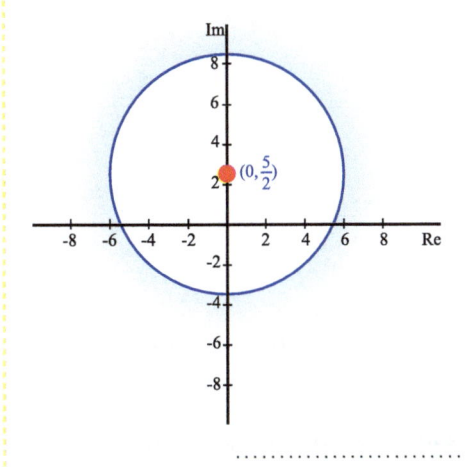

7. Problem:
Give the geometrical interpretation of the equation $|2z - 4 - 3i| = 5$

Solution:
Remember that $|z| = \sqrt{x^2 + y^2}$, where $z = x + iy$
Substituting $z = x + iy$ in $|2z - 4 - 3i| = 5$: $|2(x + iy) - 4 - 3i| = 5$
Grouping the real and imaginary parts together: $|(2x - 4) + i(2y - 3)| = 5$

Since we do not want any square roots we will square both sides:
$|(2x - 4) + i(2y - 3)|^2 = 5^2$
Expanding the left hand side, we use the definition $|z|^2 = x^2 + y^2$:
$(2x - 4)^2 + (2y - 3)^2 = 5^2$
The general equation of a circle is: $(x - a)^2 + (y - b)^2 = r^2$
To find the centre we equate $x - a = 0$ and $y - b = 0$: $2x - 4 = 0 \Rightarrow x = 2$
$2y - 3 = 0 \Rightarrow y = \frac{3}{2}$
This is a circle with centre $(2, \frac{3}{2})$ and a radius of 5.
= means that we are interested in the points on the circle.
Argand diagram: $|2z - 4 - 3i| = 5$

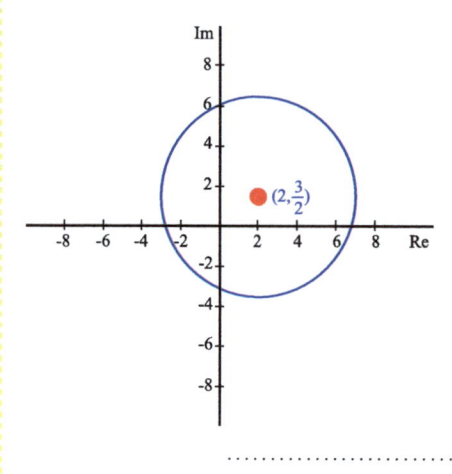

....

8. Problem:
Give the equation whose solution set is a circle centre $2i$ and radius 5 in the complex plane.
Solution:
A circle has the general formula: $(x - a)^2 + (y - b)^2 = r^2$
With centre (a, b) and radius r, where x is the real part and y is the imaginary part.
The centre is $(0, 2)$ and the radius is 5.
Thus, the equation is: $x^2 + (y - 2)^2 = 5^2$
Putting back into modulus notation we work backwards:
$|x + i(y - 2)|^2 = 5^2$
$\quad |x + iy - 2|^2 = 5^2$
$\quad\quad |z - 2|^2 = 5^2$
$\quad\quad\quad |z - 2| = 5$
....

9. Problem:

Give the equation whose solution set is a circle centre -4 and radius 3 in the complex plane.

Solution:

A circle has the general formula: $(x - a)^2 + (y - b)^2 = r^2$

With centre (a, b) and radius r, where x is the real part and y is the imaginary part.

The centre is $(-4, 0)$ and the radius is 3.

Thus, the equation is: $(x + 4)^2 + y^2 = 3^2$

Putting back into modulus notation we work backwards:

$|(x + 4) + iy|^2 = 3^2$

$|x + iy + 4|^2 = 3^2$

$|z + 4|^2 = 3^2$

$|z + 4| = 3$

10. Problem:

Give the equation whose solution set is a circle centre $(-1, 5)$ and radius 6 in the complex plane.

Solution:

A circle has the general formula: $(x - a)^2 + (y - b)^2 = r^2$

With centre (a, b) and radius r, where x is the real part and y is the imaginary part.

The centre is $(-1, 5)$ and the radius is 6.

Thus, the equation is: $(x + 1)^2 + (y - 5)^2 = 6^2$

Putting back into modulus notation we work backwards:

$|(x + 1) + i(y - 5)|^2 = 6^2$

$|x + iy + 1 - 5i|^2 = 6^2$

$|z + 1 - 5i|^2 = 6^2$

$|z + 1 - 5i| = 6$

Key point

Remember:

- z can be replaced by $x + iy$;
- from here you can group the real and imaginary parts;
- $|x + iy + (a + ib)|^2 = r2 \Rightarrow (x - a)^2 + (y - b)^2 = r^2$ which is a circle with centre (a, b) and radius r.

Geometric interpretations: Circle representations exercise

Give a geometrical interpretation of the following equations:

Q102: $|z - 3| > 2$

Q103: $|z + 1| = 3$

Q104: $|z - 2 + 3i| \leq 4$

Q105: $|2z - 3i| > 2$

Give a geometrical interpretation of the following equations by stating the:

a) centre;

b) radius;

c) shaded region (choose from: inside, on, on and inside, outside, on and outside).

Q106: $|z - 3i| > 5$

Q107: $|z - 5 + 2i| \leq 4$

Q108: $|3z - 2 - 5i| \geq 2$

6.5.2 Geometric interpretations: Straight line representations

Straight line representations

Finally, look at equations of the form $|z - a| = |z - b|$

The following example shows that the solution set of such equations is a straight line.

Example

Problem:

Find a geometrical interpretation for the equation $|z - 3| = |z - 4i|$

Solution:

The complex numbers 3 and $4i$ are represented by the points $(3, 0)$ and $(0, 4)$ in an Argand

TOPIC 6. COMPLEX NUMBERS

diagram.

Hence, $|z - 3| = |z - 4i|$ if and only if the distance from $(3, 0)$ to z is equal to the distance from $(0, 4)$ to z.

$\Leftrightarrow$ z lies on the perpendicular bisector of the line joining $(3, 0)$ and $(0, 4)$.

The next activity will demonstrate the fact used above that the perpendicular bisector of a straight line consists of all the points that are equidistant from the ends of the line.

Geometric interpretations: Straight line representation

- Plot the points $(3, 0)$ and $(0, 4)$, which represent the complex numbers 3 and $4i$, on an Argand diagram.
- Plot four more points that are, in turn, equidistant from these.
- Join these new points, which will lie in a straight line.

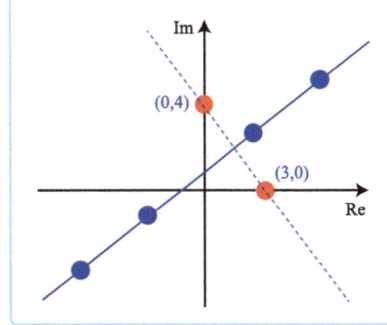

The following is a diagram which shows the construction of a perpendicular bisector of the line joining two points.

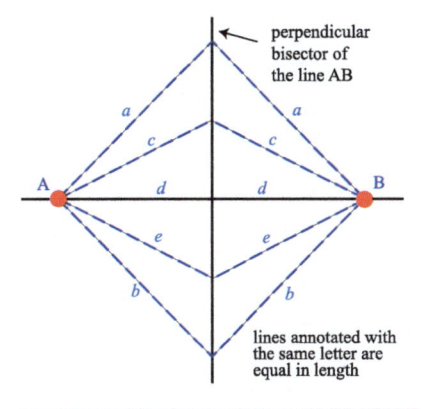

perpendicular bisector of the line AB

lines annotated with the same letter are equal in length

© HERIOT-WATT UNIVERSITY

The equation of the perpendicular bisector in terms of x and y can be found as follows:

Let $z = x + iy$

$$|z - 3| = |z - 4i|$$
$$|x + iy - 3| = |x + iy - 4i|$$
$$|(x - 3) + iy| = |x + i(y - 4)|$$
$$(x - 3)^2 + y^2 = x^2 + (y - 4)^2$$
$$x^2 - 6x + 9 + y^2 = x^2 + y^2 - 8y + 16$$
$$-6x + 9 = -8y + 16$$
$$8y = 6x + 7$$

Examples

1. Problem:

Find the equation of the locus for $|z - 2| = |z + 3|$ and draw this on an Argand diagram.

Solution:

Let $z = x + iy$: $|z - 2| = |z + 3|$
Replace z with $x + iy$: $|x + iy - 2| = |x + iy + 3|$
Group the imaginary and real parts together: $|(x - 2) + iy| = |(x + 3) + iy|$
Square both sides:
$$|(x - 2) + iy|^2 = |(x + 3) + iy|^2$$
$$(x - 2)^2 + y^2 = (x + 3)^2 + y^2$$
Expand out and simplify:
$$x^2 - 4x + 4 + y^2 = x^2 + 6x + 9 + y^2$$
$$-4x + 4 = 6x + 9$$
$$-10x = 5$$
$$x = -\frac{1}{2}$$

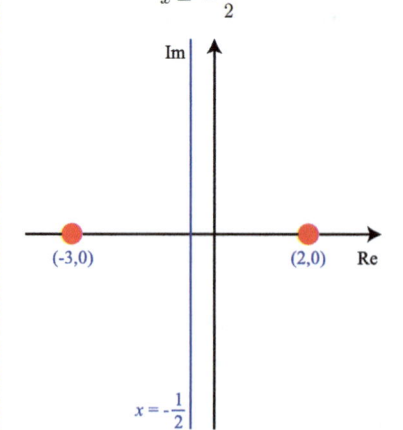

2. Problem:

Find the equation of the locus for $|z + 3i| = |z - 1|$ and draw this on an Argand diagram.

Solution:

Let $z = x + iy$: $|z + 3i| = |z - 1|$

Replace z with $x + iy$: $|x + iy + 3i| = |x + iy - 1|$

Group the imaginary and real parts together: $|x + i(y + 3)| = |(x - 1) + iy|$

Square both sides:

$|x + i(y + 3)|^2 = |(x - 1) + iy|^2$

$x^2 + (y + 3)^2 = (x - 1)^2 + y^2$

Expand out and simplify:

$x^2 + y^2 + 6y + 9 = x^2 - 2x + 1 + y^2$

$ 6y + 9 = -2x + 1$

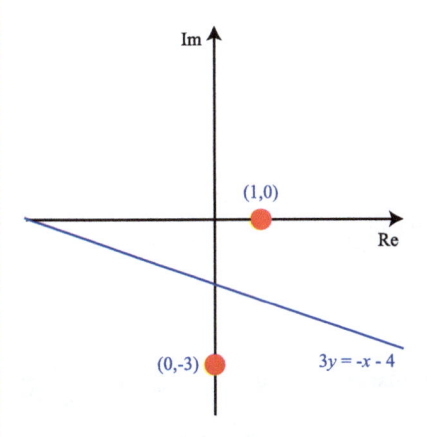

$ 6y = -2x - 8$

$ 3y = -x - 4$

3. Problem:

Give a geometric interpretation of $|z + 1 - 3i| = |z - 2 + i|$ and state the answer in the form $Ax + By + C = 0$.

Solution:

Let $z = x + iy$: $|z + 1 - 3i| = |z - 2 + i|$

Replace z with $x + iy$: $|x + iy + 1 - 3i| = |x + iy - 2 + i|$

Group the imaginary and real parts together: $|(x + 1) + i(y - 3)| = |(x - 2) + i(y + 1)|$

Square both sides:

$$|(x+1)+i(y-3)|^2 = |(x-2)+i(y+1)|^2$$
$$(x+1)^2+(y-3)^2 = (x-2)^2+(y+1)^2$$

Expand out and simplify:

$$x^2+2x+1+y^2-6y+9 = x^2-4x+4+y^2+2y+1$$
$$2x+1-6y+9 = -4x+4+2y+1$$
$$6x-8y+5 = 0$$

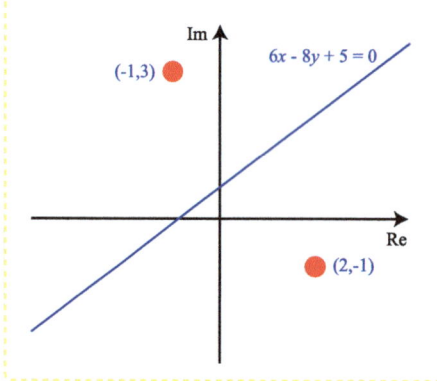

Key point

Remember that z can be replaced by $x+iy$

From here, you can group the real and imaginary parts:

$$|(x-a)+i(y-b)|^2 = |(x-c)+i(y-d)|^2$$
$$\Rightarrow (x-a)^2+(y-b)^2 = (x-c)^2+(y-d)^2$$

Geometric interpretations: Straight line representation exercise Go online

Q109: Find the equation of the locus for $|z-1| = |z-2|$ and draw this on an Argand diagram.

....

Q110: Find the equation of the locus for $|z+i| = |z-3|$ and draw this on an Argand diagram.

....

Q111: Find the equation of the locus for $|z+1-2i| = |z-2+3i|$ and draw this on an Argand diagram.

....

TOPIC 6. COMPLEX NUMBERS

Q112: Find the equation of the locus for $|z + 1| = |z - 2|$

....

Q113: Find the equation of the locus for $|z - 3i| = |z + i|$

....

Q114: Find the equation of the locus for $|z - 1 + 2i| = |z + 5 - i|$
Give your answer in the form $Ax + By + C = 0$

....

Q115: Give a geometric interpretation of $|z + 2i| = |z + 3|$ and find its equation using x, y coordinates.

The following examples give some general rules that can be used to find a geometric interpretation in the complex plane of an equation.

$|z| = r$	The solutions of the equation form a circle with centre $(0, 0)$ and radius r.
$|z - a| = r$	The solutions of the equation form a circle of radius r units with centre a, where a is a point in the complex plane.
$|z - a| > r$	The solutions of the equation are the set of all points that lie outside the circle with radius r units with centre a, where a is a point in the complex plane.
$|z - a| < r$	The solutions of the equation are the set of all points that lie inside the circle with radius r units with centre a, where a is a point in the complex plane.
$|z - a| = |z - b|$	The solutions of the equation are the set of points that lie on the perpendicular bisector of the line joining a and b.

Finally, it is worth mentioning another property of complex numbers involving the modulus, which has an interesting geometric interpretation, the **triangle inequality**. This is seen clearly in the following diagram.

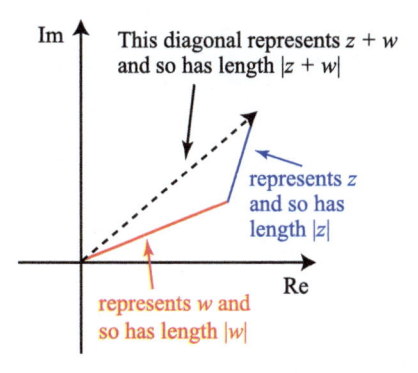

© HERIOT-WATT UNIVERSITY

The geometric interpretation of this is that there are no occasions where it is possible in a triangle for one side of a triangle to be larger than the sum of the lengths of the other two sides.

6.6 Fundamental theorem of algebra and solving complex equations

Fundamental theorem of algebra

A lot of the time when modelling a problem, the solution comes down to solving an equation of some form. However, how do we know that we have all the possible solutions and that we have not missed a vital answer? When solving polynomials, the fundamental theorem of algebra solves this problem for us. It tells us that the number of solutions is equivalent to the degree of the polynomial that is being solved.

Solving complex equations

Now that we know the number of solutions to a polynomial is equal to the degree of the polynomial, we can use previous methods to solve these polynomials. We can use synthetic or long division to find an initial factor and continue in this way, or use the quadratic formula where appropriate, to find remaining factors which could be real or complex.

6.6.1 Fundamental theorem of algebra

If the equation $x^2 + ax + b = 0$ has solutions $x = \alpha$ and $x = \beta$, then $x^2 + ax + b = (x - \alpha)(x - \beta)$

Thus, the factors of the polynomial can be found from the roots and vice versa.

By using the set of complex numbers, every quadratic equation can be solved and so every quadratic equation can be factorised.

Examples

1. Consider the equation $x^2 + 1 = 0$

The equation has solutions $x = i$ and $x = -i$

It follows that $x^2 + 1 = (x + i)(x - i)$ and this can be easily checked.

....

2. Consider $x^2 + x + 1 = 0$

Using the quadratic formula:

$$x = \frac{-b \pm \sqrt{b^2 - 4ac}}{2a}$$

$$x = \frac{-1 \pm \sqrt{1^2 - 4(1)(1)}}{2(1)}$$

$$x = \frac{-1 \pm \sqrt{-3}}{2}$$

$$x = \frac{-1 + \sqrt{-3}}{2} \quad \text{or} \quad x = \frac{-1 - \sqrt{-3}}{2}$$

The equation has solutions $x = -\frac{1}{2} \pm i\frac{\sqrt{3}}{2}$

It follows that $x^2 + x + 1 = \left(x + \frac{1}{2} - i\frac{\sqrt{3}}{2}\right)\left(x + \frac{1}{2} + i\frac{\sqrt{3}}{2}\right)$

....

3. Consider $3x^2 - 5x = -6$

First rearrange into the form $ax^2 + bx + c = 0$ we have:

$3x^2 - 5x + 6 = 0$

Using the quadratic formula:

$z = \dfrac{5 \pm \sqrt{(-5)^2 - 4(3)(6)}}{2(3)}$

$z = \dfrac{5 \pm \sqrt{-47}}{6}$

$z = \dfrac{5 \pm i\sqrt{47}}{6}$

$z = \dfrac{5 + i\sqrt{47}}{6}$ or $z = \dfrac{5 - i\sqrt{47}}{6}$

The equation has solutions $x = \frac{5}{6} \pm i\frac{\sqrt{47}}{6}$

It follows that $3x^2 - 5x + 6 = \left(x - \frac{5}{6} - i\frac{\sqrt{47}}{6}\right)\left(x - \frac{5}{6} + i\frac{\sqrt{47}}{6}\right)$

Similar results for solving polynomial equations and factorising polynomial expressions hold for polynomials of higher order.

Mathematicians find this a very striking result - hence the name of the theorem.

Fundamental theorem of algebra

Let $P(z) = a_n z^n + a_{n-1} z^{n-1} + \ldots + a_1 z + a_0$ be a polynomial of degree n (with real or complex coefficients).

The fundamental theorem of algebra states that:

- $P(z) = 0$ has n solutions and that $\alpha_1, \ldots, \alpha_n$ in the complex numbers;
- $P(z) = (z - \alpha_1)(z - \alpha_2)\ldots(z - \alpha_n)$

6.6.2 Solving complex equations

Now that we know that the number of solutions to a polynomial is equal to the degree of the polynomial, we can use previous methods to solve these polynomials. We can use synthetic or long division to find an initial factor and continue in this way or use the quadratic formula, where appropriate, to find the remaining factors, which could be real or complex.

Quadratic equations can always be solved by using the quadratic formula and hence the factors can be found.

For higher order equations, the solutions are much harder to find.

112 TOPIC 6. COMPLEX NUMBERS

In some cases, however, it is possible to find easy solutions by inspection simply by trying $x = \pm 1, \pm 2, \pm 3, \ldots$

Example
Problem:
Find a solution of $x^3 + x - 2 = 0$
Solution:
Try $x = 1$, then: $x^3 + x - 2 = 1^3 + 1 - 2 = 0$
Thus $x = 1$ is a solution.

Having found one solution, it is sometimes possible to find other solutions.

Example
Problem:
Find all the solutions of the equation $x^3 - x^2 - x - 2 = 0$
Solution:
Let $f(x) = x^3 - x^2 - x - 2$
Then $f(1) = -3$, $f(-1) = -3$, $f(2) = 0$
Since $f(x) = 0$, then $x - 2$ is a factor.
Using long division or synthetic division gives:
$x^3 - x^2 - x - 2 = (x - 2)(x^2 + x + 1) = 0$
But: $(x^2 + x + 1) = 0 \Rightarrow x = -\frac{1}{2} + i\frac{\sqrt{3}}{2}$
Hence the equation has solutions $x = 2$, $x = -\frac{1}{2} + i\frac{\sqrt{3}}{2}$ and $x = -\frac{1}{2} - i\frac{\sqrt{3}}{2}$

Given one complex root, another can be found by considering the conjugate.

It is easy to see from the quadratic formula that if $z = \alpha$ is a solution of a quadratic equation, then so is $z = \bar{\alpha}$

In fact this is true in general.

Example
Problem:
Solve $x^2 + x + 3$
Solution:
$x^2 + x + 3 = 0$

© HERIOT-WATT UNIVERSITY

Using the quadratic formula:
$$x = \frac{-b \pm \sqrt{b^2 - 4ac}}{2a}$$
$$x = \frac{-1 \pm \sqrt{1^2 - 4(1)(3)}}{2(1)}$$
$$x = \frac{-1 \pm \sqrt{11i^2}}{2}$$

The equation has solutions $x = -\frac{1}{2} \pm i\frac{\sqrt{11}}{2}$

These solutions are complex conjugates.

The conjugate roots property states that if $P(x)$ is a polynomial with real coefficients and $z = \alpha$ is a solution of $P(x) = 0$, then so is $z = \overline{\alpha}$

The roots $z = \alpha$ and $z = \overline{\alpha}$ are said to be *conjugate pairs*.

For a brief look at the proof of the conjugate roots go to the extra study activity now. It can be found in the section headed Proofs at the end of this topic.

Proof 1: Conjugate roots property

Proof 1

Suppose that α is a root of the polynomial.
$P(z) = a_n z^n + a_{n-1} z^{n-1} + \ldots + a_2 z^2 + a_1 z + a_0$

Then $P(\alpha) = 0$
i.e. $a_n \alpha^n + a_{n-1} \alpha^{n-1} + \ldots + a_2 \alpha^2 + a_1 \alpha + a_0 = 0$

Hence

$$\begin{aligned}\overline{P(\alpha)} &= \overline{a_n \alpha^n + a_{n-1}\alpha^{n-1} + \cdots + a_2 \alpha^2 + a_1 \alpha + a_0} \\ &= \overline{a_n \alpha^n} + \overline{a_{n-1}\alpha^{n-1}} + \cdots + \overline{a_2 \alpha^2} + \overline{a_1 \alpha} + \overline{a_0} \\ &= \overline{a_n}\left(\overline{\alpha^n}\right) + \overline{a_{n-1}}\left(\overline{\alpha^{n-1}}\right) + \cdots + \overline{a_2}\left(\overline{\alpha^2}\right) + \overline{a_1 \alpha} + \overline{a_0} \\ &= a_n (\overline{\alpha})^n + a_{n-1}(\overline{\alpha})^{n-1} + \cdots + a_2 (\overline{\alpha})^2 + a_1 \overline{\alpha} + a_0 \\ &= P(\overline{\alpha})\end{aligned}$$

However, $\overline{P(\alpha)} = \overline{0} = 0$ and so $P(\overline{\alpha}) = 0$

So, starting from the point that alpha (a complex number) is a root of the polynomial and applying conjugate properties to this, we have shown that if alpha is a root, then its conjugate is also a root.

Therefore, $\overline{\alpha}$ is a root of the polynomial.

With this property, it is possible to find conjugate roots and hence quadratic factors.

If $z = \alpha$ and $z = \overline{\alpha}$ are complex roots of $P(x)$, then $P(x)$ has factors $(x - \alpha)$ and $(x - \overline{\alpha})$ $P(x)$ then factorises to $(x - \alpha)(x - \overline{\alpha}) P_1(x)$ for some other polynomial $P_1(x)$

$Q(x) = (x - \alpha)(x - \overline{\alpha})$ will always have real coefficients and so provides a quadratic factor of $P(x)$ with real coefficients.

> **Example**
>
> Suppose $x = i$ is one root of a polynomial.
>
> By the conjugate roots property, $x = -i$ is also a root.
>
> Then, $(x - i)$ and $(x + i)$ are factors.
>
> Hence, $(x - i)(x + i) = x^2 + 1$ is a quadratic factor with real coefficients.

Strategy for solving a cubic

- Find one solution α by inspection.
- Using long division, or synthentic division where appropriate, obtain a factorisation of the form $P(x) = (x - \alpha) Q(x)$, where $Q(x)$ is a quadratic.
- Solve $Q(x) = 0$

Strategy for solving a quartic given a complex solution $z = \alpha$, where $\alpha \notin \mathbb{R}$

- Take $z = \overline{\alpha}$ also as a solution.
- Find a quadratic factor $Q_1(x) = (x - \alpha)(x - \overline{\alpha})$ of $P(x)$
- Obtain a factorisation of $P(x)$ of the form $P(x) = Q_1(x) Q_2(x)$, where $Q_2(x)$ is another quadratic, by using long division.
- Solve $Q_2(x) = 0$

> **Examples**
>
> **1. Problem:**
>
> Find the roots of the polynomial $2x^3 + x^2 + x + 2$
>
> **Solution:**
>
> By the fundamental theorem of algebra, there will be three roots.
>
> Try to find a real root by inspection. There is one, namely $x = -1$
>
> $f(-1) = -2 + 1 - 1 + 2 = 0$ so $x = -1$ is a root.
>
> Use long or synthetic division to divide $2x^3 + x^2 + x + 2$ by $x + 1$
>
> When $x = -1$, synthetic division gives:
>
> ```
> -1 | 2    1    1    2
>    |      0   -2    1   -2
>    |_____
>      2   -1    2    0
> ```

So $2x^3 + x^2 + x + 2 = (x + 1)(2x^2 - x + 2)$

Thus, the polynomial factorises to $(x + 1)(2x^2 - x + 2)$

Since $2x^2 - x + 2$ does not factorise readily, use the quadratic formula to find the other two roots:

$$x = \frac{-b \pm \sqrt{b^2 - 4ac}}{2a}$$

$$x = \frac{1 \pm \sqrt{(-1)^2 - 4(2)(2)}}{2(2)}$$

$$x = \frac{1 \pm \sqrt{-15i^2}}{4}$$

The quadratic equation has solutions $x = \frac{1}{4} \pm i\frac{\sqrt{15}}{4}$

Therefore, the three roots of this polynomial are:

- $x = -1$
- $x = \frac{1}{4} + i\frac{\sqrt{15}}{4}$
- $x = \frac{1}{4} - i\frac{\sqrt{15}}{4}$

....

2. Problem:

Find the other roots of the quartic $x^4 + x^3 + 2x^2 + x + 1$ given that one root is $x = i$

Solution:

By the fundamental theorem of algebra, there will be four roots.

By the conjugate roots property, complex roots occur in conjugate pairs.

Thus $x = -i$ must also be a root since $x = i$ is given as a root.

Hence $x - i$ and $x + i$ are factors and so $(x - i)(x + i) = x^2 + 1$ is a factor.

Using long division:

```
              x² + x + 1
        ┌─────────────────────────
x² + 1  │  x⁴ + x³ + 2x² + x + 1
           -x⁴        -x²
           ──────────────────────
            0 + x³ + x² + x + 1
              - x³         - x
              ──────────────────
                0 + x² + 0 + 1
                  - x²       - 1
                  ────────────────
                              0
```

So $x^4 + x^3 + 2x^2 + x + 1 = (x^2 + 1)(x^2 + x + 1)$

Since $x^2 + x + 1$ does not factorise readily, use the quadratic formula to find the other two roots:

$$x = \frac{-b \pm \sqrt{b^2 - 4ac}}{2a}$$

$$x = \frac{-1 \pm \sqrt{1^2 - 4(1)(1)}}{2(1)}$$

$$x = \frac{-1 \pm \sqrt{3i^2}}{2}$$

The quadratic equation has solutions $x = \frac{-1}{2} \pm i\frac{\sqrt{3}}{2}$

Therefore, the four roots of $x^4 + x^3 + 2x^2 + x + 1$ are:

- $x = i$
- $x = -i$
- $x = \frac{-1}{2} + \frac{\sqrt{3}}{2}i$
- $x = \frac{-1}{2} - \frac{\sqrt{3}}{2}i$

....

3. Problem:

Given that $(-2 - i)$ and $(3 + 2i)$ are the roots of a quadratic $x^2 - bx - c$ where $b, c \in \mathbb{C}$, find the complex numbers b and c.

Solution:

The factors of the quadratic are: $(x - (-2 - i)) = (x + 2 + i)$ and $(x - (3 + 2i)) = (x - 3 - 2i)$

Thus, the quadratic is:

$$(x + 2 + i)(x - 3 - 2i) = x^2 + -3x - 2ix + 2x - 6 - 4i + ix - 3i - 2i^2$$

$$= x^2 + (-3 - 2i + 2 + i)x - 6 - (4 + 3)i - 2i^2$$

$$= x^2 + (-1 - i)x - 4 - 7i$$

$$= x^2 - (1 + i)x - (4 + 7i)$$

Since the quadratic is in the form $ax^2 - bx - c$, equate the coefficients of the same powers of x to give $b = (1 + i)$ and $c = (4 + 7i)$.

....

4. Problem:

Verify that $z = 1 + i$ is a root of the equation $z^4 + 3z^2 - 6z + 10 = 0$
Hence find all the roots of the equation.

Solution:

Substitute $z = 1 + i$ into the equation. If it equals zero it is a root.

TOPIC 6. COMPLEX NUMBERS

Use the binomial theorem to expand the brackets for $(1+i)^4$:

$$(1+i)^4 = \binom{4}{0}1^4 + \binom{4}{1}1^3 i + \binom{4}{2}1^2 i^2 + \binom{4}{3}1 i^3 + \binom{4}{4}i^4$$
$$= 1 \times 1 + 4 \times 1 \times i + 6 \times 1 \times i^2 + 4 \times 1 \times i^3 + 1 \times i^4$$
$$= 1 + 4i - 6 - 4i + 1$$
$$= -4$$

$(1+i)^2 = 2i$

So
$$z^4 + 3z^2 - 6z + 10 = -4 + 3 \times 2i - 6(1+i) + 10$$
$$= -4 + 6i - 6 - 6i + 10$$
$$= 0$$

Therefore, $z = 1+i$ is a root.

Since the roots come in complex conjugate pairs, we have the roots: $z = 1+i$ and $z = 1-i$

This gives the factors: $(z - (1+i))$ and $(z - (1-i))$

If we multiply these factors together we get another factor, which is real:

$$(z - (1+i))(z - (1-i)) = z^2 - (1-i)z - (1+i)z + (1+i)(1-i)$$
$$= z^2 - (1-i+1+i)z + 1 - i + i - i^2$$
$$= z^2 - 2z + 2$$

Using long division:

```
                     z² + 2z  + 5
           ─────────────────────────────
z² - 2z + 2 │ z⁴ + 0z³ + 3z² - 6z + 10
             -z⁴ + 2z³ - 2z²
             ─────────────────
                  0 + 2z³ + z² - 6z + 10
                    - 2z³ + 4z² - 4z
                    ─────────────────
                       0 + 5z² -10z + 10
                         - 5z² +10z - 10
                         ─────────────────
                                       0
```

So $z^4 + 3z^2 - 6z + 10 = (z^2 - 2z + 2)(z^2 + 2z + 5)$

Using the quadratic formula to solve $z^2 + 2z + 5 = 0$:

$$z = \frac{-b \pm \sqrt{b^2 - 4ac}}{2a}$$
$$= \frac{-2 \pm \sqrt{2^2 - 4(1)(5)}}{2(1)}$$
$$= \frac{-2 \pm \sqrt{-16}}{2}$$

The quadratic equation has solutions $z = -1 \pm 2i$

© HERIOT-WATT UNIVERSITY

Therefore, the four roots of $z^4 + 3z^2 - 6z + 10 = 0$ are:

- $z = 1 + i$
- $z = 1 - i$
- $z = -1 + 2i$
- $z = -1 - 2i$

Top tip

- If $z = a + ib$ is a root, then $f(z) = 0$.
- If z is a root, then $\bar{z}$ is also a root.
- Use synthetic division or long division to find the other factors.
- Use the quadratic formula, is necessary, to find the remaining factors.

Solving complex equations exercise   Go online

Verify that each term is a root of the corresponding equation, hence find all of the roots.

Q116: $z = 3 - 3i$ is a root of $z^4 - 8z^3 + 32z^2 - 48z + 36 = 0$

Q117: $x = 2$ is a root of $x^3 - 6x^2 + 13x - 10 = 0$

Q118: $z = -1$ is a root of $z^3 - 5z^2 + 7z + 13 = 0$

Q119: $x - 2i$ is a root of $x^4 + 2x^3 + x^2 + 8x - 12 = 0$

Q120: $x = 1 - 3i$ is a root of $x^4 - 2x^3 + 26x^2 - 32x + 160 = 0$

Q121: $z = 4 + 3i$ is a root of $z^4 - 12z^3 + 70z^2 - 204z + 325 = 0$

Q122: $z = 2 + 5i$ is a root of $z^4 - 10z^3 + 63z^2 - 214z + 290 = 0$

TOPIC 6. COMPLEX NUMBERS

6.7 De Moivre's theorem

Thus far, all of the number work involving complex numbers has been done in Cartesian form (i.e. in the form $x + iy$).

However, it is sometimes easier and faster to transform the complex number into polar form and then carry out the addition, subtraction, multiplication and division. The following section will explore this through the use of de Moivre's theorem.

6.7.1 Multiplication using polar form

The multiplication of two complex numbers becomes much easier using the polar form.

Take two complex numbers, $z = r_1(\cos\theta + i\sin\theta)$ and $w = r_2(\cos\phi + i\sin\phi)$, and multiply them.

> **Key point**
>
> **Remember:**
> $$\cos(\theta + \varphi) = \cos\theta\cos\varphi - \sin\theta\sin\varphi$$
> $$\sin(\theta + \varphi) = \cos\theta\sin\varphi + \sin\theta\cos\varphi$$

$$zw = r_1(\cos\theta + i\sin\theta) \times r_2(\cos\varphi + i\sin\varphi)$$
$$= r_1 r_2 \left(\cos\theta\cos\varphi + i\cos\theta\sin\varphi + i\sin\theta\cos\varphi + i^2\sin\theta\sin\varphi\right)$$
$$= r_1 r_2 \left(\cos\theta\cos\varphi - \sin\theta\sin\varphi + i(\cos\theta\sin\varphi + \sin\theta\cos\varphi)\right)$$
$$= r_1 r_2 \left(\cos(\theta + \varphi) + i\sin(\theta + \varphi)\right)$$

Therefore, to multiply two complex numbers:

- multiply the moduli: $|zw| = r_1 r_2$
- add the arguments: $\arg(zw) = \theta + \phi$

> **Examples**
>
> **1. Problem:**
> If $z = 3(\cos(30°) + i\sin(30°))$ and $w = 4(\cos(60°) + i\sin(60°))$, find zw.
>
> **Solution:**
>
> Modulus:
> $|zw| = r_1 r_2$
> $\quad\quad = 3 \times 4$
> $\quad\quad = 12$
>
> Argument:
> $\arg(zw) = \theta + \varphi$
> $\quad\quad\quad = 30° + 60°$
> $\quad\quad\quad = 90°$
>
> So:
> $zw = 12(\cos(90°) + i\sin(90°))$
> $\quad\, = 12i$

© HERIOT-WATT UNIVERSITY

2. Problem:

If $z = 4\left(\cos\left(\frac{\pi}{6}\right) - i\sin\left(\frac{\pi}{6}\right)\right)$ and $w = 6\left(\cos\left(\frac{5\pi}{6}\right) + i\sin\left(\frac{5\pi}{6}\right)\right)$, find zw.

Solution:

We need to rewrite z in the form $\cos\theta + i\sin\theta$.

Remember that $\cos\theta - i\sin\theta = \cos(-\theta) + i\sin(-\theta)$ so:

$z = 4\left(\cos\left(-\frac{\pi}{6}\right) + i\sin\left(-\frac{\pi}{6}\right)\right)$

Modulus:

$|zw| = r_1 r_2$
$= 4 \times 6$
$= 24$

Argument:

$\arg(zw) = \theta + \varphi$
$= -\frac{\pi}{6} + \frac{5\pi}{6}$
$= \frac{2\pi}{3}$

So:

$zw = 24\left(\cos\left(\frac{2\pi}{3}\right) + i\sin\left(\frac{2\pi}{3}\right)\right)$

$= 24\left(-\frac{1}{2} + i\frac{\sqrt{3}}{2}\right)$

$= -12 + 12i\sqrt{3}$

Multiplication using polar form exercise   Go online

Remember: when using polar form, the argument is expected to be the principle argument. This is standard practice. That is θ is between $-\pi$ and π. This can make calculations without a calculator easier and more manageable.

Q123: If $z = 2\left(\cos\left(\frac{\pi}{3}\right) + i\sin\left(\frac{\pi}{3}\right)\right)$ and $w = 5\left(\cos\left(\frac{\pi}{2}\right) + i\sin\left(\frac{\pi}{2}\right)\right)$, find zw.

....

Q124: If $z = \left(\cos\left(\frac{\pi}{3}\right) - i\sin\left(\frac{\pi}{3}\right)\right)$ and $w = 4\left(\cos\left(\frac{2\pi}{3}\right) + i\sin\left(\frac{2\pi}{3}\right)\right)$, find wz.

....

Q125: If $z = 8\left(\cos\left(\frac{\pi}{2}\right) - i\sin\left(\frac{\pi}{2}\right)\right)$ and $w = 6\left(\cos\left(\frac{4\pi}{3}\right) - i\sin\left(\frac{4\pi}{3}\right)\right)$, find wz.

....

Q126: If $z = 3\left(\cos\left(\frac{\pi}{4}\right) + i\sin\left(\frac{\pi}{4}\right)\right)$, find z^3.

6.7.2 Division using polar form

In a similar way, division can be discussed using polar form.

Take two complex numbers, $z = r_1(\cos\theta + i\sin\theta)$ and $w = r_2(\cos\phi + i\sin\phi)$, and divide them.

TOPIC 6. COMPLEX NUMBERS

> **Key point**
>
> **Remember:**
>
> $$\cos(\theta - \varphi) = \cos\theta\cos\varphi + \sin\theta\sin\varphi$$
> $$\sin(\theta - \varphi) = \cos\theta\sin\varphi - \sin\theta\cos\varphi$$

$$\begin{aligned}
\frac{z}{w} &= \frac{r_1(\cos\theta + i\sin\theta)}{r_2(\cos\varphi + i\sin\varphi)} \\
&= \frac{r_1}{r_2} \times \frac{(\cos\theta + i\sin\theta)(\cos\varphi - i\sin\varphi)}{(\cos\varphi + i\sin\varphi)(\cos\varphi - i\sin\varphi)} \\
&= \frac{r_1}{r_2} \times \frac{\cos\theta\cos\varphi - i\cos\theta\sin\varphi + i\sin\theta\cos\varphi - i^2\sin\theta\sin\varphi}{(\cos^2\varphi + \sin^2\varphi)} \\
&= \frac{r_1}{r_2} \times \frac{\cos\theta\cos\varphi + \sin\theta\sin\varphi + i(\sin\theta\cos\varphi - \cos\theta\sin\varphi)}{(\cos^2\varphi + \sin^2\varphi)} \\
&= \frac{r_1}{r_2}(\cos(\theta - \varphi) + i\sin(\theta - \varphi))
\end{aligned}$$

Therefore, to divide two complex numbers:

- divide the moduli: $\left|\frac{z}{w}\right| = \frac{r_1}{r_2}$
- subtract the arguments: $arg\left(\frac{z}{w}\right) = \theta - \varphi$

Examples

1. Problem:
If $z = 18\left(\cos\left(\frac{\pi}{3}\right) + i\sin\left(\frac{\pi}{3}\right)\right)$ and $w = 6\left(\cos\left(\frac{\pi}{6}\right) + i\sin\left(\frac{\pi}{6}\right)\right)$, find $\frac{z}{w}$

Solution:

Modulus:
$$\frac{r_1}{r_2} = \frac{18}{6}$$
$$= 3$$

Argument:
$$\theta - \varphi = \frac{\pi}{3} - \frac{\pi}{6}$$
$$= \frac{\pi}{6}$$

So:
$$\frac{z}{w} = 3\left(\cos\left(\frac{\pi}{6}\right) + i\sin\left(\frac{\pi}{6}\right)\right)$$
$$= \frac{3\sqrt{3}}{2} + i\frac{3}{2}$$

....

2. Problem:
If $z = 10(cos(60°) - i\sin(60°))$ and $w = 2(cos(-15°) + i\sin(-15°))$, find zw.

Solution:

We need to rewrite z in the form $\cos\theta + i\sin\theta$.

© HERIOT-WATT UNIVERSITY

Remember that $\cos\theta - i\sin\theta = \cos(-\theta) + i\sin(-\theta)$ so: $z = 10(\cos(-60°) + i\sin(-60°))$

Modulus:
$$\frac{r_1}{r_2} = \frac{10}{2} = 5$$

Argument:
$$\theta - \varphi = -60° - (-15)°$$
$$= -45°$$

So:
$$zw = 5(\cos(-45°) + i\sin(-45°))$$
$$= 5(\cos(45°) - i\sin(45°))$$
$$= 5\left(\frac{\sqrt{2}}{2} - i\frac{\sqrt{2}}{2}\right)$$
$$= \frac{5\sqrt{2}}{2} - i\frac{5\sqrt{2}}{2}$$

Division using polar form exercise   Go online

Remember: when using polar form, the argument is expected to be the principle argument. This is standard practice. That is θ is between $-\pi$ and π. This can make calculations without a calculator easier and more manageable.

Q127: If $z = 3(\cos(75°) + i\sin(75°))$ and $w = 6(\cos(15°) + i\sin(15°))$, find $\frac{z}{w}$.

....

Q128: If $z = 10\left(\cos\left(\frac{\pi}{2}\right) + i\sin\left(\frac{\pi}{2}\right)\right)$ and $w = 2\left(\cos\left(\frac{\pi}{3}\right) + i\sin\left(\frac{\pi}{3}\right)\right)$, find $\frac{z}{w}$.

....

Q129: If $z = 7\left(\cos\left(\frac{\pi}{7}\right) - i\sin\left(\frac{\pi}{7}\right)\right)$ and $w = 2\left(\cos\left(\frac{2\pi}{7}\right) - i\sin\left(\frac{2\pi}{7}\right)\right)$, find $\frac{z}{w}$.

6.7.3 De Moivre's theorem

Multiplying two complex numbers in polar form is easily undertaken by using the rules set out in the last section. This leads to a very simple formula for calculating powers of complex numbers which is known as de Moivre's theorem.

Take the complex number $z = r(\cos\theta + i\sin\theta)$ and consider z^2.

Key point

Remember:

- Multiply the moduli.
- Add the arguments.

We have:
$z^2 = r(\cos\theta + i\sin\theta) \times r(\cos\theta + i\sin\theta)$
$= r \times r(\cos(\theta + \theta) + i\sin(\theta + \theta))$
$= r^2(\cos(2\theta) + i\sin(2\theta))$

and

$z^3 = r^2(\cos(2\theta) + i\sin(2\theta)) \times r(\cos\theta + i\sin\theta)$
$= r^2 \times r(\cos(2\theta + \theta) + i\sin(2\theta + \theta))$
$= r^3(\cos(3\theta) + i\sin(3\theta))$

There is a pattern which can be extended to the nth power and is known as de Moivre's theorem.

> **Key point**
>
> **De Moivre's theorem**
>
> If $z = r(\cos\theta + i\sin\theta)$, then $z^n = r^n(\cos(n\theta) + i\sin(n\theta))$ for all $n \in \mathbb{N}$.

This holds for the argument θ in radians $(-\pi, \pi)$ and degrees $(-180°, 180°)$.

For a brief look at the proof of de Moivre's theorem go to the extra study activity now. It can be found in the section headed Proofs at the end of this topic.

Proof 2: Proof of de Moivre's theorem

Proof 2

The theorem states that if $z = r(\cos\theta + i\sin\theta)$, then:
$z^n = r^n(\cos(n\theta) + i\sin(n\theta))$ for all $n \in \mathbb{N}$.

This is a proof by induction. The technique is explained in greater detail in the topic on methods of proof.

For $n = 1$:
$(r(\cos\theta + i\sin\theta))^1 = r(\cos\theta + i\sin\theta)$
$\qquad\qquad\qquad\qquad\quad = r^1(\cos(1\theta) + i\sin(1\theta))$

And so it is true for $n = 1$

Now suppose that the result is true for $n = k$, then:
$(r(\cos\theta + i\sin\theta))^k = r^k(\cos(k\theta) + i\sin(k\theta))$

Consider $n = k + 1$, then:

$$\begin{aligned}(r\left(\cos\theta + i\sin\theta\right))^{k+1} &= (r\left(\cos\theta + i\sin\theta\right))(r\left(\cos\theta + i\sin\theta\right))^{k}\\ &= (r\left(\cos\theta + i\sin\theta\right))\left(r^{k}\left(\cos\left(k\theta\right) + i\sin\left(k\theta\right)\right)\right)\\ &= r^{k+1}\left(r\left(\cos\theta\cos\left(k\theta\right) - \sin\theta\sin\left(k\theta\right)\right) + i\left(\sin\theta\cos\left(k\theta\right) + \cos\theta\sin\left(k\theta\right)\right)\right)\\ &= r^{k+1}\left(\cos\left(\theta + \left(k\theta\right)\right) + i\sin\left(\theta + \left(k\theta\right)\right)\right)\\ &= r^{k+1}\left(\cos\left(\left(k+1\right)\theta\right) + i\sin\left(\left(k+1\right)\theta\right)\right)\end{aligned}$$

So the result is true for $n = k + 1$ if it is true for $n = k$.
Since it is also true for $n = 1$, then it is true for all $n \in \mathbb{N}$.

This is a very useful result as it makes it simple to find z^n once z is expressed in polar form.

Examples

1.

Problem:

Calculate $\left(2\left(\cos\left(\frac{\pi}{5}\right) + i\sin\left(\frac{\pi}{5}\right)\right)\right)^{5}$

Solution:

If $z = r(\cos\theta + i\sin\theta)$, then $z^n = r^n(\cos(n\theta) + i\ \sin(n\theta))$ for all $n \in \mathbb{N}$.
Thus, using de Moivre's theorem with $r = 2$ and $\theta = \frac{\pi}{5}$:

$$\begin{aligned}\left(2\left(\cos\left(\frac{\pi}{5}\right) + i\sin\left(\frac{\pi}{5}\right)\right)\right)^{5} &= 2^{5}\left(\cos\left(5\times\frac{\pi}{5}\right) + i\sin\left(5\times\frac{\pi}{5}\right)\right)\\ &= 32\left(\cos\pi + i\sin\pi\right)\\ &= -32\end{aligned}$$

(To reach this result, multiply the moduli to give 2^5 and add the arguments to give $\frac{5\pi}{5} = \pi$)

....

2. Problem:

Calculate $\left(1 - i\sqrt{3}\right)^{6}$

Solution:

$z = 1 - i\sqrt{3}$, but we need it in the form $z = r(\cos\theta + i\sin\theta)$ to use de Moivre's theorem.
Consider its representation on an Argand diagram:

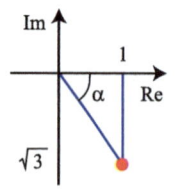

Modulus:

$$|z| = \sqrt{1^2 + \left(-\sqrt{3}\right)^2}$$
$$= 2$$

Argument:

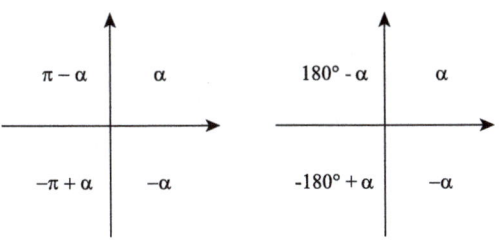

As z is in the fourth quadrant, $\arg(z) = -\alpha$, where $\tan \alpha = \left|\frac{-\sqrt{3}}{1}\right|$, i.e. $\alpha = \frac{\pi}{3}$

Using de Moivre's theorem with $r = 2$ and $\theta = -\frac{\pi}{3}$:

$$z^6 = 2^6 \left(\cos\left(-\frac{6\pi}{3}\right) + i\sin\left(-\frac{6\pi}{3}\right)\right)$$
$$= 64\left(\cos\left(-2\pi\right) + i\sin\left(-2\pi\right)\right)$$
$$= 64$$

....

3. Problem:

Calculate $(-1 + i)^5$

Solution:

$z = -1 + i$, but we need it in the form $z = r(\cos\theta + i\sin\theta)$ to use de Moivre's theorem. Consider its representation on an Argand diagram:

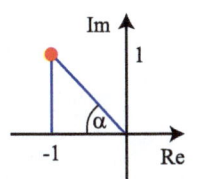

Modulus:

$$|z| = \sqrt{(-1)^2 + 1^2}$$
$$= \sqrt{2}$$

Argument:

As the argument is in the second quadrant, $\arg(z) = 180° - \alpha$, where $\tan \alpha = \left|\frac{1}{-1}\right|$, i.e. $\alpha = 45°$.

Using de Moivre's theorem with $r = \sqrt{2}$ and $135°$:

$z^5 = \left(\sqrt{2}\right)^5 (\cos(5 \times 135°) + i\sin(5 \times 135°))$
$= 4\sqrt{2}(\cos(675°) + i\sin(675°))$

The angle $675°$ is a clockwise rotation. It is equivalent to: $360° + 315°$.
On a diagram this would look like:

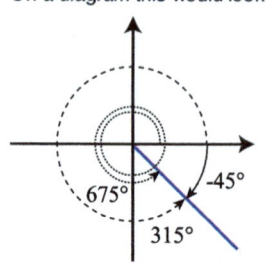

So the equivalent angle to $675°$ is $-45°$.

$z^5 = 4\sqrt{2}(\cos(-45°) + i\sin(-45°))$
$= 4\sqrt{2}(\cos(45°) - i\sin(45°))$

De Moivre's theorem and the rule for dividing complex numbers in polar form can be used to simplify fractions involving powers.

Example

Problem:

Simplify $\dfrac{(1+i)^6}{(1-i\sqrt{3})^4}$

Solution:

First express $1 - i\sqrt{3}$ and $1 + i$ in polar form.

From the previous example, $1 - i\sqrt{3} = 2\left(\cos\left(\dfrac{-\pi}{3}\right) + i\sin\left(\dfrac{-\pi}{3}\right)\right)$

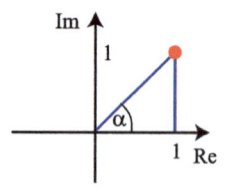

From an Argand diagram, $1 + i = \sqrt{2}\left(\cos\left(\dfrac{\pi}{4}\right) + i\sin\left(\dfrac{\pi}{4}\right)\right)$

By de Moivre's theorem:

$$\left(1 - i\sqrt{3}\right)^4 = 2^4 \left(\cos\left(4 \times \frac{-\pi}{3}\right) + i\sin\left(4 \times \frac{-\pi}{3}\right)\right)$$
$$= 16\left(\cos\left(\frac{-4\pi}{3}\right) + i\sin\left(\frac{-4\pi}{3}\right)\right)$$

and

$$(1+i)^6 = \sqrt{2}^6 \left(\cos\left(6 \times \frac{\pi}{4}\right) + i\sin\left(6 \times \frac{\pi}{4}\right)\right)$$
$$= 8\left(\cos\left(\frac{3\pi}{2}\right) + i\sin\left(\frac{3\pi}{2}\right)\right)$$

Hence:

$$\frac{(1+i)^6}{\left(1-i\sqrt{3}\right)^4} = \frac{8}{16}\left(\cos\left(\frac{3\pi}{2} - \left(-\frac{4\pi}{3}\right)\right) + i\sin\left(\frac{3\pi}{2} - \left(-\frac{4\pi}{3}\right)\right)\right)$$
$$= \frac{1}{2}\left(\cos\left(\frac{17\pi}{6}\right) + i\sin\left(\frac{17\pi}{6}\right)\right)$$

Principal argument:

$$\frac{17\pi}{6} = \left(2 + \frac{5}{6}\right)\pi$$
$$= \frac{5\pi}{6}$$

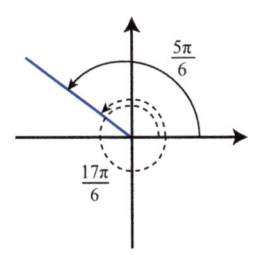

So:

$\frac{(1+i)^6}{(1-i\sqrt{3})^4} = \frac{1}{2}\left(\cos\left(\frac{5\pi}{6}\right) + i\sin\left(\frac{5\pi}{6}\right)\right)$

(To divide complex numbers in polar form, divide the moduli and subtract the arguments.)

Top tip

Remember that to evaluate cos θ and sin θ use the CAST diagram to find the equivalent acute angle. Then we decide if cos θ and sin θ are positive or negative in those quadrants.

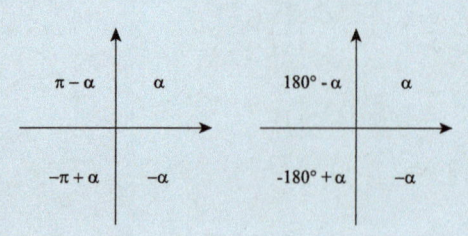

De Moivre's theorem exercise 1   Go online

Q130: Calculate $\left(2\left(\cos\left(\frac{3\pi}{4}\right) + i\sin\left(\frac{3\pi}{4}\right)\right)\right)^4$

Q131: Calculate $\left(4\left(\cos\left(\frac{\pi}{3}\right) + i\sin\left(\frac{\pi}{3}\right)\right)\right)^3$

Q132: Calculate $2\left(\cos\left(\frac{3\pi}{4}\right) + i\sin\left(\frac{3\pi}{4}\right)\right)^{10}$.

Q133: Calculate $(3(\cos(30°) + i\sin(30°)))^8$, giving your answer to 1 decimal place.

Q134: Calculate $(6(\cos(42°) + i\sin(42°)))^4$, giving your answer to 1 decimal place.

Q135: Calculate $(-1 - i)^5$

Q136: Calculate $(-4 + 5i)^7$, giving your answer to 1 decimal place.

Q137: Calculate $\left(\sqrt{3} - 2i\right)^3$, giving your answer to 1 decimal place.

Q138: Calculate $\left(\sqrt{3} + i\right)^4$

Q139: Simplify $\frac{(3-3i)^4}{\left(\sqrt{3}+i\right)^3}$

Q140: Simplify $\frac{\left(\sqrt{3}+i\right)^4}{(1-i)^3}$, giving your answer to 1 decimal place.

TOPIC 6. COMPLEX NUMBERS

Q141: Simplify $\dfrac{(-\sqrt{2}+i\sqrt{2})^5}{(-\sqrt{3}-i)^4}$, giving your answer to 1 decimal place.

De Moivre's theorem for $n \in \mathbb{Z}$

Now that we have seen how de Moivre's theorem works for positive whole numbers, it can be extended to include negative whole numbers.

By considering $z = \cos\theta + i\sin\theta$, calculate z^{-n} (that is $\frac{1}{z^n}$).

Consider n to be a negative integer.

Let $n = -p$

So if $z^n = (\cos\theta + i\sin\theta)^n$.

Then:

$$(\cos\theta + i\sin\theta)^n = (\cos\theta + i\sin\theta)^{-p}$$

$$= \frac{1}{(\cos\theta + i\sin\theta)^p}$$

$$= \frac{1}{\cos(p\theta) + i\sin(p\theta)} \times \left(\frac{\cos(p\theta) - i\sin(p\theta)}{\cos(p\theta) - i\sin(p\theta)}\right)$$

$$= \frac{\cos(p\theta) - i\sin(p\theta)}{\cos^2(p\theta) + i\sin^2(p\theta)}$$

$$= \cos(p\theta) - i\sin(p\theta)$$

$$= \cos(-p\theta) + i\sin(-p\theta)$$

$$= \cos(n\theta) + i\sin(n\theta)$$

which can be simplified to $\cos(p \times \theta) - i\sin(p \times \theta)$ since cosine is an even function and sine is an odd function.

So $z^n = \cos(n\theta) + i\sin(n\theta)$ where $n \in \mathbb{Z}$.

Example

Problem:

Calculate $\left(2\left(\cos\left(\frac{\pi}{4}\right) + i\sin\left(\frac{\pi}{4}\right)\right)\right)^{-3}$

Solution:

Using de Moivre's theorem, $z^n = r^n(\cos(n\theta) + i\sin(n\theta))$ for all $n \in \mathbb{Z}$.

We have:

$$\left(2\left(\cos\left(\frac{\pi}{4}\right) + i\sin\left(\frac{\pi}{4}\right)\right)\right)^{-3} = 2^{-3}\left(\cos\left(-3 \times \frac{\pi}{4}\right) + i\sin\left(-3 \times \frac{\pi}{4}\right)\right)$$

$$= \frac{1}{8}\left(\cos\left(-\frac{3\pi}{4}\right) + i\sin\left(-\frac{3\pi}{4}\right)\right)$$

$$= \frac{1}{8}\left(-\frac{1}{\sqrt{2}} - \frac{1}{\sqrt{2}}i\right)$$

© HERIOT-WATT UNIVERSITY

De Moivre's theorem exercise 2

Go online

Q142: Calculate $(4(cos(125°) + i\ sin(125°)))^{-4}$, giving your answer to three decimal places.

....

Q143: Convert $\left(5\left(\cos\frac{7\pi}{4} + i\sin\frac{7\pi}{4}\right)\right)^{-3}$ into the form $x + iy$.

....

Q144: Evaluate $\left(\sqrt{3} - i\right)^{-6}$ in the form $x + iy$.

....

Q145: Evaluate $\left(1 + i\sqrt{2}\right)^{-7}$ in the form $x + iy$.

6.7.4 De Moivre's theorem and multiple angle formulae

De Moivre's theorem is extremely useful in deriving trigonometric formulae. The examples which follow demonstrate the strategy to obtain these.

Examples

1. Problem:
Find a formula for $\cos(3\theta)$ in terms of powers of $\cos\theta$ by using de Moivre's theorem.
Hence express $\cos^3\theta$ in terms of $\cos\theta$ and $\cos(3\theta)$.

Solution:
Use de Moivre's theorem: $\cos(3\theta) + i\sin(3\theta) = (\cos\theta + i\sin\theta)^3$
By the binomial theorem:
$(\cos\theta + i\sin\theta)^3 = \cos^3\theta + 3i\cos^2\theta\sin\theta - 3\cos\theta\sin^2\theta - i\sin^3\theta$
Hence: $\cos(3\theta) + i\sin(3\theta) = \cos^3\theta + 3i\cos^2\theta\sin\theta - 3\cos\theta\sin^2\theta - i\sin^3\theta$
The question only asks for an expression for $\cos(3\theta)$. This is a real term. To get the equivalent expression on the right hand side, we need to equate real parts on both sides.
Reminder: the terms with i are all imaginary terms. Those without are real parts.
Equating real parts: $\cos(3\theta) = \cos^3\theta - 3\cos\theta\sin^2\theta$
Hence:
$\cos(3\theta) = \cos^3\theta - 3\cos\theta\left(1 - \cos^2\theta\right)$
$= \cos^3\theta - 3\cos\theta + 3\cos^3\theta$
$= 4\cos^3\theta - 3\cos\theta$

Rearranging gives:
$\cos^3\theta = \frac{\cos(3\theta) + 3\cos(3\theta)}{4}$

....

2. Problem:

Using De Moivre's and the binomial theorem, express $\cos(4\theta)$ as a polynomial in $\cos\theta$.

Solution:

Use de Moivre's theorem: $(cos\ \theta\ +\ i\sin\theta)^4\ =\ \cos(4\theta)\ +\ i\sin(4\theta)$

By the binomial theorem:

$(cos\ \theta\ +\ i\sin\theta)^4\ =\ \cos^4\theta\ +\ 4i\cos^3\theta\sin\theta\ -\ 6\cos^2\theta\sin^2\theta\ -\ 4i\cos\theta\sin^3\theta\ +\ \sin^4\theta$

Hence:

$\cos(4\theta)\ +\ i\sin(4\theta)\ =\ \cos^4\theta\ +\ 4i\cos^3\theta\sin\theta\ -\ 6\cos^2\theta\sin^2\theta\ -\ 4i\cos\theta\sin^3\theta\ +\ \sin^4\theta$

The question only asks for an expression for $\cos(4\theta)$. This is a real term. To get the equivalent expression on the right hand side we need to equate real parts on both sides.

Reminder: the terms with i are all imaginary terms. Those without are real parts.

Equating real parts: $\cos(4\theta)\ =\ \cos^4\theta\ -\ 6\cos^2\theta\sin^2\theta\ +\ \sin^4\theta$

Hence:

$$\cos(4\theta) = \cos^4\theta - 6\cos^2\theta\left(1 - \cos^2\theta\right) + \left(1 - \cos^2\theta\right)^2$$
$$= \cos^4\theta - 6\cos^2\theta + 6\cos^4\theta + 1 - 2\cos^2\theta + \cos^4\theta$$
$$= 8\cos^4\theta - 8\cos^2\theta + 1$$

....

3. Problem:

Express $\tan(5\theta)$ in terms of $\tan\theta$.

Solution:

We know that $\tan(5\theta) = \frac{\sin(5\theta)}{\cos(5\theta)}$ so we need to find expressions for $\sin(5\theta)$ and $\cos(5\theta)$ in terms of powers of cosine and sine.

So: $\cos(5\theta)\ +\ i\sin(5\theta)\ =\ (cos\ \theta\ +\ i\sin\theta)^5$

Expanding this using the binomial theorem:

$(cos\ \theta\ +\ i\sin\theta)^5\ =\ \cos^5\theta\ +\ 5i\cos^4\theta\sin\theta\ -\ 10\cos^3\theta\sin^2\theta\ -\ 10i\cos^2\theta\sin^3\theta\ +\ 5\cos\theta\sin^4\theta\ +\ i\sin^5\theta$

In the question we are asked to find expressions for $\cos(5\theta)$ and $\sin(5\theta)$. Cosine is real and sine is imaginary. To find expressions for these, we need to equate the real terms on both sides and the imaginary terms on both sides.

Reminder: the terms with i are all imaginary terms. Those without are real parts.

Equating real and imaginary parts:

$$\cos(5\theta) = \cos^5\theta - 10\cos^3\theta\sin^2\theta + 5\cos\theta\sin^4\theta$$

$$\sin(5\theta) = 5\cos^4\theta\sin\theta - 10\cos^2\theta\sin^3\theta + \sin^5\theta$$

Putting this into the expression for $\tan(5\theta)$:

$$\tan(5\theta) = \frac{5\cos^4\theta\sin\theta - 10\cos^2\theta\sin^3\theta + \sin^5\theta}{\cos^5\theta - 10\cos^3\theta\sin^2\theta + 5\cos\theta\sin^4\theta}$$

We divide through by $\cos^5\theta$ because we only want expressions in terms of $\tan\theta$ and the highest power of sine and cosine is 5.

$$\tan(5\theta) = \frac{\frac{5\cos^4\theta\sin\theta}{\cos^5\theta} - \frac{10\cos^2\theta\sin^3\theta}{\cos^5\theta} + \frac{\sin^5\theta}{\cos^5\theta}}{\frac{\cos^5\theta}{\cos^5\theta} - \frac{10\cos^3\theta\sin^2\theta}{\cos^5\theta} + \frac{5\cos\theta\sin^4\theta}{\cos^5\theta}}$$

$$= \frac{5\frac{\sin\theta}{\cos\theta} - 10\frac{\sin^3\theta}{\cos^3\theta} + \frac{\sin^5\theta}{\cos^5\theta}}{1 - 10\frac{\sin^2\theta}{\cos^2\theta} + 5\frac{\sin^4\theta}{\cos^4\theta}}$$

$$= \frac{5\tan\theta - 10\tan^3\theta + \tan^5\theta}{1 - 10\tan^2\theta + 5\tan^4\theta}$$

De Moivre's theorem and multiple angle formulae exercise   Go online

Q146: Express $\sin(3\theta)$ as a polynomial in $\sin\theta$ and hence express $\sin^3\theta$ in terms of $\sin\theta$ and $\sin(3\theta)$.

....

Q147: Express $\cos(5\theta)$ as a polynomial in $\cos\theta$.

....

Q148: Express $\frac{\sin(5\theta)}{\sin\theta}$ as a polynomial in $\sin\theta$.

....

Q149: Express $\tan(4\theta)$ in terms of $\tan\theta$.

....

Q150: Show that $\frac{\cos(3\theta)}{\cos^2\theta} = 4\cos\theta - 3\sec\theta$

6.7.4.1 De Moivre's theorem with integers

It can be shown that de Moivre's theorem holds for all integers ($n \in \mathbb{Z}$).

For a brief look at the proof that de Moivre's theorem holds for integers go to the extra study activity now. It can be found in the section headed Proofs at the end of this topic.

Proof 3: Proof of de Moivre's theorem for integers

Proof 3

Consider n to be a negative integer.
Let $n = -p$
If $z^n = (cos\,\theta + i\,sin\theta)^n$, then:

$$(\cos\theta + i\sin\theta)^n = (\cos\theta + i\sin\theta)^{-p}$$
$$= \frac{1}{(\cos\theta + i\sin\theta)^p}$$
$$= \frac{1}{\cos(p\theta) + i\sin p\theta} \times \left(\frac{\cos(p\theta) - i\sin(p\theta)}{\cos(p\theta) - i\sin(p\theta)}\right)$$
$$= \frac{\cos(p\theta) - i\sin(p\theta)}{\cos^2(p\theta) + \sin^2(p\theta)}$$
$$= \cos(p\theta) - i\sin(p\theta)$$
$$= \cos(-p\theta) + i\sin(-p\theta)$$
$$= \cos(n\theta) + i\sin(n\theta)$$

So $z^n = \cos(n\theta) + i\,\sin(n\theta)$ where $n \in \mathbb{Z}$.

Using this we can create some new expressions that allows us to state powers of sine and cosine in terms of $\sin(n\theta)$ and $\cos(n\theta)$.

Suppose that $z = \cos\theta + i\sin\theta$ and $\overline{z} = \cos\theta - i\sin\theta$

We also have:
$$z^{-1} = (\cos\theta + i\sin\theta)^{-1}$$
$$= \cos(-\theta) + i\sin(-\theta)$$
$$= \cos\theta - i\sin\theta$$

So we have: $\overline{z} = z^{-1}$

Adding gives:
$$z + z^{-1} = \cos\theta + i\sin\theta + \cos\theta - i\sin\theta$$
$$= 2\cos\theta$$

i.e. $z + \frac{1}{z} = 2\cos\theta$

so: $\cos\theta = \frac{1}{2}\left(z + \frac{1}{z}\right)$

Also:
$$z^n = \cos(n\theta) + i\sin(n\theta)$$
$$z^n + z^{-n} = 2\cos(n\theta)$$
$$\cos(n\theta) = \frac{1}{2}\left(z^n + \frac{1}{z^n}\right)$$

Subtracting gives:
$$z - z^{-1} = \cos\theta + i\sin\theta - (\cos\theta - i\sin\theta)$$
$$= 2i\sin\theta$$

© HERIOT-WATT UNIVERSITY

i.e. $z - \frac{1}{z} = 2i\sin\theta$

so: $\sin\theta = \frac{1}{2i}\left(z - \frac{1}{z}\right)$

Also:
$$z^{-n} = \cos(n\theta) - i\sin(n\theta)$$
$$z^n - z^{-n} = 2i\sin\theta$$
$$\sin(n\theta) = \frac{1}{2i}\left(z^n - \frac{1}{z^n}\right)$$

Key point

$$2\cos\theta = z + \frac{1}{z} \qquad 2i\sin\theta = z - \frac{1}{z}$$
$$2\cos(n\theta) = z^n + \frac{1}{z^n} \qquad 2i\sin(n\theta) = z^n - \frac{1}{z^n}$$

Examples

1. Problem:

Express $\cos^3\theta$ in the form of a series of cosines and multiples of θ.

Solution:

Let $z = \cos\theta + i\sin\theta$

We know we can give $\cos(n\theta)$ as: $2\cos(n\theta) = z^n + \frac{1}{z^n}$

In this case $n = 1$ since we wish to find an expression for $(\cos(1 \times \theta))^3$ (rewritten from $\cos^3\theta$)
so we have: $(2\cos\theta)^3 = \left(z + \frac{1}{z}\right)^3$

From here we expand using the binomial: $8\cos^3\theta = z^3 + 3z^2\frac{1}{z} + 3z\frac{1}{z^2} + \frac{1}{z^3}$

Simplifying this expression: $8\cos^3\theta = z^3 + \frac{1}{z^3} + 3\left(z + \frac{1}{z}\right)$

We should be able to notice that we can group terms into the form $z^n + \frac{1}{z^n}$ which we can then replace with $2\cos(n\theta)$:

$$8\cos^3\theta = z^3 + \frac{1}{z^3} + 3\left(z + \frac{1}{z}\right)$$
$$= 2\cos(3\theta) + 3(2\cos\theta)$$
$$= 2\cos(3\theta) + 6\cos\theta$$

We have: $\cos^3\theta = \frac{1}{4}(\cos(3\theta) + 3\cos\theta)$

....

2. Problem:

Express $\sin^5\theta$ as a series of sines and multiples of θ.

Solution:

Let $z = \cos\theta + i\sin\theta$

We will use the form: $2i\sin(n\theta) = z^n - \frac{1}{z^n}$

In this case $n = 1$ since we wish to find an expression for $(\sin(1 \times \theta))^5$ (rewritten from $\sin^5\theta$)
so we have: $(2i\sin\theta)^5 = \left(z - \frac{1}{z}\right)^5$

TOPIC 6. COMPLEX NUMBERS

Using the binomial theorem to expand the RHS:
$32i\sin^5\theta = z^5 - 5z^4\frac{1}{z} + 10z^3\frac{1}{z^2} - 10z^2\frac{1}{z^3} + 5z\frac{1}{z^4} - \frac{1}{z^5}$
(Remember that $i^5 = i$)
Simplifying: $32i\sin^5\theta = z^5 - 5z^3 + 10z - \left(\frac{10}{z}\right) + \left(\frac{5}{z^3}\right) - \left(\frac{1}{z^5}\right)$
Grouping terms together into the form $z^n - \frac{1}{z^n}$, we can then replace them with $2i\sin(n\theta)$:

$$32i\sin^5\theta = z^5 - \frac{1}{z^5} - 5\left(z^3 - \frac{1}{z^3}\right) + 10\left(z - \frac{1}{z}\right)$$
$$= 2i\sin(5\theta) - 5(2i\sin(3\theta)) + 10(2i\sin\theta)$$
$$= 2i\sin(5\theta) - 10i\sin(3\theta) + 20i\sin\theta$$

We have: $\sin^5\theta = \frac{1}{16}(\sin(5\theta) - 5\sin(3\theta) + 10\sin\theta)$

....

3. Problem:

Express $\cos^4(2\theta)$ in the form of a series of cosines and multiples of θ.

Solution:

Let $z = \cos\theta + i\sin\theta$

We will use the form: $2\cos(n\theta) = z^n + \frac{1}{z^n}$

In this case n = 2 since we wish to find an expression for $(cos(2\theta))^4$ (rewritten from $\cos^4(2\theta)$) so we have: $(2\cos(2\theta))^4 = \left(z^2 + \frac{1}{z^2}\right)^4$

Using the binomial theorem to expand the RHS:
$16\cos^4(2\theta) = (z^2)^4 + 4(z^2)^3\frac{1}{z^2} + 6(z^2)^2\frac{1}{(z^2)^2} + 4z^2\frac{1}{(z^2)^3} + \frac{1}{(z^2)^4}$

Simplifying:
$$16\cos^4(2\theta) = z^8 + 4z^6\frac{1}{z^2} + 6z^4\frac{1}{z^4} + 4z^2\frac{1}{z^6} + \frac{1}{z^8}$$
$$= z^8 + 4z^4 + 6 + 4\frac{1}{z^4} + \frac{1}{z^8}$$

Grouping terms together into the form $z^n + \frac{1}{z^n}$, we can then replace them with $2\cos(n\theta)$:

$$16\cos^4(2\theta) = z^8 + \frac{1}{z^8} + 4\left(z^4 + \frac{1}{z^4}\right) + 6$$
$$= 2\cos(8\theta) + 4(2\cos(4\theta)) + 6$$
$$= 2\cos(8\theta) + 8\cos(4\theta) + 6$$

We have: $\cos^4(2\theta) = \frac{1}{8}(\cos(8\theta) + 4\cos(4\theta) + 3)$

De Moivre's theorem with integers exercise   Go online

Q151: Express $\sin^7\theta$ as a series of sines of multiples of θ.
....

Q152: Express $\cos^6\theta$ as a series of cosines of multiples of θ.
....

Q153: Express $\sin^6\theta$ as a series of cosines of multiples of θ.

....

Q154: Show that $\cos^2\theta \sin\theta = \frac{1}{8i}\left(z^3 - \frac{1}{z^3} + z - \frac{1}{z}\right)$.

6.7.5 De Moivre's theorem with fractional powers

De Moivre's theorem is not only true for the integers but can be extended to fractions.

Key point

De Moivre's theorem for fractional powers gives the equation:

$$(r(\cos\theta + i\sin\theta))^{\frac{p}{q}} = r^{\frac{p}{q}}\left(\cos\left(\frac{p}{q}\theta\right) + i\sin\left(\frac{p}{q}\theta\right)\right)$$

Examples

1. Problem:

Calculate $\left(\cos\left(\frac{\pi}{4}\right) + i\sin\left(\frac{\pi}{4}\right)\right)^{\frac{1}{3}}$

Solution:

By de Moivre's theorem for fractional powers:

$$\left(\cos\left(\frac{\pi}{4}\right) + i\sin\left(\frac{\pi}{4}\right)\right)^{\frac{1}{3}} = \left(\cos\left(\frac{1}{3} \times \frac{\pi}{4}\right) + i\sin\left(\frac{1}{3} \times \frac{\pi}{4}\right)\right)$$

$$= \left(\cos\left(\frac{\pi}{12}\right) + i\sin\left(\frac{\pi}{12}\right)\right)$$

....

2. Problem:

Using de Moivre's theorem, calculate $(16\cos(140°) + 16i\sin(140°))^{\frac{3}{2}}$

Leave the answer in polar form.

Solution:

This can be rewritten as:

$$16^{\frac{3}{2}}(\cos(140°) + i\sin(140°))^{\frac{3}{2}} = 16^{\frac{3}{2}}\left(\cos\left(\frac{3}{2} \times 140°\right) + i\sin\left(\frac{3}{2} \times 140°\right)\right)$$

$$= 64(\cos(210°) + i\sin(210°))$$

The argument is not within the range of the principal argument, namely $(-180°, 180°)$.

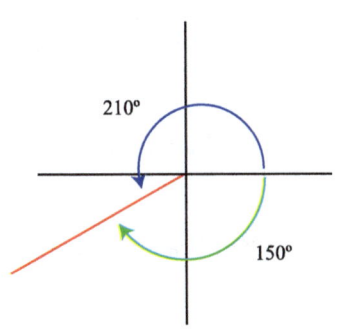

TOPIC 6. COMPLEX NUMBERS

$$16^{\frac{3}{2}} \left(\cos\left(140°\right) + i \sin\left(140°\right)\right) = 64 \left(\cos\left(-150°\right) + i \sin\left(-150°\right)\right)$$
$$= 64 \left(\cos\left(150°\right) - i \sin\left(150°\right)\right)$$

....

3. Problem:

Using de Moivre's theorem, calculate $\left(16 \left(\cos\left(\frac{\pi}{2}\right) + i \sin\left(\frac{\pi}{2}\right)\right)\right)^{-\frac{3}{4}}$

Leave the answer in polar form.

Solution:

This can be rewritten as:

$$\left(16 \left(\cos\left(\frac{\pi}{2}\right) + i \sin\left(\frac{\pi}{2}\right)\right)\right)^{-\frac{3}{4}} = 16^{-\frac{3}{4}} \left(\cos\left(\frac{\pi}{2}\right) + i \sin\left(\frac{\pi}{2}\right)\right)^{-\frac{3}{4}}$$
$$= 16^{-\frac{3}{4}} \left(\cos\left(-\frac{3}{4} \times \frac{\pi}{2}\right) + i \sin\left(-\frac{3}{4} \times \frac{\pi}{2}\right)\right)^{-\frac{3}{4}}$$
$$= \frac{1}{8} \left(\cos\left(-\frac{3\pi}{8}\right) + i \sin\left(-\frac{3\pi}{8}\right)\right)$$
$$= \frac{1}{8} \left(\cos\left(\frac{3\pi}{8}\right) - i \sin\left(\frac{3\pi}{8}\right)\right)$$

De Moivre's theorem and fractional powers exercise   Go online

Q155: Use de Moivre's theorem to calculate $\left(27 \left(\cos\left(\frac{\pi}{2}\right) + i \sin\left(\frac{\pi}{2}\right)\right)\right)^{\frac{1}{3}}$ leaving your answer in polar form.

....

Q156: Use de Moivre's theorem to calculate $\left(8 \left(\cos\left(\frac{\pi}{4}\right) + i \sin\left(\frac{\pi}{4}\right)\right)\right)^{\frac{2}{3}}$ leaving your answer in polar form.

....

Q157: Use de Moivre's theorem to calculate $\left(81 \left(\cos\left(36°\right) + i \sin\left(36°\right)\right)\right)^{\frac{1}{4}}$ leaving the answer in polar form.

....

© HERIOT-WATT UNIVERSITY

Q158: Use de Moivre's theorem to calculate $\left(32\left(\cos\left(\frac{\pi}{3}\right) + i\sin\left(\frac{\pi}{3}\right)\right)\right)^{\frac{3}{5}}$ leaving the answer in polar form.

....

Q159: Use de Moivre's theorem to calculate $\left(64\left(\cos\left(\frac{3\pi}{2}\right) + i\sin\left(\frac{3\pi}{2}\right)\right)\right)^{\frac{4}{3}}$ leaving the answer in polar form.

....

Q160: Use de Moivre's theorem to calculate $\left(16\left(\cos\left(\frac{5\pi}{3}\right) + i\sin\left(\frac{5\pi}{3}\right)\right)\right)^{\frac{3}{2}}$ leaving the answer in polar form.

....

Q161: Use de Moivre's theorem to calculate $\left(343\left(\cos\left(243°\right) + i\sin\left(243°\right)\right)\right)^{-\frac{1}{3}}$ leaving the answer in polar form.

....

Q162: Using de Moivre's theorem calculate $\left(25\left(\cos\left(184°\right) + i\sin\left(184°\right)\right)\right)^{-\frac{5}{2}}$ leaving the answer in polar form.

6.7.6 De Moivre's theorem and n^{th} roots

A previous worked example showed that $\left(\cos\left(\frac{\pi}{4}\right) + i\sin\left(\frac{\pi}{4}\right)\right)^{\frac{1}{3}} = \left(\cos\left(\frac{\pi}{12}\right) + i\sin\left(\frac{\pi}{12}\right)\right)$

That is, $\cos\left(\frac{\pi}{12}\right) + i\sin\left(\frac{\pi}{12}\right)$ is a cube root of $\cos\left(\frac{\pi}{4}\right) + i\sin\left(\frac{\pi}{4}\right)$

This cube root is obtained by dividing the argument of the original number by 3.

However, the cube roots of $\cos\left(\frac{\pi}{4}\right) + i\sin\left(\frac{\pi}{4}\right)$ are complex numbers z which satisfy $z^3 = 1$ and so, by the fundamental theorem of algebra, since this equation is of degree 3, there should be three roots. That is, in general, a complex number should have three cube roots.

Given a complex number, these three cube roots can always be found.

> **Top tip**
>
> **Strategy for finding the cube roots of a complex number**
>
> - State the complex number in polar form $z = r(\cos\theta + i\sin\theta)$
> - State z in two more equivalent alternative ways by adding 2π to the argument each time:
>   - $z = r(\cos(\theta + 2\pi) + i\sin(\theta + 2\pi))$
>   - $z = r(\cos(\theta + 4\pi) + i\sin(\theta + 4\pi))$
> - Determine the cube roots of z by taking the cube root of r and dividing each of the arguments by 3

TOPIC 6. COMPLEX NUMBERS

Note that the strategy for finding the cube roots of a complex number gives the three cube roots as:

$$r^{\frac{1}{3}} \left(\cos\left(\frac{\theta}{3}\right) + i \sin\left(\frac{\theta}{3}\right) \right)$$

$$r^{\frac{1}{3}} \left(\cos\left(\frac{\theta}{3} + \frac{2\pi}{3}\right) + i \sin\left(\frac{\theta}{3} + \frac{2\pi}{3}\right) \right)$$

$$r^{\frac{1}{3}} \left(\cos\left(\frac{\theta}{3} + \frac{4\pi}{3}\right) + i \sin\left(\frac{\theta}{3} + \frac{4\pi}{3}\right) \right)$$

If $z = r(\cos\theta + i \sin\theta)$ is written in any further alternative ways, such as:

$z = r(\cos(\theta + 6\pi) + i \sin(\theta + 6\pi))$

This gives a cube root of: $r^{\frac{1}{3}} \left(\cos\left(\frac{\theta}{3} + \frac{6\pi}{3}\right) + i \sin\left(\frac{\theta}{3} + \frac{6\pi}{3}\right) \right) = r^{\frac{1}{3}} \left(\cos\left(\frac{\theta}{3}\right) + i \sin\left(\frac{\theta}{3}\right) \right)$,

which is the same as one of the previously mentioned roots.

It is impossible to find any more.

Example

Problem:

Find the cube roots of $1 + i$

Solution:

First express $1 + i$ in polar form:

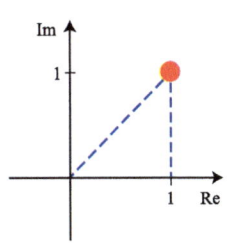

$|1 + i| = \sqrt{2}$ and $\arg(1 + i) = \frac{\pi}{4}$

Hence $1 + i$ can be expressed as: $1 + i = \sqrt{2} \left(\cos\left(\frac{\pi}{4}\right) + i \sin\left(\frac{\pi}{4}\right) \right)$

But $1 + i$ can also be expressed as:

$$1 + i = \sqrt{2} \left(\cos\left(\frac{\pi}{4} + 2\pi\right) + i \sin\left(\frac{\pi}{4} + 2\pi\right) \right)$$
$$= \sqrt{2} \left(\cos\left(\frac{9\pi}{4}\right) + i \sin\left(\frac{9\pi}{4}\right) \right)$$

and

$$1 + i = \sqrt{2} \left(\cos\left(\frac{\pi}{4} + 4\pi\right) + i \sin\left(\frac{\pi}{4} + 4\pi\right) \right)$$
$$= \sqrt{2} \left(\cos\left(\frac{17\pi}{4}\right) + i \sin\left(\frac{17\pi}{4}\right) \right)$$

© HERIOT-WATT UNIVERSITY

Hence, taking the cube root of the modulus and dividing the argument by 3, the cube roots of $1 + i$ are:

$$z_1 = \left(2^{\frac{1}{2}}\right)^{\frac{1}{3}} \left(\cos\left(\frac{\pi}{4}\right) + i\sin\left(\frac{\pi}{4}\right)\right)^{\frac{1}{3}}$$
$$= \left(\frac{1}{64}\right)\left(\cos\left(\frac{\pi}{12}\right) + i\sin\left(\frac{\pi}{12}\right)\right)$$

$$z_2 = \left(2^{\frac{1}{2}}\right)^{\frac{1}{3}} \left(\cos\left(\frac{9\pi}{4}\right) + i\sin\left(\frac{9\pi}{4}\right)\right)^{\frac{1}{3}}$$
$$= \left(\frac{1}{64}\right)\left(\cos\left(\frac{3\pi}{4}\right) + i\sin\left(\frac{3\pi}{4}\right)\right)$$

$$z_3 = \left(2^{\frac{1}{2}}\right)^{\frac{1}{3}} \left(\cos\left(\frac{17\pi}{4}\right) + i\sin\left(\frac{17\pi}{4}\right)\right)^{\frac{1}{3}}$$
$$= \left(\frac{1}{64}\right)\left(\cos\left(\frac{17\pi}{12}\right) + i\sin\left(\frac{17\pi}{12}\right)\right)$$

In this way, the nth roots of any complex number can be found.

Example

Problem:

Find the fifth roots of $z = 243(cos(25°) + i\sin(25°))$

Solution:

For the fifth roots, we would expect to find five solutions:

$z_1 = 243^{\frac{1}{5}}(\cos(25°) + i\sin(25°))^{\frac{1}{5}}$
$= 3(\cos(5°) + i\sin(5°))$

$z_2 = 243^{\frac{1}{5}}(\cos(385°) + i\sin(385°))^{\frac{1}{5}}$
$= 3(\cos(77°) + i\sin(77°))$

$z_3 = 243^{\frac{1}{5}}(\cos(745°) + i\sin(745°))^{\frac{1}{5}}$
$= 3(\cos(149°) + i\sin(149°))$

$z_4 = 243^{\frac{1}{5}}(\cos(1105°) + i\sin(1105°))^{\frac{1}{5}}$
$= 3(\cos(221°) + i\sin(221°))$ convert to principal argument
$= 3(\cos(-139°) + i\sin(-139°))$
$= 3(\cos(139°) - i\sin(139°))$

$z_5 = 243^{\frac{1}{5}}(\cos(1465°) + i\sin(1465°))^{\frac{1}{5}}$
$= 3(\cos(293°) + i\sin(293°))$ convert to principal argument
$= 3(\cos(-67°) + i\sin(-67°))$
$= 3(\cos(67°) - i\sin(67°))$

De Moivre's theorem and n^{th} roots exercise

Go online

Q163: Find the cube roots of $z = 64(cos(30°) + i\sin(30°))$

....

Q164: Find the fourth roots of $81i$, that is $81\left(\cos\left(\frac{\pi}{2}\right) + i\sin\left(\frac{\pi}{2}\right)\right)$

....

Q165: Find the sixth roots of $\sqrt{3} + i$ where $(-\pi, \pi)$.

6.7.7 De Moivre's theorem and roots of unity

It is easy and important to find the nth roots of 1, i.e. complex numbers such that $z^n = 1$. Such numbers are often referred to as the nth roots of unity.

Since $1 = \cos(2\pi) + i\sin(2\pi)$, it follows that $\cos\left(\frac{2\pi}{n}\right) + i\sin\left(\frac{2\pi}{n}\right)$ is an nth root of unity.

But 1 can be written using different arguments as follows:
$1 = \cos(2\pi) + i\sin(2\pi)$
$1 = \cos(4\pi) + i\sin(4\pi)$
$1 = \cos(6\pi) + i\sin(6\pi)$
$1 = \cos(2n\pi) + i\sin(2n\pi)$

Hence, dividing the argument in each case by n gives the following n^{th} roots of unity:

$z = \cos\left(\frac{2\pi}{n}\right) + i\sin\left(\frac{2\pi}{n}\right)$

$z = \cos\left(\frac{4\pi}{n}\right) + i\sin\left(\frac{4\pi}{n}\right)$

$z = \cos\left(\frac{6\pi}{n}\right) + i\sin\left(\frac{6\pi}{n}\right)$

$z = \cos\left(\frac{2n\pi}{n}\right) + i\sin\left(\frac{2n\pi}{n}\right)$

Note that arguments increase by $\frac{2\pi}{n}$ each time. The roots of unity are regularly spaced on an Argand diagram.

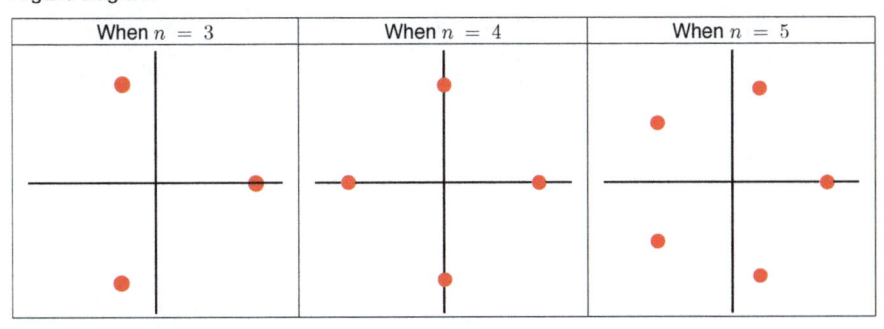

When $n = 3$	When $n = 4$	When $n = 5$

When $n = 6$	When $n = 7$	When $n = 8$
(dots arranged as 6th roots of unity on unit circle)	(dots arranged as 7th roots of unity on unit circle)	(dots arranged as 8th roots of unity on unit circle)

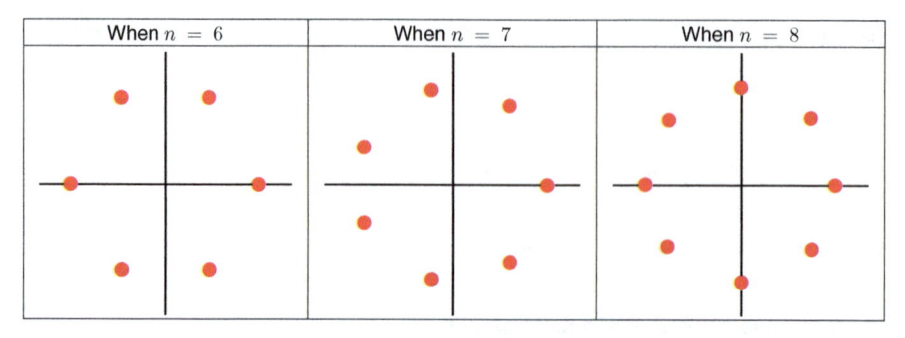

Roots of unity

The notation for the roots are given in the form $e^{i\theta}$ because $e^{i\theta} = \cos\theta + i\sin\theta$. You do not need to know this notation and it is not examinable. If you were to state the roots in this form you would not be given the marks. It is merely for shorthand and for interest.

Examples

1. Problem:

Find the cube roots of unity and plot them on an Argand diagram.

Solution:

The cube roots of unity are the solutions to $z^3 = 1$

1 can be written in polar form as:

$1 = \cos(2\pi) + i\sin(2\pi)$
$1 = \cos(4\pi) + i\sin(4\pi)$
$1 = \cos(6\pi) + i\sin(6\pi)$

So:

$z^3 = \cos(2\pi) + i\sin(2\pi)$
$z^3 = \cos(4\pi) + i\sin(4\pi)$
$z^3 = \cos(6\pi) + i\sin(6\pi)$

Taking the cube roots gives:

$z = (\cos(2\pi) + i\sin(2\pi))^{\frac{1}{3}}$
$= \cos\left(\frac{2\pi}{3}\right) + i\sin\left(\frac{2\pi}{3}\right)$
$= \frac{1}{2}\left(-1 + i\sqrt{3}\right)$

$z = (\cos(4\pi) + i\sin(4\pi))^{\frac{1}{3}}$
$= \cos\left(\frac{4\pi}{3}\right) + i\sin\left(\frac{4\pi}{3}\right)$
$= \frac{1}{2}\left(-1 - i\sqrt{3}\right)$

$z = (\cos(6\pi) + i\sin(6\pi))^{\frac{1}{3}}$
$= \cos\left(\frac{6\pi}{3}\right) + i\sin\left(\frac{6\pi}{3}\right)$
$= 1$

Plotted on an Argand diagram:

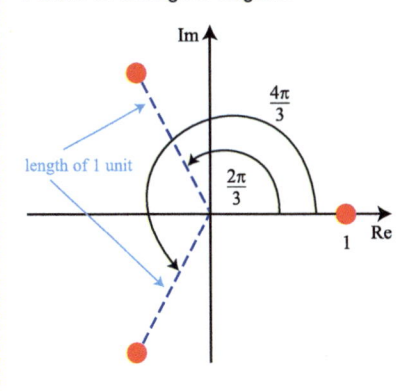

....

2. Cube root of a complex number
Problem:
Find the solutions of $z^3 + 1 = 0$ and plot them on an Argand diagram.
Solution:
$z^3 + 1 = 0 \rightarrow z^3 = -1$
-1 can be written in polar form as:
$-1 = \cos \pi + i \sin \pi$
$-1 = \cos(3\pi) + i \sin(3\pi)$
$-1 = \cos(5\pi) + i \sin(5\pi)$
So:
$$z^3 = \cos \pi + i \sin \pi$$
$$z^3 = \cos(3\pi) + i \sin(3\pi)$$
$$z^3 = \cos(5\pi) + i \sin(5\pi)$$
Taking the cube roots gives:
$z = (\cos \pi + i \sin \pi)^{\frac{1}{3}}$
$ = \cos\left(\frac{\pi}{3}\right) = i \sin\left(\frac{\pi}{3}\right)$
$ = \frac{1}{2} + \frac{\sqrt{3}}{2} i$
$z = (\cos(3\pi) + i \sin(3\pi))^{\frac{1}{3}}$
$ = \cos \pi + i \sin \pi$
$ = -1$
$z = (\cos(5\pi) + i \sin(5\pi))^{\frac{1}{3}}$
$ = \cos\left(\frac{5\pi}{3}\right) + i \sin\left(\frac{5\pi}{3}\right)$
$ = \frac{1}{2} - \frac{\sqrt{3}}{2} i$

Plotted on an Argand diagram:

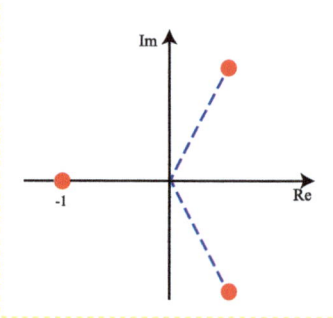

Notice that the complex solutions are conjugates of each other and the roots are equally spaced on the Argand diagram.

Example

Problem:

Solve the equation $z^5 = 2 - i2\sqrt{3}$

Solution:

Change $2 - i2\sqrt{3}$ into polar form first.

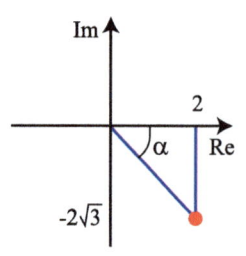

Modulus:

$|z| = \sqrt{2^2 + \left(-2\sqrt{3}\right)^2}$

$= 4$

Argument:

$\tan \alpha = \left| \dfrac{-2\sqrt{3}}{2} \right|$

$\alpha = \dfrac{\pi}{3}$

Since $2 - i2\sqrt{3}$ is in the fourth quadrant, $\arg(z) = -\alpha$

$\arg(z) = -\dfrac{\pi}{3}$

We have: $z^5 = 4\left(\cos\left(-\dfrac{\pi}{3}\right) + i\sin\left(-\dfrac{\pi}{3}\right)\right)$

For the fifth roots we have: $z^5 = 4\left(\cos\left(-\dfrac{\pi}{3} + 2k\pi\right) + i\sin\left(-\dfrac{\pi}{3} + 2k\pi\right)\right)$

TOPIC 6. COMPLEX NUMBERS

We do not need to change $-\frac{\pi}{3}$ to the equivalent clockwise angle $\frac{5\pi}{3}$ because both will give the same answer when evaluated. Hence, we will still get all the same roots. Using $-\frac{\pi}{3}$ is just more convenient because we do not have to do extra work to find this equivalent clockwise angle.

Evaluate for $k = 0, 1, 2, 3, 4$:

$k = 0:$ $z^5 = 4\left(\cos\left(-\frac{\pi}{3}\right) + i\sin\left(-\frac{\pi}{3}\right)\right)$

$k = 1:$ $z^5 = 4\left(\cos\left(-\frac{\pi}{3} + 2\pi\right) + i\sin\left(-\frac{\pi}{3} + 2\pi\right)\right)$
$ = 4\left(\cos\left(\frac{5\pi}{3}\right) + i\sin\left(\frac{5\pi}{3}\right)\right)$

$k = 2:$ $z^5 = 4\left(\cos\left(-\frac{\pi}{3} + 4\pi\right) + i\sin\left(-\frac{\pi}{3} + 4\pi\right)\right)$
$ = 4\left(\cos\left(\frac{11\pi}{3}\right) + i\sin\left(\frac{11\pi}{3}\right)\right)$

$k = 3:$ $z^5 = 4\left(\cos\left(-\frac{\pi}{3} + 6\pi\right) + i\sin\left(-\frac{\pi}{3} + 6\pi\right)\right)$
$ = 4\left(\cos\left(\frac{17\pi}{3}\right) + i\sin\left(\frac{17\pi}{3}\right)\right)$

$k = 4:$ $z^5 = 4\left(\cos\left(-\frac{\pi}{3} + 8\pi\right) + i\sin\left(-\frac{\pi}{3} + 8\pi\right)\right)$
$ = 4\left(\cos\left(\frac{23\pi}{3}\right) + i\sin\left(\frac{23\pi}{3}\right)\right)$

Using de Moivre's theorem, the roots are:

$k = 0:$ $z = 4^{\frac{1}{5}}\left(\cos\left(-\frac{\pi}{15}\right) + i\sin\left(-\frac{\pi}{15}\right)\right)$
$ = 4^{\frac{1}{5}}\left(\cos\left(\frac{\pi}{15}\right) - i\sin\left(\frac{\pi}{15}\right)\right)$

$k = 1:$ $z = 4^{\frac{1}{5}}\left(\cos\left(\frac{5\pi}{15}\right) + i\sin\left(\frac{5\pi}{15}\right)\right)$
$ = 4^{\frac{1}{5}}\left(\cos\left(\frac{\pi}{3}\right) + i\sin\left(\frac{\pi}{3}\right)\right)$

$k = 2:$ $z = 4^{\frac{1}{5}}\left(\cos\left(\frac{11\pi}{15}\right) + i\sin\left(\frac{11\pi}{15}\right)\right)$

$k = 3:$ $z = 4^{\frac{1}{5}}\left(\cos\left(\frac{17\pi}{15}\right) + i\sin\left(\frac{17\pi}{15}\right)\right)$
$ = 4^{\frac{1}{5}}\left(\cos\left(\frac{13\pi}{15}\right) - i\sin\left(\frac{13\pi}{15}\right)\right)$

$k = 4:$ $z = 4^{\frac{1}{5}}\left(\cos\left(\frac{23\pi}{15}\right) + i\sin\left(\frac{23\pi}{15}\right)\right)$
$ = 4^{\frac{1}{5}}\left(\cos\left(\frac{7\pi}{15}\right) - i\sin\left(\frac{7\pi}{15}\right)\right)$

© HERIOT-WATT UNIVERSITY

Evaluating the roots:

$k = 0:$ $\quad z = 1 \cdot 29 - 0 \cdot 27i$
$k = 1:$ $\quad z = 0 \cdot 66 + 1 \cdot 14i$
$k = 2:$ $\quad z = -0 \cdot 88 + 0 \cdot 98i$
$k = 3:$ $\quad z = -1 \cdot 21 - 0 \cdot 54i$
$k = 4:$ $\quad z = 0 \cdot 14 - 1 \cdot 31i$

Key point

The general solution for a root of unity is given by:

if $z^n = r(cos(\theta + 2k\pi) + i \sin(\theta + 2k\pi))$

then $z = r^{\frac{1}{n}} \left(\cos \left(\frac{\theta + 2k\pi}{n} \right) + i \sin \left(\frac{\theta + 2k\pi}{n} \right) \right)$

De Moivre's theorem and roots of unity exercise   Go online

Q166: Find the fourth roots of unity and plot them on an Argand diagram.

....

Q167: Find the solutions of the equation $z^6 - 1 = 0$. Plot the answers on an Argand diagram.

....

Q168: Solve $z^4 = i$. Plot the answers on an Argand diagram. Compare this to the solution of $z^4 = 1$. What do you notice?

....

Q169: How would you expect the solutions of $z^4 = 16$ and $z^4 = -16$ to differ?

....

Q170: In the equation $z^4 = 1$, take the four solutions to be denoted by z_1, z_2, z_3, z_4. What is the value of $z_1 + z_2 + z_3 + z_4$? Explain why this result is to be expected.

....

Q171: Solve the equation $z^3 = -64$

TOPIC 6. COMPLEX NUMBERS

6.8 Conjugate properties

There are special identities which apply to complex numbers and their conjugates. Some of these properties are listed below and will be used in this section. They should be learned.

> **Key point**
>
> If $z = a + ib$, then its conjugate is $\bar{z} = a - ib$
>
> If $w = u + iv$, then its conjugate is $\bar{w} = u - iv$

Let the two complex numbers be z and w with conjugates $\bar{z}$ and $\bar{w}$ respectively.

1. $\bar{\bar{z}} = z$ i.e., the conjugate of the conjugate of z is equal to z.
2. $z = \bar{z}$ where $z \in \mathbb{R}$, i.e. the real part of z is equal to the real part of the conjugate of z.
3. $z + \bar{z} = 2\text{Re}(z)$ i.e., adding the real part of z and the real part of the conjugate with give twice the real part of z.
4. $|\bar{z}| = |z|$ i.e., the magnitude, or modulus, of the conjugate of z is equivalent to the magnitude, or modulus, of z.
5. $z \times \bar{z} = |z|^2$ i.e., z multiplied by the conjugate of z is the same as the modulus of z squared.
6. $\overline{z + w} = \bar{z} + \bar{w}$ i.e., taking the conjugate of the sum of z and w is the same as adding the conjugate of z to the conjugate of w.
7. $\overline{z \times w} = \bar{z} \times \bar{w}$ i.e., taking the conjugate of the multiple of z and w is the same as multiplying the conjugate of z by the conjugate of w.

These properties of conjugates lead easily to the following properties of moduli.

Let z and w be complex numbers. Then:

1. $|\bar{z} \times \bar{w}| = |z| \times |w|$
2. $|z + w| \leq |z| + |w|$
   This is commonly known as the triangle inequality.

For a brief look at the proofs go to the extra study activities now. They can be found in the section headed Proofs at the end of this topic.

> **Proof 4:** $\bar{\bar{z}} = z$
>
> **Proof 4**
>
> Let $z = a + bi$ for a and $b \in \mathbb{R}$
> By the definition of a conjugate:
> $\bar{z} = a - bi$
> $\phantom{\bar{z}} = a + (-b)i$

Again, using the conjugate definition:
$$\overline{\overline{z}} = \overline{a + (-b)i}$$
$$= a - (-b)i$$
$$= a + bi$$
$$= z$$

Proof 5: $z = \overline{z}$ where $z \in \mathbb{R}$

Proof 5

If $z \in \mathbb{R}$, then:

$$z = a + 0 \times i$$
$$= a$$

(since $0 \times i = 0$) for some $a \in \mathbb{R}$

By the definition of a conjugate: $\bar{z} = a - 0 \times i$
But $0 \times i = 0$ so:
$$\bar{z} = a$$
$$= z$$

Proof 6: $z + \overline{z} = 2\text{Re}(z)$

Proof 6

Let $z = a + bi$, then $\bar{z} = a - bi$

Hence:
$$z + \bar{z} = a + bi + a - bi$$
$$= 2a$$
$$= 2\,\text{Re}(z)$$

Proof 7: $|\bar{z}| = |z|$

Proof 7

Let $z = a + bi$ and $|z| = \sqrt{(a^2 + b^2)}$

Hence:
$$|\bar{z}| = \sqrt{(a^2 + (-b^2))}$$
$$= \sqrt{(a^2 + b^2)}$$
$$= |z|$$

Proof 8: $z \times \bar{z} = |z|^2$

Proof 8

Let $z = a + bi$, then $\bar{z} = a - bi$
Hence:
$$z \times \bar{z} = (a + bi)(a - bi)$$
$$= a^2 + b^2$$

But $|z| = \sqrt{a^2 + b^2}$ so: $|z|^2 = a^2 + b^2$
Thus:
$$z \times \bar{z} = a^2 + b^2$$
$$= |z|^2$$

Proof 9: $\overline{z + w} = \bar{z} + \bar{w}$

Proof 9

Let $z = a + bi$ and $w = c + di$, then:
$z + w = (a + c) + i(b + d)$ and so:
$\overline{z + w} = (a + c) - i(b + d)$

Also $\bar{z} = a - bi$ and $\bar{w} = c - di$, and so:

$$\bar{z} + \bar{w} = (a - bi) + (c - di)$$
$$= (a + c) - i(b + d)$$
$$= \overline{z + w}$$

Proof 10: $\overline{z \times w} = \bar{z} \times \bar{w}$

Proof 10

Let $z = a + bi$ and $w = c + di$, then:
$z \times w = (a + bi)(c + di)$
$ = (ac - bd) + i(bc + ad)$
So: $\overline{z \times w} = (ac - bd) - i(bc + ad)$

Also $\bar{z} = a - bi$ and $\bar{w} = c - di$, and so:
$\bar{z} \times \bar{w} = (a - bi)(c - di)$
$\phantom{\bar{z} \times \bar{w}} = (ac - bd) - i(bc + ad)$
$\phantom{\bar{z} \times \bar{w}} = \overline{z \times w}$

Proof 11: $|z \times w| = |z| \times |w|$

Proof 11

$$\begin{aligned}|z \times w|^2 &= (z \times w) \times \overline{z \times w} &&\text{by Proof 8}\\ &= (z \times \overline{z}) \times (w \times \overline{w}) &&\text{by Proof 10}\\ &= |z|^2 \times |w|^2 &&\text{by Proof 8}\end{aligned}$$

Hence, taking the square root of each side:
$|z \times w| = |z| \times |w|$

Proof 12: Triangle inequality

Proof 12

This proof uses another result: $\text{Re}(z) \leq |z|$ for all $z \in \mathbb{C}$.

(If time permits, try this proof.)

The triangle inequality states that if z and $w \in \mathbb{C}$, then $|z + w| \leq |z| + |w|$
So let z and $w \in \mathbb{C}$

$$\begin{aligned}|z+w|^2 &= (z+w)\overline{(z+w)} &&\text{from Proof 8}\\ &= (z+w)(\overline{z}+\overline{w}) &&\text{from Proof 9}\\ &= z\overline{z} + w\overline{w} + z\overline{w} + \overline{z}w\\ &= |z|^2 + |w|^2 + z\overline{w} + \overline{z\overline{w}} &&\text{from Proof 4}\\ &= |z|^2 + |w|^2 + 2\,\text{Re}(z\overline{w}) &&\text{from Proof 6}\\ &\leq |z|^2 + |w|^2 + 2|z\overline{w}| &&\text{from result given at start}\\ &\leq |z|^2 + |w|^2 + 2|z||\overline{w}| &&\text{from Proof 11}\\ &\leq |z|^2 + |w|^2 + 2|z||w| &&\text{from Proof 7}\\ &\leq (|z| + |w|)^2\end{aligned}$$

Since $|z + w| \geq 0$ and $|z| + |w| \geq 0$

Taking the square root of each side gives $|z + w| \leq |z| + |w|$

Examples

1. Problem:
If $z = 3 - 2i$ and $w = -4 + 3i$, show that $\overline{z + w} = \overline{z} + \overline{w}$
Solution:
$$\begin{aligned}z + w &= 3 - 2i + (-4 + 3i)\\ &= -1 + i\end{aligned}$$
So $\overline{z + w} = -1 - i$

TOPIC 6. COMPLEX NUMBERS

$$\bar{z} + \bar{w} = 3 + 2i + (-4 - 3i)$$
$$= -1 - i$$

Therefore $\overline{z + w} = \bar{z} + \bar{w}$

---

2. Problem:
If $z = -1 + 4i$ and $w = 5 - 2i$, show that $|\bar{z} \times \bar{w}| = |z| \, |w|$

Solution:
$\bar{z} = -1 - 4i$ and $\bar{w} = 5 + 2i$

So:
$$\bar{z}\bar{w} = (-1 - 4i)(5 + 2i)$$
$$= -5 - 22i - 8i^2$$
$$= 3 - 22i$$
$$|\bar{z}\bar{w}| = \sqrt{3^2 + (-22)^2}$$
$$= \sqrt{493}$$

Now:
$$|z| = \sqrt{(-1)^2 + (-4)^2}$$
$$= \sqrt{17}$$
$$|w| = \sqrt{(5)^2 + (2)^2}$$
$$= \sqrt{29}$$

So $|z| \, |w| = \sqrt{17}\sqrt{29}$
$$= \sqrt{493}$$

Therefore $|\bar{z}\bar{w}| = |z| \, |w|$

---

3. Problem:
Given the equation $z\bar{z} + 3iz = 11 - 3i$, express z in the form $a + ib$

Solution:
Let $z = a + ib$ and $\bar{z} = a - ib$ then:
$$(a + ib)(a - ib) + 3i(a + ib) = 11 - 3i$$
$$a^2 - i^2b^2 + 3ia + 3i^2b = 11 - 3i$$
$$a^2 + b^2 - 3b + 3ia = 11 - 3i$$

Equating real and imaginary parts:
$$3a = -3 \Rightarrow a = -1$$
$$a^2 + b^2 - 3b = 11 \Rightarrow 1 + b^2 - 3b = 11$$
$$\Rightarrow b^2 - 3b - 10 = 0$$
$$\Rightarrow (b - 5)(b + 2) = 0$$
$$\Rightarrow b = 5 \text{ or } b = -2$$

Therefore, $z = -1 - 2i$ or $z = -1 + 5i$

© HERIOT-WATT UNIVERSITY

Conjugate properties exercise   Go online

Q172: If $z = 4 - 5i$, show that $\overline{\overline{z}} = z$

Q173: If $z = -4 + \frac{1}{2}i$, show that $\overline{z} = z$ where $z \in \mathbb{R}$

Q174: If $z = -\frac{\sqrt{3}}{2} + 10i$, show that $\overline{z} + z = 2\text{Re}(z)$

Q175: If $z = 3 + 7i$, show that $|\overline{z}| = |z|$

Q176: If $z = 6 - 2i$, show that $z\overline{z} = |z|^2$

Q177: If $z = 5 - 2i$ and $w = 3 + 7i$, show that $\overline{z + w} = \overline{z} + \overline{w}$

Q178: If $z = 6 + i$ and $w = 2 - 3i$, show that $\overline{z}\,\overline{w} = \overline{zw}$

Q179: If $z = 9 - 5i$ and $w = 1 - 4i$, show that $|\overline{zw}| = |z|\,|w|$

Q180: Given the equation $z + 2i\overline{z} = -5 + 2i$, express z in the form $a + ib$

6.9 Learning points

Complex numbers
Introduction to Imaginary numbers

- Natural numbers (denoted by $\mathbb{N}$) are positive whole numbers, i.e. 1, 2, 3...
- Integers (denoted by $\mathbb{Z}$) are positive and negative whole numbers.
- Rational numbers (denoted by $\mathbb{Q}$) are fractions where the numerator and denominator are integers.
- Real numbers (denoted by $\mathbb{R}$) include all the previous numbers as well as surds and numbers that can be written as infinite decimals.
- $i^2 = -1$
- Complex numbers (denoted $\mathbb{C}$) include all the previous numbers as well as the imaginary part of numbers denoted by i, e.g. $3 + 5i$

Complex numbers and the complex plane

- If the discriminant is less than zero ($b^2 - 4ac < 0$) in the quadratic formula, replace -1 with i^2, e.g.
  if $b^2 - 4ac = -4$, then $b^2 - 4ac = 4i^2$
- If $z = a + ib$, then $\text{Re}(a + ib) = a$ and $\text{Im}(a + ib) = b$
- Complex numbers are plotted on the complex plane or an Argand diagram; the x-axis is the real axis and the y-axis is the imaginary axis:

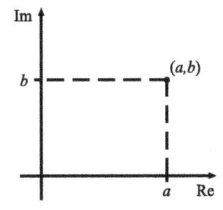

The arithmetic of complex numbers

- To add two complex numbers, add the *real* parts and *imaginary* part separately:
  $(a + ib) + (c + id) = (a + c) + i(b + d)$
- To subtract two complex numbers, subtract the *real* parts and *imaginary* part separately:
  $(a + ib) - (c + id) = (a - c) + i(b - d)$
- To multiply complex numbers, use the technique for multiplying out two brackets:
  $(a + ib)(c + id) = (ac - bd) + i(bc + ad)$
- To divide by a complex number, we multiply the numerator and denominator by the complex conjugate of the denominator:
  $\frac{(a+ib)}{(c+id)} \times \frac{(c-id)}{(c-id)}$

154　TOPIC 6.　COMPLEX NUMBERS

- To square root a complex number:
  - we know that the answer must be of the form $a + ib$, e.g.
    $(a + ib)^2 = 5 + 12i$
  - now multiply out the LHS and equate the real and imaginary parts, e.g.
    $a^2 + 2iab + i^2b^2 = 5 + 12i$
    $a^2 + 2iab - b^2 = 5 + 12i$
    Real parts　$a^2 - b^2 = 5$
    Imaginary parts　$2ab = 12$
  - solve the two resulting equations, e.g.
    $$a^2 - b^2 = 5 \quad (1)$$
    $$2ab = 12 \quad (2)$$
    Rearranging (2):　$b = \dfrac{6}{a} \quad (3)$
    Substituting (3) into (1):　$a^2 - \left(\dfrac{6}{a}\right)^2 = 5$
    Simplifying:　$a^2 - \dfrac{36}{a^2} = 5$
    $$a^4 - 5a^2 - 36 = 0$$
    Factorising and solving:　$(a^2 + 4)(a^2 - 9) = 0$
    $$a^2 = -4 \text{ or } a^2 = 9$$
    $a \in \mathbb{R}$ so $a^2 = -4$ is impossible
    so $a = \pm 3$

    Substituting into (2):
    when $a = 3$, then $b = \dfrac{6}{3} \Rightarrow b = 2$; when $a = -3$, then $b = \dfrac{6}{-3} \Rightarrow b = -2$
    The square roots of $5 + 12i$ are $3 + 2i$ and $-3 - 2i$

The modulus, argument and polar form of a complex number

- Given the Cartesian form of a complex number, $z = a + ib$, the polar form is given by $z = r\cos\theta + i\,r\sin\theta$ where:
  - the modulus, r, is given by $r = \sqrt{a^2 + b^2}$
  - the principal argument, θ, is between $-\pi$ and π
  - θ is calculated in the following way in each quadrant where $\tan\alpha = \left|\dfrac{b}{a}\right|$

```
       |
  π − α |    α
  ──────┼──────→
 −π + α |   −α
        |
```

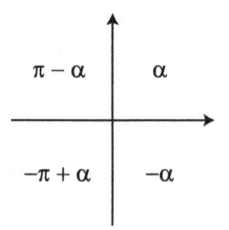

© HERIOT-WATT UNIVERSITY

TOPIC 6. COMPLEX NUMBERS

Geometric interpretations

- **Equations of the form** $|z| = r$
  The solutions of the equation form a circle with centre $(0,0)$ and radius r.

- **Equations of the form** $|z - a| = r$
  The solutions of the equation form a circle of radius r units with centre a, where a is a point in the complex plane.

- **Equations of the form** $|z - a| > r$
  The solutions of the equation are the set of all points that lie outside the circle with radius r units with centre a, where a is a point in the complex plane.

- **Equations of the form** $|z - a| < r$
  The solutions of the equation are the set of all points that lie inside the circle with radius r units with centre a, where a is a point in the complex plane.

- **Equations of the form** $|z - a| = |z - b|$
  The solutions of the equation are the set of points that lie on the perpendicular bisector of the line joining a and b.

- **Triangle inequality** $|z + w| \leq |z| + |w|$
  Geometrically, this shows that there are no occasions where one side is bigger than the other two sides of a triangle added together.

Fundamental theorem of algebra

- Let $P(z) = z^n + a_{(n-1)}z^{(n-1)} + \ldots + a_1 z + a_0$ be a polynomial of degree n (with real or complex coefficients).
  The fundamental theorem of algebra states that $P(z) = 0$ has n solutions α_1, ..., α_n in the complex numbers and $P(z) = (z - \alpha_1)(z - \alpha_2)\ldots(z - \alpha_n)$

Solving equations

- If $z = a + ib$ is a root, then its conjugate, $\bar{z} = a - ib$, is also a root.
- Evaluating $(z - (a + ib))(z - (a - ib))$ will give a factor which is real.
- Use this factor and long division, or synthetic division, to find the remaining factors.
- Apply the quadratic formula if necessary to factorise the resulting factor.

De Moivre's theorem

- If $z = r_1(cos\,\theta + i\,sin\,\theta)$ and $w = r_2(cos\,\phi + i\,sin\,\phi)$, then zw has:
  - $|zw| = r_1 r_2$
  - $arg(\theta + \phi)$ (give the principal argument)
- If $z = r_1(cos\,\theta + i\,sin\,\theta)$ and $w = r_2(cos\,\phi + i\,sin\,\phi)$, then $\frac{z}{w}$ has:

© HERIOT-WATT UNIVERSITY

- $\left|\frac{z}{w}\right| = \frac{r_1}{r_2}$
- $arg(\theta - \phi)$ (give the principal argument)

- If $z = r(cos\,\theta + i\,sin\,\theta)$, then $z^n = r^n(cos(n\theta) + i\,sin(n\theta))$ for all $n \in \mathbb{Z}$
  This can also be extended to fractional powers where:
  $(r\,(\cos\theta + i\sin\theta))^{\frac{p}{q}} = r^{\frac{p}{q}}\left(\cos\left(\frac{p}{q}\theta\right) + i\sin\left(\frac{p}{q}\theta\right)\right)$

- To state $\cos(n\theta)$ in terms of powers of $\cos\theta$ and/or $\sin\theta$:
  - expand $(cos\,\theta + i\,sin\,\theta)^n$ using the binomial theorem;
  - expand $(cos\,\theta + i\,sin\,\theta)^n$ using de Moivre's theorem;
  - equate the real parts of each expansion;
  - to state solely in terms of powers of $\cos\theta$, replace $\sin^2\theta$ with $1 - \cos^2\theta$;
  - to state solely in terms of powers of $\sin\theta$, replace $\cos^2\theta$ with $1 - \sin^2\theta$.

- To state $\sin(n\theta)$ in terms of powers of $\cos\theta$ and/or $\sin\theta$:
  - do the same as for $\cos(n\theta)$, but equate the imaginary parts of each expansion.

- To state $\tan(n\theta)$ in terms of powers of $\tan\theta$:
  - use $\tan(n\theta) = \frac{\sin(n\theta)}{\cos(n\theta)}$;
  - go through the previous processes to state $\sin(n\theta)$ and $\cos(n\theta)$ in terms of powers of $\sin\theta$ and $\cos\theta$;
  - divide through by the highest power of $\cos\theta$.

- To state a power of $\cos\theta$ in terms of multiple angles of $\cos\theta$:
  - rewrite $2\cos\theta = \left(z + \frac{1}{z}\right)$ as $(2\cos\theta)^n = \left(z + \frac{1}{z}\right)^n$
    (note: we are taking $2\cos(n\theta) = \left(z^n + \frac{1}{z^n}\right)$, where $n = 1$);
  - expand the RHS using the binomial theorem;
  - gather expressions together of the form $z^n + \frac{1}{z^n}$ and replace with $2\cos(n\theta)$.

- To state a power of $\sin\theta$ in terms of multiple angles of $\sin\theta$:
  - rewrite $2i\sin\theta = \left(z - \frac{1}{z}\right)$ as $(2i\sin\theta)^n$
    (note: we are taking $2i\sin(n\theta) = \left(z^n - \frac{1}{z^n}\right)$, where $n = 1$);
  - expand the RHS using the binomial theorem;
  - gather expressions of the form $z^n + \frac{1}{z^n}$ together and replace with $2\cos(n\theta)$ or gather expressions of the form $z^n - \frac{1}{z^n}$ and replace with $2i\sin(n\theta)$ where appropriate.

- To find the roots, rewrite $z^n = r(cos\,\theta + i\,sin\,\theta)$ as $z = (r\,(\cos\theta + i\sin\theta))^{\frac{1}{n}}$:
  - since the arguments are not unique, we can add $2k\pi$:
    $z = r^{\frac{1}{n}}((\cos(\theta + 2k\pi)) + i\sin(\theta + 2k\pi))^{\frac{1}{n}}$;
  - apply de Moivre's theorem to the RHS;
  - by the fundamental theorem of algebra, there will be n solutions which will be equally spaced when drawn on an Argand diagram.

Conjugate properties

- Let the two complex numbers be z and w with conjugates $\bar{z}$ and $\bar{w}$ respectively.
    1. $\bar{\bar{z}} = z$
    2. $z = \bar{z}$ where $z \in \mathbb{R}$
    3. $z + \bar{z} = 2\text{Re}(z)$
    4. $|\bar{z}| = |z|$
    5. $z \times \bar{z} = |z|^2$
    6. $\overline{z + w} = \bar{z} + \bar{w}$
    7. $\overline{z \times w} = \bar{z} \times \bar{w}$
    8. $|\bar{z} \times \bar{w}| = |z| \times |w|$

6.10 Proofs

Proof 1: Conjugate roots property

Suppose that α is a root of the polynomial.
$P(z) = a_n z^n + a_{n-1} z^{n-1} + \ldots + a_2 z^2 + a_1 z + a_0$
Then $P(\alpha) = 0$
i.e. $a_n \alpha^n + a_{n-1} \alpha^{n-1} + \ldots + a_2 \alpha^2 + a_1 \alpha + a_0 = 0$
Hence

$$\begin{aligned}\overline{P(\alpha)} &= \overline{a_n \alpha^n + a_{n-1}\alpha^{n-1} + \cdots + a_2 \alpha^2 + a_1 \alpha + a_0}\\ &= \overline{a_n \alpha^n} + \overline{a_{n-1}\alpha^{n-1}} + \cdots + \overline{a_2 \alpha^2} + \overline{a_1 \alpha} + \overline{a_0}\\ &= \overline{a_n}\left(\overline{\alpha^n}\right) + \overline{a_{n-1}}\left(\overline{\alpha^{n-1}}\right) + \cdots + \overline{a_2}\left(\overline{\alpha^2}\right) + \overline{a_1}\overline{\alpha} + \overline{a_0}\\ &= a_n (\overline{\alpha})^n + a_{n-1}(\overline{\alpha})^{n-1} + \cdots + a_2 (\overline{\alpha})^2 + a_1 \overline{\alpha} + a_0\\ &= P(\overline{\alpha})\end{aligned}$$

However, $\overline{P(\alpha)} = \overline{0} = 0$ and so $P(\overline{\alpha}) = 0$
So, starting from the point that alpha (a complex number) is a root of the polynomial and applying conjugate properties to this, we have shown that if alpha is a root, then its conjugate is also a root.
Therefore, $\overline{\alpha}$ is a root of the polynomial.

Proof 2: Proof of de Moivre's theorem

The theorem states that if $z = r(\cos\theta + i\sin\theta)$, then:
$z^n = r^n(\cos(n\theta) + i\sin(n\theta))$ for all $n \in \mathbb{N}$.
This is a proof by induction. The technique is explained in greater detail in the topic on methods of proof.

For $n = 1$:
$$\begin{aligned}(r(\cos\theta + i\sin\theta))^1 &= r(\cos\theta + i\sin\theta)\\ &= r^1(\cos(1\theta) + i\sin(1\theta))\end{aligned}$$
And so it is true for $n = 1$

Now suppose that the result is true for $n = k$, then:
$(r(\cos\theta + i\sin\theta))^k = r^k(\cos(k\theta) + i\sin(k\theta))$

Consider $n = k + 1$, then:
$$\begin{aligned}(r(\cos\theta + i\sin\theta))^{k+1} &= (r(\cos\theta + i\sin\theta))(r(\cos\theta + i\sin\theta))^k\\ &= (r(\cos\theta + i\sin\theta))\left(r^k(\cos k\theta + i\sin k\theta)\right)\\ &= r^{k+1}(r(\cos\theta\cos k\theta - \sin\theta\sin k\theta) + i(\sin\theta\cos k\theta + \cos\theta\sin k\theta))\\ &= r^{k+1}(\cos(\theta + k\theta) + i\sin(\theta + k\theta))\\ &= r^{k+1}(\cos((k+1)\theta) + i\sin((k+1)\theta))\end{aligned}$$

So the result is true for $n = k + 1$ if it is true for $n = k$.
Since it is also true for $n = 1$, then it is true for all $n \in \mathbb{N}$.

Proof 3: Proof of de Moivre's theorem for integers

Consider n to be a negative integer
Let $n = -p$
If $z^n = (\cos\theta + i\sin\theta)^n$, then:

$$(\cos\theta + i\sin\theta)^n = (\cos\theta + i\sin\theta)^{-p}$$
$$= \frac{1}{(\cos\theta + i\sin\theta)^p}$$
$$= \frac{1}{\cos(p\theta) + i\sin p\theta} \times \left(\frac{\cos(p\theta) - i\sin(p\theta)}{\cos(p\theta) - i\sin(p\theta)}\right)$$
$$= \frac{\cos(p\theta) - i\sin(p\theta)}{\cos^2(p\theta) + \sin^2(p\theta)}$$
$$= \cos(p\theta) - i\sin(p\theta)$$
$$= \cos(-p\theta) + i\sin(-p\theta)$$
$$= \cos(n\theta) + i\sin(n\theta)$$

So $z^n = \cos(n\theta) + i\sin(n\theta)$ where $n \in \mathbb{Z}$.

Proof 4: $\bar{\bar{z}} = z$

Let $z = a + bi$ for a and $b \in \mathbb{R}$.
By the definition of a conjugate:
$\bar{z} = a - bi$
$\phantom{\bar{z}} = a + (-b)i$

Again, using the conjugate definition:
$\bar{\bar{z}} = \overline{a + (-b)i}$
$\phantom{\bar{\bar{z}}} = a - (-b)i$
$\phantom{\bar{\bar{z}}} = a + bi$
$\phantom{\bar{\bar{z}}} = z$

Proof 5: $z = \bar{z}$ where $z \in \mathbb{R}$

If $z \in \mathbb{R}$, then:

$z = a + 0 \times i$
$ = a$

(since $0 \times i = 0$) for some $a \in \mathbb{R}$
By the definition of a conjugate:
$\bar{z} = a - 0 \times i$
But $0 \times i = 0$ so:
$\bar{z} = a$
$\phantom{\bar{z}} = z$

Proof 6: $z + \bar{z} = 2\mathbf{Re}(z)$

Let $z = a + bi$, then $\bar{z} = a - bi$

Hence:
$z + \bar{z} = a + bi + a - bi$
$\phantom{z + \bar{z}} = 2a$
$\phantom{z + \bar{z}} = 2\operatorname{Re}(z)$

Proof 7: $|\bar{z}| = |z|$

Let $z = a + bi$ and $|z| = \sqrt{(a^2 + b^2)}$

Hence:
$|\bar{z}| = \sqrt{(a^2 + (-b^2))}$
$= \sqrt{(a^2 + b^2)}$
$= |z|$

Proof 8: $z \times \bar{z} = |z|^2$

Let $z = a + bi$, then $\bar{z} = a - bi$
Hence:
$z \times \bar{z} = (a + bi)(a - bi)$
$= a^2 + b^2$
But $|z| = \sqrt{a^2 + b^2}$ so:
$|z|^2 = a^2 + b^2$
Thus:
$z \times \bar{z} = a^2 + b^2$
$= |z|^2$

Proof 9: $\overline{z + w} = \bar{z} + \bar{w}$

Let $z = a + bi$ and $w = c + di$, then:
$z + w = (a + c) + i(b + d)$ and so:
$\overline{z + w} = (a + c) - i(b + d)$

Also $\bar{z} = a - bi$ and $\bar{w} = c - di$, and so:

$\bar{z} + \bar{w} = (a - bi) + (c - di)$
$= (a + c) - i(b + d)$
$= \overline{z + w}$

Proof 10: $\overline{z \times w} = \bar{z} \times \bar{w}$

Let $z = a + bi$ and $w = c + di$, then:
$z \times w = (a + bi)(c + di)$
$= (ac - bd) + i(bc + ad)$
So:
$\overline{z \times w} = (ac - bd) - i(bc + ad)$
Also $\bar{z} = a - bi$ and $\bar{w} = c - di$, and so:
$\bar{z} \times \bar{w} = (a - bi)(c - di)$
$= (ac - bd) - i(bc + ad)$
$= \overline{z \times w}$

Proof 11: $|z \times w| = |z| \times |w|$

$|z \times w|^2 = (z \times w) \times \overline{z \times w}$ by Proof 8
$= (z \times \bar{z}) \times (w \times \bar{w})$ by Proof 10
$= |z|^2 \times |w|^2$ by Proof 8
Hence, taking the square root of each side:
$|z \times w| = |z| \times |w|$

Proof 12: Triangle inequality

This proof uses another result: $\text{Re}(z) \leq |z|$ for all $z \in \mathbb{C}$.

(If time permits, try this proof.)

The triangle inequality states that if z and $w \in \mathbb{C}$, then $|z + w| \leq |z| + |w|$
So let z and $w \in \mathbb{C}$

$$
\begin{aligned}
|z+w|^2 &= (z+w)\overline{(z+w)} & \text{from Proof 8} \\
&= (z+w)(\bar{z}+\bar{w}) & \text{from Proof 9} \\
&= z\bar{z} + w\bar{w} + z\bar{w} + w\bar{z} \\
&= |z|^2 + |w|^2 + z\bar{w} + \overline{z\bar{w}} & \text{from Proof 4} \\
&= |z|^2 + |w|^2 + 2\,\text{Re}(z\bar{w}) & \text{from Proof 6} \\
&\leq |z|^2 + |w|^2 + 2|z\bar{w}| & \text{from result given at start} \\
&\leq |z|^2 + |w|^2 + 2|z||\bar{w}| & \text{from Proof 11} \\
&\leq |z|^2 + |w|^2 + 2|z||w| & \text{from Proof 7} \\
&\leq (|z|+|w|)^2
\end{aligned}
$$

Since $|z + w| \geq 0$ and $|z| + |w| \geq 0$

Taking the square root of each side gives $|z + w| \leq |z| + |w|$

6.11 Extended information

There are links online which give a selection of interesting web sites to visit. These sites can lead to an advanced study of the topic, but there are many areas which will be of passing interest.

Cardano

In 1545, Cardano published the 'Ars magna', also known as 'Artis Magnae' (The Rules of Algebra), which includes the use of negative numbers. He described them as 'ficticious' but said that imaginary numbers were 'sophistic'.

Bombelli

He was probably the first mathematician to have a clear idea of a complex number. He discussed imaginary and complex numbers in a treatise written in 1572 called 'L'Algebra'.

Descartes

He first called numbers involving the square root of a negative number 'imaginary', but assumed from this that the problem was insoluble.

Euler

In 1748, he found the formula $e^{i\theta} = \cos\theta + i\sin\theta$ (now known as Euler's formula) where $i = \sqrt{-1}$. This is the exponential form of a complex number.

Argand

Although the idea of representing a complex number by a point in a plane had been suggested by several mathematicians earlier, it was Argand's proposal that was accepted.

Gauss

He was the first mathematician to mention a complex number plane. He also gave the first proof of the fundamental theorem of algebra in his Ph.D. thesis.

Hamilton

In 1833, Hamilton, an Irish mathematician, introduced the complex number notation $a + bi$ and made the connection with the point (a, b) in the plane although many mathematicians argued that they had found this earlier. He is well known for his work on graph theory and networks.

De Moivre

A French mathematician and statistician who discovered an easy way to take the power of a complex number. He is probably best known for his work on statistics.

TOPIC 6. COMPLEX NUMBERS

6.12 End of topic test

End of topic 6 test   Go online

Q181: Three complex numbers are defined as follows:

- $u = 3 - i$
- $v = 3 + 2i$
- $w = m - 3i$

Express $2u - w + 3v$ in the form $a + ib$ where a and b are real numbers.

....

Q182: Three complex numbers are defined as follows:

- $u = 2 + 5i$
- $v = -1 - 2i$
- $w = 6 - 2i$

Express $3iu + 2\overline{w} - 4v$ in the form $a + ib$ where a and b are real numbers.

....

Q183: Two complex numbers are defined as follows:

- $u = 5 + i$
- $v = 2 - 3i$

Express $\frac{u}{v}$ in the form $a + ib$ where a and b are real numbers.

....

Q184: Two complex numbers are defined as follows:

- $u = 4 - 5i$
- $v = 1 + 2i$

Express uv in the form $a + ib$ where a and b are real numbers.

....

Q185: Find the square root of the complex number $16 - 30i$ where the real part is greater than the imaginary part.

....

Q186: Divide $3 + 4i$ by $2 - 5i$

....

Q187: Convert $z = -3 + 3\sqrt{3}i$ into polar form, giving your answer in radians.

....

Q188: Convert $z = 2\sqrt{3} - 2i$ into polar form, giving your answer in radians.

....

© HERIOT-WATT UNIVERSITY

164 TOPIC 6. COMPLEX NUMBERS

Q189: Convert $z = -5 - 3i$ into polar form, giving your answer to three decimal places using radians.

....

Q190: Convert $z = -8i$ into polar form, giving your answer in radians.

....

Q191: Convert $z = 9$ into polar form, giving your answer in radians.

....

Q192: The complex number z has modulus 3 and argument $-\frac{3\pi}{4}$
Determine z in Cartesian form using exact values.

....

Q193: The complex number z has modulus 2 and argument $\frac{4\pi}{3}$
Determine z in Cartesian form using exact values.

....

Q194: Expand $(5(cos\ y + i \sin y))^3$ using de Moivre's theorem, giving your answer in the same form.

....

Q195: Expand $(3(cos\ x + i \sin x))^{-4}$ using de Moivre's theorem, giving your answer in the same form.

....

Q196: Expand $(8(\cos x + i \sin x))^{\frac{2}{3}}$ using de Moivre's theorem, giving your answer in the same form.

....

Q197: Expand $(-3 - 2i)^6$ using de Moivre's theorem, giving your answer in the same form.

....

Q198: State the centre and radius of the circle represented by $|z - 5| = 8$

....

Q199: State the centre and radius of the circle represented by $|z + 3i| = 4$

....

Q200: State the centre and radius of the circle represented by $|z - 4 + 2i| = 6$

....

Q201: Give the equation of the line represented by $|z + 2i| = |z - 3i|$. State the right hand side of the equation $y =$.

....

Q202: Give the equation of the line represented by $|z + 2i| = |z - 5|$. State the right hand side of the equation $y =$.

....

© HERIOT-WATT UNIVERSITY

TOPIC 6. COMPLEX NUMBERS

Q203: Give the equation of the line represented by $|z - 4| = |z - 7|$. State the right hand side of the equation $x =$.

....

Q204: If the roots of a quadratic $x^2 + bx + c$ are $-3 + i$ and $2 - 5i$, find complex numbers b and c.

....

Q205: Find the roots of the cubic $x^3 - 7x^2 + 25x - 39 = 0$ given that one of them is $2 + 3i$.

....

Q206: Find the roots of the quartic $z^4 - 2z^3 - z^2 + 2z + 10 = 0$ given that one of them is $-1 - i$.

....

Q207: Calculate $\dfrac{(\sqrt{3}-i)^5}{(1+i)^3}$ using de Moivre's theorem.

....

Q208:
Using de Moivre's theorem, simplify $\dfrac{(\cos(3\theta)+i\sin(3\theta))^5}{(\cos\theta+i\sin\theta)^2}$ giving your answer in the same form.

....

Q209:
Using de Moivre's theorem, simplify $\dfrac{\left(\cos\left(\frac{\pi}{3}\right)-i\sin\left(\frac{\pi}{3}\right)\right)^3}{\left(\cos\left(\frac{\pi}{5}\right)+i\sin\left(\frac{\pi}{5}\right)\right)^4}$ giving your answer in the form $\cos\theta + i\sin\theta$.

....

Q210: Express $\cos(6\theta)$ as a polynomial in $\sin\theta$.
Let $z = \cos\theta + i\sin\theta$

....

Q211: Express $\sin(3\theta)$ as a polynomial in $\sin\theta$ and $\cos\theta$. Let $z = \cos\theta + i\sin\theta$

....

Q212: Express $\tan(4\theta)$ as a polynomial in $\tan\theta$.

....

Q213: Express $\cos^4\theta$ as a polynomial in terms of $\cos(n\theta)$.

....

Q214: Express $\cos^6\theta$ as a polynomial in terms of $\cos(n\theta)$.

....

Q215: Express $\cos^3\theta$ as a polynomial in terms of $\sin(n\theta)$.

....

© HERIOT-WATT UNIVERSITY

Q216: Find the cube roots of $5 - 2i$

Give the answer in the form $u + iv$, with u and v to three decimal places (working in radians).

....

Q217: Solve $z^4 = 27 \left(\cos \left(\frac{3\pi}{4} \right) + i \sin \left(\frac{3\pi}{4} \right) \right)$.

Give your answer in the form $u + iv$, where u and v are to three decimal places (working in radians).

....

Q218: Solve $z^5 = 32(cos(135°) + i \sin(135°))$.

Give your answer in the form $u + iv$, where u and v are to three decimal places (working in degrees).

....

Q219: Solve $z^3 - 1 = 0$

Give your answer in the form $u + iv$, where u and v are to three decimal places (working in radians).

....

Q220: Solve $3\bar{z} + 2iz = 22 + 23i$ where $z = a + ib$

Topic 7

Sequences and series

Contents

- 7.1 Looking back . . . 171
  - 7.1.1 Simple recurrence relations . . . 173
  - 7.1.2 Finding a limit . . . 180
  - 7.1.3 Solving recurrence relations . . . 185
- 7.2 Sequences and recurrence relations . . . 188
  - 7.2.1 What is a sequence? . . . 189
  - 7.2.2 Sequences and recurrence relations . . . 191
- 7.3 Arithmetic and geometric sequences . . . 193
  - 7.3.1 Arithmetic sequences . . . 193
  - 7.3.2 Geometric sequences . . . 197
- 7.4 Fibonacci and other sequences . . . 203
- 7.5 Convergence and limits . . . 207
- 7.6 Definitions of e and π as limits of sequences . . . 212
  - 7.6.1 The definition of e as a limit of a sequence . . . 213
  - 7.6.2 The definition of π as a limit of a sequence . . . 214
- 7.7 Series and sums . . . 216
- 7.8 Arithmetic and geometric series . . . 219
  - 7.8.1 Arithmetic series . . . 219
  - 7.8.2 Geometric series . . . 225
- 7.9 Sums to infinity . . . 228
  - 7.9.1 Convergent geometric series . . . 228
  - 7.9.2 Binomial theorem and the geometric series . . . 230
  - 7.9.3 Numeric expansion using the power of -1 . . . 233
  - 7.9.4 Sums to infinity exercise . . . 235
- 7.10 Power series . . . 236
- 7.11 Maclaurin series for simple functions . . . 237
- 7.12 Maclaurin's theorem . . . 239
- 7.13 The Maclaurin series for $\tan^{-1}(x)$ . . . 244
- 7.14 Maclaurin's series expansion to a given number of terms . . . 245
- 7.15 Composite Maclaurin's series expansion . . . 247
  - 7.15.1 Composite function examples using standard results . . . 251
- 7.16 Learning points . . . 256
- 7.17 Extended information . . . 259

7.18 End of topic test 262

TOPIC 7. SEQUENCES AND SERIES

Learning objective

At points it may be necessary to draw graphs and explore these.
The maths is fun website has a graphing tool that will allow you to explore these graphs.

http://www.mathsisfun.com/data/function-grapher.php

By the end of this topic, you should be able to:

- identify:
    - a geometric sequence;
    - an arithmetic sequence;
- determine a linear recurrence relation;
- use a recurrence relation to calculate terms;
- identify when a recurrence relation has a limit;
- find the limit of a converging recurrence relation;
- solve recurrence relations;
- interpret recurrence relations in context.
- define a:
    - sequence;
    - finite and infinite sequence;
- know the meaning of nth term, element, general term;
- define a first order recurrence relation;
- use the recurrence relation to calculate the nth term in the sequence;
- define an arithmetic sequence using a recurrence relation and a general form;
- know the meaning of common difference;
- define a geometric sequence using a recurrence relation and a general form;
- know the meaning of common ratio;
- define the:
    - Fibonacci sequence;
    - triangular number sequence;
- give an example of:
    - an alternating sequence;
    - a uniform sequence;
- know what it means for a sequence to converge to a limit;

© HERIOT-WATT UNIVERSITY

Learning objective continued

- identify a null sequence;
- use rules for convergent sequences to determine the limit of more complicated sequences;
- identify bounded and unbounded sequences;
- state the limit of the sequence $\left\{\left(1+\frac{1}{n}\right)^n\right\}$;
- state the limit of the sequence $\left\{n\tan\left(\frac{180}{n}\right)\right\}$;
- know what the following are:
  - a series;
  - a partial sum;
  - the sum and multiple rule for convergent series.
- know and use the nth partial sum of:
  - an arithmetic progression;
  - a geometric progression;
- know the nth partial sum for $\sum_{r=1}^{n} r$, $\sum_{r=1}^{n} r^2$ and $\sum_{r=1}^{n} r^3$;
- use the sum to infinity of a geometric series and know its conditions of use;
- know that the Binomial expansion of $(1-x)^{-1}$ represents an infinite geometric series;
- know and use the Binomial expansion of $(1-x)^{-1}$ to write geometric series for $(a-x)^{-1}$.
- recognise the form of a power series;
- generate Maclaurin series for various functions using the Maclaurin series expansion formula;
- appreciate that given certain conditions the Macluarin series generated by a function is actually equal to that function;
- know the:
  - interval of convergence for an infinite geometric series;
  - range of convergence for the Maclaurin series of $\ln(1+x)$, $\frac{1}{1+x}$ and $\tan^{-1}x$;
- apply both Macluarin's Theorem and Binomial's Theorem to obtain a power series for $\tan^{-1}x$;
- determine the Maclaurin series:
  - expansion for a given function to a specified number of terms;
  - for composite functions by using Maclaurin's Theorem and standard results.

TOPIC 7. SEQUENCES AND SERIES

7.1 Looking back

Pre-requisites from Advanced Higher

You should have covered the following in the *Differentiation* topic. If you need to reinforce your learning go back and study this topic.

Conditions for differentiability

- A function is:
    - continuous if $\lim_{h \to 0} f(x) = f(a)$, otherwise it is discontinuous, in other words if you can draw a curve without lifting the pen from the paper it is continuous.;
    - differentiable at a point if it has a tangent at that point.
- $f(x)$ is differentiable over the interval $[a, b]$ if $f'(x)$ exists for each $x \in [a, b]$
- Higher derivatives are denoted using either $f^{(n)}(x)$ or $\frac{d^n y}{dx^n}$ where n is the nth derivative.

The product rule

- When $k(x) = f(x) g(x)$, then $k'(x) = f'(x) g(x) + f(x) g'(x)$

The quotient rule

- When $k(x) = \frac{f(x)}{g(x)}$, then $k'(x) = \frac{f'(x)g(x) - f(x)g'(x)}{[g(x)]^2}$

Differentiating cot(x), sec(x), cosec(x) and tan(x)

Secant Function	Cosecant Function	Cotangent Function
$\sec x = \frac{1}{\cos x}$	$\operatorname{cosec} x = \frac{1}{\sin x}$ or $\csc x = \frac{1}{\sin x}$	$\cot x = \frac{1}{\tan x}$

- $\frac{d}{dx}(\sec x) = \sec x \tan x$
- $\frac{d}{dx}(\csc x) = -\csc x \cot x$
- $\frac{d}{dx}(\cot x) = -\csc^2 x$
- $\frac{d}{dx}(\tan x) = \sec^2 x$

Differentiate exp(x)

- $\frac{d}{dx} e^{ax} = a e^{ax}$

Logarithmic differentiation

- The natural logarithm is the logarithm with base e.
- The notation is $\log_e |x| = \ln |x|$

- $\frac{dy}{dx} = \frac{1}{\frac{dx}{dy}}$
- $\frac{d}{dx} \ln(x) = \frac{1}{x}$
- $\frac{d}{dx} \ln |f(x)| = f'(x) \times \frac{1}{f(x)}$
- When differentiating a function that is either a product or quotient of two or more functions or has a variable in the power:
    - apply the natural logarithm to both the LHS and the RHS;
    - logarithmic rules may need to be applied to simplify before differentiation at this stage;
    - differentiate the LHS and RHS implicitly;
    - re-arrange to give $\frac{dy}{dx} =$.
    - replace y with the original function to give the final solution in terms of only x.

Summary of prior knowledge
Recurrence relations: Higher

- A sequence is a series of numbers or terms with a definite pattern. A sequence can be defined by a rule or a formula for the nth term.
- A recurrence relation describes a sequence in which each term is a function of the previous term or terms.
- A geometric sequence takes the form $u_{n+1} = au_n$
    - The nth term can be written as $u_n = a^n u_0$
- An arithmetic sequence takes the form $u_{n+1} = u_n + b$
    - The nth term can be written as $u_n = u_0 + nb$
- A linear recurrence relation is a sequence defined by $u_{n+1} = au_n + b, a \neq 0$
- For the linear recurrence relation $u_{n+1} = au_n + b$, a limit exists if $-1 < a < 1$
- The limit is given by the formula $L = \frac{b}{1-a}$ or $L = \frac{c}{1-m}$

TOPIC 7. SEQUENCES AND SERIES

7.1.1 Simple recurrence relations

A **sequence** is a series of terms with a definite pattern. A sequence can be defined by a rule or a formula for the nth term.

A **recurrence relation** describes a sequence in which each term is a function of the previous term or terms.

Examples

1.

2, 5, 8, 11, ...

The next two terms would be 14 and 17.

A rule for this sequence would be to start with 2 then add on 3 each time.

A formula to find the nth term would be $3n - 1$.

This is because our sequence goes up in 3s so we think of the multiples of 3, i.e. 3, 6, 9, 12, ...

You should be able to see that if we subtract 1 from each term we would get our sequence 2, 5, 8, 11, ...

....

2.

1, 1, 2, 3, 5, 8, ...

This is a **Fibonacci** sequence.

The next two terms would be 13 and 21.

A rule for this sequence would be to add the previous two terms to get the next term.

A formula to find the nth term for this sequence is much harder and would be $\approx \frac{1}{\sqrt{5}}\left(\frac{1+\sqrt{5}}{2}\right)^{n+1}$, where the "$\approx$" is close enough that you can round to the nearest integer.

You may have also spotted that $\frac{1+\sqrt{5}}{2}$ is the golden ratio.

It can be difficult to describe a formula to find the nth term so we will focus mainly on constructing a recurrence relation.

Just as we did when describing a formula for a sequence, we need some notation to describe recurrence relations.

> **Key point**
>
> u_0 defines the initial value.
>
> u_1 defines the first term.
>
> u_n defines the nth term.
>
> u_{n-1} defines the term before the nth term.
>
> u_{n+1} defines the term after the nth term.

© HERIOT-WATT UNIVERSITY

Simple recurrence relations

Consider the sequence of numbers:

$u_1, u_2, u_3, u_4, u_5, u_6, \ldots$
↓   ↓   ↓   ↓   ↓   ↓
5, 8, 11, 14, 17, 20, ...

Notice that given $u_1 = 5$ it is possible to calculate $u_2, u_3, u_4, u_5, u_6, \ldots$ by repeatedly adding 3.

$u_1 = 5$
$u_2 = u_1 + 3 = 8$
$u_3 = u_2 + 3 = 11$
$u_4 = u_3 + 3 = 14$
$u_5 = u_4 + 3 = 17$
$u_6 = u_5 + 3 = 20$

This sequence can then be defined in another way as $u_{n+1} = u_n + 3$

A sequence defined in this way is known as a recurrence relation because the pattern +3 recurs.

Geometric sequence

Under certain laboratory conditions a bamboo plant grows at a rate of 20% per day. At the start of the experiment the height of the bamboo plant is $B_0 = 30\ cm$.

a) Write down a recurrence relation that describes the growth of the plant.

b) Calculate how tall the plant is after five days.

c) Find a formula for the height of the plant after n days giving your answer in terms of B_0

(Notice that the notation is slightly different in this question B_0 is used here for the initial height of the plant. Indeed, it makes sense that B_1 represents the height of the plant after one day and similarly for $B_2, B_3, B_4, \ldots$).

a) $B_0 = 30$
Since the plant grows by 20% each day, the height will be $120\% = 1 \cdot 2$ times its height from the previous day.
The recurrence relation is therefore $B_{n+1} = 1 \cdot 2 B_n$ with $B_0 = 30$.

b) $B_0 = 30$
$B_1 = 1 \cdot 2 \times B_0 = 1 \cdot 2 \times 30 = 36$
$B_2 = 1 \cdot 2 \times B_1 = 1 \cdot 2 \times 36 = 43 \cdot 2$
$B_3 = 1 \cdot 2 \times B_2 = 1 \cdot 2 \times 43 \cdot 2 = 51 \cdot 84$

TOPIC 7. SEQUENCES AND SERIES

$B_4 = 1 \cdot 2 \times B_3 = 1 \cdot 2 \times 51 \cdot 84 = 62 \cdot 208$
$B_5 = 1 \cdot 2 \times B_4 = 1 \cdot 2 \times 62 \cdot 208 = 74 \cdot 6496$

Thus after five days the plant will be 75 cm tall (to the nearest cm).

c) The preceding working can be written in another way as shown.

$B_0 = 30$
$B_1 = 1 \cdot 2 \times B_0$
$B_2 = 1 \cdot 2 \times B_1 = 1 \cdot 2 \times (1 \cdot 2 \times B_0) = 1 \cdot 2^2 \times B_0$
$B_3 = 1 \cdot 2 \times B_2 = 1 \cdot 2 \times (1 \cdot 2^2 \times B_0) = 1 \cdot 2^3 \times B_0$
$B_4 = 1 \cdot 2 \times B_3 = 1 \cdot 2 \times (1 \cdot 2^3 \times B_0) = 1 \cdot 2^4 \times B_0$
$B_5 = 1 \cdot 2 \times B_4 = 1 \cdot 2 \times (1 \cdot 2^4 \times B_0) = 1 \cdot 2^5 \times B_0$

A pattern develops and in general:

$B_n = 1 \cdot 2^n B_0$

Example

Problem:

Bacteria increase at a rate of 30% every hour.

Write down a recurrence relation to describe the growth of bacteria.

Solution:

Each hour the number of bacteria will be 130% of the previous hour where u_n is the number of bacteria after n hours.

$u_{n+1} = 1 \cdot 3 u_n$

Key point

A geometric sequence is a special type of recurrence relation that takes the form: $u_{n+1} = au_n$, where a is a scalar or multiple.

It is worth noting that a formula for the nth term is given by $u_n = a^n u_0$

Example

Problem:

A car depreciates at a rate of 15% per annum. In 2010 a car cost £9500.

a) Describe the price each year by a recurrence relation in the form $V_{n+1} = aV_n$ where V_n is the value of the car after n years.

b) Calculate V_1, V_2, and V_3.

c) Write down a formula for V_n.

d) How long will it take for the car to half its value?

© HERIOT-WATT UNIVERSITY

Solution:

a)
The value is 85% of its value the previous year.
$V_{n+1} = 0 \cdot 85 V_n, \; V_0 = 9500$

b)
Use the recurrence relation from part (a).
$V_1 = 0 \cdot 85 \times 9500 = 8075$
$V_2 = 0 \cdot 85 \times 8075 = 6863 \cdot 75$
$V_3 = 0 \cdot 85 \times 6863 \cdot 75 = 5834 \cdot 19$

c)
This example uses the method for compound interest.
$V_n = (0 \cdot 85)^n \times 9500$

d)
Using trial and error is often the easiest way.

Half of 9500 = £4750
$V_3 = 5834 \cdot 19$
$V_4 = 0 \cdot 85 \times 5834 \cdot 19 = 4959 \cdot 06$
This value is close but not yet half.

$V_5 = 0 \cdot 85 \times 4959 \cdot 06 = 4215 \cdot 20$
This value is below half so it will take five years to half its value.

> **Key point**
>
> Remember a recurrence relation describes how to find the next term in a sequence whereas a rule describes how to find the nth term.

Example

Problem:

John saves £15 each month. Initially he had £35.

a) If a_n is the amount he has saved after n months, construct a recurrence relation for a_{n+1} and state the value of a_0.

b) Calculate a_1, a_2, a_3.

c) Write down a formula for a_n.

d) How much had John saved after 10 months?

Solution:

a)
$a_{n+1} = a_n + 15, \; a_0 = 35$

b)
$a_1 = 35 + 15 = 50$

TOPIC 7. SEQUENCES AND SERIES

$a_2 = 50 + 15 = 65$
$a_3 = 65 + 15 = 80$

c)
The sequence so far is 50, 65, 80.
It is easier to ignore the initial value when $n = 0$.

Our sequence goes up in 15s so we think of the multiples of 15, i.e. 15, 30, 45, ...
You should be able to see that if we add 35 to each term we would get our sequence 50, 65, 80, ...

The formula is $a_n = 15n + 35$

d)
When $n = 10$, $a_{10} = 15 \times 10 + 35 = 185$
He has saved £185.

An alternative solution would be to use the recurrence relation to find a_4, a_5, a_6, a_7, a_8, a_9 and finally a_{10}.

Key point

An **arithmetic sequence** is a special type of recurrence relation that takes the form
$u_{n+1} = u_n + b$
A formula for the nth term is given by $u_n = u_0 + bn$

Example

Problem:

Given the recurrence relation: $u_{n+1} = {}^1/_2 u_n - 3$ and $u_1 = 10$

List the first four terms of the sequence.

Solution:

$u_1 = 10$
$u_2 = {}^1/_2 \times 10 - 3 = 5 - 3 = 2$
$u_3 = {}^1/_2 \times 2 - 3 = 1 - 3 = -2$
$u_4 = {}^1/_2 \times (-2) - 3 = -1 - 3 = -4$

Key point

A linear recurrence relation is defined by:
$u_{n+1} = au_n + b, a \neq 0$

Notice that the equation looks like the equation of a straight line i.e. $y = mx + c$.

© HERIOT-WATT UNIVERSITY

Example

Problem:

During an epidemic a hospital claims that a new drug will cure 40% of patients with a virus each day.

Currently 3 new patients will be admitted with the virus each day.

If there were 22 patients with the virus last night find a recurrence relation for the number of patients in the hospital with the virus.

Solution:

We need to know about the number of patients with the virus so if 40% are cured then 60% still have the virus and 3 new patients must be added each day.

$v_{n+1} = 0 \cdot 6 v_n + 3$, $v_0 = 22$ where v_n is the number of patients with the virus on day n.

> **Top tip**
>
> Remember the initial value is part of the recurrence relation and must be included.

Example

Problem:

A farmer grows a variety of plum tree which ripens during the months of July and August.

On the last day in July there was 2000 kg of ripe fruit ready to be picked.

At the beginning of August the farmer hires some fruit pickers who manage to pick 75% of the ripe fruit each day.

Also, each day 60 kg more of the plums become ripe.

a) Find a recurrence relation for the weight of ripe plums left in the orchard.

b) What is the estimated weight of ripe plums left in the orchard at the end of the day on the 7th August ?

Solution:

a) Let P_0 represent the weight of ripe plums available at the start then $P_0 = 2000$
P_1 represents the amount of ripe plums left in the orchard at the end of the day on the 1st August and similarly for $P_2, P_3, P_4, \ldots$

Since 75% of the ripe fruit is picked each day then 25% of the ripe fruit is left in the orchard for the next day.
Also, each day 60 kg more of the plums ripen.
Thus the recurrence relation is: $P_{n+1} = 0 \cdot 25 P_n + 60$

b) With $P_0 = 2000$ and $P_{n+1} = 0 \cdot 25 P_n + 60$ then, using a calculator, you can check that
$P_1 = 0 \cdot 25 \times 2000 + 60 = 560$
$P_2 = 200$
$P_3 = 110$
$P_4 = 87 \cdot 5$
$P_5 = 81 \cdot 875$
$P_6 = 80 \cdot 46875$
$P_7 = 80 \cdot 1171875$

Thus after the 7th of August there is approximately 80 kg of ripe fruit in the orchard.

Occasionally it is useful to be able to identify specific terms in a recurrence relation.

Further linear recurrence relations — Go online

A sequence is defined by the recurrence relation $u_{n+1} = 0 \cdot 8 u_n + 500$, $u_0 = 10$. Calculate the value of u_4 and find the smallest value of n for which $u_n > 2000$.

The recurrence relation gives (to two decimal places):
$u_0 = 10$
$u_1 = 0 \cdot 8 \times 10 + 500 = 508$
$u_2 = 0 \cdot 8 \times 508 + 500 = 906 \cdot 4$
$u_3 = 0 \cdot 8 \times 906 \cdot 4 + 500 = 1225 \cdot 12$
$u_4 = 0 \cdot 8 \times 1225 \cdot 12 + 500 = 1480 \cdot 10$
$u_5 = 0 \cdot 8 \times 1480 \cdot 10 + 500 = 1684 \cdot 08$
$u_6 = 0 \cdot 8 \times 1684 \cdot 08 + 500 = 1847 \cdot 26$
$u_7 = 0 \cdot 8 \times 1847 \cdot 26 + 500 = 1977 \cdot 81$
$u_8 = 0 \cdot 8 \times 1977 \cdot 81 + 500 = 2082 \cdot 25$
$u_9 = 0 \cdot 8 \times 2082 \cdot 81 + 500 = 2165 \cdot 80$

So $u_4 = 1480 \cdot 10$ (to 2 d.p.) and $n = 8$ for $u_n > 2000$.

Key point

You may round your answers to write them down but you must keep and use all decimal places to calculate the next value in the sequence.

Simple recurrence relations exercise

Q1: Write down a recurrence relation for the sequence 100, 50, 25, ...

Q2: A lorry driver travels 80 miles per day. At the start of the week his speedometer reads 14200 miles.

Write down a recurrence relation that describes the reading on the speedometer M_{n+1} and state the value of M_0.

Q3: Given the recurrence relation $u_{n+1} = 2u_n + 1$ and $u_1 = 3$, what is u_5?

7.1.2 Finding a limit

Some recurrence relations tend toward a limit or settle around a particular value.

Remember the orchard example from a previous sub-topic, it looks like the amount of fruit may settle around 80 kg.

Compare the following two recurrence relations.

Investigating recurrence relations

$u_{n+1} = 0 \cdot 6 u_n + 20$, $u_0 = 10$ generates the following sequence (to 2 decimal places).

$u_0 = 10$
$u_1 = 26$
$u_2 = 35 \cdot 6$
$u_3 = 41 \cdot 36$
$u_4 = 44 \cdot 82$
$u_5 = 46 \cdot 89$
$u_6 = 48 \cdot 13$
$u_7 = 48 \cdot 88$
$u_8 = 49 \cdot 33$
$u_9 = 49 \cdot 60$
$u_{10} = 49 \cdot 76$
$u_{11} = 49 \cdot 85$
$u_{12} = 49 \cdot 91$

TOPIC 7. SEQUENCES AND SERIES

This can be represented on a graph as shown:

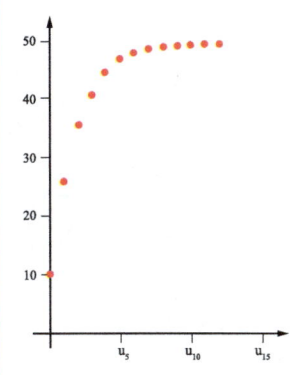

This shows that $u_n \to 50$ as $n \to \infty$. The terms become closer and closer to 50. This is a **convergent sequence**.

$u_{n+1} = 2u_n + 20$, $u_0 = 10$ generates the following sequence (to 2 decimal places).

$u_0 = 10$
$u_1 = 40$
$u_2 = 100$
$u_3 = 220$
$u_4 = 460$
$u_5 = 940$
$u_6 = 1900$
$u_7 = 3820$
$u_8 = 7660$
$u_9 = 15340$

This is represented on a graph as shown here.

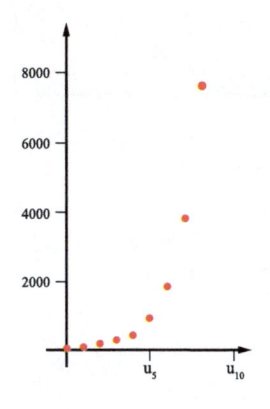

Notice that this time the **sequence is divergent**. Subsequent terms continue getting bigger and bigger.

> **Key point**
>
> A convergent recurrence relation tends toward a limit, L as $n \to \infty$.
>
> A divergent recurrence relation does not have a limit.

> **Proof: The limit of a recurrence relation**
>
> **Proof**
>
> The formula for a limit L can be derived from the following.
> In general if $u_{n+1} \to L$ as $n \to \infty$ then we also have that $u_n \to L$ as $n \to \infty$
> Thus, as $n \Rightarrow \infty$, the formula $u_{n+1} = au_n + b$ tends to the following:
>
> $L = aL + b$
> $L - aL = b$
> $L(1 - a) = b$
> $L = \dfrac{b}{1 - a}$

> **Key point**
>
> For the linear recurrence relation $u_{n+1} = au_n + b$
>
> A limit exists if $-1 < a < 1$ and the limit L is given by the formula:
>
> $$L = \dfrac{b}{1-a}$$
>
> Some students find it difficult to remember this formula so there is an alternative.
> For the recurrence relation $u_{n+1} = mu_n + c$:
>
> A limit exists if $-1 < m < 1$ and the limit L is given by the formula:
>
> $$L = \dfrac{c}{1-m}$$

Examples

1. **Problem:**

    a) Find the first six terms of the recurrence relation $u_{n+1} = 0 \cdot 8u_n + 4$ with $u_0 = 2$
    b) Give a reason why this recurrence relation generates a sequence which has a limit.
    c) Calculate the value of the limit.

Solution:

a) The first six terms in the sequence are:
$u_0 = 2$
$u_1 = 5 \cdot 6$
$u_2 = 8 \cdot 48$
$u_3 = 10 \cdot 784$
$u_4 = 12 \cdot 6272$
$u_5 = 14 \cdot 10176$

b) $-1 < 0 \cdot 8 < 1$ so a limit exists.
Note that it is essential that you make this statement to gain full marks.

c) As $n \Rightarrow \infty$ the recurrence relation $u_{n+1} = 0 \cdot 8 u_n + 4$ tends to

$$L = \frac{b}{1-a}$$
$$L = \frac{4}{1 - 0 \cdot 8}$$
$$L = \frac{4}{0 \cdot 2}$$
$$L = \frac{40}{2} = 20$$

Thus the sequence tends towards 20.

This type of question is often asked in the non-calculator paper. Be sure that you can calculate the limit without a calculator.

....

2. Problem:

Fish, like all animals need oxygen to survive. The fish in a certain tank use up 15% of the oxygen in the water each hour.
However, due to the action of a pump, oxygen is added to the water at a rate of 1 part per metre^3 each hour.
The oxygen level in the tank should be between 5 and 7 parts per metre^3 for the survival of the fish.
Initially the concentration of oxygen in the tank is 6 ppm^3.

a) Write down a recurrence relation to describe the oxygen level in the water.

b) Say whether or not a limit exists, giving a reason.

c) Determine, in the long term, whether the fish will survive.

Solution:

a) Let F_n represent the oxygen level in the water after n hours then
$F_{n+1} = 0 \cdot 85 F_n + 1$ with $F_0 = 6$

b) $-1 < 0.85 < 1$ so a limit exists.

c) The fish will survive because when $n \Rightarrow \infty$ then $u_n \Rightarrow L$ and $u_{n+1} \Rightarrow L$ thus

$$L = \frac{b}{1-a}$$
$$= \frac{1}{1-0.85}$$
$$= \frac{1}{0.15}$$
$$L = 6.667$$

so oxygen levels in the tank will tend towards 6·667 ppm^3.

Since $5 < 6.667 < 7$ the fish will survive.

....

3. Problem:
A farmer grows a variety of plum tree which ripens during the months of July and August.
On the last day in July there was 2000 kg of ripe fruit ready to be picked.
At the beginning of August the farmer hires some fruit pickers who manage to pick 75% of the ripe fruit each day.
Also, each day 60 kg more of the plums become ripe.

a) Find a recurrence relation for the weight of ripe plums left in the orchard.

b) What is the estimated weight of ripe plums left in the orchard in the long term during the picking season?

Solution:

a)
The recurrence relation is $P_{n+1} = 0.25 P_n + 60$

b)
$$L = \frac{c}{1-m}$$
$$= \frac{60}{1-0.25}$$
$$= \frac{60}{0.75}$$
$$= 60 \div \frac{3}{4}$$
$$= 60 \times 4 \div 3$$
$$L = 80$$

During the picking season the amount of fruit left unpicked in the orchard will settle around 80 kg.

TOPIC 7. SEQUENCES AND SERIES

> **Key point**
>
> It is important to remember that the limit is the value that a recurrence relation will tend towards or settle around. It will never actually be that value it will just get very close to it as $n \to \infty$.

Finding a limit exercise   Go online

Q4: Does a limit exist for the recurrence relation $u_{n+1} = 2 \cdot 4u_n + 3$?
Answer yes or no.

....

Q5: Does a limit exist for the recurrence relation $u_{n+1} = -u_n - 5$?
Answer yes or no.

....

Q6: Does a limit exist for the recurrence relation $u_{n+1} = 0 \cdot 9u_n + 10$?
Answer yes or no.

7.1.3 Solving recurrence relations

Example Problem:

The recurrence relation $A_{n+1} = kA_n + 6$ has the same limit as the recurrence relation $B_{n+1} = -0 \cdot 8B_n + 18$ as $n \Rightarrow \infty$.

Find the value of k.

Solution:

We know that a limit exists so we can find the limit for B because $-1 < -0 \cdot 8 < 1$.

$L_B = \dfrac{b}{1-a}$

$L_B = \dfrac{18}{1-(-0 \cdot 8)}$

$L_B = \dfrac{18}{1 \cdot 8} = 10$

The formula for the limit of A is: $L_A = \dfrac{6}{1-k}$, but we all know that $L_A = 10$ so

$$10 = \dfrac{6}{1-k}$$
$10(1-k) = 6$
$10 - 10k = 6$
$-10k = -4$
$k = 0 \cdot 4$

Be sure you can do this calculation without a calculator.

Notice that $-1 < k < 1$, so we know our answer lies in the range of values for which there is a limit.

Examples

1. Problem:
Given the recurrence relation $u_{n+1} = au_n + b$ with $u_1 = 2$, $u_2 = 7$ and $u_3 = 17$:

a) Find the values of a and b.
b) Calculate u_5.

Solution:

a) From the recurrence relation $u_{n+1} = au_n + b$ we can write an equation for u_2 using u_1 and an equation for u_3 using u_2.

$u_2 = au_1 + b$ $\qquad\qquad u_3 = au_2 + b$

$7 = 2a + b$ $\qquad\qquad 17 = 7a + b$

This gives us a pair of simultaneous equations to solve.

$2a + b = 7 \quad \times -1$
$7a + b = 17$

$\begin{array}{r} -2a - b = -7 \\ + \quad 7a + b = 17 \\ \hline 5a = 10 \\ a = 2 \end{array}$

Let $a = 2$ in $2a + b = 7$ gives,
$2 \times 2 + b = 7$
$4 + b = 7$
$b = 3$

Hence, the recurrence relation is $u_{n+1} = 2u_n + 3$

b) Given the recurrence relation $u_{n+1} = 2u_n + 3$ and $u_3 = 17$ then

$u_4 = 2 \times 17 + 3 = 37$
$u_5 = 2 \times 37 + 3 = 77$
Thus $u_5 = 77$

....

2. Problem:
Garry saved the same amount each month to a fixed rate savings account where no withdrawals were allowed during a fixed term.

The fixed rate of interest was paid monthly and the amount in his bank at the end of three consecutive months were £402, £642·20, £906·32.

What was the interest rate and the amount that he saved each month?

Solution:

The recurrence relation can be expressed as $u_{n+1} = au_n + b$ where a is the compound interest as a decimal and b is his monthly savings.
We can make $u_1 = 402$, $u_2 = 642 \cdot 20$ and $u_3 = 906 \cdot 42$.

From the recurrence relation $u_{n+1} = au_n + b$ we can write an equation for u_2 using u_1 and an equation for u_3 using u_2.

$$u_2 = au_1 + b \qquad\qquad u_3 = au_2 + b$$
$$642 \cdot 20 = 402a + b \qquad\qquad 906 \cdot 42 = 642 \cdot 20a + b$$

This gives us a pair of simultaneous equations.

$$402a + b = 642 \cdot 20 \quad \times \; -1$$
$$642 \cdot 20a + b = 906 \cdot 42$$

$$\begin{aligned}-402a - b &= -642 \cdot 20 \\ +\quad 642 \cdot 20a + b &= 906 \cdot 42 \\ \hline 240 \cdot 20a &= 264 \cdot 22 \\ a &= 1 \cdot 1\end{aligned}$$

Let $a = 1 \cdot 1$ in $402a + b = 642 \cdot 20$ gives,
$$402 \times 1 \cdot 1 + b = 642 \cdot 20$$
$$442 \cdot 20 + b = 642 \cdot 20$$
$$b = 200$$

Since $a = 1 \cdot 1$ that equates to compound interest of 110% and an increase of 10%.

Hence his monthly savings are £200 and the interest rate is 10%.

Solving recurrence relations exercise — Go online

Given the recurrence relation $u_{n+1} = au_n + b$ with $u_1 = -2$, $u_2 = -17$ and $u_3 = -92$ find a and b.

Q7: Let $n = 1$ then $u_{n+1} = au_n + b$ becomes $u_2 = au_1 + b$ hence $-17 = ?$
Give your answer in terms of a and b.

....

Q8: Let $n = 2$ then $u_{n+1} = au_n + b$ becomes $u_3 = au_2 + b$ hence $-92 = ?$

....

Q9: Using your answers to the previous questions solve the simultaneous equations to find a and b.

Given the recurrence relation $u_{n+1} = au_n + b$

Q10: Find a and b when $u_1 = 9200$, $u_2 = 7400$ and $u_3 = 6050$.

....

Q11: Hence what is $u_0 = ?$

....

Q12: What is the limit L of the sequence as n tends to infinity?

....

Q13:
A recurrence relation is given as $u_{n+1} = 2u_n + b$ with $u_0 = 1$ and $u_2 = 58$.
What is the value of b?

A mushroom bed contains 6000 mushrooms on the first morning.
Each day x percent of the mushrooms are picked and each night another y mushrooms are ready for picking.
On the second and third days there are 1600 then 720 mushrooms ready to be picked.

Q14:
The recurrence relation that describes the number of mushrooms that are ready to be picked is $u_{n+1} = au_n + b$
What are the values of a and b?

....

Q15: If x percent of mushrooms are picked each day.
What is the value of x?

....

Q16: Each night another y mushrooms become ready to be picked.
What is the value of y?

Two sequences are defined by the recurrence relations:

1. $u_{n+1} = 0 \cdot 5u_n + x$ with $u_0 = 20$

2. $v_{n+1} = 0 \cdot 1v_n + y$ with $v_0 = 10$

Q17: What is the limit of sequence 1 in terms of x?

....

Q18: What is the limit of sequence 2 in terms of y?

....

Q19: Express x in terms of y.

7.2 Sequences and recurrence relations

The terms 'sequence' and 'series' are used in a wide variety of contexts. There are, for example, film sequences, television series, a sequence of events and so on. The two terms are often interchangeable and in fact most dictionaries will give similar definitions of both terms.

There are however mathematical sequences and series and these have distinctly different meanings although there is a strong relationship between them.

TOPIC 7. SEQUENCES AND SERIES

7.2.1 What is a sequence?

> **Key point**
>
> A sequence is an ordered list of terms.

Example

The following are sequences:

- x, x^2, x^3, x^4
- $\frac{1}{2}, \frac{1}{4}, \frac{1}{8}, \frac{1}{16}, \frac{1}{32}$
- 1, 2, 3, 4, 5, ..., 50, ..., 99, 100
- 2, 6, 12, 20, 30, 42,

In the last example the dots between the terms in the sequence represent missing terms. This sequence is the list of natural numbers between 1 and 100.

These examples demonstrate that not all sequences have an obvious pattern. In this topic the emphasis will be on numeric sequences that have a clear relationship between the terms but it is important to remember that not all sequences have an obvious pattern.

Each number in a sequence is called a **term** or an **element**. The nth term (or general term) is often denoted by u_n.

Example Problem:

Find the 3rd term in the sequence 2, 3, 5, 8, 12.

Solution:

Count from the left. The 3rd term is 5.

Q20: What is the 5th term of the sequence 2, 4, 6, 8, 10, 12?

Finite sequence

All the sequences mentioned previously are finite sequences. Each sequence specifically states the first and last terms.

The general example of a finite sequence is $u_1, u_2, u_3, \ldots, u_n$.

Where u_1 is the first term and u_n is the last term.

> **Key point**
>
> A finite sequence is one which has a last term.

Infinite sequence

The general example of a sequence which does not have a last term is $u_1, u_2, u_3, \ldots$.

Where the three dots at the end indicate that the sequence continues indefinitely.

Such a sequence is called an infinite sequence.

> **Key point**
>
> An infinite sequence is one which continues indefinitely.

> **Example**
>
> These are all infinite sequences:
>
> - 1, 2, 3, 4, 5, ..., 50, ..., 99, 100, ... (the natural numbers listed in order).
> - 2, 3, 5, 7, 11, 13, 17, ... (the prime numbers listed in ascending order).
> - $1, \frac{1}{2}, \frac{1}{3}, \frac{1}{4}, \frac{1}{5}, \ldots$ (the reciprocals of the natural numbers listed in order).

If a sequence can be defined by a general term such as u_n (or a_n, x_n, etc.) then it is common for the sequence to be written as $\{u_n\}$, where $n \in \mathbb{N}$ or $n \in \mathbb{W}$.

> **Key point**
>
> The first term of a sequence can be defined as u_0 or u_1 where $n = 0$, or $n = 1$. This can change the sequence slightly.
>
> e.g. The sequence $\{n + 1\}$ where $n \in \mathbb{N}$ gives 2, 3, 4, 5, ...
>
> But the sequence $\{n + 1\}$ where $n \in \mathbb{W}$ gives 1, 2, 3, 4, 5, ...
>
> For some sequences, the way it is defined dictates whether the first term is u_0 or u_1.
>
> For example, for the sequence $\{\frac{1}{n}\}$ where $n \in \mathbb{N}$ gives $1, \frac{1}{2}, \frac{1}{3}, \frac{1}{4}, \frac{1}{5}, \frac{1}{6}, \ldots$
>
> But if $n \in \mathbb{W}$ then we have $\frac{1}{0}, 1, \frac{1}{2}, \frac{1}{3}, \frac{1}{4}, \frac{1}{5}, \frac{1}{6}, \ldots$. The first term is undefined. So if the sequence is defined as $\{\frac{1}{n}\}$ then $n \in \mathbb{N}$.

> **Example**
>
> Here are some sequences defined by a general term (or the nth term):
>
> - $\{2n + 1\}$ where $n \in \mathbb{N}$ denotes the sequence 3, 5, 7, 9, 11, 13, 15, ...
>   or
> $\{2n + 1\}$ where $n \in \mathbb{W}$ denotes the sequence 1, 3, 5, 7, 9, 11, 13, ...

TOPIC 7. SEQUENCES AND SERIES

- $\{n^2 + n - 2\}$ where $n \in \mathbb{N}$ denotes the sequence 0, 4, 10, 18, ...

  or

  $\{n^2 + n - 2\}$ where $n \in \mathbb{W}$ denotes the sequence -2, 0, 4, 10, ...

What is a sequence exercise   Go online

Q21: Write out the first five terms of the sequence $\{(n-2)^3 + 2\}$, where $n \in \mathbb{N}$.

....

Q22: A sequence is defined by $u_n = \frac{2n}{n+1}$, $n \in \mathbb{W}$. Find the first four terms.

....

Q23: A sequence is defined by $u_n = \frac{nx^n}{1-x}$, where $n \in \mathbb{W}$. Find the first five terms.

....

Q24: Write out the first four terms of the sequence $\{11 - 3n\}$, where $n \geqslant 1$.

7.2.2 Sequences and recurrence relations

Recall that a linear recurrence relation can be defined by:

$u_{n+1} = au_n + b$, where $a \neq 0$ and the starting value u_0 or u_1 is specified.

It is a linear recurrence relation since the equation is of the form $y = mx + c$.

Examples

1. The sequence 12, 22, 32, 42, 52, ... can be expressed as an nth **term formula** or as a **recurrence relation**.

n^{th} formula: $\{10n + 2\}$, where $n \in \mathbb{N}$ or $\{10n + 12\}$, where $n \in \mathbb{W}$.

Recurrence relation: $u_{n+1} = u_n + 10$, where the initial value $u_1 = 12$ or $u_0 = 12$.

In some cases it is easier to find a formula for the recurrence relation than it is for the nth term.

....

2. Consider the sequence 5, 12, 21, 32, 45, ...

The nth term formula is $\{n^2 + 2n - 3\}$, where $n \geqslant 2$

The recurrence relation is:

5,   12,   21,   32,   45, ...   Increase by odd numbers starting with 7. This is the sequence $\{2n + 5\}$, $n \in \mathbb{N}$

 +7   +9   +11   +13

So the recurrence relation is: $u_{n+1} = u_n + 2n + 5$, where $u_1 = 5$.

It is also possible for a sequence to be defined by a second order recurrence relation in which the first two terms are specified.

Example

Problem:

Consider the sequence 3, 6, 15, 33, 78, ...

This can be defined by the recurrence relation $u_{n+2} = u_{n+1} + 3u_n$, where $u_1 = 3$ and $u_2 = 6$.

Solution:

$u_{n+2} = u_{n+1} + 3u_n$, where $u_1 = 3$ and $u_2 = 6$

$u_1 = 3$
$u_2 = 6$
$u_3 = u_2 + 3u_1 \Rightarrow u_3 = 6 + 3(3) = 15$
$u_4 = u_3 + 3u_2 \Rightarrow u_4 = 15 + 3(6) = 33$
$u_5 = u_4 + 3u_3 \Rightarrow u_5 = 33 + 3(15) = 78$

This leads back to the general term of a sequence. As already stated, another way of describing it is the nth term of a sequence. Thus if the general term or n^{th} term has a formula it is possible to find any specific term in the sequence.

In summary, a specific term of a sequence can be found from the recurrence relation $u_{n+1} = au_n + b$ or a general formula $\{u_n\}$. Which one is used depends on the information you have.

Examples

1. Using the recurrence relation to find a named term

Problem:

Find the 4th term of the sequence defined by the recurrence relation.
$u_{n+1} = 0.5u_n + 3$, where $u_1 = 2$.

Solution:

$u_2 = 0 \cdot 5(2) + 3 \Rightarrow u_2 = 4$
$u_3 = 0 \cdot 5(4) + 3 \Rightarrow u_3 = 7$
$u_4 = 0 \cdot 5(7) + 3 \Rightarrow u_4 = 5 \cdot 5$

The fourth term is 5·5.

....

2.

Problem:

Find the 8th term of the sequence defined by $\{n^2 - 2n\}$, $n \in \mathbb{N}$.

Solution:

8th term is when $n = 8$ so $8^2 - 2(8) = 48$

The 8th term is 48.

TOPIC 7. SEQUENCES AND SERIES

Sequences and recurrence relations exercise Go online

Q25: Find the first three terms of the sequence defined by $\left\{\frac{n^2}{n+1}\right\}$, $n \in \mathbb{W}$.

....

Q26: Find the fourth term of a sequence given by the recurrence relation $u_{n+1} = \frac{1}{3}u_n + 1$, $u_1 = 15$.

....

Q27: Find the fifth term of the sequence $\left\{3^{n-1} + 1\right\}$, $n \in \mathbb{N}$.

7.3 Arithmetic and geometric sequences

The analysis of sequence data allows the extrapolation of existing data to forecast future trends, spreading of disease, population growth, component failure, reliability and much more.

If we did not have this analysis then the Centre for Disease Control would not be able to predict the flu season, its severity and how many vaccines to produce for instance. A manufacturer would be not be able to extrapolate how much of a product to produce etc.

Trend analysis is therefore important and this section will start by looking at two specific sequences in more detail: arithmetic and geometric sequences.

7.3.1 Arithmetic sequences

Definition of an arithmetic sequence

An arithmetic sequence is one which takes the form:

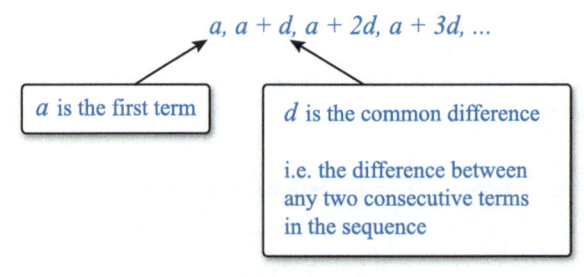

a, a + d, a + 2d, a + 3d, ...

a is the first term

d is the common difference

i.e. the difference between any two consecutive terms in the sequence

The link with recurrence relations still remains.

Consider the general linear recurrence relation $u_{n+1} = bu_n + c$, where $b, c \in \mathbb{R}$ and $n \in \mathbb{N}$.

When we let $b = 1$ and $c \neq 0$ then $c = d$.

We get $u_{n+1} = u_n + d$, where $d \in \mathbb{R}$ and $n \in \mathbb{N}$.

© HERIOT-WATT UNIVERSITY

When we let $u_1 = a$, we get:

$u_1 = a$
$u_2 = a + d$
$u_3 = a + 2d$

Which is an arithmetic sequence.

Example
Problem:

Consider the arithmetic sequence defined by the recurrence relation $u_{n+1} = u_n + c$ with $c = -2$ and $u_1 = 3$.

State a, d and hence the first 4 terms of the sequence.

Solution:

Arithmetic sequence takes the form $u_{n+1} = u_n + d$, where $u_1 = a$.

We have: $u_{n+1} = u_n - 2$, where $u_1 = 3$.

So $a = 3$ and $d = -2$

$u_1 = 3$
$u_2 = 3 - 2 \Rightarrow u_2 = 1$
$u_3 = 1 - 2 \Rightarrow u_3 = -1$
$u_4 = -1 - 2 \Rightarrow u_4 = -3$

Finding the nth term of an arithmetic sequence

An arithmetic sequence is one which takes the form: $u_{n+1} = u_n + d$, where $u_1 = a$.

This idea of finding the terms one by one until the required term is reached is fine for terms near the beginning of a sequence. However it would be rather laborious to do this to find, say, the 40th term in a sequence.

There is a way round this which comes directly from the definition.

$a, a + d, a + 2d, a + 3d, \ldots$

If we write this in relation to their term number we can define a formula for the nth term.

$u_1 = a$
$u_2 = a + 1d$
$u_3 = a + 2d$
$u_4 = a + 3d$
$\ldots$
$u_n = a + (n - 1)d$

TOPIC 7. SEQUENCES AND SERIES

> **Key point**
>
> So the nth term of an arithmetic sequence can be written as: $u_n = \{a + (n-1)d\}$, where $n \in \mathbb{N}$.
>
> Where:
>
> - a is the first term
> - d is the common difference
> - n is the term number

> **Top tip**
>
> Note that if the values of the terms of an arithmetic sequence are plotted on a graph against the term number the relationship can clearly be seen as a linear one.

Arithmetic sequence graphs   Go online

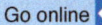

To demonstrate that the geometric sequence does indeed graph a straight line we will use the following example.

Take the recurrence relation with:

- Common ratio $d = -4$
- Starting value $a = 3$

Recurrence relation is therefore given by $u_{n+1} = u_n - 4$, $u_1 = 3$

This gives the terms:

$u_1 = 3$, $u_2 = -1$, $u_3 = -5$, $u_4 = -9$, $u_5 = -13, \ldots$

Plotting these on a graph we can see that they form a straight line.

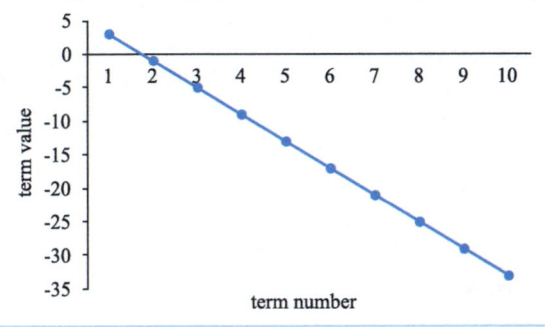

© HERIOT-WATT UNIVERSITY

Examples

1. Identifying arithmetic sequences
Problem:
Using the definition, which of the following are arithmetic sequences?

a) -12, -9, -6, -3
b) 5, 15, 45 ,55, 85, 95, 125, ...
c) 9, 14, 20, 27, ..., 64, 76, ...
d) 0, 1, 2, 3, 4, 5, ..., 45, ..., 90, 91
e) 0·01, 0·001, 0·0001, 0·00001, 0·000001, ...
f) 1, 8, 27, 64, 125, ...

Solution:

a) This is an arithmetic sequence with $a = -12$ and $d = 3$.
b) This is not an arithmetic sequence. The difference between terms 1 and 2 is 10 but between terms 2 and 3 the difference is 30. There is no common difference.
c) This is not an arithmetic sequence. There is no common difference. For example, the first difference is 5 and the second difference is 6.
d) This has $a = 0$ and $d = 1$ and is an arithmetic sequence.
e) This has no common difference and is not an arithmetic sequence.
f) This is not an arithmetic sequence. It has no common difference.

....

2. Finding the nth term of an arithmetic sequence
Problem:
Find the 18th term of the arithmetic sequence 10, 6, 2, -2, -6, ...
Hints: The nth term of an arithmetic sequence takes the form: $u_n = a + (n-1)d$

- a is the 1st term
- d is the common difference between consecutive terms
- n is the term number

Solution:
$$u_n = a + (n-1)d$$
$$= 10 + (18-1) - 4$$
$$= 78$$

....

3. Defining the sequence from two terms
Problem:
If the 6th term of an arithmetic sequence is -22 and the third term is -10 define the sequence.

TOPIC 7. SEQUENCES AND SERIES

Solution:
For the 6th term $-22 = a + 5d$ and for the 3rd term $-10 = a + 2d$.
Solving these simultaneously gives $d = -4$.
So $a = -2$ and the sequence is defined by $\{-2 - 4(n - 1)\} = \{-4n + 2\}$.

Arithmetic sequence exercise Go online

Q28: If the 5th term of an arithmetic sequence is 8 and the 14th term is 35, define the sequence.

....

Q29: If the 3rd term of an arithmetic sequence is -7 and the 10th term is -49, define the sequence.

....

Q30: Find the value of n when $a = -2$, $d = 5$ and $u_n = 23$.

....

Q31: If the 4th, 5th and 6th terms of an arithmetic sequence are $2x$, $x + 8$ and $3x + 1$ respectively.
What are the values of the 4th, 5th and 6th terms?

7.3.2 Geometric sequences

Definition of a geometric sequence

A geometric sequence is one which has the form:

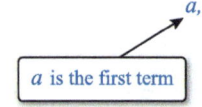

$$a,\ ar,\ ar^2,\ ar^3, \ldots$$

a is the first term

r is the common ratio
i.e. the ratio of two consecutive terms

$$r = \frac{u_{n+1}}{u_n}$$

Example

5, 15, 45, 135, 405, ... is a geometric sequence with $a = 5$ and $r = 3$.
The common ratio is: $r = \frac{u_{n+1}}{u_n}$ $\Rightarrow$ $r = \frac{15}{5} = 3$
This is the same for all consecutive terms.

A geometric sequence can be linked to the general linear recurrence relation.

$u_{n+1} = bu_n + c$, where $b, c \in \mathbb{R}$ and $n \in \mathbb{N}$.

Consider the outcome when $b \neq c$ with $c = 0$.

This gives: $u_{n+1} = bu_n$

This represents a geometric sequence.

The first term is $a = u_1$ and the sequence has a common ratio of $r = b$.

Example
Problem:

Consider the geometric sequence defined by the recurrence relation $u_{n+1} = bu_n$ with $b = -3$ and $u_1 = -1$.

State a, r and hence the first 4 terms of the sequence.

Solution:

A Geometric sequence takes the form: $u_{n+1} = ru_n$

In this case $r = -3$ and $a = -1$ so we have: $u_{n+1} = -3u_n$

Terms:

$u_1 = -1$
$u_2 = -3(-1) \Rightarrow u_2 = 3$
$u_3 = -3(3) \Rightarrow u_3 = -9$
$u_4 = -3(-9) \Rightarrow u_4 = 27$

As with arithmetic sequences it would be rather labourious to use this approach to find, say, the 20th term in a sequence.

Again there is an alternative approach which comes directly from the definition.

Definition of a geometric sequence: Alternative approach

A geometric sequence is one which takes the form: $a, ar, ar^2, ar^3, \ldots$

If we write this in relation to their term number we can define a formula for the nth term.

$u_1 = a$
$u_2 = ar^1$
$u_3 = ar^2$
$u_4 = ar^3$
$\ldots$
$u_n = ar^{(n-1)}$

TOPIC 7. SEQUENCES AND SERIES

> **Key point**
>
> So the nth term of a geometric sequence can be written as:
>
> $u_n = ar^{(n-1)}$, where $n \in \mathbb{N}$.
>
> Where:
>
> - a is the first term
> - r is the common ratio
> - n is the term number

The terms in a geometric sequence demonstrate an exponential relationship with the equation in the form $u_n = ab^n$.

Geometric sequence graphs   Go online

To demonstrate that the geometric sequence does indeed graph a straight line we will use the following example.

Take the recurrence relation with:

Common ratio $r = 3$

Starting value $a = 2$

Recurrence relation is therefore given by $u_{n+1} = 3u_n$, $u_1 = 2$

This gives the terms:

$u_1 = 2$, $u_2 = 6$, $u_3 = 18$, $u_4 = 54$, $u_5 = 162$, ...

Plotting these on a graph we can see that they form an exponential curve.

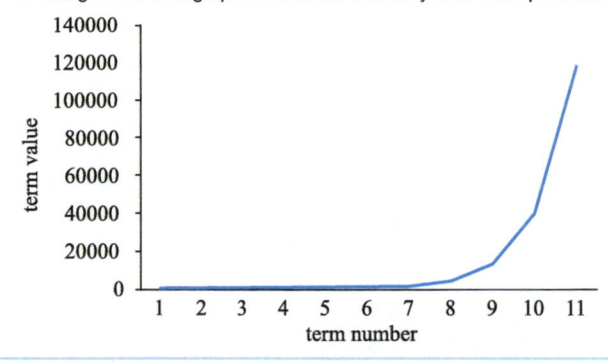

Geometric sequence graph: Alternating sequence

If the common ratio is a negative number we end up with an alternating sequence.

If $r = -1$ it will alternate between two values.

TOPIC 7. SEQUENCES AND SERIES

Take the recurrence relation with:

Common ratio $r = -1$

Starting value $a = 2$

Recurrence relation is therefore given by $u_{n+1} = -u_n$, $u_1 = 2$

This gives the terms:

$u_1 = 2$, $u_2 = -2$, $u_3 = 2$, $u_4 = -2$, $u_5 = 2$, ...

Plotting these on a graph we can see that they form an alternating sequence.

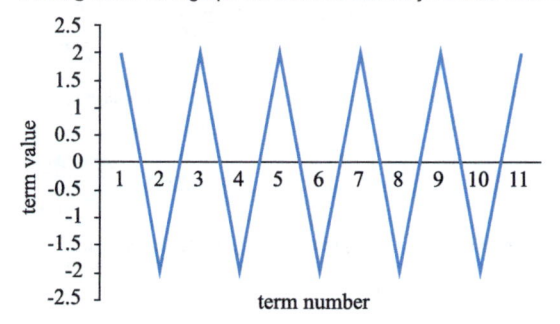

If $r < -1$ then it will be an alternating sequence where the positive numbers will get bigger and the negative numbers smaller.

Take the recurrence relation with:

Common ratio $r = -3$

Starting value $a = 2$

Recurrence relation is therefore given by $u_{n+1} = -3u_n$, $u_1 = 2$

This gives the terms:

$u_1 = 2$, $u_2 = -6$, $u_3 = 18$, $u_4 = -54$, $u_5 = 162$, ...

Plotting these on a graph we can see that they form a growing alternating sequence.

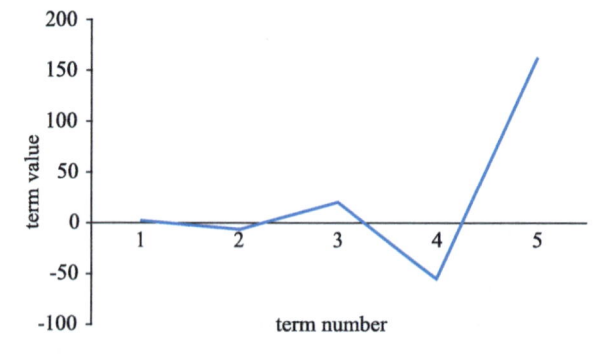

TOPIC 7. SEQUENCES AND SERIES

Examples

1. Identifying geometric sequences
Problem:
Using the definition, which of the following are geometric sequences?

a) -1, 1, -1, 1, -1, 1, -1, 1, -1, ...
b) 5, -10, 20, -40, 80, ...
c) 1, 4, 9, 16, 25, 36, 49, ...
d) 2·2, 3·3, 4·4, 5·5, ..., 9·9
e) 0·01, 0·001, ..., 0·0000001, 0·00000001, ...
f) 64, 32, 16, 8, 4

Solution:

a) This has $a = -1$ and $r = -1$ and is a geometric sequence.
b) This has $a = 5$ and $r = -2$ and is a geometric sequence.
c) This is not a geometric sequence since the ratio of two consecutive terms is not the same throughout the sequence.
d) This is not a geometric sequence since there is no common ratio.
e) This has $a = 0 \cdot 01$ and $r = 0 \cdot 1$ and is a geometric sequence.
f) This has $a = 64$ and $r = 0 \cdot 5$ and is a geometric sequence.

....

2. Finding the nth term of a geometric sequence 1
Problem:
Find the 12th term of the geometric sequence 4, 1·2, 0·36, 0·108, 0·0324, ...
Solution:
The nth term of a geometric sequence is $u_n = ar^{n-1}$
The first term is: $a = 4$
The common ratio is: $r = \frac{u_{n+1}}{u_n}$ ⇒ $r = \frac{1 \cdot 2}{4}$ ⇒ $r = 0 \cdot 3$
The 12th term is:
$u_{12} = 4 \times 0 \cdot 3^{11}$
$u_{12} = 7 \cdot 08588 \times 10^{-6}$

....

3. Finding the nth term of a geometric sequence 2
Problem:
Find the 17th term of the geometric sequence $6, 3, \frac{3}{2}, \frac{3}{4}, \frac{3}{8}, ...$

© HERIOT-WATT UNIVERSITY

Hints: The nth term of a geometric sequence takes the form: $u_n = ar^{(n-1)}$

- a is the 1st term.
- r is the common ratio between consecutive terms.
- n is the term number.

Solution:
$a = 6$
Common ratio:
$r = \dfrac{3}{6}$
$= \dfrac{1}{2}$
so
$u_{17} = 6 \times \left(\dfrac{1}{2}\right)^{16}$
$= \dfrac{3}{32768}$

....

4. Defining the sequence from two terms

Problem:
If the 6th term of a geometric sequence is -486 and the 3rd term is 18, define the sequence.

Solution:
For the 6th term $-486 = ar^5$ and for the 3rd term $18 = ar^2$.
Dividing one equation by the other gives:
$\dfrac{-486}{18} = \dfrac{ar^5}{ar^2}$
$-27 = r^3$
So, $-27 = r^3 \Rightarrow r = -3$
Substituting $r = -3$ into $18 = ar^2$: $18 = a(-3)^2 \Rightarrow a = \dfrac{18}{9} \Rightarrow a = 2$
Therefore the sequence is defined by $\{2 \times (-3)^{n-1}\}$.

Geometric sequence exercise   Go online

Q32: If the 4th term of a geometric sequence is 192 and the 7th term is 12288, define the sequence.

....

Q33: If the 3rd term of an geometric sequence is -0·02 and the 6th term is -0·00002, define the sequence.

....

TOPIC 7. SEQUENCES AND SERIES

Q34: The population of an endangered bat increases annually by 2%. Initially there are 1000 bats.

a) Write down the general geometric sequence for successive populations.
b) The species will be taken off the danger list when the population exceeds 4000. After how many years will this happen?

7.4 Fibonacci and other sequences

There are many interesting sequences. Here are a few.

> **Key point**
>
> An alternating sequence is any sequence which has alternate positive and negative terms.

This type of sequence was mentioned in the geometric sequence section. The example given was -1, 1, -1, 1, -1, ...

The Fibonacci sequence is formed by first stating two starting values. The most common example given is taking the first two terms to be 1 and 1. Each subsequent term is formed by adding the two terms immediately before this. Note that the first two terms do not have to be 1 and 1.

> **Key point**
>
> The Fibonacci sequence is defined by a second order recurrence relation of the form
> $$u_{n+2} = u_{n+1} + u_n$$
> An example is: 1, 1, 2, 3, 5, 8, 13, ...

The Fibonacci sequence can be shown using squares, where the length of each side represents a term in the Fibonacci sequence.

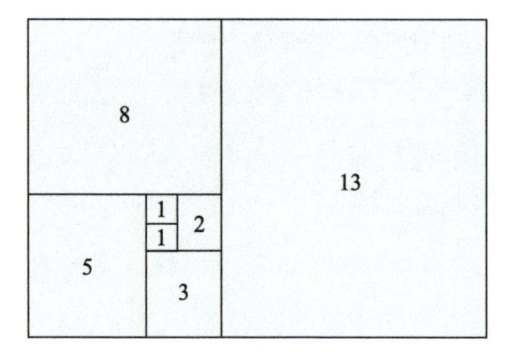

The Fibonacci sequence occurs in many natural phenomena such as the shape of snail shells and

© HERIOT-WATT UNIVERSITY

the geometry of sunflower heads.

The following image demonstrates the structure of the snail shell using the Fibonacci sequence. It is obtained from arcs connecting the opposite corners of each square starting at the first term of 1 and continuing with an uninterrupted line.

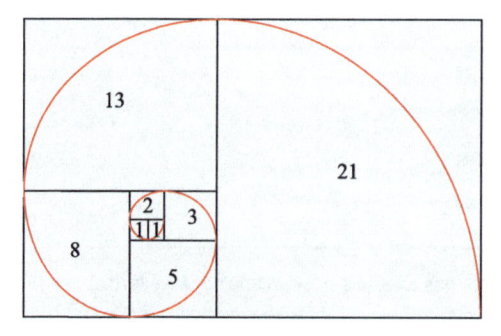

Triangular number sequence

The triangular number sequence comprises the natural numbers which can be drawn as dots in a triangular shape.

The first four terms are represented as:

> **Example**
>
> **Problem:**
>
> The triangular number sequence begins 1, 3, 6, 10, 15, ...
>
> Find a formula for the nth term.
>
> **Solution:**
>
> To help understand the formula we can re-draw the triangular numbers as follows:
>
> $n =$ 1   2   3   4   5

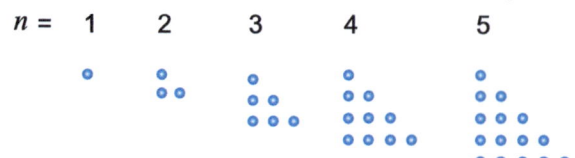

TOPIC 7. SEQUENCES AND SERIES

Now we can add dots to create rectangles:

$n = 1 \quad 2 \quad 3 \quad 4 \quad 5$

The area of the rectangles is given by $n(n+1)$.

However, we want the area of triangles: $\frac{n(n+1)}{2}$.

So the general term for the triangular numbers is given by: $u_n = \frac{n(n+1)}{2}$.

Types of sequences

Go online

Uniform sequence

Example: 3, 3, 3, 3, 3, 3, ...

Alternating sequence

Example: 2, -2, 2, -2, 2, -2, ...

Arithmetic sequence

Example: 2, 5, 8, 11, 14, ...

Geometric sequence

Example: 3, 6, 12, 24, 48, ...

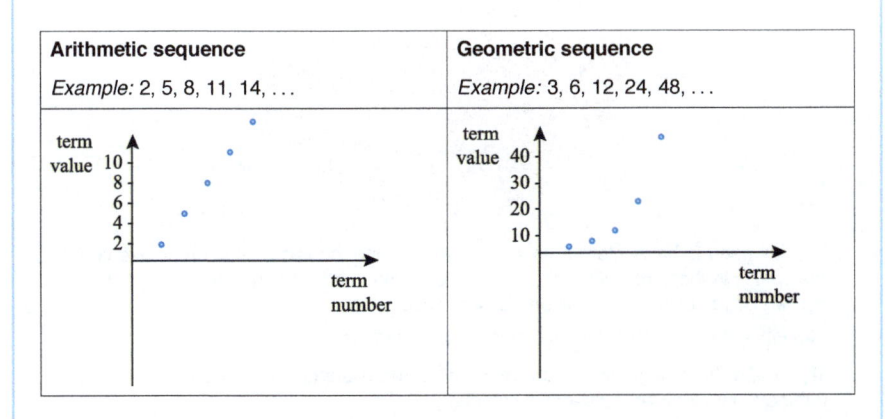

© Heriot-Watt University

Fibonacci sequence

Example: 1, 1, 2, 3, 5, 8, 13, ...

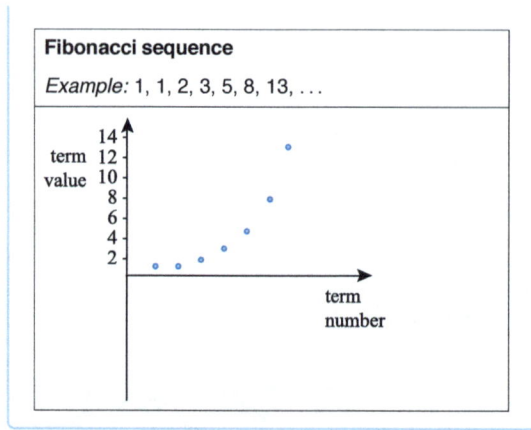

Sequence activities

Here are a variety of sequence activities.

1. Take the image of a sunflower which follows and identify the Fibonacci sequence in it using this rectangular construction.

   You will need to have internet access in order to find the information on how to count the spirals in the sunflower (use the url below). No matter which way you count them, you will get a number in the Fibonacci sequence.

   http://momath.org/home/fibonacci-numbers-of-sunflower-seed-spirals/

2. Try to find three sequences with geometrical representations such as the triangular numbers (for example, pentagonal numbers).

3. Draw on squared paper, the Fibonacci square representation of the first 12 terms.

TOPIC 7. SEQUENCES AND SERIES

Fibonacci and other sequences exercise   Go online

Q35: Write the next three numbers in this linear alternating sequence: $2, -2, 2, -2, \ldots$

Q36: Write the next three numbers in this linear alternating sequence: $-2, 4, -6, 8, -10, \ldots$

Q37: Write the next three numbers in this Fibonacci sequence: $3, 4, 7, 11, 18, \ldots$

Q38: Write the next three numbers in this Fibonacci sequence: $-2, -5, -7, -12, \ldots$

7.5 Convergence and limits

Convergence

We have already met both **finite** and **infinite** sequences.

For finite sequences the last term is explicitly stated. For an infinite sequence $\{u_n\}$, there is no last term but it may be possible to say something about the behaviour of the sequence for large values of n.

Consider the sequence $\{\frac{1}{n}\}$ with terms $\frac{1}{1}, \frac{1}{2}, \frac{1}{3}, \frac{1}{4}, \frac{1}{5}, \ldots$

The terms in this sequence become smaller and smaller.

This can be clearly seen by plotting some of the terms on a graph.

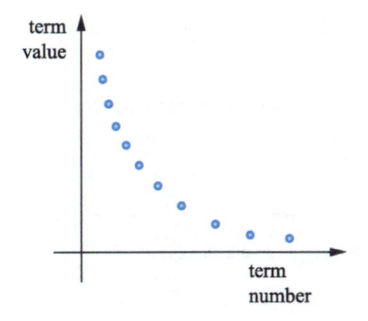

From this evidence it appears that the sequence tends towards the value 0 for large values of n.

In a similar fashion, it appears that the sequence $\{1 + \frac{1}{n}\}$ with terms $\frac{2}{1}, \frac{3}{2}, \frac{4}{3}, \frac{5}{4}, \frac{6}{5}, \ldots$ approaches the value 1 for large n.

© HERIOT-WATT UNIVERSITY

208 TOPIC 7. SEQUENCES AND SERIES

This intuitive approach suggests a mathematical rule for determining how sequences behave.

An infinite sequence $\{u_n\}$ tends to the limit k if for all positive numbers p there is an integer N such that $k - p < u_n < k + p$ for all $n > N$.

If the condition is satisfied then $\lim_{n \to \infty} u_n = k$ (The limit of the sequence $\{u_n\}$ as n tends to infinity is k).

Graphically this means that if the terms of the sequence are plotted on a graph then at some value along the x-axis all the subsequent terms of the sequence (to the right) lie between horizontal lines drawn at $x = k - p$ and $x = k + p$.

This is what is meant by $k - p < u_n < k + p$, for a value of p. i.e. the nth term in the sequence lies between $k - p$ and $k + p$.

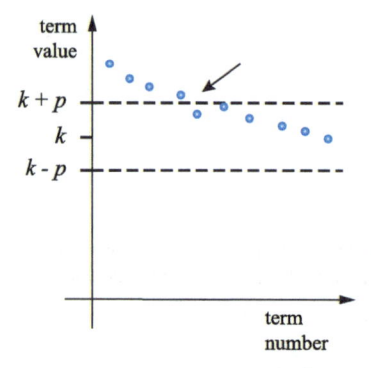

In the above example, after the 4th term (4th blue dot) all terms in the sequence lie between the lines $x = k - p$ and $x = k + p$. Looking at the rule defining the behaviour of sequences it follows that $N = 4$, since for $n = 5$ and above the difference between the term value and the limit is less than p ($|u_n - k| < p$).

Mathematically we can write this as $|u_n - k| < p$. This means the distance between the term number and the limit is between $-p$ and p.

By letting p become smaller and smaller it is apparent that the limiting value of the sequence must be k.

A sequence which has such a limit is called convergent.

> **Key point**
>
> An infinite sequence $\{u_n\}$ for which $\lim_{n \to \infty} u_n = k$ is called a convergent sequence with limit k.

Example

Problem:

Take the sequence $u_n = \{\frac{1}{n}\}$ and confirm that it is convergent with limit 0.

© HERIOT-WATT UNIVERSITY

TOPIC 7. SEQUENCES AND SERIES

Solution:

For the infinite sequence $\left\{\frac{1}{n}\right\}$ to tend to the limit 0 there must exist for all positive numbers p an integer N such that $|u_n| < p$ for all $n > N$.

First consider some example for fixed p such that $|u_n| < p$.

Suppose $p = \frac{2}{11}$, then for each term from u_6 onwards $|u_n| < p$ (e.g. $u_6 = \frac{1}{6} < \frac{2}{11}$).

In this case $N = 5$, so there exists a p such that $|u_n| < p$ when $n > 5$.

In other words, the absolute value of the sequence u_n is less than this fixed value for p which was $\frac{2}{11}$.

Similarly, if $p = \frac{2}{97}$ then when $N = 48$, u_{49} and all subsequent values in the sequence satisfy $|u_n| < p$ (e.g. $u_{49} = \frac{1}{49} < \frac{2}{97}$).

Algebraically, $\frac{1}{n} < p \Rightarrow \frac{1}{p} < n$ so for any p if N is chosen as the first integer $\geqslant \frac{1}{p}$ the convergence condition holds and hence the sequence converges to 0.

Algebraically, the sequence term $\frac{1}{n} < p \Rightarrow \frac{1}{p} < n$. Remember that for the limit to exist we want p to be as small as possible. When we evaluate $\frac{1}{p}$ then this means that n, the nth term and the value of n will be very large tending towards infinity. This means that the sequence term $\frac{1}{n}$ is heading towards zero. So if for any value of p we can choose an N as the first integer $\geqslant \frac{1}{p}$ (very large number heading towards infinity) the convergence conditions holds and hence the sequence, $\left\{\frac{1}{n}\right\}$, converges to 0.

Null sequence

Sequences which converge to the value 0 are given a special name.

> **Key point**
>
> A convergent sequence which converges to the limit 0 is called a null sequence.

Example

Problem:

Are the following null sequences?

a) $\left\{\frac{1}{n^3}\right\}$

b) $\left\{3 + \frac{1}{n^3}\right\}$

c) $\{2n\}$

Solution:

a) This is a null sequence.

b) This is not a null sequence because $u_n > 3$ for all n.

c) This is not a null sequence because $2n > 1$ for all n.

© HERIOT-WATT UNIVERSITY

Rules for convergent sequences

There are a variety of rules which help to determine limits on more complicated sequences. Here are some of the more common ones.

If $\lim_{n \to \infty} a_n = k$ and $\lim_{n \to \infty} b_n = m$ then the following can be applied.

- **The sum rule for convergent sequences**
$$\lim_{n \to \infty} (a_n + b_n) = k + m$$
- **The multiple rule for convergent sequences**
$$\lim_{n \to \infty} (\lambda a_n) = \lambda k \quad \text{for} \quad \lambda \in \mathbb{R}$$
- **The product rule for convergent sequences**
$$\lim_{n \to \infty} (a_n b_n) = km$$
- **The quotient rule for sequences**
$$\lim_{n \to \infty} \left(\frac{a_n}{b_n} \right) = \frac{k}{m} \quad \text{provided that} \quad m \neq 0$$

Applying rules for convergent sequences

Determine whether the sequence $\left\{ 1 + \frac{1}{n} \right\}$ converges.

Use $\lim_{n \to \infty} (a_n + b_n) = k + m$ to evaluate $\lim_{n \to \infty} \left\{ 1 + \frac{1}{n} \right\}$.

So

$\lim_{n \to \infty} \left\{ 1 + \frac{1}{n} \right\} = \lim_{n \to \infty} \{1\} + \lim_{n \to \infty} \left\{ \frac{1}{n} \right\}$

$\lim_{n \to \infty} \{1\} = 1$, since this is a uniform sequence

$\lim_{n \to \infty} \left\{ \frac{1}{n} \right\} = 0$, since this is a null sequence

So

$$\lim_{n \to \infty} \left\{ 1 + \frac{1}{n} \right\} = 1 + 0 = 1$$

Examples

1. The sequence $\{2^n\}$ is not convergent. For any given p and k there is no term in the sequence which can be chosen such that every subsequent term lies between $k + p$ and $k - p$.

However, for any chosen value m there is a point in the sequence where all subsequent terms are larger than m. The sequence is **unbounded**.

The sequence $\{2^n\}$ is an example of a sequence which tends to infinity.

....

TOPIC 7. SEQUENCES AND SERIES

2. The alternating sequence $\{(-1)^n\}$ with terms -1, 1, -1, 1, -1, ... is not convergent. For $p = \frac{1}{2}$ and any k there is no term in the sequence which can be chosen such that every subsequent term lies between $k + p$ and $k - p$.

However, unlike the previous example $\{2^n\}$ this sequence is not unbounded since all terms lie between -1 and +1.

Complicated sequences

Example

Consider the sequence $\left\{\frac{(2n-1)(3n+3)}{4n(n+2)}\right\}$.

This seems a very complicated sequence. The limit, if it exists, is not clear from the expression as it stands. However with some algebraic manipulation the sequence can be transformed into a state in which it is easier to determine if it is convergent.

In this case the following steps make the problem much easier to address.

Multiply out the brackets $\left\{\frac{(2n-1)(3n+3)}{4n(n+2)}\right\} = \left\{\frac{6n^2+3n-3}{4n^2+8n}\right\}$.

The dominant term is n^2 so divide top and bottom by this.

The new expression is $\left\{\frac{6+\frac{3}{n}-\frac{3}{n^2}}{4+\frac{8}{n}}\right\}$.

Now the rules can be used. The terms $\frac{3}{n}$, $\frac{8}{n}$ and $\frac{3}{n^2}$ all generate null sequences and the sequence $\left\{\frac{(2n-1)(3n+3)}{4n(n+2)}\right\}$ has a limit of $\frac{6}{4}$.

Finding the limit of complicated function   Go online

Strategy for finding the limit of a complicated quotient:

1. Simplify by removing any brackets.
2. Identify the dominant term.
3. Divide top and bottom by this dominant term.
4. Use the rules to determine the limit, if it exists.

© HERIOT-WATT UNIVERSITY

> **Convergence and limits exercise**   Go online
>
> Using a graphics calculator explore whether the following sequences have a limit.
>
> (The maths is fun website has a graphing tool that will allow you to explore these graphs.)
>
> http://www.mathsisfun.com/data/function-grapher.php
>
> **Q39:** $\left\{\frac{1}{(3n-1)}\right\}$
>
> **Q40:** $\left\{\frac{n^2}{(3n-1)}\right\}$
>
> **Q41:** $\left\{10 - \frac{1}{n}\right\}$
>
> **Q42:** $\left\{\frac{(-1)^n}{n}\right\}$

7.6 Definitions of e and π as limits of sequences

When thinking of π, we immediately think of calculating the area and circumference of circles. Because of this link with circles π is also found in geometry, trigonometry, complex numbers as well as mechanics, statistics and many more areas of study. It describes the approximate period of a pendulum, it appears in the formula for the normal distribution (bell curve) in statistics, calculating the volume of a sphere and cylinders, tracking the path of an elliptical orbit and much more.

Similarly, the number e appears in many areas of study. For example, it appears in the study of compound interest, natural growth and decay, gambling, particularly Bernoulli trials which is linked to the Binomial Theorem. It is also known as Napier or Euler's constant.

Both of these numbers have such wide applications and often require accuracy to a number of decimal places. Since both of these numbers are irrational, there needs to be a way of calculating the value of these numbers. This section will look at how these numbers can be calculated by creating a sequence, which when the limit is taken will tend towards e and π.

TOPIC 7. SEQUENCES AND SERIES

7.6.1 The definition of e as a limit of a sequence

Consider the sequence $\left\{\left(1+\frac{1}{n}\right)^n\right\}$.

The definition of e as a limit of a sequence

Q43: Using a calculator find and plot the values of the first twenty terms on a graph using the x-axis for the term number and the y-axis for these term values.

Suggest a limit for this sequence.

Give your answers to 4 decimal places.

n	$\left\{\left(1+\frac{1}{n}\right)^n\right\}$	n	$\left\{\left(1+\frac{1}{n}\right)^n\right\}$
1		11	
2		12	
3		13	
4		14	
5		15	
6		16	
7		17	
8		18	
9		19	
10		20	

....

Q44:

Investigate the values of the terms in this sequence with a graphics calculator.

Examine a selection of the terms up to the number 10000 and hence suggest a limit for this sequence.

Write you answers to 4 decimal places.

n	1000	2000	3000	4000	5000	6000	7000	8000	9000	10000
$\left\{\left(1+\frac{1}{n}\right)^n\right\}$										

....

Q45: What is the value of e?

....

Q46: Compare the last three answers and make a statement connecting e and the sequence $\left\{\left(1+\frac{1}{n}\right)^n\right\}$.

The results of the questions are important and worth stating again.

© HERIOT-WATT UNIVERSITY

The sequence $\left\{\left(1+\frac{1}{n}\right)^n\right\}$ converges with a limit equal to e.
That is:

Key point

$$\lim_{n\to\infty}\left(1+\frac{1}{n}\right)^n = e$$

7.6.2 The definition of π as a limit of a sequence

π can be defined as the limit of a sequence. This can be done by trying to calculate the perimeter of polygons using simple trigonometry.

Examples

1. Deriving the formula for the perimeter

Take a circle of unit radius and circumscribe a hexagon on it.

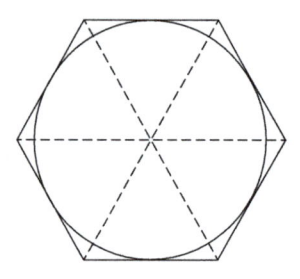

By elementary trigonometry in degrees it can be shown that the length of each side of the hexagon is given by the formula $2\tan\left(\frac{180}{6}\right)$.

The perimeter of the hexagon is therefore $6 \times 2\tan\left(\frac{180}{6}\right)$.

Considering one of the six triangles we have:

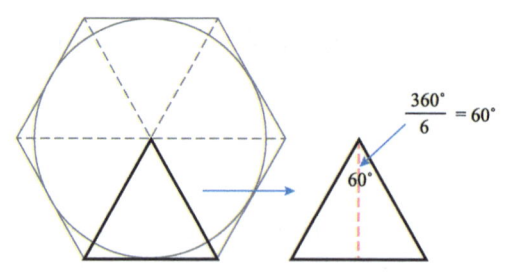

$$\frac{360°}{6} = 60°$$

Now we cut this triangle into a right angled triangle so that we can use right angled trigonometry.

TOPIC 7. SEQUENCES AND SERIES

Unit circle has radius 1

$$\frac{360°}{12} = \frac{180°}{6} = 30°$$

$\tan\left(\frac{180}{6}\right) = \frac{opp}{1} \Rightarrow opp = \tan\left(\frac{180}{6}\right)$

So the length of one side of the triangle is $2\tan\left(\frac{180}{6}\right)$.

So the perimeter of the hexagon is $6 \times 2\tan\left(\frac{180}{6}\right)$.

....

2. Problem:

Suggest a formula for the perimeter of a polygon with twelve sides. From this deduce a formula for the perimeter of a polygon with n sides.

Solution:

A reasonable formula for the perimeter of a polygon with twelve sides is $12 \times 2\tan\left(\frac{180}{12}\right)$ and for a polygon with n sides is $n \times 2\tan\left(\frac{180}{n}\right)$.

Thus the perimeters of polygons around a unit circle generate a sequence $n \times 2\tan\left(\frac{180}{n}\right)$ for $n > 3$.

The definition of π as a limit of a sequence exercise   Go online

Q47: Using a graphics calculator take this function and investigate the sequence of perimeters of polygons around a unit circle for increasing values of n.

Explore the values and suggest a limit.

Write you answers to 4 decimal places

(The maths is fun website has a graphing tool that will allow you to explore these graphs.)

http://www.mathsisfun.com/data/function-grapher.php

Now consider geometrically the effect of circumscribing a polygon around a unit circle. As the polygon has more and more sides the polygon itself tends closer and closer to the shape of the circle. In other words the limit of the perimeter of polygons circumscribed around a circle is the circumference of the circle.

Q48: Suggest a value for the circumference of the unit circle from the work done here on polygons and hence give a value for π.

....

Q49: If $\{a_n\}$ is the sequence of perimeter values of polygons around a unit circle, suggest a definition for π based on this sequence.

This is another result worth stating as π can be defined as a limit of a sequence.

> **Key point**
>
> $\pi = \lim\limits_{n \to \infty} \left[n \tan\left(\frac{180}{n}\right) \right]$

7.7 Series and sums

The link between series and sequences is a simple one.

> **Key point**
>
> A series is the sum of the terms in an infinite sequence.

Consider the sequence $u_1, u_2, u_3, u_4, \ldots$

Let

$u_1 + u_2 = S_2$
$u_1 + u_2 + u_3 = S_3$
and so on.

In this way, $u_1 + u_2 + u_3 + \ldots + u_n = S_n$.

Such a sum is known as a **partial sum** or the **sum to n terms** of a sequence and is denoted by S_n.

It is customary to use the sigma sign $\sum$ when describing a series. It gives a useful, accurate shorthand way of writing down a series without having to specify each term.

Here are some examples:

- $\sum\limits_{r=1}^{4} (r\{r+2\}) = 3 + 8 + 15 + 24 = 50$ this is a partial sum

- $\sum\limits_{n=2}^{5} \left(\frac{1}{n}\right) = \frac{1}{2} + \frac{1}{3} + \frac{1}{4} + \frac{1}{5} = \frac{77}{60}$ this is a partial sum

- $\sum\limits_{r=1}^{\infty} \left(2 + \frac{2}{r}\right)$ this is a sum to infinity

The upper and lower limits on the sigma sign indicate the range of values which should be used to construct the series.

The partial sum is the sum of the terms from 1 to n where $n \in \mathbb{N}$. It is denoted by S_n and represented as $S_n = \sum\limits_{r=1}^{n} u_r$.

In sigma notation the series $\sum\limits_{r=1}^{\infty} u_r$ is the sum to infinity of the terms of the sequence $\{u_r\}$. It is denoted by S_∞. This concept of sums to infinity for geometric series will be explained later in a separate subsection.

Example

Problem:

Find S_4 for $\sum\limits_{r=1}^{\infty} (2 + r)$.

Solution:

S_4 means the partial sum to four terms. So r = 1, 2, 3 and 4 .
This gives (2 + 1) + (2 + 2) + (2 + 3) + (2 + 4) = 18.

There is another aid to finding the sum to n terms. The two combination rules which follow apply to sums to infinity provided that the series a_n and b_n are convergent. The actual sums to infinity will be explored in one of the following subsections but at present the rules can be adapted and used for sums to n terms.

Key point

If a_n and b_n form convergent series then:

- **The sum rule for convergent series**

$$\sum_{n=1}^{\infty} (a_n + b_n) = \sum_{n=1}^{\infty} a_n + \sum_{n=1}^{\infty} b_n$$

- **The multiple rule for convergent series**

$$\sum_{n=1}^{\infty} \lambda a_n = \lambda \sum_{n=1}^{\infty} a_n \text{ for } \lambda \in \mathbb{N}$$

Examples

1.

Problem:

Find $\sum\limits_{n=1}^{6} (2n + 3)$.

Solution:

$$\sum_{n=1}^{6} (2n + 3) = \sum_{n=1}^{6} (2n) + \sum_{n=1}^{6} (3)$$
$$= 2 \sum_{n=1}^{6} n + 3 \times 6$$
$$= 2 (1 + 2 + 3 + 4 + 5 + 6) + 18$$
$$= 42 + 18 = 60$$

2. Problem:

Find the value of $\sum\limits_{n=1}^{\infty} \left(\frac{-3}{n^2 + 5n + 4} \right)$

Solution:

This is more difficult since this is a sum to infinity. One way to tackle this is to write it as partial fractions and then see if any terms cancel each other out.

Using partial fractions:
$$\frac{-3}{n^2 + 5n + 4} = \frac{-3}{(n+4)(n+1)}$$
$$= \frac{A}{(n+4)} + \frac{B}{(n+1)}$$
$$-3 = A(n+1) + B(n+4)$$

Substituting $n = -1$: $B = -1$

$$\frac{-3}{n^2 + 5n + 4} = \frac{1}{(n+4)} - \frac{1}{(n+1)}$$

$$\sum_{n=1}^{\infty} \left(\frac{-3}{n^2 + 5n + 4}\right) = \sum_{n=1}^{\infty} \left(\frac{1}{(n+4)} - \frac{1}{(n+1)}\right)$$

Now expand out:

$$= \left(\frac{1}{5} + \frac{1}{6} + \frac{1}{7} + \frac{1}{8} + \cdots\right) - \left(\frac{1}{2} + \frac{1}{3} + \frac{1}{4} + \frac{1}{5} + \frac{1}{6} + \frac{1}{7} + \cdots\right)$$

$$= -\left(\frac{1}{2} + \frac{1}{3} + \frac{1}{4}\right)$$

$$= \frac{13}{12}$$

Series and sums exercise　　　　　　　　　　　　　　　　　　　　　　　Go online

Q50: Find $\sum_{n=1}^{4} \left(\frac{1}{2}n + 3n^2\right)$ using the combination rules.

....

Q51: Find the value of $\sum_{n=1}^{\infty} \left(\frac{-2}{n^2 + 8n + 15}\right)$.

....

Q52: Find $\sum_{n=1}^{5} (3 - n)$ using the combination rules.

....

Q53: Find $\sum_{n=1}^{4} \left(2n^2 - 4n + 1\right)$ using the combination rules.

....

Q54: Find $\sum_{n=1}^{\infty} \left(\frac{-2}{n^2 + 2n}\right)$ using the combination rules.

....

Q55: Find $\sum_{n=1}^{\infty} \left(\frac{-3}{n^2 + 5n + 4}\right)$ using the combination rules.

7.8 Arithmetic and geometric series

When Gauss was a young boy in school it was said that he was asked to find the sum of all the numbers from 1 to 20. The teacher expected this to take his students a while, but Gauss came up with the answer almost immediately. What Gauss did was to realise that 1 + 20 is 21 and 2 +19 is 21 and so on. Counting the number of pairs like this and finding the total he gave the answer 210. What Gauss did here was to find the sum of an arithmetic sequence. In the following sections we will look at this technique as well as others to find formulae for arithmetic and geometric series.

7.8.1 Arithmetic series

Recall that an **arithmetic sequence** is one which has the form $a, a + d, a + 2d, \ldots$ where a is the first term in the sequence and d is the **common difference**.
The nth term of an arithmetic sequence was also defined earlier.
It is $u_n = a + (n - 1)d$.

With this information it is possible to find a formula which can be used to provide the sum to n terms of an arithmetic series.

Arithmetic series

> **Key point**
>
> An arithmetic series (or arithmetic progression) is the sum of the terms of an arithmetic sequence.

Given the arithmetic sequence: $a, a+d, a+2d, a+3d, \ldots \ldots$ and the formula for the nth term $u_n = a + (n-1)d$, where d is the **common difference**, we can generate a **formula** for an arithmetic series as follows.

$$S_n = \sum_{n=1}^{\infty} (a + (n-1)d)$$
$$= a + (a+d) + (a+2d) + \ldots\ldots + (a + (n-2)d) + (a + (n-1)d)$$

We can reverse this:

$$S_n = (a + (n-1)d) + (a + (n-2)d) + \ldots\ldots + (a+2d) + (a+d) + a$$

Now add the two expressions together:

$$2S_n = (2a + (n-1)d) + (2a + (n-1)d) + \ldots + (2a + (n-1)d) + (2a + (n-1)d)$$
$$2S_n = n(2a + (n-1)d)$$

The sum to n terms for an arithmetic series (n^{th} partial sum) is given by:

> **Key point**
>
> $$S_n = \tfrac{n}{2}(2a + (n-1)d)$$
>
> - a is the first **term**
> - d is the **common difference**
> - $n \in \mathbb{N}$

Examples

1. Problem:
Find the sum of the arithmetic series $1 + 3 + 5 + 7 + \ldots + 37$.
Solution:
Let the nth term be equal to 37 so $a + (n - 1)d = 37$.
But $d = 2$ and $a = 1$ so $1 + 2(n - 1) = 37 \Rightarrow n = 19$.
$S_{19} = \frac{19}{2}(2 + 18 \times 2) = 361$

....

2. Problem:
If the third term of an arithmetic sequence is 11 and $S_6 = 78$, find S_{22}.
Solution:
From the first piece of information $u_3 = 11$.
So using the formula $u_n = a + (n - 1)d$ gives the equation $11 = a + 2d$.
The second piece of information $S_6 = 78$ gives another equation in a and d, namely,
$78 = 3(2a + (5)d) = 6a + 15d$.
These two equations, $11 = a + 2d$ and $78 = 6a + 15d$ can be solved simultaneously to give $d = 4$ and $a = 3$.
Using the sum formula $S_n = \frac{n}{2}(2a + (n-1)d)$ with these values for S_{22} gives:
$S_{22} = \frac{22}{2}(2 \times 3 + (22 - 1)4) = 990$

The sum of the first n integers   Go online

The values for the sum of the n integers between 1 and 10 are:

n	S_n
1	1
2	3
3	6
4	10
5	15
6	21
7	28
8	36
9	45
10	55

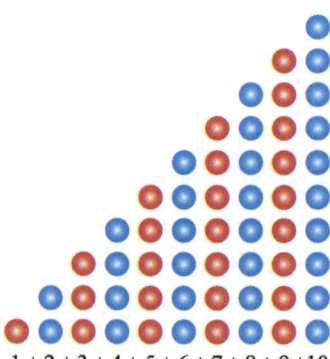

$1 + 2 + 3 + 4 + 5 + 6 + 7 + 8 + 9 + 10$

Can you find an expression for the sum of the first n integers?

TOPIC 7. SEQUENCES AND SERIES

Arithmetic series exercise Go online

Q56: Find the sum of the arithmetic series 12 + 19 + 26 + ... + 355.

....

Q57: Find the sum of the arithmetic series 0·01 + 0·32 + 0·63 + ... + 2·8.

....

Q58: Find the sum of the arithmetic series 1 + 2 + 3 + 4 + ...

....

Q59: The sum of the first 60 terms of an arithmetic series is 5190. The common difference is 3. What is the first term?

....

Q60: The first term of an arithmetic progression is 5 and the last term is 45. The sum of the terms is 525. How many terms are in the progression?

....

Q61: Given that $u_{17} = 20$ and $S_{12} = 114$, calculate S_{18}.

Infinite series

The infinite series 1 + 2 + 3 + 4 + 5 + ... is an arithmetic series of particular interest and use.

This series has $a = 1$ and $d = 1$.

Thus

$$S_n = \frac{n}{2}[2a + (n-1)d]$$

$$S_n = \frac{n}{2}[2 + n - 1] = \frac{n(n+1)}{2}$$

Key point

This is another important result:

$$\sum_{r=1}^{n} r = \tfrac{1}{2}n(n+1)$$

We can use this new found formula to derive formulae for $\sum_{r=1}^{n} r^2$ and $\sum_{r=1}^{n} r^3$.

A formula for $\sum_{r=1}^{n} r^2 = 1^2 + 2^2 + 3^2 + \ldots + n^2$

We start with the set up which will allow us to have a term of the form $\sum_{r=1}^{n} r^2$ on the LHS.

TOPIC 7. SEQUENCES AND SERIES

By substituting in numbers write the terms for each summation:

$$\sum_{r=1}^{n}(1+r)^3 - \sum_{r=1}^{n}r^3 = \left\{2^3 + 3^3 + 4^3 + 5^3 + \ldots + n^3 + (1+n)^3\right\}$$
$$- \left\{1^3 + 2^3 + 3^3 + 4^3 + 5^3 + \ldots + n^3\right\}$$

Expand the LHS and simplify the RHS: $\sum_{r=1}^{n}1 + \sum_{r=1}^{n}3r + \sum_{r=1}^{n}3r^2 + \sum_{r=1}^{n}r^3 - \sum_{r=1}^{n}r^3 = (1+n)^3 - 1^3$

Evaluate the summations we know: $\sum_{r=1}^{n}r = \frac{1}{2}n(n+1)$

$n + 3 \times \frac{1}{2}n(n+1) + \sum_{r=1}^{n}3r^2 = 1 + 3n + 3n^2 + n^3 - 1$

Multiply by 2 to get rid of the fraction: $2n + 3n(n+1) + 2\sum_{r=1}^{n}3r^2 = 2(3n + 3n^2 + n^3)$

Rearrange so that summation of r^2 is on the LHS and everything else in on the RHS:
$6\sum_{r=1}^{n}r^2 = 2(3n + 3n^2 + n^3) - 2n - 3n(n+1)$

Take out factor of n

$6\sum_{r=1}^{n}r^2 = n(6 + 6n + 2n^2 - 2 - 3n - 3)$

$6\sum_{r=1}^{n}r^2 = n(2n^2 + 3n + 1)$

Factorise: $\sum_{r=1}^{n}r^2 = \frac{n(n+1)(2n+1)}{6}$

$\sum_{r=1}^{n}r^3$ can be evaluated in a similar way, where $\sum_{r=1}^{n}r^3 = \frac{n^2(n+1)^2}{4}$.

> **Key point**
>
> Important results:
>
> $$\sum_{r=1}^{n}r^2 = \frac{n(n+1)(2n+1)}{6}$$
>
> $$\sum_{r=1}^{n}r^3 = \frac{n^2(n+1)^2}{4}$$

Examples

1. Problem:

Evaluate $\sum_{n=1}^{9}(3n+2)$

TOPIC 7. SEQUENCES AND SERIES

Solution:
By the combination rules:
$$\sum_{n=1}^{9}(3n+2) = \sum_{n=1}^{9}(3n) + \sum_{n=1}^{9}(2)$$
$$= 3\sum_{n=1}^{9} n + \sum_{n=1}^{9}(2)$$
$$= \frac{3 \times 9(10)}{2} + (2 \times 9)$$
$$= \frac{3(90)}{2} + 18$$
$$= 153$$

....

2. Problem:
Evaluate $\sum_{r=2}^{5}\left(2r^2 - r^3\right)$

Solution:
$$\sum_{r=2}^{5}\left(2r^2 - r^3\right) - \sum_{r=1}^{1}\left(2r^2 - r^3\right)$$
$$= 2\sum_{r=1}^{5} r^2 - \sum_{r=1}^{5} r^3 - \left(2\sum_{r=1}^{1} r^2 - \sum_{r=1}^{1} r^3\right)$$

Notice the lower and upper limits. To find $\sum_{r=2}^{5}$ we need to subtract off $\sum_{r=1}^{1}$. We do not subtract off $\sum_{r=1}^{2}$ since we are including the second term in the summation.

$$= 2\frac{5(6)(11)}{6} - \frac{5^2(6)^2}{4} - \left(2(1)^2 - (1)^3\right)$$
$$= 110 - 225 - 1$$
$$= -116$$

....

3. Problem:
Express the arithmetic progression $10 + 13 + 16 + 19 + \cdots + 31$ in sigma notation.

Solution:
We know that $u_n = a + (n-1)d$. Where $a = 10$, $d = 3$ and the last term is 31.
$31 = 10 + (n-1)3 \quad \Rightarrow \quad n = 8$
In summation notation:
$$\sum_{r=1}^{n} a + (r-1)d = \sum_{r=1}^{8} 10 + (r-1)3$$
$$= \sum_{r=1}^{8} 7 + 3r$$

....

© HERIOT-WATT UNIVERSITY

4. Problem:

Show that $\sum_{r=1}^{n}(4-5r) = \frac{3n-5n^2}{6}$

Solution:

$$\sum_{r=1}^{n}(4-5r) = \sum_{r=1}^{n}4 - \sum_{r=1}^{n}5r$$
$$= \sum_{r=1}^{n}4 - 5\sum_{r=1}^{n}r$$
$$= 4n - 5 \times \frac{n(n+1)}{2}$$
$$= \frac{8n - 5n(n+1)}{2}$$
$$= \frac{8n - 5n^2 - 5n}{2}$$
$$= \frac{3n - 5n^2}{6}$$

Infinite series exercise

Go online

Q62: Calculate $\sum_{n=1}^{6}(2n-4)$ using the techniques of the previous example.

....

Q63: Calculate $\sum_{n=1}^{7}\left(3n - \frac{1}{2}\right)$.

....

Q64: Evaluate $\sum_{n=1}^{5}\left(5n^2 - 6n\right)$.

....

Q65: Evaluate $\sum_{n=4}^{8}(3-2n)$.

....

Q66: Evaluate $\sum_{n=3}^{5}(4n-1)$.

....

Q67: Express the arithmetic progression $14 + 18 + 22 + 26 + 30 + \ldots + 90$ in sigma notation.

....

Q68: Express the arithmetic progression $30 + 24 + 18 + 12 + 6 + \ldots + (-78)$ in sigma notation.

....

Q69: Show that $\sum_{r=1}^{n}(3r+2) = \frac{3}{2}\left(\frac{n}{2}\right)(7n+3)$.

....

Q70: Show that $\sum_{r=4}^{n}(6-8r) = -4n^2 + 2n - 30$.

7.8.2 Geometric series

You may recall that a **geometric sequence** is one which has the form $a, ar, ar^2, ar^3, ar^4, \ldots$, where a is the first term in the sequence and r is the **common ratio**.

The nth term of a geometric sequence is $u_n = ar^{n-1}$.

Using this information it is possible to find a formula which can be used to provide the sum to n terms of a geometric series.

Geometric series

A geometric series (or progression) is the sum of the terms of a geometric sequence.

Given the geometric sequence: $a, ar, ar^2, ar^3, ar^4, \ldots$ and the formula for the nth term $u_n = ar^{n-1}$, where r is the common ratio, defined by $r = \frac{u_{n+1}}{u_n}$, we can generate a **formula** for the geometric series as follows:

$$S_n = \sum_{k=1}^{n} ar^{k-1}$$
$$= a + ar + ar^2 + \ldots + ar^{n-2} + ar^{n-1}$$

Now we multiply this by r: $rS_n = ar + ar^2 + \ldots + ar^{n-2} + ar^{n-1} + ar^n$

Now we subtract: $S_n - rS_n$

$S_n - rS_n = a - ar^n$
$(1-r)S_n = a(1-r^n)$

The sum to n terms for a geometric series (n^{th} partial sum) is given by:

$S_n = \frac{a(1-r^n)}{1-r}, r \neq 1$

- a is the first **term**
- r is the **common difference**
- $n \in \mathbb{N}$

Key point

This may be written in two different ways depending on the value of r:

- $r < 1$ then $S_n = \frac{a(1-r^n)}{1-r}$
- $r > 1$ then $S_n = \frac{a(r^n-1)}{r-1}$

The only reason for this is that it prevents a negative number on the numerator and denominator. This would not affect the overall answer as these would cancel with each other anyway.

Either formula can be used regardless of the size of r.

> **Key point**
>
> Note that if $r = 1$ then:
> $$S_n = a + a + a + \ldots + a + a$$
> $$= an$$

As an aside, at this point it is worth summing up the possibilities which can arise with a general geometric series dependant on the values of r. The range $-1 < r < 1$ will be discussed in the next section.

Behaviour of a general geometric series for values of r, where $r \geq 1$ and $r \leq -1$

$r > 1$	$S_n \to \pm \infty$ as $n \to \infty$
$r < -1$	S_n alternates between +ve and -ve values, $|S_n| \to \infty$
$r = 1$	$S_n = na$
$r = -1$	$S_n = 0$ for even n and $S_n = a$ for odd n

Example Problem:

Find S_9 of the geometric series $1 + 3 + 9 + 27 + 81 + \ldots$.

Solution:

Using the formula $\frac{a(1-r^n)}{1-r}$

We have:

$a = 1$

$r = \dfrac{u_{n+1}}{u_n} \quad \Rightarrow \quad r = \dfrac{3}{1} = 3$

$n = 9$

gives:

$S_9 = \dfrac{1(3^9 - 1)}{3 - 1} = 9841$

The techniques of finding a specific term and summing a geometric series can be combined to solve more complex problems where only limited information is available.

Example Problem:

Find S_6 for the geometric series which has $u_2 = 3$ and $u_5 = 1/9$.

Solution:

From the first part of the information $u_2 = 3$.

So using the formula $u_n = a\, r^{n-1}$ gives the equation $3 = ar$.

The second part, $u_5 = 1/9$ gives another equation in a and r, namely, $1/9 = a\, r^4$.

TOPIC 7. SEQUENCES AND SERIES

Thus, $\frac{1/9}{3} = \frac{ar^4}{ar}$ so $\frac{1}{27} = r^3 \Rightarrow r = \frac{1}{3}$

Substitute $r = \frac{1}{3}$ into $ar = 3$: $a \times \frac{1}{3} = 3 \Rightarrow a = 9$

Using the sum formula $S_n = \frac{a(1-r^n)}{1-r}$ with these values and $n = 6$ gives S_6:

$S_6 = \frac{9\left(1-\left(\frac{1}{3}\right)^6\right)}{1-\frac{1}{3}} = \frac{364}{27}$

Geometric series exercise   Go online

Q71: Find the sum of the geometric sequence 4, -8, 16, -32, ..., 1024.

....

Q72: Find the sum of the geometric series $4 - \frac{1}{3} + \frac{1}{36} - \ldots$ to 12 terms.

....

Q73: How many terms must be added for the geometric series 3 + 9 + 27 + ... to equal 9840?

....

Q74: The geometric series has a common ratio of 4. Its sum for the first four terms is -850. Calculate the first term.

....

Q75: Find S_4 for the geometric series which has $u_3 = \frac{3}{2}$ and $u_5 = \frac{3}{8}$, where $r > 0$.

....

Q76: The first two terms of a geometric series add up to 4. The fourth and fifth terms add up to 108. Write down the geometric series using sigma notation.

Achilles and the tortoise   Go online

A Greek philosopher called Zeno put forward a now famous paradox concerning Achilles and the Tortoise. He suggested that a tortoise and Achilles were in a race. Achilles could travel at say, 10 times the rate of the tortoise but the tortoise had a head start of say 300 metres. He argued that when Achilles travelled the 300 metres, the tortoise would have moved ahead by 30 metres; when Achilles travelled the next 30 metres, the tortoise would have travelled 3 metres; when Achilles had travelled the next 3 metres the tortoise would have travelled a further 0·3 metres and so on. Thus Achilles could never catch the tortoise. Discuss and find out the flaw in this argument (which looks like a geometric series).

This video gives a discussion of this paradox: https://www.youtube.com/watch?v=i_1jBGRmS0U

The demonstration of this paradox is available online.

© HERIOT-WATT UNIVERSITY

7.9 Sums to infinity

It is straightforward to find the limit of a convergent sequence. It is also easy to calculate sums to n terms of an infinite series. In some cases however, it would be useful to know whether an infinite series actually has a sum.

Consider the infinite geometric series 27 + 2·7 + 0·27 + 0·027 + 0·0027 + ...

$S_1 = 27$
$S_2 = 29 \cdot 7$
$S_3 = 29 \cdot 97$
and so on.

The values of these partial sums form a sequence 27, 29·7, 29·97, 29·997, 29·9997, ... Clearly the sequence approaches 30.
That is, the sum to infinity is the limit of the sequence of partial sums.
In this case $S_n \to 30$ as $n \to \infty$ (S_n tends towards a limiting value of 30 as n approaches infinity).

If a limit exists, the series is convergent.

A convergent series is one for which the limit of partial sums exists. This limit is called the **sum** and is denoted by S_∞ or $\sum_{n=1}^{\infty} u_n$.

A divergent series is one which is not convergent. e.g. 1 + 2 + 3 + ...

7.9.1 Convergent geometric series

Consider the general infinite geometric series $a + ar + ar^2 + ...$.

The formula for S_n is given by: $S_n = \frac{a(1-r^n)}{1-r}$, and $r^n \to 0$ as $n \to \infty$.

Since r is a number less than 1, when r is raised to a large power it will approach zero.

For example, $\frac{1}{2^n} \to 0$ as $n \to \infty$.

Consider the general infinite geometric series $a + ar + ar^2 + ...$.

Suppose that $-1 < r < 1$ i.e. $|r| < 1$.

The formula for S_n then becomes: $S_\infty = \frac{(a-r^n)}{(1-r)}$

Since $r^n \to 0$ as $n \to \infty$, then: $S_\infty = \frac{a}{1-r}$

Suppose that $-1 < r < 1$ i.e. $|r| < 1$.

Then start with the formula for S_n: $S_n = \frac{a(1-r^n)}{1-r}$

Multiplying out the top line: $S_n = \frac{a-ar^n}{1-r}$

Separating this into two fractions: $S_n = \frac{a}{1-r} - \frac{ar^n}{1-r}$

Now remembering that $|r| < 1$, let $n \to \infty$. As this happens $r^n \to 0$. Now we are left with the formula for the sum to infinity being:

$S_\infty = \frac{a}{1-r}$

This means that a geometric sequence is convergent when $|r| < 1$.

TOPIC 7. SEQUENCES AND SERIES

> **Key point**
>
> For a convergent geometric series $S_\infty = \frac{a}{1-r}$, where $|r| < 1$

To conclude, the sum to infinity of a series only exists if the sequence of partial sums is convergent. In the case of geometric series, this only occurs when $|r| < 1$.

Examples

1. Finding the sum to infinity

Problem:

Find S_∞ of the geometric series 625, 125, 25, 5, 1, ...

Solution:

$a = 625$ and $r = \frac{1}{5}$.

So

$S_\infty = \frac{625}{1 - \frac{1}{5}} = \frac{625 \times 5}{4} = 781.25$

---

2. Problem:

Express the recurring decimal 0·14141... as an infinite geometric series, and hence as a vulgar fraction in its simplest form.

Solution:

We can express 0·14141 ... as 0·14 + 0·0014 + 0·000014 + ...

This is a geometric series where $a = 0 \cdot 14$ and $r = 0 \cdot 01$, since $|r| < 1$ then S_∞ will exist.

Substituting into: $S_\infty = \frac{a}{1-r}$

$S_\infty = \frac{0.14}{1 - 0.01}$

$= \frac{14}{99}$

Convergent geometric series exercise

Go online

Q77: State which of the following geometric series are convergent and for those that are convergent find the sum to infinity.

a) 2 + 4 + 8 + 16 + 32 + ...
b) 32 + 16 + 8 + 4 + 2 + ...
c) 32 - 16 + 8 - 4 + 2 - ...
d) $\frac{1}{3} + \frac{1}{6} + \frac{1}{12} + \frac{1}{24} + \frac{1}{48} + \ldots$
e) $-\frac{1}{3} + 1 - 3 + 9 - 27 + 81 \ldots$

Q78: Find the sum to infinity of the geometric series $1 + \frac{1}{2} + \frac{1}{4} + \frac{1}{8} + \frac{1}{16} + \ldots$

Q79: Find the first term of a geometric series which has a sum to infinity of 9 and a common ratio of $\frac{2}{3}$.

Q80: Find the sixth term of a geometric series which has a common ratio of $\frac{5}{6}$ and a sum to infinity of 72.

Q81: Find the sum to infinity of the geometric series with all terms positive in which $u_2 = 4$ and $u_4 = 1$.

Q82: Express the recurring decimal 0·370370... as an infinite geometric series in its simplest form.

7.9.2 Binomial theorem and the geometric series

Consider the infinite geometric series:

$S_\infty = \sum\limits_{n=1}^{\infty} ar^{n-1}$, where $a = 1$ and $|r| < 1$

This can be expanded to: $S_\infty = 1 + r + r^2 + \ldots$

This will converge and give a limit given by $S_\infty = \frac{1}{1-r}$, which is equivalent to $S_\infty = (1-r)^{-1}$.

The binomial theorem is given by: $(x+y)^n = \sum\limits_{r=0}^{n} \binom{n}{r} x^{n-r} y^r$

This can be expanded to:

$(x+y)^n = \frac{n!}{0!n!}x^n + \frac{n!}{1!(n-1)!}x^{n-1}y + \frac{n!}{2!(n-2)!}x^{n-2}y^2 + \frac{n!}{3!(n-3)!}x^{n-3}y^3 + \ldots$

Which simplifies to:

$(x+y)^n = x^n + nx^{n-1}y + \frac{n(n-1)}{2!}x^{n-2}y^2 + \frac{n(n-1)(n-2)}{3!}x^{n-3}y^3 + \ldots$

Now if we replace $x = 1$, $y = -r$, where $|r| < 1$ and $n = -1$ we have the following:

$(1+(-r))^{-1} = 1 + (-1)(-r) + \frac{(-1)(-2)}{2!}(-r)^2 + \frac{(-1)(-2)(-3)}{3!}(-r)^3 + \ldots$

Which simplifies to: $(1-r)^{-1} = 1 + r + r^2 + r^3 \ldots$

That is: $S_\infty = \frac{1}{1-r} = 1 + r + r^2 + r^3 \ldots$

Which is equivalent to the infinite sum for a geometric series.

TOPIC 7. SEQUENCES AND SERIES

Example

Consider the geometric series $\sum_{n=1}^{\infty} ar^{n-1}$, where $a = 1$ and $r = \frac{1}{2}$.

This series takes the form $1 + \frac{1}{2} + \frac{1}{4} + \frac{1}{8} + \frac{1}{16} + \ldots$.

Using $S_\infty = \frac{1}{1-r}$ gives the sum to infinity for this series as $\frac{1}{1-\frac{1}{2}}$ which of course can be written as $\left(1 - \frac{1}{2}\right)^{-1}$.

Using the binomial theorem gives:

$1 + \left(\frac{-1}{1!}\right)\left(-\frac{1}{2}\right) + \left(\frac{-1 \times -2}{2!}\right)\left(-\frac{1}{2}\right)^2 + \left(\frac{-1 \times -2 \times -3}{3!}\right)\left(-\frac{1}{2}\right)^3 + \ldots$

$= 1 + \frac{1}{2} + \frac{1}{4} + \frac{1}{8} + \frac{1}{16} + \ldots$ as required.

The binomial theorem can express $(1-r)^{-1}$ as an infinite series:

$(1-r)^{-1} = 1 + r + r^2 + r^3 \ldots$

However, this only works for expressions in the form $(1+x)^n$, where $n \in \mathbb{R}$.

The technique can be adapted though to cope with expressions of the form $(a+x)^n$.

These will contain algebraic terms as well as numbers.

To transform $(a+x)^n$ into $(1+x)^n$, where $|r| < 1$ we have two scenarios:

1. **Take out a factor of a^n**

$$(a+x)^n = \left(a\left(1 + \frac{x}{a}\right)\right)^n$$
$$= a^n \left(1 + \frac{x}{a}\right)^n$$
$$= a^n \left(1 + (-1)\frac{x}{a} + \frac{(-1)(-2)}{2!}\left(\frac{x}{a}\right)^2 + \frac{(-1)(-2)(-3)}{3!}\left(\frac{x}{a}\right)^3 + \ldots\right)$$
$$= a^n \left(1 - \frac{x}{a} + \left(\frac{x}{a}\right)^2 - \left(\frac{x}{a}\right)^3 + \ldots\right)$$

Where $\left|\frac{x}{a}\right| < 1$ for the series to converge and have an infinite sum.

Which can be simplified.

2. **Take out a factor of x^n**

$$(a+x)^n = \left(x\left(1 + \frac{a}{x}\right)\right)^n$$
$$= x^n \left(1 + \frac{a}{x}\right)^n$$
$$= x^n \left(1 + (-1)\frac{a}{x} + \frac{(-1)(-2)}{2!}\left(\frac{a}{x}\right)^2 + \frac{(-1)(-2)(-3)}{3!}\left(\frac{a}{x}\right)^3 + \ldots\right)$$
$$= x^n \left(1 - \frac{a}{x} + \left(\frac{a}{x}\right)^2 - \left(\frac{a}{x}\right)^3 + \ldots\right)$$

Where $\left|\frac{a}{x}\right| < 1$ for the series to converge and have an infinite sum.

Which can be simplified.

© HERIOT-WATT UNIVERSITY

Examples

1. Problem:
Show that the expansion of $(3-x)^{-1}$ is the geometric series with first term $a = \frac{1}{3}$ and common ratio $r = \frac{x}{3}$.

Solution:
$(3-x)^{-1} = 3^{-1}(1-\frac{x}{3})^{-1} = \frac{1}{3}(1+(-\frac{x}{3}))^n$

By the binomial theorem the expansion is:

$\frac{1}{3}\left\{1 + \left(\frac{-1}{1!}\right)\left(-\frac{x}{3}\right) + \left(\frac{-1 \times -2}{2!}\right)\left(-\frac{x}{3}\right)^2 + \left(\frac{-1 \times -2 \times -3}{3!}\right)\left(-\frac{x}{3}\right)^3 + \ldots\right\}$

This simplifies to $\frac{1}{3}\left\{1 + \frac{x}{3} + \frac{x^2}{9} + \frac{x^3}{27} + \ldots\right\} = \frac{1}{3} + \frac{x}{9} + \frac{x^2}{27} + \frac{x^3}{81} + \ldots$

This is a geometric series with first term $a = \frac{1}{3}$ and common ratio $r = \frac{x}{3}$ and converges for $\left|-\frac{x}{3}\right| < 1$ i.e. for $|x| < 3$.

....

2. Problem:
As mentioned, for $(1-r)^{-1}$ to be expanded into an infinite series, then $|r| < 1$. This restriction could drastically alter the way the expansion looks.

Lets consider the expansion of $\frac{1}{2x-3} = (2x-3)^{-1}$ stating the conditions under which the expansion is valid.

Solution 1:

Take out a factor of 3^{-1}

For the expansion to be valid $\left|\frac{2x}{3}\right| < 1$, i.e. $|x| < \frac{3}{2}$

$(-3)^{-1}\left(1 + \left(-\frac{2x}{3}\right)\right)^{-1}$

$= -\frac{1}{3}\left(1 + (-1)\left(-\frac{2x}{3}\right) + \frac{(-1)(-2)}{2!}\left(-\frac{2x}{3}\right)^2 + \frac{(-1)(-2)(-3)}{3!}\left(-\frac{2x}{3}\right)^3 + \ldots\right)$

$= -\frac{1}{3} - \frac{2x}{9} - \frac{4x^2}{27} - \frac{8x^3}{81} + \ldots$

or

Solution 2:

Take out a factor of $(2x)^{-1}$

For the expansion to be valid $\left|\frac{3}{2x}\right| < 1$, i.e. $x > \frac{3}{2}$ or $x < -\frac{3}{2}$

$(2x)^{-1}\left(1 + \left(-\frac{3}{2x}\right)\right)^{-1}$

$= \frac{1}{2x}\left(1 + (-1)\left(-\frac{3}{2x}\right) + \frac{(-1)(-2)}{2!}\left(-\frac{3}{2x}\right)^2 + \frac{(-1)(-2)(-3)}{3!}\left(-\frac{3}{2x}\right)^3 + \ldots\right)$

$= \frac{1}{2x} + \frac{3}{4x^2} + \frac{9}{8x^3} + \frac{27}{16x^4} + \ldots$

TOPIC 7. SEQUENCES AND SERIES

Binomial theorem and the geometric series exercise   Go online

Q83: Use the binomial theorem to expand $(a+x)^{-1}$ to four terms where $\left|\frac{x}{a}\right| < 1$.

....

Q84: Find the geometric series which relates to the expression $(2+3x)^{-1}$ when $|x| < \frac{2}{3}$.

....

Q85: Find the geometric series which relates to the expression $\frac{1}{x-5}$ stating the conditions under which the expansion is valid.

....

Q86: Find the geometric series which relates to the expression $\frac{1}{4x-3}$ stating the conditions under which the expansion is valid.

7.9.3 Numeric expansion using the power of -1

The following example shows a numeric expansion using the power of -1. Recall that in the section on the binomial theorem there were similar examples and questions using positive powers.

Examples

1. Problem:

Express $(0 \cdot 99)^{-1}$ as a geometric series and give an approximate value for it to four decimal places.

Solution:

$(0 \cdot 99)^{-1} = (1 + (-0 \cdot 01))^{-1}$

By the binomial theorem the expansion is:

$(1 - 0 \cdot 01)^{-1} = 1 + \dfrac{(-1)}{1!}(-0 \cdot 01) + \dfrac{(-1) \times (-2)}{2!}(-0 \cdot 01)^2 + \dfrac{(-1) \times (-2) \times (-3)}{3!}(-0 \cdot 01)^3 + \ldots$

$= 1 + 0 \cdot 01 + 0 \cdot 0001 + 0 \cdot 000001 + \ldots$

Which is a geometric series with the first term $a = 1$ and the common ratio $r = 0 \cdot 01$.

The series approximates to a value of $1 \cdot 0101$.

Note: the exact value is given by the repeating decimal $1 \cdot 010101 \ldots$

....

2.

Problem:

Express $(4 \cdot 05)^{-1}$ as a geometric series and give an approximate value to for it to four decimal places.

Solution:

$(4 \cdot 05)^{-1} = (4 + 0 \cdot 05)^{-1} = \frac{1}{4}(1 + 0 \cdot 0125)^{-1}$

We have to take out a factor of 4 to put in into the standard form $(1+r)^{-1}$, remembering that the factor is chosen so that $|r| < 1$.

© HERIOT-WATT UNIVERSITY

Using the expansion $(1+r)^{-1} = 1 - r + r^2 - r^3 + \ldots$
We have:
$$\frac{1}{4}(1 + 0 \cdot 0125)^{-1} = \frac{1}{4}\left(1 - (0 \cdot 0125) + (0 \cdot 0125)^2 - (0 \cdot 0125)^3 + \ldots\right)$$
$$= \frac{1}{4}(1 - 0 \cdot 0125 + 0 \cdot 000125 - 0 \cdot 00000125 + \ldots)$$
$$= 0 \cdot 2469 \text{ to 4 d.p.}$$

Numeric expansion using the power of -1 exercise   Go online

Q87: Express $(1 \cdot 05)^{-1}$ as a geometric series and give an approximate value for it to four decimal places.

....

Q88: Express $(1 \cdot 21)^{-1}$ as a geometric series and give an approximate value to for it to four decimal places.

....

Q89: Express $(0 \cdot 81)^{-1}$ as a geometric series and give an approximate value to for it to four decimal places.

....

Q90: Express $(0 \cdot 65)^{-1}$ as a geometric series and give an approximate value to for it to four decimal places.

....

Q91: Express $(1 \cdot 99)^{-1}$ as a geometric series and give an approximate value to for it to four decimal places.

....

Q92: Express $(2 \cdot 1)^{-1}$ as a geometric series and give an approximate value to for it to four decimal places.

....

Q93: Express $(3 \cdot 02)^{-1}$ as a geometric series and give an approximate value to for it to four decimal places.

Partial sums on two common series

The series $\sum_{n=1}^{\infty} \frac{1}{n}$ is divergent. (This is called the harmonic series.)

$$\sum_{n=1}^{\infty}\frac{1}{n} = 1 + \frac{1}{2} + \left(\frac{1}{3}+\frac{1}{4}\right) + \left(\frac{1}{5}+\frac{1}{6}+\frac{1}{7}+\frac{1}{8}\right) + \left(\frac{1}{9}+\ldots+\frac{1}{16}\right) + \ldots$$
$$= 1 + \frac{1}{2} + \frac{2}{4} + \frac{4}{8} + \frac{8}{16} + \ldots$$
$$= \infty$$

Calculate the partial sums $S_2, S_4, S_8, S_{16}, \ldots$ and confirm that $S_{2n} > 1 + \frac{n}{2}$.

For additional interest the series $\sum_{n=1}^{\infty}\frac{1}{n^2}$ is convergent. It converges to $\frac{\pi^2}{6}$.

7.9.4 Sums to infinity exercise

Sums to infinity exercise — Go online

Q94: State which of the following geometric series are convergent and for those that are convergent find the sum to infinity.

a) $2 + 4 + 8 + 16 + 32 + \ldots$
b) $32 + 16 + 8 + 4 + 2 + \ldots$
c) $32 - 16 + 8 - 4 + 2 - \ldots$
d) $\frac{1}{3} + \frac{1}{6} + \frac{1}{12} + \frac{1}{24} + \frac{1}{48} + \ldots$
e) $-\frac{1}{3} + 1 - 3 + 9 - 27 + 81 - \ldots$

....

Q95: Find the sum to infinity of the geometric series
$1 + \frac{1}{2} + \frac{1}{4} + \frac{1}{8} + \frac{1}{16} + \ldots$

....

Q96: Find the first term of a geometric series which has a sum to infinity of 9 and a common ratio of $^2/_3$

....

Q97: Find the sixth term of a geometric series which has a common ratio of $^5/_6$ and a sum to infinity of 72

....

Q98: Find the sum to infinity of the geometric series with all terms positive in which $u_2 = 4$ and $u_4 = 1$

....

Q99: Use the Binomial theorem to expand $(a + x)^{-1}$ to four terms where $\left|\frac{x}{a}\right| < 1$

....

Q100: Find the geometric series which relates to the expression $(2 + 3x)^{-1}$ when $|x| < {}^2/_3$

....

Q101: Find the geometric series which relates to the expression $(2 + 3x)^{-1}$ when $|x| < {}^2/_3$

7.10 Power series

A power series is an expression of the form:

$$\sum_{n=0}^{\infty} a_n x^n = a_0 + a_1 x + a_2 x^2 + a_3 x^3 + \ldots + a_n x^n + \ldots$$

Where $a_0, a_1, a_2, a_3, \ldots, a_n, \ldots$ are constants and x is a variable. It is called a power series as it is made up of a sequence of powers of x with coefficients $a_0, a_1, a_2, a_3, \ldots, a_n, \ldots$.

Power series are useful in solving differential equations that occur in physics, including the equations that describe motion of a simple pendulum, vibrating strings, heat flow and electrical current.

In numerical analysis, power series can be used to determine how many decimal places are required in a computation to guarantee a specified accuracy.

It is also useful to express some simple functions such as e^x, $\sin x$, $\cos x$, $\tan^{-1} x$, $(1 + x)^n$ and $\ln(1 + x)$ in terms of power series.

Examples

1. $1 + x + x^2 + x^3 + x^4 + \ldots x^n + \ldots$ is an example of a power series with $1 = a_0 = a_1 = a_2 = a_3 = a_4 = \ldots = a_n = \ldots$.

Note that this is a geometric series.

....

2. Another important example of a power series is:

$\frac{1}{0!} + \frac{x}{1!} + \frac{x^2}{2!} + \frac{x^3}{3!} + \frac{x^4}{4!} + \frac{x^5}{5!} + \ldots$

This time $a_r = \frac{1}{r!}$

This power series converges for all values of $x \in \mathbb{R}$

Recall the factorial:
$n! = n \times (n-1) \times (n-2) \times \ldots \times 2 \times 1$ for $n \in \mathbb{N}$
e.g.
$5! = 5 \times 4 \times 3 \times 2 \times 1$
$= 120$

Power series exercise   Go online

Q102: Substitute $x = 1$ into the power series $\frac{1}{0!} + \frac{x}{1!} + \frac{x^2}{2!} + \frac{x^3}{3!} + \frac{x^4}{4!} + \frac{x^5}{5!} + \ldots$ and calculate the sum of the first ten terms, S_{10}, to 6 decimal places.

....

Q103: Write down a power series up to the fourth term where $a_0 = 3$ and each coefficient after is doubled.

....

Q104: Write down a power series up to the seventh term where $a_n = \frac{1}{n}$ except for a_0 which is 1.

7.11 Maclaurin series for simple functions

Suppose that $f(x)$ is a function which can be differentiated as often as necessary and that there is no problem differentiating when $x = 0$. Functions like this do exist, for example e^x and $\sin x$ are functions that satisfy these conditions. These conditions provide a special type of power series called the Maclaurin series.

Calculators use a more general version of the Maclaurin series expansion (The Taylor series expansion) to evaluate trigonometric functions about $x = 0$. It also allows for integration of functions which have non-standard results such as e^{x^2}. This is because the Maclaurin series generates a polynomial approximation to the original function.

> **Key point**
>
> The Maclaurin series generated by the function $f(x)$ is:
>
> $$\sum_{r=0}^{\infty} f^{(r)}(0) \frac{x^r}{r!} = f(0) + f^{(1)}(0) \frac{x}{1!} + f^{(2)}(0) \frac{x^2}{2!} + f^{(3)}(0) \frac{x^3}{3!} + \ldots$$
> $$+ f^{(n)}(0) \frac{x^n}{n!} + \ldots$$

For the following example we need to recall the derivative notation $f^{(n)}(x)$ where n tells us the number of times we must differentiate. For example $f^{(3)}(x)$ describes the third derivative i.e. we must differentiate three times.

Example

Problem:

Find the Maclaurin series generated by $f(x) = e^x$.

Solution:

When $f(x) = e^x$ is repeatedly differentiated we obtain:

$f(x) = e^x$ $f(0) = e^0 = 1$
$f^{(1)}(x) = e^x$ $f^{(1)}(0) = e^0 = 1$
$f^{(2)}(x) = e^x$ $f^{(2)}(0) = e^0 = 1$
$f^{(3)}(x) = e^x$ $f^{(3)}(0) = e^0 = 1$
$f^{(4)}(x) = e^x$ $f^{(4)}(0) = e^0 = 1$
$f^{(5)}(x) = e^x$ $f^{(5)}(0) = e^0 = 1$

Therefore the Maclaurin series generated by $f(x) = e^x$ becomes:

$$\sum_{r=0}^{\infty} \frac{x^r}{r!} f^{(r)}(0) = f(0) + f^{(1)}(0) \frac{x}{1!} + f^{(2)}(0) \frac{x^2}{2!} + f^{(3)}(0) \frac{x^3}{3!} + f^{(4)}(0) \frac{x^4}{4!} + f^{(5)}(0) \frac{x^5}{5!} + \ldots$$

$$e^x = 1 + (1) \frac{x}{1!} + (1) \frac{x^2}{2!} + (1) \frac{x^3}{3!} + (1) \frac{x^4}{4!} + (1) \frac{x^5}{5!} + \ldots$$

$$e^x = 1 + x + \frac{x^2}{2!} + \frac{x^3}{3!} + \frac{x^4}{4!} + \frac{x^5}{5!} + \ldots$$

> **Key point**
>
> The Maclaurin series expansion for e^x is given by:
>
> $e^x = 1 + x + \frac{x^2}{2!} + \frac{x^3}{3!} + \frac{x^4}{4!} + \frac{x^5}{5!} + \ldots, \forall x \in \mathbb{R}$

The general rule for the Maclaurin series expansion will be given on the formula sheet in the exam. You will either have to derive the expansion for e^x or remember it. If you remember it, it will save you time in the exam. This will be the case for all other Maclaurin series expansions explored in this topic.

Pay close attention to those expansions referenced in the key points.

Example Problem:

Find the Maclaurin series generated by $f(x) = \sin(x)$.

Solution:

When $f(x) = \sin(x)$ is repeatedly differentiated we obtain:

$f(x) = \sin(x)$ $\qquad$ $f(0) = 0$
$f^{(1)}(x) = \cos(x)$ $\qquad$ $f^{(1)}(0) = 1$
$f^{(2)}(x) = -\sin(x)$ $\qquad$ $f^{(2)}(0) = 0$
$f^{(3)}(x) = -\cos(x)$ $\qquad$ $f^{(3)}(0) = -1$
$f^{(4)}(x) = \sin(x)$ $\qquad$ $f^{(4)}(0) = 0$
$f^{(5)}(x) = \cos(x)$ $\qquad$ $f^{(5)}(0) = 1$

Therefore the Maclaurin series generated by $f(x) = \sin(x)$ becomes:

$$\sum_{r=0}^{\infty} f^{(r)}(0) \frac{x^r}{r!} = f(0) + f^{(1)}(0)\frac{x}{1!} + f^{(2)}(0)\frac{x^2}{2!} + f^{(3)}(0)\frac{x^3}{3!} + f^{(4)}(0)\frac{x^4}{4!} + f^{(5)}(0)\frac{x^5}{5!} + \ldots$$

$$= 0 + (1)\frac{x}{1!} + (0)\frac{x^2}{2!} + (-1)\frac{x^3}{3!} + (0)\frac{x^4}{4!} + (1)\frac{x^5}{5!} + \ldots$$

$$\sin(x) = x - \frac{x^3}{3!} + \frac{x^5}{5!} - \ldots$$

> **Key point**
>
> The Maclaurin series expansion for $\sin(x)$ is given by:
>
> $\sin(x) = x - \frac{x^3}{3!} + \frac{x^5}{5!} - \frac{x^7}{7!} + \ldots, \forall x \in \mathbb{R}$

Maclaurin series for simple functions exercise $\qquad$ Go online

Q105: Find the Maclaurin series generated by $f(x) = \cos(x)$.

TOPIC 7. SEQUENCES AND SERIES

Q106: Find the Maclaurin series generated by $f(x) = \ln(1+x)$, where $-1 < x \leqslant 1$.

....

Q107: Find the Maclaurin series generated by $f(x) = (1+x)^n$.

7.12 Maclaurin's theorem

What is the relationship between the Maclaurin series of a function and the function itself?

Take a function $f(x)$, which can be differentiated as often as required and that there is no problem differentiating when $x = 0$. Also suppose that it is possible to write this function as a series expansion so that:

$$f(x) = a_0 + a_1 x + a_2 x^2 + a_3 x^3 + a_4 x^4 + a_5 x^5 + \ldots$$

Then the following definition applies.

> **Key point**
>
> Maclaurin's theorem states that:
>
> $$f(x) = \sum_{r=0}^{\infty} f^{(r)}(0) \frac{x^r}{r!}$$
>
> $$= f(0) + f^{(1)}(0) \frac{x}{1!} + f^{(2)}(0) \frac{x^2}{2!} + f^{(3)}(0) \frac{x^3}{3!} + \ldots + f^{(n)}(0) \frac{x^n}{n!} + \ldots$$

Notice that this theorem claims that the function $f(x)$ is actually equal to its infinite power series expansion. The theorem is named after the Scottish mathematician Colin Maclaurin (1698-1746) who first proposed this result in his publication *Treatise of fluxions*.

This is true for the following reasoning.

Let $f(x) = a_0 + a_1 x + a_2 x^2 + a_3 x^3 + a_4 x^4 + a_5 x^5 + \ldots$.

Substitute $x = 0$ into the expansion to give $f(0) = a_0$.

Differentiate with respect to x to obtain:

$f'(x) = a_1 + 2a_2 x + 3a_3 x^2 + 4a_4 x^3 + 5a_5 x^4 + \ldots$.

Now substituting $x = 0$ into this equation gives $f'(0) = a_1 = 1!a_1$.

Repeat this process again.

Differentiating again with respect to x gives:

$f''(x) = (2 \times 1) a_2 + (3 \times 2) a_3 x + (4 \times 3) a_4 x^2 + (5 \times 4) a_5 x^3 + \ldots$.

Therefore $f''(0) = (2 \times 1) a_2 = 2!a_2$.

© HERIOT-WATT UNIVERSITY

Differentiating once more with respect to x gives:

$f'''(x) = (3 \times 2 \times 1) a_3 + (4 \times 3 \times 2) a_4 x + (5 \times 4 \times 3) a_5 x^2 + \ldots$

Therefore $f'''(0) = (3 \times 2 \times 1) a_3 = 3! a_3$.

Remember that for higher derivatives it is often more convenient to replace a series of dashes with a number.

In other words, for example, f''' can be rewritten as $f^{(3)}$.

Example

Problem:

Find $f^{(4)}(0)$ for the function given above.

Solution:

$f^{(4)}(0) = (4 \times 3 \times 2 \times 1) a_4 = 4! a_4$

Rearranging the previous results the coefficients are as follows,

$a_0 = f(0)$
$a_1 = \dfrac{f^{(1)}(0)}{1!}$
$a_2 = \dfrac{f^{(2)}(0)}{2!}$
$a_3 = \dfrac{f^{(3)}(0)}{3!}$
$a_4 = \dfrac{f^{(4)}(0)}{4!}$

Since the assumption was that $f(x)$ could be differentiated indefinitely then:

$a_n = \dfrac{f^{(n)}(0)}{n!}$

Now substituting these expressions for a_r back into the power series:

$f(x) = a_0 + a_1 x + a_2 x^2 + a_3 x^3 + a_4 x^4 + a_5 x^5 \ldots$

gives the Maclaurin series expansion:

$f(x) = f(0) + f^{(1)}(0) \dfrac{x}{1!} + f^{(2)}(0) \dfrac{x^2}{2!} + f^{(3)}(0) \dfrac{x^3}{3!} + \ldots$

Note that this result depends on being able to differentiate the infinite series term-by-term and is only valid within an interval of convergence.

The following example should help to explain this.

A convergent series is one for which the limit of partial sums exists. This limit is called the sum and is denoted by S_∞.

Example

Consider the power series:

$$1 + x + x^2 + x^3 + x^4 + \ldots + x^n + \ldots$$

In the topic sequences and series, we learned that for a convergent geometric series:

$$S_\infty = a + ar + ar^2 + ar^3 + ar^4 + \ldots = \frac{a}{1-r}, \text{ where } -1 < r < 1$$

The interval $-1 < r < 1$ is known as the interval of convergence for the previous series. For values of r outside this interval the series does not converge and we say that the series is a **divergent series** for $r \leq -1$ and for $r \geq 1$

Therefore comparing the power series:

$$1 + x + x^2 + x^3 + x^4 + \ldots + x^n + \ldots$$

to a geometric series it is clear that this series is convergent for $-1 < x < 1$ and from the formula for S_∞, taking $a = 1$ and $r = x$ gives:

$$1 + x + x^2 + x^3 + x^4 + \ldots + x^n + \ldots = \frac{1}{1-x}, \text{ for } -1 < x < 1$$

$-1 < x < 1$ is the interval of convergence for this series. Outside the interval of convergence the power series is divergent.

For example, when $x = 1$ the series is simply $1 + 1 + 1 + 1 + \ldots + 1 + \ldots$ which is obviously divergent as the sum of the series will continue increasing the more terms we add. It will not tend towards a limit.

It is also possible to perform a binomial expansion on $\frac{1}{1-x}$ which of course can be written as $(1-x)^{-1}$. Remember that when the binomial expansion is used with negative powers the expansion is infinite and to ensure that the infinite series converges then $-1 < x < 1$, just as in the geometric series formula.

This was discussed in the section Binomial theorem and the geometric series.

The Binomial expansion gives: $(x+y)^n = \sum_{r=0}^{n} \binom{n}{r} x^{n-r} y^r$

This can be expanded to:

$$(x+y)^n = \frac{n!}{0!n!}x^n + \frac{n!}{1!(n-1)!}x^{n-1}y + \frac{n!}{2!(n-2)!}x^{n-2}y^2 + \frac{n!}{3!(n-3)!}x^{n-3}y^3 + \ldots$$

Which simplifies to:

$$(x+y)^n = x^n + nx^{n-1}y + \frac{n(n-1)}{2!}x^{n-2}y^2 + \frac{n(n-1)(n-2)}{3!}x^{n-3}y^3 + \ldots$$

Now if we replace $x = 1$, $y = -x$, where $|x| < 1$ and $n = -1$ we have the following:

$(1 + (-x))^{-1} = 1 + (-1)(-x) + \frac{(-1)(-2)}{2!}(-x)^2 + \frac{(-1)(-2)(-3)}{3!}(-x)^3 + \ldots$

Which simplifies to: $(1 - x)^{-1} = 1 + x + x^2 + x^3 + \ldots$

Key point

The Maclaurin series expansion for $f(x) = (1-x)^{-1}$ is given by:
$(1-x)^{-1} = 1 + x + x^2 + x^3 + x^4 + \ldots$, for $|x| < 1$

From the questions in the previous section we also know the following results:

Key point

The Maclaurin series expansion for $f(x) = \cos(x)$ is given by:
$\cos(x) = 1 - \frac{x^2}{2!} + \frac{x^4}{4!} - \frac{x^6}{6!} + \frac{x^8}{8!} - \ldots, \forall x \in \mathbb{R}$

Key point

The Maclaurin series expansion for $f(x) = \ln(1+x)$ is given by:
$\ln(1+x) = x - \frac{x^2}{2} + \frac{x^3}{3} - \frac{x^4}{4} + \frac{x^5}{5} - \frac{x^6}{6} + \ldots$, for $-1 < x \leq 1$

It will be useful to remember the following information.

Key point

A Maclaurin series will normally converge to its generating function in an interval about the origin. For many functions this is the entire x-axis.

For example, the Maclaurin series for functions such as $\sin x$, $\cos x$, e^x all converge for $x \in \mathbb{R}$.

The series below however have restricted intervals of convergence.

$\ln(1+x)$ converges for $-1 < x \leq 1$

$\frac{1}{1+x}$ converges for $|x| < 1$

$\tan^{-1} x$ converges for $|x| \leq 1$

TOPIC 7. SEQUENCES AND SERIES

The Maclaurin series for sin x

The Maclaurin series generated by $\sin x$ is: $x - \frac{x^3}{3!} + \frac{x^5}{5!} - \frac{x^7}{7!} + \frac{x^9}{9!} - \frac{x^{11}}{11!} + \ldots$

This series has partial sums as below.

$S_1 = x$

$S_2 = x - \frac{x^3}{3!}$

$S_3 = x - \frac{x^3}{3!} + \frac{x^5}{5!}$

$S_4 = x - \frac{x^3}{3!} + \frac{x^5}{5!} - \frac{x^7}{7!}$

...

...

$S_n = x - \frac{x^3}{3!} + \frac{x^5}{5!} - \frac{x^7}{7!} + \frac{x^9}{9!} - \frac{x^{11}}{11!} + \ldots + \frac{(-1)^n x^{2n-1}}{(2n-1)!}$

Maclaurin's theorem exercise

Q108: Write down the Maclaurin series expansion for $f(x) = \sin(x)$.

....

Q109: Evaluate the Maclaurin series of $\sin(x)$ to the fourth term when $x = 1$.

....

Q110: Write down the Maclaurin series expansion for $f(x) = \cos(x)$.

....

Q111: Evaluate the Maclaurin series of $\cos(x)$ to the fifth term when $x = 0 \cdot 5$.

....

Q112: Write down the Maclaurin series expansion for $f(x) = \ln|1 + x|$.

....

Q113: Evaluate the Maclaurin series of $\ln|1 + x|$ to the third term when $x = 0 \cdot 3$.

7.13 The Maclaurin series for $\tan^{-1}(x)$

We can find the Maclaurin series for $\tan^{-1} x$ in the same way as for the previous functions.

Example

Problem:

Find the Maclaurin series generated by $f(x) = \tan^{-1} x$ as far as the third derivative.

Solution:

$f(x) = \tan^{-1} x$ $\qquad$ $f(0) = 0$

$f^{(1)}(x) = \frac{1}{1+x^2}$ $\qquad$ $f^{(1)}(0) = 1$

$f^{(2)}(x) = \frac{-2x}{(1+x^2)^2}$ $\qquad$ $f^{(2)}(0) = 0$

$f^{(3)}(x) = \frac{6x^2-2}{(1+x^2)^3}$ $\qquad$ $f^{(3)}(0) = -2$

From this only the first two terms in the Maclaurin series can be found.

$$\tan^{-1} x = \sum_{r=0}^{\infty} \frac{x^r}{r!} f^{(r)}(0)$$

$$= f(0) + f^{(1)}(0)\frac{x}{1!} + f^{(2)}(0)\frac{x^2}{2!} + f^{(3)}(0)\frac{x^3}{3!} + f^{(4)}(0)\frac{x^4}{4!} + f^{(5)}(0)\frac{x^5}{5!}\ldots$$

$$= 0 + (1)\frac{x}{1!} + (0)\frac{x^2}{2!} + (-2)\frac{x^3}{3!} + \ldots$$

$$= x - \frac{x^3}{3} + \ldots$$

To differentiate further and so obtain more terms in the series soon becomes quite tedious. The following method gives another way to obtain more terms.

Maclaurin series for $\tan^{-1}(x)$: Alternative method

Find the Maclaurin series generated by $f(x) = \tan^{-1}(x)$.

$f(x) = \tan^{-1}(x)$

Differentiate $f(x)$ and rearrange: $f'(x) = \frac{1}{1+x^2} = (1+x^2)^{-1}$

Provided $|x^2| < 1$ we can now use the binomial expansion (see Maclaurin's Theorem) to give:

$$f'(x) = 1 + \frac{(-1)}{1!}x^2 + \frac{(-1)(-2)}{2!}(x^2)^2 + \frac{(-1)(-2)(-3)}{3!}(x^2)^3 + \frac{(-1)(-2)(-3)(-4)}{4!}(x^2)^4 + \ldots$$

$$= 1 - x^2 + x^4 - x^6 + x^8 - \ldots$$

Integrating this will take us back to $f(x) = \tan^{-1}(x)$.

$$\int \frac{1}{1+x^2} dx = \int (1 - x^2 + x^4 - x^6 + x^8 - \ldots) dx$$

$$\tan^{-1}(x) = C + x - \frac{x^3}{3} + \frac{x^5}{5} - \frac{x^7}{7} + \frac{x^9}{9} - \ldots$$

TOPIC 7. SEQUENCES AND SERIES

C is the constant of integration. However when $x = 0$, $\tan^{-1}(x) = 0$ and therefore $C = 0$.

So we can now write: $\tan^{-1}(x) = x - \frac{x^3}{3} + \frac{x^5}{5} - \frac{x^7}{7} + \frac{x^9}{9} - \ldots$

> **Key point**
>
> The Maclaurin series expansion for $\tan^{-1}(x)$ is given by:
>
> $$\tan^{-1}(x) = x - \frac{x^3}{3} + \frac{x^5}{5} - \frac{x^7}{7} + \frac{x^9}{9} - \ldots, \text{ for } |x| \leqslant 1$$

The series for $\tan^{-1}(x)$ actually converges for $|x| \leqslant 1$ and is used to obtain an approximation for π in the following way.

Let $x = 1$, then

$\tan^{-1} 1 = \frac{\pi}{4} = 1 - \frac{1}{3} + \frac{1}{5} - \frac{1}{7} + \frac{1}{9} - \frac{1}{11} + \ldots$

and so,

$\pi = 4 \left(1 - \frac{1}{3} + \frac{1}{5} - \frac{1}{7} + \frac{1}{9} - \frac{1}{11} + \ldots \right)$

The previous series is known as the Leibniz formula for π. However, this series converges very slowly so in practice it is not used in approximating π to many decimal places. In fact 1000 terms are needed before it gives a value accurate to 4 decimal places.

There is more information about different methods for calculating π in the extended information chapter.

> **The Maclaurin series for $\tan^{-1}(x)$ exercise**   Go online
>
> Try to obtain the Maclaurin series for the following functions and explain what happens.
>
> **Q114:** $f(x) = \ln x$
>
> ....
>
> **Q115:** $f(x) = \sqrt{x}$
>
> ....
>
> **Q116:** $f(x) = \cot x$

7.14 Maclaurin's series expansion to a given number of terms

Often you may be required to find a specific number of terms in a Maclaurin series expansion.

We proceed as before as you can see in the following example.

Examples

1. Problem:
Use Maclaurin's theorem to write down the expansion of $(1 + x)^{-3}$ as far as the term in x^3.

Solution:
When $(1 + x)^{-3}$ is repeatedly differentiated we obtain:

$f(x) = (1 + x)^{-3}$ $f(0) = 1$
$f^{(1)}(x) = -3(1 + x)^{-4}$ $f^{(1)}(0) = -3$
$f^{(2)}(x) = (-3)(-4)(1 + x)^{-5}$ $f^{(2)}(0) = (-3)(-4)$
$f^{(3)}(x) = (-3)(-4)(-5)(1 + x)^{-6}$ $f^{(3)}(0) = (-3)(-4)(-5)$

As the series is only needed as far as the term in x^3, the calculations stop.
From this the Maclaurin series is:

$$\begin{aligned}(1 + x)^{-3} &= \sum_{r=0}^{\infty} \frac{x^r}{r!} f^{(r)}(0) \\ &= f(0) + f^{(1)}(0)\frac{x}{1!} + f^{(2)}(0)\frac{x^2}{2!} + f^{(3)}(0)\frac{x^3}{3!} + \ldots \\ &= 1 + \frac{x}{1!}(-3) + \frac{x^2}{2!}(-3)(-4) + \frac{x^3}{3!}(-3)(-4)(-5) + \ldots \\ &= 1 - 3x + 6x^2 - 10x^3 + \ldots\end{aligned}$$

Thus the Maclaurin series for $(1 + x)^{-3}$, as far as the term in x^3, is $1 - 3x + 6x^2 - 10x^3$. This is also known as the Maclaurin series to third order because the highest power of x in the expansion is 3.

....

2. Problem:
Be careful if asked to find the Maclaurin series generated by, for example $(1 + x)^5$.
Since this expression has a positive integer power it will have a finite number of terms when expanded.

Solution:
When $(1 + x)^5$ is repeatedly differentiated:

$f(x) = (1 + x)^5$ $f(0) = 1$
$f^{(1)}(x) = 5(1 + x)^4$ $f^{(1)}(0) = 5$
$f^{(2)}(x) = (5)(4)(1 + x)^3$ $f^{(2)}(0) = 20$
$f^{(3)}(x) = (5)(4)(3)(1 + x)^2$ $f^{(3)}(0) = 60$
$f^{(4)}(x) = (5)(4)(3)(2)(1 + x)^1$ $f^{(4)}(0) = 120$
$f^{(5)}(x) = (5)(4)(3)(2)$ $f^{(5)}(0) = 120$
$f^{(6)}(x)$ and further derivatives $= 0$ $f^{(6)}(0)$ and further derivatives $= 0$

TOPIC 7. SEQUENCES AND SERIES

From this the following Maclaurin series is:

$$(1+x)^5 = \sum_{r=0}^{\infty} \frac{x^r}{r!} f^{(r)}(0)$$

$$= f(0) + f^{(1)}(0)\frac{x}{1!} + f^{(2)}(0)\frac{x^2}{2!} + f^{(3)}(0)\frac{x^3}{3!} + f^{(4)}(0)\frac{x^4}{3!} + f^{(5)}(0)\frac{x^5}{3!}$$

$$= 1 + \frac{x}{1!}(5) + \frac{x^2}{2!}(20) + \frac{x^3}{3!}(60) + \frac{x^4}{4!}(120) + \frac{x^5}{5!}(120)$$

$$= 1 + 5x + 10x^2 + 10x^3 + 5x^4 + x^5$$

Note that this is the same as the binomial expansion for $(1+x)^5$.

Maclaurin's series expansion to a given number of terms exercise Go online

Q117: Find the Maclaurin expansion for $f(x) = e^{x^2}$ as far as the term in x^6.

....

Q118: Find the Maclaurin expansion for $f(x) = \frac{1}{2}\sin(2x)$ as far as the first four non-zero terms.

....

Q119: Find the Maclaurin expansion for $f(x) = \frac{2}{3}\tan^{-1}(3x)$ as far as the first three non-zero terms, where $|x| \leqslant \frac{1}{3}$.

....

Q120: Find the Maclaurin expansion for $f(x) = \ln(1-3x)$ as far as the first four non-zero terms, where $|x| < \frac{1}{3}$.

....

Q121: Find the Maclaurin series expansion for $f(x) = \ln(\cos(x))$ as far as the term in x^4, where $0 \leqslant x < \frac{\pi}{2}$.

7.15 Composite Maclaurin's series expansion

It is also possible to find Maclaurin series for functions such as e^{-3x} and $\sin 2x$. The following example shows how this can be done.

Example

Problem:

Find the Maclaurin series expansion for e^{-3x} as far as the term in x^5.

Solution:

When $f(x) = e^{-3x}$ is repeatedly differentiated we obtain:

$f(x) = e^{-3x}$ $\qquad\qquad$ $f(0) = e^0 = 1$
$f^{(1)}(x) = -3e^{-3x}$ $\qquad\qquad$ $f^{(1)}(0) = -3e^0 = -3$
$f^{(2)}(x) = (-3)^2 e^{-3x}$ $\qquad\qquad$ $f^{(2)}(0) = (-3)^2 e^0 = (-3)^2$
$f^{(3)}(x) = (-3)^3 e^{-3x}$ $\qquad\qquad$ $f^{(3)}(0) = (-3)^3 e^0 = (-3)^3$
$f^{(4)}(x) = (-3)^4 e^{-3x}$ $\qquad\qquad$ $f^{(4)}(0) = (-3)^4 e^0 = (-3)^4$
$f^{(5)}(x) = (-3)^5 e^{-3x}$ $\qquad\qquad$ $f^{(5)}(0) = (-3)^5 e^0 = (-3)^5$

Therefore the Maclaurin series for $f(x) = e^{-3x}$ becomes:

$$e^{-3x} = \sum_{r=0}^{\infty} \frac{x^r}{r!} f^{(r)}(0)$$

$$= f(0) + f^{(1)}(0)\frac{x}{1!} + f^{(2)}(0)\frac{x^2}{2!} + f^{(3)}(0)\frac{x^3}{3!} + f^{(4)}(0)\frac{x^4}{4!} + f^{(5)}(0)\frac{x^5}{5!}...$$

$$= 1 + \frac{x}{1!}(-3) + \frac{x^2}{2!}(-3)^2 + \frac{x^3}{3!}(-3)^3 + \frac{x^4}{4!}(-3)^4 + \frac{x^5}{5!}(-3)^5...$$

$$= 1 - \frac{3x}{1!} + \frac{(3x)^2}{2!} - \frac{(3x)^3}{3!} + \frac{(3x)^4}{4!} - \frac{(3x)^5}{5!} + ...$$

$$= 1 - 3x + \frac{9}{2}x^2 - \frac{9}{2}x^3 + \frac{27}{8}x^4 - \frac{81}{40}x^5 + ...$$

Composite Maclaurin series expansion exercise $\qquad$ Go online

Q122: Find the Maclaurin series generated by $f(x) = \sin(3x)$.

....

Q123: Find the Maclaurin series generated by $f(x) = \cos(2x)$.

....

Q124: Find the Maclaurin series generated by $f(x) = e^{-x}$.

....

Q125: Find the Maclaurin series generated by $f(x) = \ln(1 - 2x)$ for $-\frac{1}{2} < x \leqslant \frac{1}{2}$.

....

Q126: Find the Maclaurin series expansion for $f(x) = ln(1 + e^x)$ for terms as far as x^4.

Top tip

Note that for example 1 we could have obtained the same result if we had substituted $(-3x)$ for x in the Maclaurin series of e^x.

This is demonstrated in the worked example below.

TOPIC 7. SEQUENCES AND SERIES

Example

Problem:

Find the Maclaurin series for e^{-3x}.

Solution:

The Maclaurin series expansion for e^x is:

$e^x = 1 + x + \frac{x^2}{2!} + \frac{x^3}{3!} + \frac{x^4}{4!} + \frac{x^5}{5!} \ldots$

Substituting $(-3x)$ for x we get:

$$\begin{aligned} e^{-3x} &= 1 + (-3x) + \frac{(-3x)^2}{2!} + \frac{(-3x)^3}{3!} + \frac{(-3x)^4}{4!} + \frac{(-3x)^5}{5!} \ldots \\ &= 1 - 3x + \frac{9x^2}{2} - \frac{27x^3}{6} + \frac{81x^4}{24} - \frac{243x^5}{120} + \ldots \\ &= 1 - 3x + \frac{9}{2}x^2 - \frac{9}{2}x^3 + \frac{27}{8}x^4 + \frac{81}{40}x^5 + \ldots \end{aligned}$$

This worked example demonstrates that it can be useful to remember some of the standard Maclaurin series expansions for e^x, $\sin(x)$, $\cos(x)$, $\tan^{-1}(x)$, $\ln(1+x)$ and $(x+1)^{-1}$. This would allow for quicker derivation when expanding the functions for expressions other than just x i.e for $-3x$ as in the example above.

Key point

To generate the Maclaurin series expansion for functions such as e^x, $\sin(x)$, $\cos(x)$, $\tan^{-1}(x)$, $\ln(1+x)$ and $(x+1)^{-1}$, where x is replaced by some other expression i.e. $-3x$ as in the worked example:

1. Write down the standard expansion which involves only x for the given function or derive it.

2. Replace x by the expression given in the question and simplify.

Standard Maclaurin series expansions

Q127: Below is a table of Maclaurin series expansions and ranges of convergence. Match the expansions and ranges of convergence to their functions in the second table.

$= x - \frac{x^3}{3!} + \frac{x^5}{5!} - \frac{x^7}{7!} + \frac{x^9}{9!} - \frac{x^{11}}{11!} + \ldots$	$-1 < x \leqslant 1$
Converges for all $x \in \mathbb{R}$	$= x - \frac{x^3}{3} + \frac{x^5}{5} - \frac{x^7}{7} + \frac{x^9}{9} - \frac{x^{11}}{11} + \ldots$
$= 1 - \frac{x^2}{2!} + \frac{x^4}{4!} - \frac{x^6}{6!} + \frac{x^8}{8!} - \frac{x^{10}}{10!} + \ldots$	$= 1 + x + x^2 + x^3 + x^4 + x^5 \ldots$
$|x| < 1$	$|x| \leqslant 1$
Converges for all $x \in \mathbb{R}$	$= 1 + x + \frac{x^2}{2!} + \frac{x^3}{3!} + \frac{x^4}{4!} + \frac{x^5}{5!} + \ldots$
$= x - \frac{x^2}{2} + \frac{x^3}{3} - \frac{x^4}{4} + \frac{x^5}{5} - \frac{x^6}{6} + \ldots$	Converges for all $x \in \mathbb{R}$

Function	Maclaurin expansion	Interval of convergence
e^x		
$\sin(x)$		
$\cos(x)$		
$\tan^{-1}(x)$		
$\ln(1+x)$		
$(1-x)^{-1}$		

Standard Maclaurin series expansions exercise

Try the following questions using the substitution method.

Q128: Using the substitution method find the Maclaurin series generated by $f(x) = e^{-4x}$ for terms as the term in x^5.

....

Q129: Using the substitution method find the Maclaurin series generated by $f(x) = \sin(5x)$ as far as the term in x^7.

....

Q130: Using the substitution method find the Maclaurin series generated by $f(x) = \cos(3x)$ as far as the term in x^6.

....

Q131: Using the substitution method find the Maclaurin series generated by $f(x) = \tan^{-1}(2x)$, where $|x| \leqslant \frac{1}{2}$ as far as the term in x^7.

TOPIC 7. SEQUENCES AND SERIES

Q132: Using the substitution method find the Maclaurin series generated by $ln(1+2x)$, where $-\frac{1}{2} < x \leqslant \frac{1}{2}$ as far as the term in x^6.

Q133: Using the substitution method find the Maclaurin series generated by $(1-3x)^{-1}$, where $|x| < \frac{1}{3}$ as far as the term in x^5.

Q134: Write down the Maclaurin expansion for e^{ix}, where $i = \sqrt{-1}$.

Q135: For the question above, can you suggest a connection between e^{ix}, $\sin x$ and $\cos x$?

7.15.1 Composite function examples using standard results

Now consider the following composite function examples using standard results.

Examples

1. Problem:
Find the Maclaurin series generated by $f(x) = e^x \cos(2x)$ as far as the term in x^4.

Solution:
Standard results:
$$e^x = 1 + x + \frac{x^2}{2!} + \frac{x^3}{3!} + \frac{x^4}{4!} + \frac{x^5}{5!} + \ldots$$
$$\cos(x) = 1 - \frac{x^2}{2!} + \frac{x^4}{4!} - \frac{x^6}{6!} + \frac{x^8}{8!} - \frac{x^{10}}{10!} + \ldots$$

Substituting $2x$ into $\cos(x)$ gives $\cos(2x)$:
$$\cos(x) = 1 - \frac{(2x)^2}{2!} + \frac{(2x)^4}{4!} - \frac{(2x)^6}{6!} + \frac{(2x)^8}{8!} - \frac{(2x)^{10}}{10!} + \ldots$$
$$= 1 - 2x^2 + \frac{2x^4}{3} - \frac{4x^6}{45} + \frac{2x^8}{315} - \frac{4x^{10}}{14175} + \ldots$$

Now multiply the expansions for e^x and $\cos(2x)$ together:
$$f(x) = e^x \cos(2x)$$
$$= \left(1 + x + \frac{x^2}{2!} + \frac{x^3}{3!} + \frac{x^4}{4!} + \frac{x^5}{5!} + \ldots\right)\left(1 - 2x^2 + \frac{2x^4}{3} - \frac{4x^6}{45} + \ldots\right)$$

© HERIOT-WATT UNIVERSITY

TOPIC 7. SEQUENCES AND SERIES

Multiply the first term in the first bracket by each term in the second bracket. Stop when the powers of x in the answer is greater than 4.

$f(x) = e^x \cos(2x)$

$= \left(\boxed{1} + x + \dfrac{x^2}{2!} + \dfrac{x^3}{3!} + \dfrac{x^4}{4!} + \dfrac{x^5}{5!} + \ldots \right) \left(\boxed{1 - 2x^2 + \dfrac{2x^4}{3} - \dfrac{4x^6}{45}} + \ldots \right)$

$= 1 - 2x^2 + \dfrac{2}{3} x^4$

Multiply the second term in the first bracket by each term in the second bracket. Stop when the powers of x in the answer is greater than 4.

$f(x) = e^x \cos(2x)$

$= \left(1 + \boxed{x} + \dfrac{x^2}{2!} + \dfrac{x^3}{3!} + \dfrac{x^4}{4!} + \dfrac{x^5}{5!} + \ldots \right) \left(\boxed{1 - 2x^2 + \dfrac{2x^4}{3}} - \dfrac{4x^6}{45} + \ldots \right)$

$= 1 - 2x^2 + \dfrac{2}{3} x^4 \boxed{+ x - 2x^3}$

Multiply the third term in the first bracket by each term in the second bracket. Stop when the powers of x in the answer is greater than 4.

$f(x) = e^x \cos(2x)$

$= \left(1 + x + \boxed{\dfrac{x^2}{2!}} + \dfrac{x^3}{3!} + \dfrac{x^4}{4!} + \dfrac{x^5}{5!} + \ldots \right) \left(\boxed{1 - 2x^2} + \dfrac{2x^4}{3} - \dfrac{4x^6}{45} + \ldots \right)$

$= 1 - 2x^2 + \dfrac{2}{3} x^4 + x - 2x^3 \boxed{+ \dfrac{x^2}{2!} - x^4}$

Multiply the fourth term in the first bracket by each term in the second bracket. Stop when the powers of x in the answer is greater than 4.

$f(x) = e^x \cos(2x)$

$= \left(1 + x + \dfrac{x^2}{2!} + \boxed{\dfrac{x^3}{3!}} + \dfrac{x^4}{4!} + \dfrac{x^5}{5!} + \ldots \right) \left(\boxed{1} - 2x^2 + \dfrac{2x^4}{3} - \dfrac{4x^6}{45} + \ldots \right)$

$= 1 - 2x^2 + \dfrac{2}{3} x^4 + x - 2x^3 + \dfrac{x^2}{2!} - x^4 \boxed{+ \dfrac{x^3}{3!}}$

Multiply the fifth term in the first bracket by each term in the second bracket. Stop when the powers of x in the answer is greater than 4.

$f(x) = e^x \cos(2x)$

$= \left(1 + x + \dfrac{x^2}{2!} + \dfrac{x^3}{3!} + \boxed{\dfrac{x^4}{4!}} + \dfrac{x^5}{5!} + \ldots \right) \left(\boxed{1} - 2x^2 + \dfrac{2x^4}{3} - \dfrac{4x^6}{45} + \ldots \right)$

$= 1 - 2x^2 + \dfrac{2}{3} x^4 + x - 2x^3 + \dfrac{x^2}{2!} - x^4 + \dfrac{x^3}{3!} \boxed{+ \dfrac{x^4}{4!}}$

We now stop multiplying further terms since we will only get terms with powers of x greater than 4.

$$f(x) = e^x \cos(2x)$$
$$= \left(1 + x + \frac{x^2}{2!} + \frac{x^3}{3!} + \frac{x^4}{4!} + \frac{x^5}{5!} + \ldots\right)\left(1 - 2x^2 + \frac{2x^4}{3} - \frac{4x^6}{45} + \ldots\right)$$
$$= 1 - 2x^2 + \frac{2}{3}x^4 + x - 2x^3 + \frac{x^2}{2!} - x^4 + \frac{x^3}{3!} + \frac{x^4}{4!}$$
$$= 1 + x - \frac{3}{2}x^2 - \frac{11}{6}x^3 + \frac{17}{24}x^4$$

....

2. Problem:

Find the Maclaurin series generated by $f(x) = \cos(\sin(x))$ as far as the term in x^4.

Solution:

Standard results:
$$\cos(x) = 1 - \frac{x^2}{2!} + \frac{x^4}{4!} - \frac{x^6}{6!} + \frac{x^8}{8!} - \frac{x^{10}}{10!} + \ldots$$
$$\sin(x) = x - \frac{x^3}{3!} + \frac{x^5}{5!} - \frac{x^7}{7!} + \frac{x^9}{9!} - \frac{x^{11}}{11!} + \ldots$$

Now substitute the expansion of $\sin(x)$ into $\cos(x)$.

$$f(x) = \cos(\sin(x))$$

We only substitute $\sin(x) = x - \frac{x^3}{3!}$ into $\cos(x)$ because the highest power in the answer is 4 and the next term in the sine expansion is of degree 5.

$$f(x) = \cos\left(x - \frac{x^3}{3!}\right)$$

First consider the expansion of $\cos(x)$, then replace x by $x - \frac{x^3}{3!}$.

$$f(x) = \cos\left(x - \frac{x^3}{3!}\right)$$
$$= 1 - \frac{\left(x - \frac{x^3}{3!}\right)^2}{2!} + \frac{\left(x - \frac{x^3}{3!}\right)^4}{4!} - \frac{\left(x - \frac{x^3}{3!}\right)^6}{6!} + \frac{\left(x - \frac{x^3}{3!}\right)^8}{8!} - \frac{\left(x - \frac{x^3}{3!}\right)^{10}}{10!} + \ldots$$

We do not need any terms with powers greater than 4 so we can eliminate all brackets with degree 5 or greater. Now expand each remaining bracket, again ignoring any terms with powers greater than 4.

$$f(x) = \cos\left(x - \frac{x^3}{3!}\right)$$
$$= 1 - \frac{\left(x - \frac{x^3}{3!}\right)^2}{2!} + \frac{\left(x - \frac{x^3}{3!}\right)^4}{4!}$$

The third term in the expansion above has a bracket to the power of 4. The question asks only for an expansion to x^4. It is therefore not necessary to expand the bracket completely, but we must ensure that all necessary terms are evaluated. The same is true for the second term when expanding the squared bracket. For demonstration purposes the brackets have been expanded in their entirety. Those terms in x with a power greater than 4 has then been eliminated on the next line of working.

$$=1-\frac{1}{2}\left(x^2-\frac{x^4}{3}+\frac{x^6}{36}\right)+\frac{1}{24}\left(x^4-\frac{2x^6}{3}+\frac{x^8}{6}-\frac{x^{10}}{54}+\frac{x^{12}}{1396}\right)$$
$$=1-\frac{1}{2}\left(x^2-\frac{x^4}{3}+\ldots\right)+\frac{1}{24}\left(x^4-\ldots\right)$$
$$=1-\frac{x^2}{2}+\frac{5x^4}{24}$$

Note that when expanding brackets to get the final answer we can again ignore all terms with powers greater than 4.

Key point

When asked to expand a series to a given term i.e. x^4, do not waste time calculating terms of higher order that will not be needed in the final answer.

Example

Problem:

Find the Maclaurin series expansion for $f(x) = (3+x)\ln(3+x)$ as far as the term in x^4, where $|x| < 3$

Solution:

First expand the brackets in the expression:

$$f(x) = (3+x)\ln(3+x)$$
$$= 3\ln(3+x) + x\ln(3+x)$$

To find the Maclaurin series expansion of $f(x)$ we need to know the expansion for $\ln(3+x)$. We do this by recalling the standard result for $\ln(1+x)$.

$$\ln(1+x) = x - \frac{x^2}{2} + \frac{x^3}{3} - \frac{x^4}{4} + \ldots, \quad -1 < x \leq 1$$

We must now rearrange $\ln(3+x)$ into this form so that we can substitute for x. We do this by taking a common factor of 3 and using the logarithmic rule $\log ab = \log a + \log b$

$$\ln(3+x) = \ln\left(3\left(1+\frac{x}{3}\right)\right)$$
$$= \ln(3) + \ln\left(1+\frac{x}{3}\right)$$

Now we expand $\ln\left(1+\frac{x}{3}\right)$ by substituting x for $\frac{x}{3}$ into the standard result for $\ln(1+x)$.

$$= \ln(3) + \frac{x}{3} - \frac{1}{2}\left(\frac{x}{3}\right)^2 + \frac{1}{3}\left(\frac{x}{3}\right)^3 - \frac{1}{4}\left(\frac{x}{3}\right)^4 + \ldots$$
$$= \ln(3) + \frac{x}{3} - \frac{1}{18}x^2 + \frac{1}{81}x^3 - \frac{1}{324}x^4 + \ldots$$

Now that we have the expansion for $\ln(3+x)$ we substitute this back into the original function $f(x) = (3+x)\ln(3+x)$.

TOPIC 7. SEQUENCES AND SERIES

$$f(x) = (3+x)\ln(3+x)$$
$$= 3\ln(3+x) + x\ln(3+x)$$
$$= 3\left(\ln(3) + \frac{x}{3} - \frac{1}{18}x^2 + \frac{1}{81}x^3 - \frac{1}{324}x^4\right) + x\left(\ln(3) + \frac{x}{3} - \frac{1}{18}x^2 + \frac{1}{81}x^3 - \frac{1}{324}x^4\right)$$
$$= 3\ln(3) + x - \frac{1}{6}x^2 + \frac{1}{27}x^3 - \frac{1}{108}x^4 + x\ln(3) + \frac{1}{3}x^2 - \frac{1}{18}x^3 + \frac{1}{81}x^4 - \ldots$$
$$= \ln(27) + (1 + \ln(3))x + \frac{1}{6}x^2 - \frac{1}{54}x^3 + \frac{1}{324}x^4$$

Composite function examples using standard results exercise   Go online

Q136:
a) Write down the Maclaurin series expansion for e^x as far as the term in x^4.
b) Obtain the Maclaurin series expansion for e^{-x^2}.
c) Deduce the Maclaurin series expansion for e^{x-x^2}.

....

Q137: Obtain the Maclaurin series expansion for $\cos^2(x)$ as far as the term in x^4.

....

Q138: Obtain the Maclaurin series expansion for $\tan^{-1}(\sin(x))$ as far as the term in x^3, where $|\sin(x)| < 1$.

7.16 Learning points

Sequences and series
Sequences and recurrence relations

- A sequence is an ordered set of numbers.
- A finite sequence has a last term.
- An infinite sequence continues indefinitely.
- An element or term is a number in a sequence.
- The nth term (or general term) is the nth number in the sequence and denoted by u_n.
- First order recurrence relation is given by $u_{n+1} = bu_n + c$, where $b \neq 0$.
- If u_1 is given we can find u_n by substituting u_1 into the recurrence relation $u_{n+1} = bu_n + c$ where b and c are known.

Arithmetic and geometric sequences

- An arithmetic sequence is defined by the sequence:
  $a, a+d, a+2d, a+3d, \ldots$
  where a is the 1st term and d is the common difference.
- The common difference in an arithmetic sequence is the difference between any two consecutive numbers in the sequence.
- An arithmetic sequence is given by the recurrence relation $u_{n+1} = u_n + d$, where $u_1 = a$ and d is the common difference.
- This leads to the general form $u_n = a + (n-1)d$.
- A geometric sequence is defined by the sequence:
  $a, ar, ar^2, ar^3, \ldots$
  where a is the 1st term and r is the common ratio.
- The common ratio in a geometric sequence is the ratio between any two consecutive numbers in the sequence and defined by $r = \frac{u_{n+1}}{u_n}$.
- A geometric sequence is given by the recurrence relation $u_{n+1} = ru_n$, where $u_1 = a$ and r is the common ratio.
- This leads to the general form $u_n = ar^{n-1}$.

Fibonacci and other sequences

- Fibonacci sequence is given by the recurrence relation $u_{n+2} = u_{n+1} + u_n$, where u_1 and u_2 are stated.
  The most commonly described Fibonacci sequence is when $u_1 = u_2 = 1$ giving:
  1, 1, 2, 3, 5, 8, 13, 21, ...
- Triangular number sequence is given by the recurrence relation $u_{n+1} = u_n + n + 1$.

TOPIC 7. SEQUENCES AND SERIES

- An alternating sequence is one in which the terms alternate between being positive and negative values e.g. -2, 2, -2, 2, ...
- An uniform sequence is one in which every element is the same and does not change e.g. 5, 5, 5, 5, ...

Convergence and limits

- An infinite sequence $\{u_n\}$ tends to the limit k if for all positive numbers p there is an integer N such that $k - p < u_n < k + p$ for all $n > N$.
- A null sequence is one which tends to a limit of zero e.g. $\lim\limits_{n \to \infty} \left\{\frac{1}{n}\right\} = 0$.
- If $\lim\limits_{n \to \infty} a_n = k$ and $\lim\limits_{n \to \infty} b_n = m$ then:
  - $\lim\limits_{n \to \infty} (a_n + b_n) = k + m$
  - $\lim\limits_{n \to \infty} (\lambda a_n) = \lambda k$ for $\lambda \in \mathbb{R}$
  - $\lim\limits_{n \to \infty} (a_n b_n) = km$
  - $\lim\limits_{n \to \infty} \left(\frac{a_n}{b_n}\right) = \frac{k}{m}$ provided $m \neq 0$.
- A sequence is bounded if there is a number M, such that all further terms in the sequence are between $\pm M$. Similarly, an unbounded sequence does not have this.

Definitions of e and π as limits of a sequence

- $\lim\limits_{n \to \infty} \left\{\left(1 + \frac{1}{n}\right)^n\right\} = e$
- $\lim\limits_{n \to \infty} \left\{n \tan\left(\frac{180}{n}\right)\right\} = \pi$

Series and sums

- A series is the sum of the terms in a sequence.
- The partial sum is the sum of the terms from 1 to n where $n \in \mathbb{N}$. It is denoted by S_n and represented as $S_n = \sum\limits_{r=1}^{n} u_r$.
- $\sum\limits_{n=1}^{\infty} (a_n + b_n) = \sum\limits_{n=1}^{\infty} a_n + \sum\limits_{n=1}^{\infty} b_n$
- $\sum\limits_{n=1}^{\infty} (\lambda a_n) = \lambda \sum\limits_{n=1}^{\infty} a_n$ for $\lambda \in \mathbb{R}$

Arithmetic and geometric series

- The nth partial sum of an arithmetic series is $S_n = \frac{n}{2}(2a + (n-1)d)$.
- $\sum\limits_{r=1}^{n} r = \frac{1}{2}n(n+1)$

© HERIOT-WATT UNIVERSITY

- $\sum_{r=1}^{n} r^2 = \frac{n(n+1)(2n+1)}{6}$

- $\sum_{r=1}^{n} r^3 = \frac{n^2(n+1)^2}{4}$

- The nth partial sum of a geometric series is $S_n = \frac{a(1-r^n)}{1-r}$, where $r \neq 1$.

Sums to infinity

- Sum to infinity of a geometric series is $S_\infty = \frac{a}{1-r}$, where $|r| < 1$.

- The Binomial expansion for $(1-r)^{-1} = 1 + r + r^2 + r^3 \ldots$ this is equivalent to the Geometric progression given by $S_\infty = \frac{1}{1-r} = 1 + r + r^2 + r^3 \ldots$.

- To expand $(a+x)^{-1}$:
    - If $\left|\frac{x}{a}\right| < 1$ then write $(a+x)^{-1} = a^{-1}\left(1+\frac{x}{a}\right)^{-1}$ and expand using binomial theorem.
    - If $\left|\frac{a}{x}\right| < 1$ then write $(a+x)^{-1} = x^{-1}\left(1+\frac{a}{x}\right)^{-1}$ and expand using binomial theorem.

Maclaurin Series

- A power series is an expression of the form:

$$\sum_{n=0}^{\infty} a_n x^n = a_0 + a_1 x + a_2 x^2 + a_3 x^3 + \ldots + a_n x^n + \ldots$$

Where $a_0, a_1, a_2, a_3, \ldots, a_n, \ldots$ are constants and x is a variable.

- The Maclaurin series generated by the function $f(x)$ is:

$$\sum_{n=0}^{\infty} f^{(r)}(0) \frac{x^r}{r!} = f(0) + f^{(1)}(0) \frac{x}{1!} + f^{(2)}(0) \frac{x^2}{2!} + f^{(3)}(0) \frac{x^3}{3!} + \ldots + f^{(n)}(0) \frac{x^n}{n!} + \ldots$$

- Standard expansions with ranges of convergence.

    Geometric Series

$(1-x)^{-1} = 1 + x + x^2 + x^3 + x^4 + x^5 \ldots$	$\|x\| < 1$
$e^x = 1 + x + \frac{x^2}{2!} + \frac{x^3}{3!} + \frac{x^4}{4!} + \frac{x^5}{5!} + \ldots$	$x \in \mathbb{R}$
$\sin(x) = x - \frac{x^3}{3!} + \frac{x^5}{5!} - \frac{x^7}{7!} + \frac{x^9}{9!} - \frac{x^{11}}{11!} + \ldots$	$x \in \mathbb{R}$
$\cos(x) = 1 - \frac{x^2}{2!} + \frac{x^4}{4!} - \frac{x^6}{6!} + \frac{x^8}{8!} - \frac{x^{10}}{10!} + \ldots$	$x \in \mathbb{R}$
$\tan^{-1}(x) = x - \frac{x^3}{3} + \frac{x^5}{5} - \frac{x^7}{7} + \frac{x^9}{9} - \frac{x^{11}}{11} + \ldots$	$\|x\| \leq 1$
$\ln(1+x) = x - \frac{x^2}{2} + \frac{x^3}{3} - \frac{x^4}{4} + \frac{x^5}{5} - \frac{x^6}{6} + \ldots$	$-1 < x \leq 1$

    These expansions will not be given to you on the formula sheet.
    It would save you time in the exam to learn these results, otherwise you must be able to derive them using the Maclaurin series formula.

- A Maclaurin series expansion can be worked out in two ways:
    - Use the Maclaurin series formula by differentiating and evaluating at $x = 0$ or;
    - substitute into the standard results.

7.17 Extended information

There are links on the web which give a variety of web sites on this topic. These sites cover the subject to a higher level and include all the convergence tests which would be useful for those who wish to study this topic further.

http://www.mathcs.org/analysis/reals/numseq/index.html
This is an excellent site on sequences with plenty of extension material.

http://www.mathcs.org/analysis/reals/numser/index.html
A sister site to the one above with the same quality materials but on series.

http://oeis.org/wiki/Welcome
This site is a must for those wishing to explore sequences and perhaps contribute to their development. There are links to other good sites too.

http://www.superkids.com/aweb/tools/logic/towers/
This is the actual Tower of Hanoi game. Play it and think about the sequence generated which depends on the number of rings.

https://www.youtube.com/watch?v=w-I6XTVZXww
She Sum of the Natural Numbers is $-\frac{1}{12}$. Impossible!
Click the link above and find out.

Zeno of Elea (ca. 490-430 BC)

This Greek philosopher and mathematician lived in the 5th century. His interests lay in paradoxes concerning motion. One of the most famous of these is 'Achilles and the Tortoise'. His arguments and logic are very interesting.

Leonardo Bonacci - known as Fibonacci (1170 - 1250)

This man was probably the greatest mathematician of his time. He was born in Pisa in 1170 and his real name was Leonardo of Pisa. He was an inspired man who produced many mathematical results. The sequence named after him was included in the third section of the book Liber Abbaci published in 1202. This section of the book contained problems on sequences, series and some number theory. Another famous discovery of Fibonacci's was Pythagorean triples.

Jean le Rond d'Alembert (1717 - 1783)

D'Alembert was a Frenchman born in 1717. He was a very argumentative man but made significant contributions in Mathematics and Physics. His ideas on limits led to the tests for convergence named after him. These are important results but are beyond the scope of this course. D'Alembert is also credited with being one of the first users of partial differential equations.

Colin Maclaurin (1698 - 1746)

Colin Maclaurin was born in Kilmodan, Scotland. He was the youngest of three brothers but never knew his father who died when he was only six weeks old. His mother also died, when Colin was nine years old, and so he was brought up by his uncle who was a minister at Kilfinan on Loch Fyne.

At the age of 19, in August 1717, Maclaurin was appointed professor of mathematics at Marischal College, Aberdeen University. Maclaurin at this time was a great supporter of Sir Isaac Newton and is reported to have travelled to London to meet him. He was elected a Fellow of the Royal Society during one of these visits.

On 3 November 1725 Maclaurin was appointed to Edinburgh University where he spent the rest of his career. He married Anne Stewart the daughter of the Solicitor General for Scotland and had seven children. His teaching at Edinburgh came in for considerable praise and he is said to have been keen to aid the understanding of his students. If they had difficulty with a concept then he was likely to try another method of explanation in order to give them a clearer understanding.

Maclaurin is best remembered for his publication the *Treatise of fluxions* where he demonstrates the special case of the Taylor series which is now named after him.

The Taylor series generated by the function $f(x)$ at $x = a$ is:

$$f(a) + f^{(1)}(a)(x - a) + f^{(2)}(a)\frac{(x - a)^2}{2!} + \ldots + f^{(n)}(a)\frac{(x-a)^n}{n!} + \ldots$$

Notice that Maclaurin series are Taylor series with $a = 0$. The Maclaurin series was not an idea discovered independently of the Taylor series and indeed Maclaurin makes acknowledgement of Taylor's influence.

Maclaurin's other interests include the annual eclipse of the sun, the structure of bees' honeycombs and actuarial studies.

'He laid sound actuarial foundations for the insurance society that has ever since helped the widows and children of Scottish ministers and professors.'

Maclaurin also became involved in the defence of Edinburgh during the Jacobite rebellion of 1745. However, when the city fell to the Jacobites he fled to England but returned when the Jacobites marched further south. Much weakened by a fall from his horse in combination with his exertions defending Edinburgh and a difficult journey through winter weather to return to his home city, he became very ill in December 1745. He died the next year and was buried in Greyfriars Churchyard, where his grave can still be seen.

Calculating π

The series for $\tan^{-1}x$ was first discovered by James Gregory in 1671.

$$\tan^{-1}x = x - \frac{x^3}{3} + \frac{x^5}{5} - \frac{x^7}{7} + \ldots$$

When $x = 1$ in the above this gives Leibniz's formula:

$$\frac{\pi}{4} = 1 - \frac{1}{3} + \frac{1}{5} - \frac{1}{7} + \frac{1}{9} - \frac{1}{11} + \ldots \frac{(-1)^{n-1}}{2n-1} + \ldots$$

$$\pi = 4\left(1 - \frac{1}{3} + \frac{1}{5} - \frac{1}{7} + \frac{1}{9} - \frac{1}{11} + \ldots \frac{(-1)^{n-1}}{2n-1} + \ldots\right)$$

This series converges very slowly and so is not used to approximate π to many decimal places. The series for $\tan^{-1}x$ converges more quickly when x is near zero. To use the series for $\tan^{-1}x$ to calculate π consider various trigonometric identities.

For example the following trigonometrical identity could be used with

$\alpha = \tan^{-1}\frac{1}{2}$ and $\beta = \tan^{-1}\frac{1}{3}$

$$\tan(\alpha + \beta) = \frac{\tan \alpha + \tan \beta}{1 - \tan \alpha \tan \beta}$$
$$= \frac{\left(\frac{1}{2}\right) + \left(\frac{1}{3}\right)}{1 - \left(\frac{1}{6}\right)}$$
$$= 1$$
$$= \tan \frac{\pi}{4}$$

Therefore,

$$\frac{\pi}{4} = \alpha + \beta = \tan^{-1}\frac{1}{2} + \tan^{-1}\frac{1}{3}$$
$$\pi = 4\left(\tan^{-1}\frac{1}{2} + \tan^{-1}\frac{1}{3}\right)$$

Then use the expansion for $\tan^{-1} x$ with $x = 1/2$ and $x = 1/3$.

Since these values for x are nearer to zero the above method will give accurate results for π more quickly than Leibniz's formula.

In our work on Maclaurin series mention was made of obtaining and estimate for π from the Maclaurin series for the inverse tan function. This link takes you to a site where you can read further information about the calculation of π.

http://www.cecm.sfu.ca/personal/pborwein/PAPERS/P159.pdf

The MacTutor History of Mathematics archive. This site provides a very comprehensive directory of biographies for hundreds of important mathematicians. This was the source for the following information on ColinMaclaurin.

http://www-history.mcs.st-andrews.ac.uk/history/index.html

7.18 End of topic test

End of topic 7 test — Go online

Arithmetic and Geometric Sequences and Series

Unit Assessment Level

Q139: Find the 25th term of the arithmetic sequence 17, 26, 35, 44, 53, .

Q140: Find S_{28} for the arithmetic sequence 13, 8, 3, -2, -7, ...

Q141: For the arithmetic sequence -11, -7, -3, 1, .
Find:

a) the 14th term
b) the sum of the first 14 terms.

Q142: Find the 8th term of the geometric sequence 2, 8, 32, 128, 512, ...

Q143: For the geometric sequence 162, 54, 18, 6, 2, ..., find an expression for the sum of the first n terms.

Q144: For the geometric sequence 625, 125, 25, 5, ...
Find:

a) the 9th term
b) the sum of the first 6 terms.

Q145: The first term of an arithmetic sequence 5 and the common difference is 2.
Find:

a) The 120th term of the arithmetic sequence.
b) The sum of the first 200 terms of this series.

Q146: Two consecutive terms in a geometric series are 1215 and 3645. The 10th term is 98415.
Find:

a) An expression for n^{th} term of this series.
b) The sum of the first 12 terms of this series.

TOPIC 7. SEQUENCES AND SERIES

Q147: An arithmetic sequence is given by 8, 30, 52, 74, ...
Find:

a) The 40th term of the arithmetic sequence.
b) The sum of the first 50 terms of this series.

Q148: A geometric sequence is given by: 7, 14, 28, 56, ...
Find:

a) The 11th term of the geometric sequence.
b) The sum of the first 9 terms of this series.

Q149: Find the sum to infinity of the geometric series $2 + \frac{6}{5} + \frac{18}{25} + \frac{54}{125} + \ldots$

Q150: A geometric series has a common ratio of $\frac{1}{6}$ and sum to infinity of $\frac{18}{5}$.
Find the third term.

Q151: The fourth term of an arithmetic sequence is 9 and the sum of the first 8 terms is 36. Find the sum of the first 21 numbers in this arithmetic sequence.

Q152: Determine the limit of the sequences if it exists of the sequence whose n^{th} term is:

a) $\frac{n^3 + 2n - 3}{n(n^4 + 5)}$
b) $\frac{(3n+1)(2n-3)}{5n(n+2)}$
c) $7 - \frac{3(n^2 - n + 1)}{n^2}$

Exam Level

Q153: The 3rd and 5th term of a geometric series are 3 and $\frac{1}{3}$ respectively.
What is the sum to infinity of this geometric series?

Q154: Express (0·81)$^{-1}$ as a geometric series and give an approximate value for it to four decimal places.

Q155: Express (1·15)$^{-1}$ as a geometric series and give an approximate value for it to four decimal places.

Q156: Express $(3·23)^{-1}$ as a geometric series and give an approximate value for it to four decimal places.

....

Q157: Express $(2·95)^{-1}$ as a geometric series showing the first four terms and give an approximate value for it correct to 6 decimal places.

....

Q158: Find the geometric series which relates to the expression $(4 + 7x)^{-1}$, when $x > \frac{4}{7}$ or $x < -\frac{4}{7}$.

....

Q159: Find the geometric series which relates to the expression $(3x - 4)^{-1}$, when $|x| < \frac{4}{3}$.

....

Q160: Find two geometric series for the expression $(3 - 4x)^{-1}$ and state the conditions for which each converges.

....

Q161: The compound interest rate is set at 7·5 % per annum. A customer banks $500 at the start of the first year and leaves it for 5 years. At the start of the 6th year the interest rate falls to 5%. He adds $500 at that point and leaves both his deposits and the interest to accumulate for 10 more years.

- Find the two geometric sequences - one for years 1 to 5 and one for years 6 to 16.
- State the first term and the common ratio for both.
- Calculate the amount of money in the account at the start of the 17th year (to the nearest cent).

....

Q162: Evaluate $\sum_{n=1}^{4} (7 - 2n)$.

....

Q163: Evaluate $\sum_{n=1}^{9} (2n^2 + 3n^3)$.

....

Q164: Evaluate $\sum_{n=4}^{10} (4 + 3n^3 - 9n)$.

....

Q165: Find the value of $\sum_{n=1}^{\infty} \left(\frac{1}{n^2+5n+6}\right)$.

....

Q166: Show that $\sum_{r=1}^{n} r(r-1) = \frac{1}{3}n(n+1)(n-1)$.

....

Q167: Show that $\sum_{r=1}^{n} r(r+1)(r+2) = \frac{1}{4}n(n+1)(n+2)(n+3)$.

....

TOPIC 7. SEQUENCES AND SERIES

Q168: Show that $\sum_{r=1}^{n} r\left(r^{2}+2\right)=\frac{1}{4}n\left(n+1\right)\left(n^{2}+n+4\right)$.

Maclaurin Series

Unit Assessment Level

Q169: Obtain the Maclaurin series for $2\sin(2x)$ for the first three non-zero terms.

....

Q170: Obtain the Maclaurin series for $2\cos(2x)$ for the first three non-zero terms.

....

Q171: Obtain the Maclaurin series for $2tan^{-1}(2x)$ for the first three non-zero terms, where $|x| \leqslant \frac{1}{2}$.

....

Q172: Obtain the Maclaurin series for $2\ln(1+2x)$ as far as the term in x^3, where $-\frac{1}{2} < x \leqslant \frac{1}{2}$.

....

Q173: Obtain the Maclaurin series for $2(1+2x)^{-1}$ as far as the term in x^3, where $|x| < \frac{1}{2}$.

....

Q174: Obtain the Maclaurin series for $2e^{2x}$ as far as the term in x^2.

....

Q175: Obtain the Maclaurin series for $\sqrt{(1+2x)}$.

....

Q176: Obtain the Maclaurin series for $(1+2x)^{-3}$.

....

Q177: Using Maclaurin expansion evaluate $\sin(0 \cdot 5)$ correct to four decimal places.

....

Q178: Using Maclaurin expansion evaluate $e^{0.5}$ correct to three decimal places.

Exam Level

Q179:
Find the Maclaurin series generated by $f(x) = e^x \sin(3x)$ as far as the term in x^4.

....

Q180: Obtain the Maclaurin series expansion for $\tan^{-1}(\sin(x))$ as far as the term in x^3, where $|\sin(x)| < 1$.

© HERIOT-WATT UNIVERSITY

Topic 8

Curve sketching

Contents

- 8.1 Looking back 270
  - 8.1.1 Determining stationary points 272
  - 8.1.2 Closed intervals 277
  - 8.1.3 Transformations of graphs 279
- 8.2 Functions 296
  - 8.2.1 Function definition 297
  - 8.2.2 One-to-one and onto functions 305
- 8.3 Inverse functions 314
- 8.4 Odd, even or neither functions 326
- 8.5 Critical and stationary points 333
- 8.6 Derivative tests 337
  - 8.6.1 The first derivative test 337
  - 8.6.2 The second derivative test 340
- 8.7 Concavity 343
- 8.8 Continuity and asymptotic behaviour 349
- 8.9 Sketching and rational functions 363
- 8.10 Type 1 rational function: Constant over linear 368
- 8.11 Type 2 rational function: Linear over linear 371
- 8.12 Type 3 rational function: Constant or linear over quadratic 375
- 8.13 Type 4 rational function: Quadratic over quadratic 383
- 8.14 Type 5 rational function: Quadratic over linear 387
- 8.15 Summary of shortcuts to sketching rational functions 391
- 8.16 Graphical relationships between functions 392
  - 8.16.1 Modulus function 392
- 8.17 Learning points 398
- 8.18 Extended Information 402
- 8.19 End of topic test 403

Learning objective

By the end of this topic, you should be able to:

- use and know the:
    - definition of a function;
    - term domain, codomain, range and image;
- state the domain and range of simple functions;
- identify functions that are:
    - one-to-one;
    - many-to-one;
    - onto.
- define an inverse and use it to restrict the domain of a function in order to have an inverse;
- draw an inverse function given the original function;
- derive the equation of the inverse of a function given the original function;
- graph the functions $\sin^{-1} x$, $\cos^{-1} x$ and $\tan^{-1} x$ and their domains;
- define and identify what an odd and even function are;
- identify:
    - algebraically if a function is odd or even;
    - critical points, stationary points, global max/min and local max/min when given a graph of a function;
- use the:
    - first derivative and the nature table to determine the nature of a turning point;
    - second derivative to determine the nature of a turning point;
- define the:
    - meaning of concave down and concave up and how to identify this given the equation of a function;
    - conditions for a non-horizontal point of inflection;
- identify continuous and discontinuous functions;
- find the vertical, horizontal and slant asymptote of a rational function if they exist;
- sketch a graph for a rational function of the form $\frac{ax^2+bx+c}{dx^2+ex+f}$;

TOPIC 8. CURVE SKETCHING

Learning objective continued

- draw a graph that has been transformed by one or a combination of the following:
    - $y = f(x) + k$
    - $y = f(x - k)$
    - $y = f(kx)$
    - $y = kf(x)$
    - $y = f^{-1}(x)$
    - $y = |f(x)|$

8.1 Looking back

Pre-requisites from Advanced Higher

You should have covered the following in the Partial fractions topic. If you need to reinforce your learning go back and study this topic.

Algebraic long division

- The method of partial fractions is applied to proper rational functions only. If the function is improper, algebraic long division must be carried out first.

$$
\begin{array}{r}
x - 2 \\
x^2 + 2x - 3 \overline{\smash{\big)}\ x^3 + 0x^2 - 2x + 5} \\
\underline{x^3 + 2x^2 - 3x} \\
-2x^2 + x + 5 \\
\underline{-2x^2 - 4x + 6} \\
5x - 1
\end{array}
$$

Divisor: $x^2 + 2x - 3$
Quotient: $x - 2$
Dividend: $x^3 + 0x^2 - 2x + 5$
Remainder: $5x - 1$

- When setting up algebraic long division each power of the variable must be accounted for in the dividend. See the example above.

The product rule

- When $k(x) = f(x)\, g(x)$, then $k'(x) = f'(x)\, g(x) + f(x)\, g'(x)$

The quotient rule

- The quotient rule gives us a method that allows us to differentiate algebraic fractions.
- It states that, in function notation:
  when $k(x) = \frac{f(x)}{g(x)}$, then $k'(x) = \frac{f'(x)g(x) - f(x)g'(x)}{[g(x)]^2}$
- In Leibnitz notation:
  when $y = \frac{u}{v}$, where u and v are functions of x, then $\frac{dy}{dx} = \frac{\frac{du}{dx}v - u\frac{dv}{dx}}{v^2}$

Summary of prior knowledge Finding stationary points and their nature: Higher

- A stationary point could be one of the following:
  - A maximum turning point.
  - A minimum turning point.
  - A rising point of inflection.
  - A falling point of inflection.

TOPIC 8. CURVE SKETCHING

- To determine stationary points and their nature:
    - find the derivative;
    - state that stationary points occur when $\frac{dy}{dx} = 0$;
    - solve for x when $\frac{dy}{dx} = 0$;
    - find the coordinates of the stationary points;
    - to determine the nature of the stationary points make a nature table;

x	$\rightarrow$	-1	$\rightarrow$	1	$\rightarrow$
$\frac{dy}{dx}$	+	0	-	0	+
shape	/	—	\	—	/

   - state the nature of the stationary points.

Identifying maximum and minimums over closed and open intervals: Higher

- To determine the maximum and minimum values in a closed interval:
    - identify the stationary points and their coordinates;
    - make a nature table to determine the shape of the function;
    - determine the coordinates of the end points;
    - make a sketch;
    - identify the maximum and minimum values.

Transformations of graphs: Higher

- $f(x) + a$ means that the graph of $f(x)$ is moved up vertically by a units.
- $f(x) - a$ means that the graph of $f(x)$ is moved down vertically by a units.
- $f(x + a)$ means that the graph of $f(x)$ is moved left horizontally by a units.
- $f(x - a)$ means that the graph of $f(x)$ is moved right horizontally by a units.
- $-f(x)$ means the graph of $f(x)$ is reflected in the x-axis.
- $f(-x)$ means the graph of $f(x)$ is reflected in the y-axis.
- $kf(x)$ means the graph of $f(x)$ is scaled vertically by a factor of k.
    - stretched for $k > 1$
    - squashed for $0 < k < 1$
- $f(kx)$ means the graph of $f(x)$ is scaled horizontally by a factor of k.
    - squashed for $k > 1$
    - stretched for $0 < k < 1$

© HERIOT-WATT UNIVERSITY

8.1.1 Determining stationary points

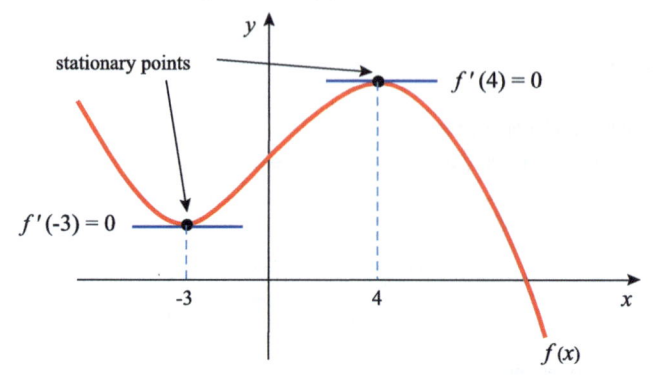

Note that there are two points on this graph where the tangent to the curve is horizontal. These are called **stationary points**.

> **Key point**
>
> At stationary points $f'(x) = 0$.

The nature of a stationary point is determined by the sign of $f'(x)$ on either side. Stationary points can be any of the following types.

Stationary points   Go online

The gradient of the tangent to the curve changes from positive to negative.

the gradient of the tangent, $f'(x)$ is positive

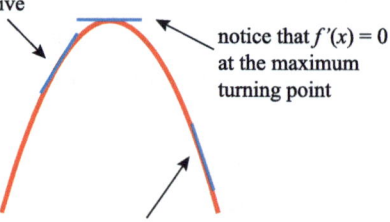

notice that $f'(x) = 0$ at the maximum turning point

the gradient of the tangent, $f'(x)$ is negative

TOPIC 8. CURVE SKETCHING

The gradient of the tangent to the curve changes from negative to positive.

the gradient of the tangent, $f'(x)$ is negative

notice that $f'(x) = 0$ at the minimum turning point

the gradient of the tangent, $f'(x)$ is positive

The gradient of the tangent to the curve is positive on either side of the point of inflection. This is a rising point of inflection.

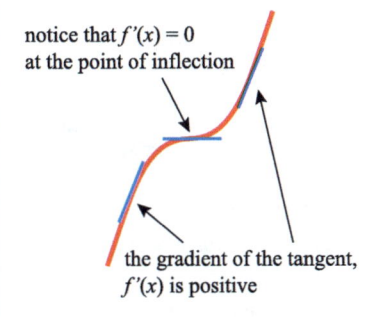

notice that $f'(x) = 0$ at the point of inflection

the gradient of the tangent, $f'(x)$ is positive

The gradient of the tangent to the curve is negative on either side of the point of inflection. This is a falling point of inflection.

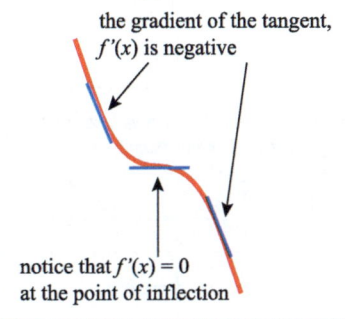

the gradient of the tangent, $f'(x)$ is negative

notice that $f'(x) = 0$ at the point of inflection

Example Problem:

Find the stationary points on the curve $f(x) = 3x^4 + 8x^3 + 6$.

TOPIC 8. CURVE SKETCHING

Solution:

$f(x) = 3x^4 + 8x^3 + 6$

$f'(x) = 12x^3 + 24x^2 = 12x^2(x+2)$

Stationary points occur when $f'(x) = 0$

It is essential that you make this statement to gain full marks.

Hence to find the stationary points we must solve:

$12x^2(x+2) = 0$

$12x^2 = 0$ or $x + 2 = 0$

$x = 0$ or $x = -2$

When $x = 0$ then $f(0) = 6$

When $x = -2$ then $f(-2) = -10$

Thus the coordinates of the stationary points are (0,6) and (-2,-10).

The nature of the stationary points in the above example is determined by making a nature table using the derivative $f'(x) = 12x^2(x+2)$. Work through the following activity to see how to make a nature table.

Making a nature table   Go online

Make a table using the factors of the derivative and the stationary points.

The arrows are for determining what the gradient of the tangents to the curve are before -2, between -2 and 0 and after 0.

x	$\rightarrow$	-2	$\rightarrow$	0	$\rightarrow$
$12x^2$					
$(x+2)$					
$f'(x)$					
shape					

It is often easier to think of a value just before -2 i.e. -3, between -2 and 0 i.e. -1 and just after 0 i.e. 1, and to uses these values to determine whether the result is positive or negative.

x	-3 $\rightarrow$	-2	-1 $\rightarrow$	0	1 $\rightarrow$
$12x^2$					
$(x+2)$					
$f'(x)$					
shape					

When $x = -3$, $12x^2 = 12 \times (-3)^2 = 108$ which is positive.
Do the same thing across the row.

TOPIC 8. CURVE SKETCHING

x	-3		-2		-1		0		1
		$\rightarrow$				$\rightarrow$			$\rightarrow$
$12x^2$	+		+		+		0		+
$(x+2)$									
$f'(x)$									
shape									

Notice that when $x = 0$, $12x^2 = 12 \times 0^2 = 0$.
Now substitute your values x into $(x+2)$.
When $x = -3$, $x + 2 = -3 + 2 = -1$ which is negative.

x	-3		-2		-1		0		1
		$\rightarrow$				$\rightarrow$			$\rightarrow$
$12x^2$	+		+		+		0		+
$(x+2)$	-		0		+		+		+
$f'(x)$									
shape									

$f'(x) = 12x^2(x+2)$ which means $12x^2$ times $(x+2)$ so to find the next row we multiply the values in the columns above.

a positive × a negative is negative
a positive × zero is zero
a positive × a positive is positive
a positive × zero is zero
a positive × a positive is positive

x	-3		-2		-1		0		1
		$\rightarrow$				$\rightarrow$			$\rightarrow$
$12x^2$	+		+		+		0		+
$(x+2)$	-		0		+		+		+
$f'(x)$	-		0		+		0		+
shape									

Now let's draw the shape of $f(x)$ depending on the value of $f'(x)$.
$f'(-3)$ is negative and a negative gradient slopes downwards from left to right giving,

x	-3		-2		-1		0		1
		$\rightarrow$				$\rightarrow$			$\rightarrow$
$12x^2$	+		+		+		0		+
$(x+2)$	-		0		+		+		+
$f'(x)$	-		0		+		0		+
shape	\		—		/		—		/

© HERIOT-WATT UNIVERSITY

This gives us a minimum turning point at (-2,-10) and a rising point of inflection at (0,6).

Example

Problem:

Find the stationary points on the curve $y = 2x^3 - 6x + 1$ and determine their nature.

Solution:

$y = 2x^3 - 6x + 1$ and $\frac{dy}{dx} = 6x^2 - 6$

Stationary points occur when $\frac{dy}{dx} = 0$.

$$6x^2 - 6 = 0$$
$$6(x^2 - 1) = 0$$
$$6(x-1)(x+1) = 0$$
$$x - 1 = 0 \quad \text{or} \quad x + 1 = 0$$
$$x = 1 \quad \text{or} \quad x = -1$$

When $x = -1, y = 2(-1)^3 - 6(-1) + 1 = 5$
When $x = 1, y = 2(1)^3 - 6(1) + 1 = -3$

Thus the stationary points occur at (-1,5) and (1,-3).

To determine their nature we draw a nature table.

x	-2 $\rightarrow$	-1	0 $\rightarrow$	1	2 $\rightarrow$
$(x-1)$	-	-	-	0	+
$(x+1)$	-	0	+	+	+
$6(x-1)(x+1)$	+	0	-	0	+
shape	/	—	\	—	/

This gives us a maximum turning point at (-1,5) and a minimum turning point at (1,-3).

Determining stationary points exercise

Go online

Q1: Find the stationary points on the curve $y = 3x^3 + 18x^2 + 27x + 1$ and determine their nature.

....

Q2: Find the stationary points on the curve $y = x^4 - 4x^3 + 1$ and determine their nature.

TOPIC 8. CURVE SKETCHING

8.1.2 Closed intervals

> **Key point**
>
> The maximum or minimum value in a closed interval occurs at either the stationary point or an endpoint.

Maximums and minimums on a closed interval

The graph has been drawn with a closed interval of $-5 \leq x \leq 7$.

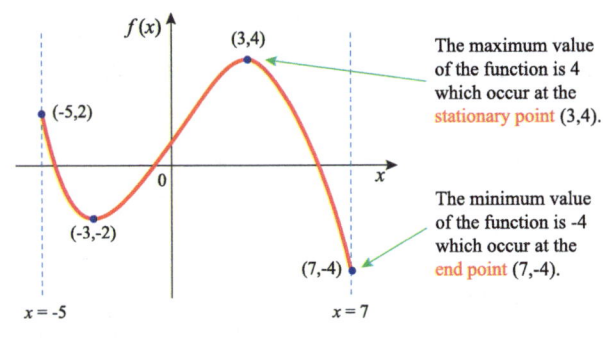

The maximum value of the function is 4 which occur at the stationary point (3,4).

The minimum value of the function is -4 which occur at the end point (7,-4).

Notice that the minimum turning point is at (-3,-2), but in this case the end point is at (7,-4) which gives a lower value for the function.

Whereas the maximum turning point is at (3,4) which gives a higher value than the end point which is at (-5,2).

Example

Problem:

Find the maximum and minimum values of $f(x) = 3x - x^3$ in the closed interval $-1 \cdot 5 \leqslant x \leqslant 3$.

Solution:

It is much easier to spot the maximum and minimum values if we make a sketch.

Step 1: Find the stationary points and their nature: $f'(x) = 3 - 3x^2$

Stationary points occur when $f'(x) = 0$.

$$\begin{aligned} 3 - 3x^2 &= 0 \\ 3\left(1 - x^2\right) &= 0 \\ 3\left(1 - x\right)\left(1 + x\right) &= 0 \\ 1 - x &= 0 \quad \text{or} \quad 1 + x = 0 \\ x &= 1 \quad \text{or} \quad x = -1 \end{aligned}$$

© HERIOT-WATT UNIVERSITY

When $x = 1$ then $y = 3 \times 1 - 1^3 = 2$.

When $x = -1$ then $y = 3 \times (-1) - (-1)^3 = -2$.

Thus the stationary points are (1,2) and (-1,-2).

x	-2 $\rightarrow$	-1	0 $\rightarrow$	1	2 $\rightarrow$
$(1-x)$	+	+	+	0	-
$(1+x)$	-	0	+	+	+
$3(1-x)(1+x)$	-	0	+	0	-
shape	\	—	/	—	\

This gives us a maximum turning point at (1,2) and a minimum turning point at (-1,-2).

Step 2: Find the end points.

When $x = -1 \cdot 5$ then $f(-1 \cdot 5) = -1 \cdot 125$.

When $x = 3$ then $f(3) = -18$.

Step 3: Make a sketch.

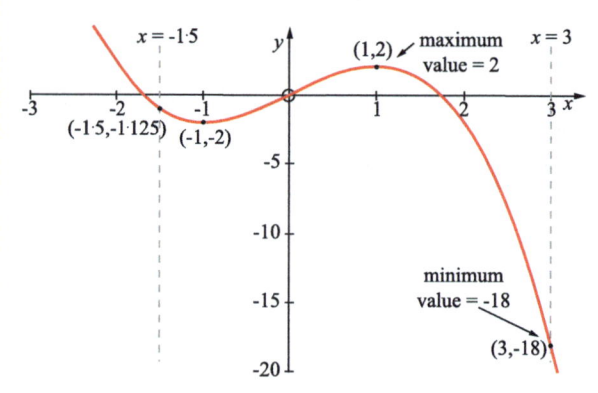

Hence the maximum value is 2 at the maximum turning point (1,2) and the minimum value is -18 at the end point (3,-18).

Closed intervals exercise

Go online

Q3:

Find the maximum and minimum values of $y = x^3 - 12x$ in the closed interval $-5 \leqslant x \leqslant 3$.

TOPIC 8. CURVE SKETCHING

8.1.3 Transformations of graphs

Sketching graphs of $f(x) + a$ **and** $f(x) - a$

> **Key point**
>
> Applying the transformation of the form $f(x) + a$ or $f(x) - a$ causes the graph of $y = f(x)$ to be moved up or down respectively by a units.

Examples

1. Problem:

Part of the graph $y = f(x)$ is shown below.

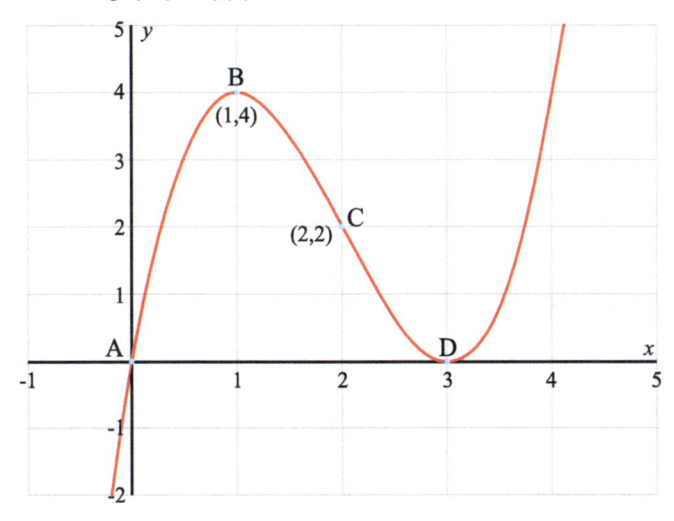

Sketch the related function $y = f(x) - 3$, identifying the coordinates of A, B, C and D.

Solution:

$f(x) - 3$ tells us that the graph of the function $f(x)$ has been moved down vertically by 3 units.

This transformation causes the y-coordinates of A, B, C and D to be reduced by 3 units giving:

A(0,0)	⇒	(0,-3)
B(1,4)	⇒	(1,1)
C(2,2)	⇒	(2,-1)
D(3,0)	⇒	(3,-3)

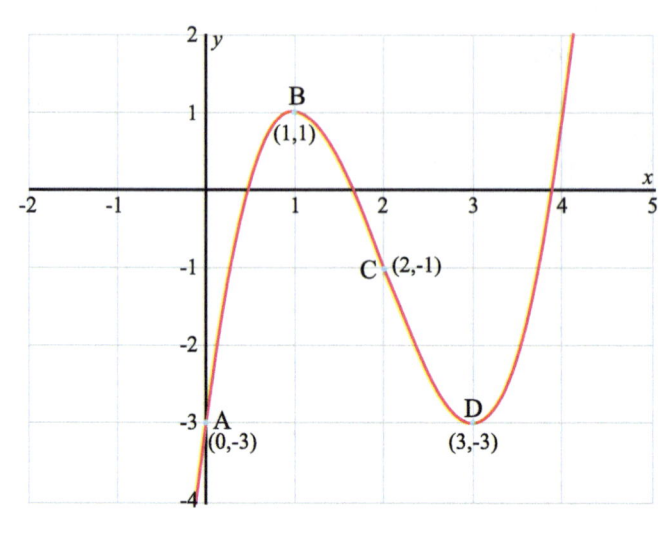

2.
Problem:

The diagram below is part of the graph of $y = f(x) - 3$.

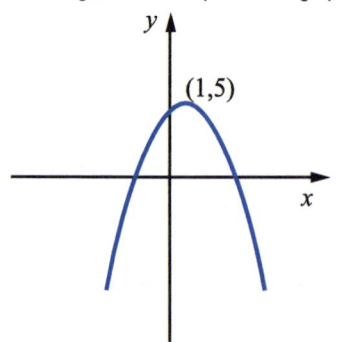

What are the coordinates of the turning point on the graph of $y = f(x)$?

Solution:

$f(x) - 3$ tells us that the graph of the function $f(x)$ has been moved down vertically by 3 units.

To undo this transformation, the graph of $f(x)$ must be 3 units up from $f(x) - 3$.

All we have to do is add 3 to the y-coordinate of the turning point. So,

f(x) − 3		**f(x)**
(1,5)	⇒	(1,8)

TOPIC 8. CURVE SKETCHING

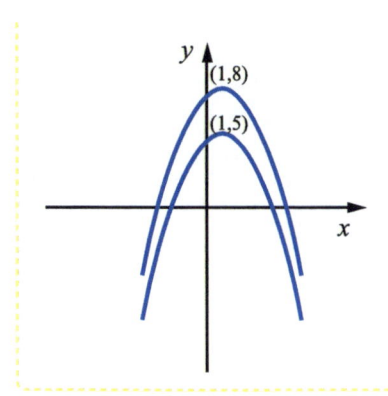

Sketching graphs of $f(x + a)$ **and** $y = f(x - a)$

> **Key point**
>
> Applying the transformation of the form $f(x + a)$ or $f(x - a)$ causes the graph of $y = f(x)$ to be moved left or right respectively by a units.

Examples

1. Problem:

The graph of $y = h(x)$ is shown below.

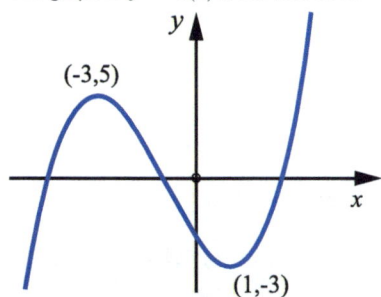

Sketch and annotate the graph of $y = h(x + 2)$.

Solution:

The transformation $h(x + 2)$ moves the graph of $h(x)$ left horizontally by 2 units. To do this we subtract 2 from the x-coordinates. The identifiable points are:

h(x)		h(x + 2)
(-3,5)	$\Rightarrow$	(-5,5)
(1,-3)	$\Rightarrow$	(-1,-3)

© HERIOT-WATT UNIVERSITY

TOPIC 8. CURVE SKETCHING

Giving,

(-5,5)
(-1,-3)

....

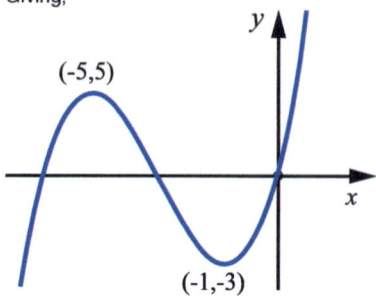

2. Problem:

The graph of $y = k(x)$ is shown below.

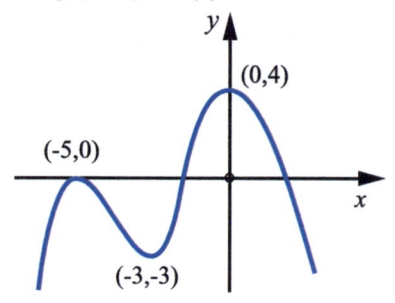

(0,4)
(-5,0)
(-3,-3)

Sketch and annotate the graph of $y = k(x - 4) + 2$.

Solution:

The transformation $k(x - 4) + 2$ moves the graph of $k(x)$ right horizontally by 4 units and up vertically by 2 units.
To do this we add 4 to the x-coordinates and add 2 to the y-coordinates.

The identifiable points are:

k(x)		**k(x − 4) + 2**
(-5,0)	⇒	(-1,2)
(-3,-3)	⇒	(1,-1)
(0,4)	⇒	(4,6)

Giving,

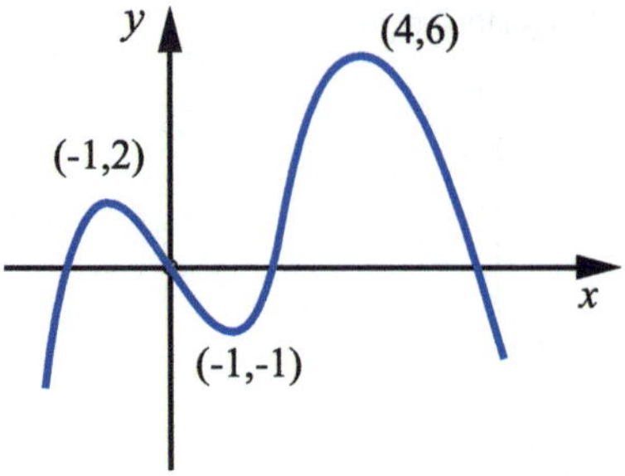

Sketching graphs of $y = -f(x)$ and $y = f(-x)$

Key point

Applying a transformation of the form $-f(x)$ causes the graph of $y = f(x)$ to be reflected in the x-axis.

Applying a transformation of the form $f(-x)$ causes the graph of $y = f(x)$ to be reflected in the y-axis.

Examples

1. Problem:

Part of the graph $y = f(x)$ is shown next.

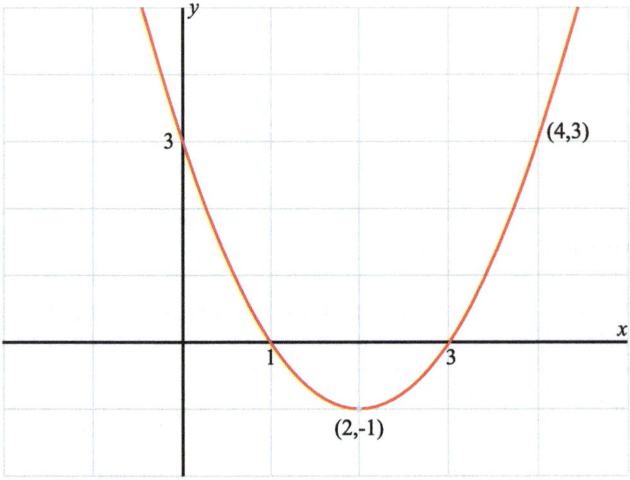

What would the graph of $y = f(-x)$ look like?

Solution:

A reflection in the y-axis causes the x-coordinates to change their sign.

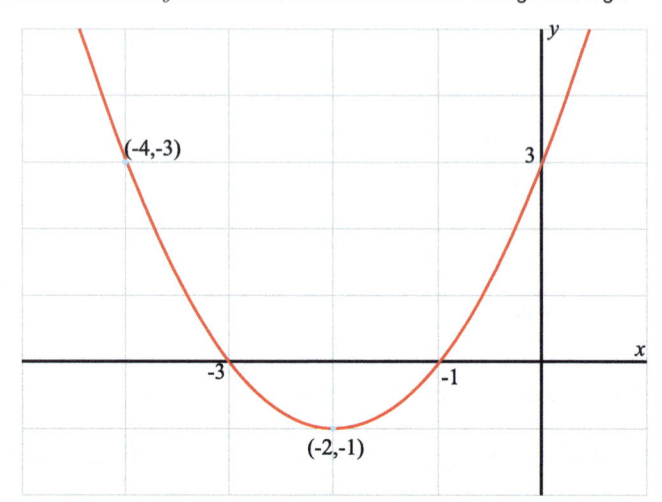

[Graph showing parabola with points (-4,-3), (-2,-1), intercepts at -3, -1, and y-intercept 3]

....

2. Problem:

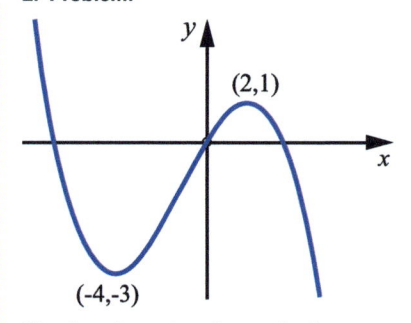

[Graph of $g(x)$ showing cubic curve with local maximum at (2,1) and point (-4,-3)]

Sketch and annotate the graph of $y = 2 - g(x)$.

Solution:

The transformation $2 - g(x) = -g(x) + 2$ and reflects the graph of $g(x)$ in the x-axis and moves the graph up vertically by 2 units.

To do this we change the sign of the y-coordinates and add 2 to the y-coordinates.

The identifiable points are:

g(x)		-g(x)		-g(x) + 2
(-4,-3)	$\Rightarrow$	(-4,3)	$\Rightarrow$	(-4,5)
(2,1)	$\Rightarrow$	(2,-1)	$\Rightarrow$	(2,1)

TOPIC 8. CURVE SKETCHING

Giving,

(Graph showing curve with points (-4,5) and (2,1))

....

3.

Problem:

The diagram below shows part of the graph of $y = g(x)$.

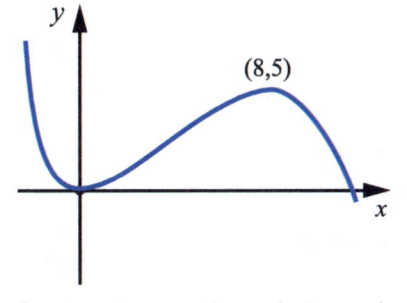

(Graph showing curve with point (8,5))

Sketch and annotate the graph of $y = g(3 - x) - 2$.

Solution:

It is important to remember that any horizontal translation happens first.

The transformation $g(3 - x) - 2 = g(-x + 3) - 2$ and moves the graph of $g(x)$ left horizontally by 3 units, reflects it in the y-axis and moves it down vertically by 2 units.

The identifiable points are:

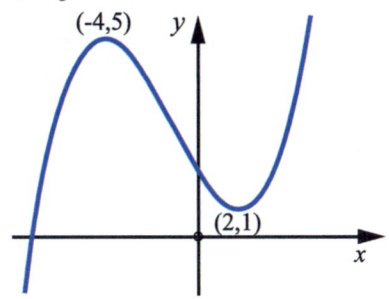

g(x)		g(x + 3)		g(−x + 3)		g(−x + 3) − 2
(0,0)	⇒	(-3,0)	⇒	(3,0)	⇒	(3,-2)
(8,5)	⇒	(5,5)	⇒	(-5,5)	⇒	(-5,3)

© HERIOT-WATT UNIVERSITY

Giving,

(-5,3)

(3,-2)

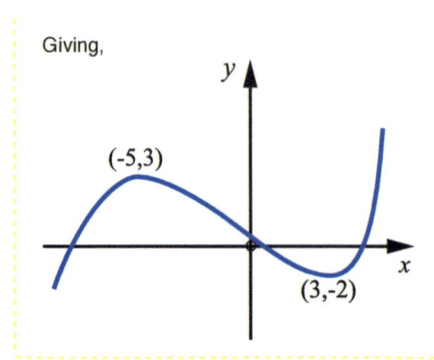

Sketching graphs of $y = k f(x)$

> **Key point**
>
> Applying the transformation of the form $kf(x)$ causes the graph of $y = f(x)$ to be scaled (stretched or squashed) vertically by a factor of k.
>
> - stretch for $k > 1$
> - squash for $0 < k < 1$

> **Top tip**
>
> It may help to remember what happens to a trig graph.
>
> - $y = \sin x$ has a maximum value of 1 and a minimum value of -1
> - $y = 5 \sin x$ has a maximum of 5 and a minimum of -5 (stretched vertically)
> - $y = 1/2 \sin x$ has a maximum of $1/2$ and a minimum of $-1/2$ (squashed vertically)

TOPIC 8. CURVE SKETCHING

Examples

1. Problem:

Part of the graph $y = h(x)$ is shown, sketch and annotate the graph of $y = -1/2h(x)$.

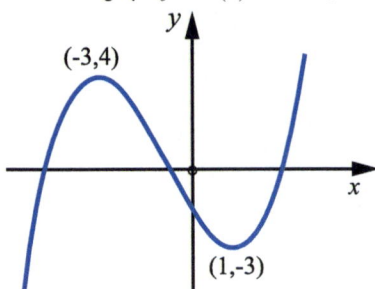

Solution:

The transformation $-1/2h(x)$ reflects the graph of $h(x)$ in the x-axis and squashes it vertically by a factor of a half.

To do this we change the sign of the y coordinates and half the y coordinates.

The identifiable points are:

$\mathbf{h(x)}$		$-1/2\mathbf{hf(x)}$
(-3,4)	⇒	(-3,-2)
(1,-3)	⇒	(1,1·5)

Giving,

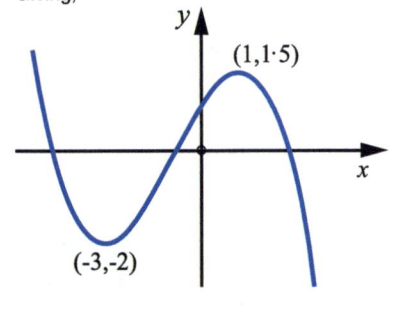

....

2.

Problem:

The diagram below shows part of the graph of $y = 5f(x) - 1$.

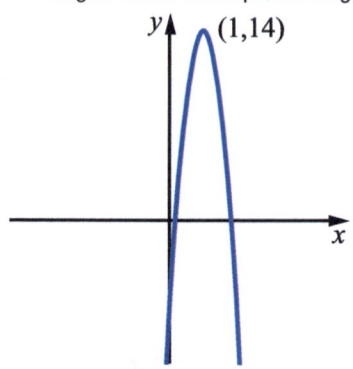

What are the coordinates of the turning point on the graph of $y = f(x)$?

Solution:

$5f(x) - 1$ tells us that the graph of the function $f(x)$ has been stretched vertically by a factor of 5 and moved down by 1 unit.

To undo this transformation, the graph must be moved up by 1 unit and squashed vertically by a factor of a fifth.

All we have to do is add 1 to the y-coordinate and divide it by 5.

So,

$$5f(x) - 1 \qquad\qquad f(x)$$
$$(1, 14) \qquad \Rightarrow \qquad (1, 3)$$

Giving,

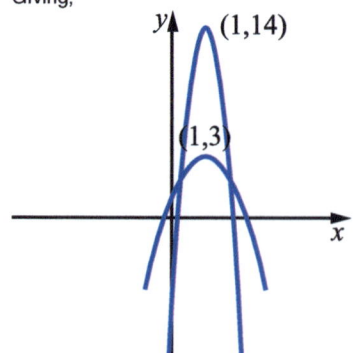

TOPIC 8. CURVE SKETCHING

Sketching graphs of $y = f(kx)$

> **Key point**
>
> Applying the transformation of the form $f(kx)$ causes the graph of $y = f(x)$ to be scaled (stretched or squashed) horizontally by a factor of k.
>
> - squash for $k > 1$
> - stretch for $0 < k < 1$

> **Top tip**
>
> It may help to remember what happens to a trig graph.
>
> - $y = \cos x$ has 1 wave or cycle or repeat in 360°
> - $y = \cos 5x$ has 5 waves in 360° (squashed horizontally)
> - $y = \cos {}^1/_2 x$ has half a wave in 360° (stretched horizontally)

Examples

1. Problem:

Part of the graph $y = h(x)$ is shown, sketch and annotate the graph of $y = h(^1/_4 x)$.

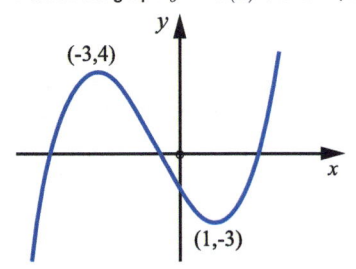

Solution:

The transformation $h(^1/_4 x)$ stretches the graph horizontally. To do this we divide the x-coordinates by $^1/_4$ which is the same as multiplying the x-coordinates by 4.

The identifiable points are:

h(x)		**h($^1/_4$x)**
(-3,4)	$\Rightarrow$	(-12,4)
(1,-3)	$\Rightarrow$	(4,-3)

Giving,

© HERIOT-WATT UNIVERSITY

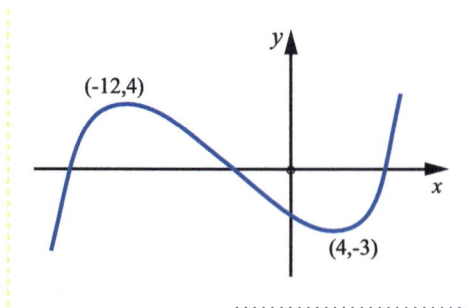

2. Problem:

The diagram shows part of the graph of $y = g(x)$, sketch and annotate the graph of $y = g(4x) + 1$.

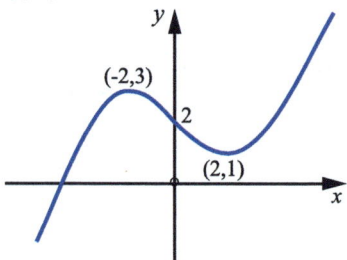

Solution:

The transformation $g(4x) + 1$ squashes the graph of $g(x)$ horizontally and moves it up vertically by 1 unit. To do this we divide the x-coordinates by 4 and add 1 to the y-coordinates.

The identifiable points are:

g(x)		g(4x) + 1
(-2,3)	$\Rightarrow$	(-1/2,4)
(0,2)	$\Rightarrow$	(0,3)
(2,1)	$\Rightarrow$	(1/2,2)

Giving,

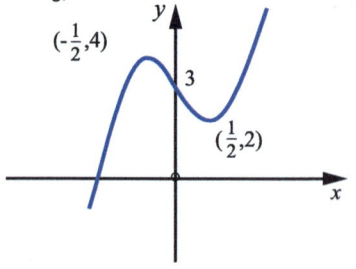

TOPIC 8. CURVE SKETCHING

3. Problem:

The diagram shows part of the graph of $y = g(x)$, sketch and annotate the graph of $y = 2g(x - 1)$.

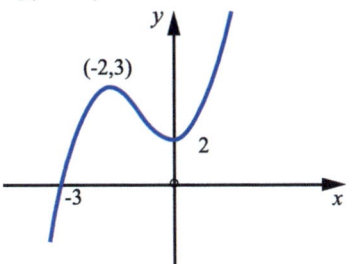

Solution:

The transformation $2g(x - 1)$ moves the graph horizontally right by 1 unit and stretches it vertically by a factor of 2. To do this we add 1 to the x coordinates and multiply the y coordinates by 2.

The identifiable points are:

g(x)		2g(x − 1)
(-3,0)	⇒	(-2,0)
(-2,3)	⇒	(-1,6)
(0,2)	⇒	(1,4)

Giving,

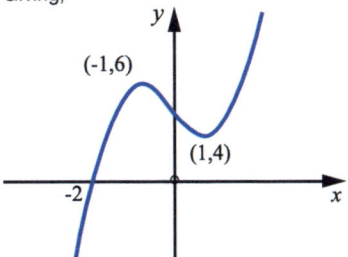

Transformations of graphs exercise

Q4: The graph of $y = f(x)$ is shown, which of the following graphs represents $y = f(x) + 2$?

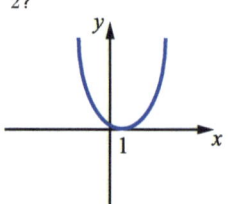

a) (1,2)

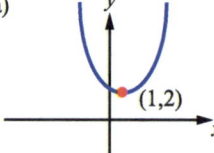

b) (-1,0)

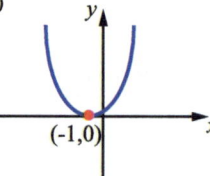

c) (3,0)

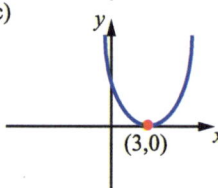

d) (1,-2)

....

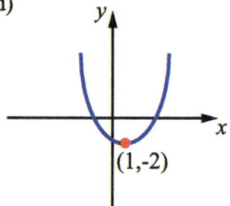

Q5: This diagram is part of the graph of $y = f(x) + 7$.

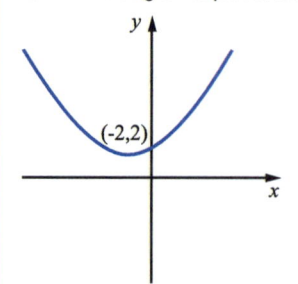

(-2,2)

What are the coordinates of the turning point on the graph of $y = f(x)$?

....

TOPIC 8. CURVE SKETCHING

Q6: The graph of $y = g(x)$ is shown, sketch and annotate the graph of $y = g(x+1) - 3$.

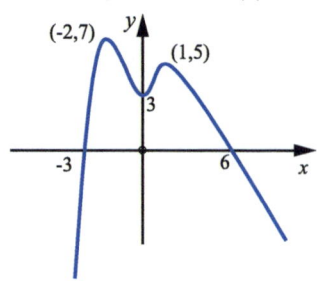

....

Q7: The graph of $y = f(x-3) + 4$ is shown. What are the coordinates of the minimum turning point on the graph of $y = f(x)$?

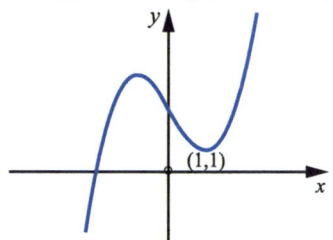

....

Q8: The graph of $y = f(x)$ is shown, which of the following graphs represents $y = -f(x)$?

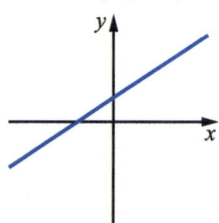

a)

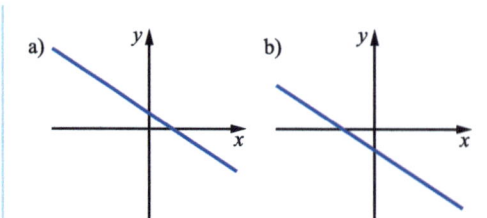

b)

c)

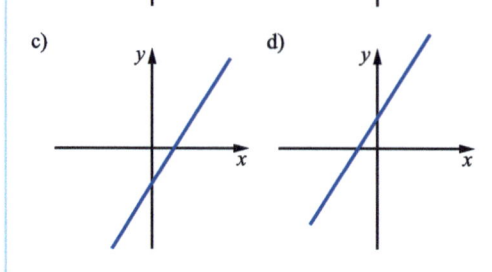

d)

Q9: The graph of $y = f(x)$ is shown, which of the following graphs represents $y = f(-x)$?

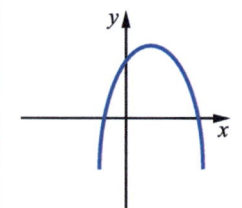

a)

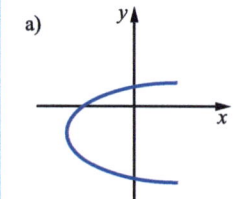

b)

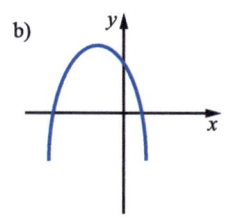

c)

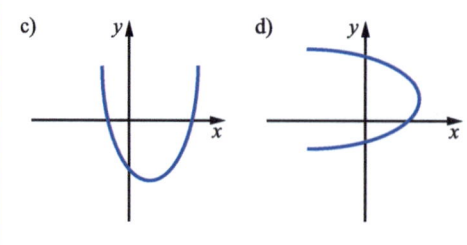

d)

TOPIC 8. CURVE SKETCHING

Q10: This diagram is part of the graph of $y = f(-x) + 3$.

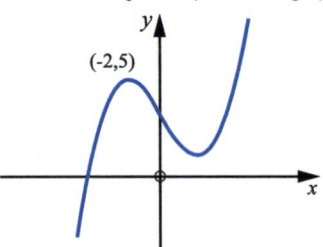

(-2,5)

What are the coordinates of the maximum turning point on the graph of $y = f(x)$.

....

Q11: $f(x) = \sin x$, which of the following represents the graph of $y = 2f(x)$?

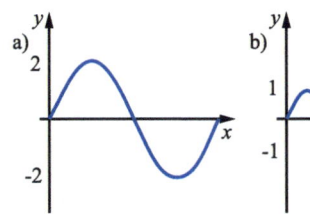

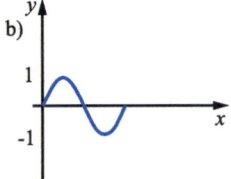

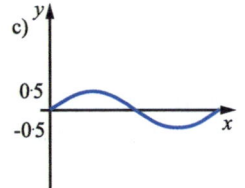

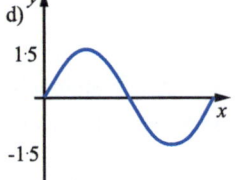

a) 2, -2

b) 1, -1

c) 0.5, -0.5

d) 1.5, -1.5

....

Q12: Part of the graph of $y = f(x)$ is shown, sketch and annotate the graph of $y = \frac{1}{2}f(x+3)$.

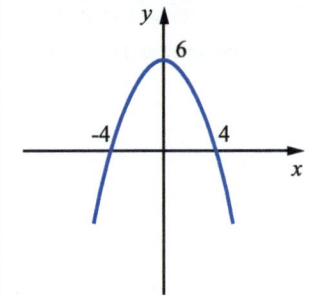

6, -4, 4

Q13: $f(x) = \cos x$, which of the following represents the graph of $y = f(3x)$?

a)

b)

c)

d)

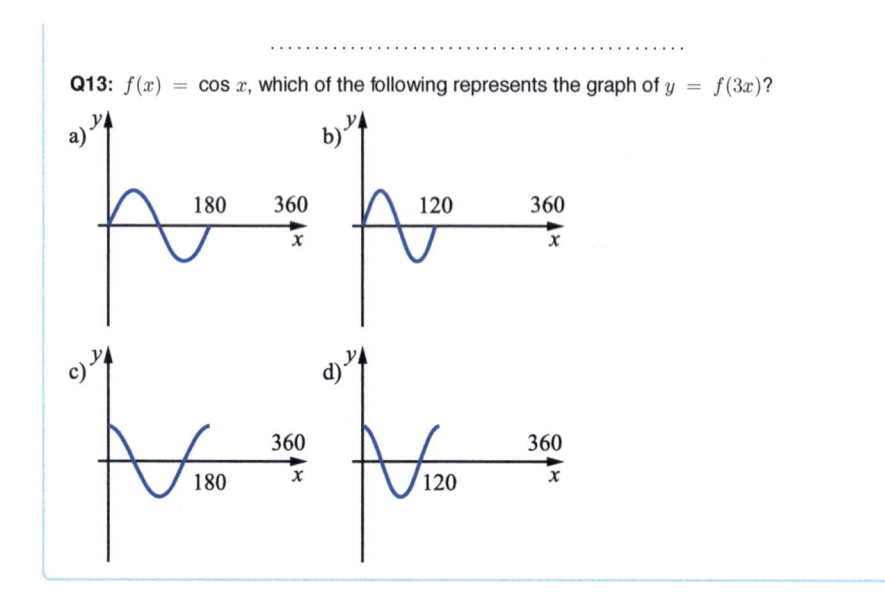

8.2 Functions

Why study functions?

Functions provide a way of describing mathematically the relationship between two quantities.

The key idea is that the values of two variables are related. For example, the price of a bag of potatoes depends on how much they weigh.

Another example is the amount of tax paid on a product depends on its cost.

This topic will explore the rules defining functions and how to sketch the graphs of functions. Graphs are a useful way to gain information about a function.

Graphs of simple functions such as $f(x) = x^2$ are easily sketched, but there are also many relationships between functions that can be used to sketch other slightly more complex graphs.

These techniques will be developed and explored in order to identify some useful ways of dealing with the graphs of rational functions.

The emphasis will be on sketching the graphs.

TOPIC 8. CURVE SKETCHING

8.2.1 Function definition

Set notation

The standard number sets shown are used in this section.

The **standard number sets** are:

- $\mathbb{N} = \{1, 2, 3, 4, 5, ...\}$ the set of natural numbers.
- $\mathbb{W} = \{0, 1, 2, 3, 4, 5, ...\}$ the set of whole numbers.
- $\mathbb{Z} = \{..., -3, -2, -1, 0, 1, 2, 3, ...\}$ the set of integers.
- $\mathbb{Q}$ = the set of all numbers which can be written as fractions, called the set of rational numbers.
- $\mathbb{R}$ = the set of rational and irrational numbers, called the set of real numbers.

These sets and their subsets (such as $\mathbb{Z}^+ = \{1, 2, 3, ...\}$ or $\mathbb{R}^+ = \{\, x \in \mathbb{R}, x > 0 \,\}$) are the only standard number sets used in this section.

Remember this notation $\in$? This symbol is a short way of saying "**is a member of**" or "**belongs to**".

- $\{x \in \mathbb{R}, x > 0\}$ - We say "x is a member of the real numbers but only when x is greater than 0". In other words, all the real numbers bigger than zero.
- $\{x \in \mathbb{R}, x \neq 3\}$ - We say "x is a member of the real numbers but x cannot be equal to 3." This means x can be any real number except 3.
- $\{x \notin \mathbb{N}\}$ - We say "x is not a member of the natural numbers". In other words x can be any number except the natural numbers.
- $\{x \in \mathbb{Z}^- : x \neq -1\}$ - We say "x is a member of the negative integers such that x cannot be equal to -1". This means x is any negative integer except 1.

Notice the " : " used in the last bullet point? In maths it is read as "such that".

Function definition

Simply, a function is a rule which connects an input to an output.

Example

If Emma saves £100 every month. Then the amount she saves is related to the number of months using the function t:

$t(\text{"months"}) = 100 \times \text{months}$

So if she saves for 6 months, the total is, $t(6) = £600$.

However, not all input values may work. For instance negative months do not make sense. This more formal definition of a function, domain, range and codomain will clear up some of these issues.

298 TOPIC 8. CURVE SKETCHING

> **Key point**
>
> A **function f** from set A to set B is a rule which assigns to each element in A exactly one element in B. This is often written as $f : A \to B$.
>
> (Read as f from A to B.)

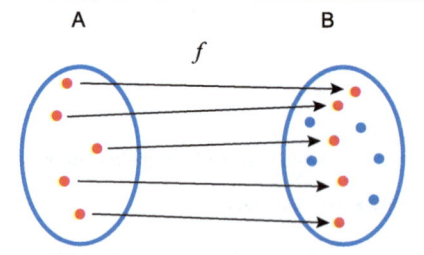

A   B

f

The collection of values that **can go into** a function is called the **domain**.

> **Key point**
>
> For a function $f : A \to B$, A is called the **domain** of the function f.

We could choose the values that go into the domain by restricting the numbers that go into the function (as long as they can go into the function).

For example, when drawing the function $f(x) = x^2$ the input values could be restricted to between -2 and 2. The function $f(x) = x^2$ though can be drawn for all values of $x \in \mathbb{R}$, but we have chosen to restrict the domain.

In this section unless otherwise stated, the domain of a function will be the largest possible set of x values for which the rule defining f (x) makes sense.

For example if $f(x) = \frac{1}{x}$ then the domain is {x $\in \mathbb{R} : x \neq 0$}

The **codomain** is the collection of all possible values that you could get out of a function.

> **Key point**
>
> For a function $f : A \to B$, B is called the **codomain** of the function f.

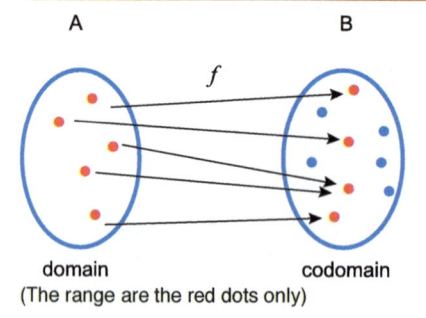

A   B

f

domain   codomain
(The range are the red dots only)

© HERIOT-WATT UNIVERSITY

TOPIC 8. CURVE SKETCHING

Lastly, the **range** or **image**, is a set within the codomain. These are the values that you get out when you have evaluated the domain.

> **Key point**
>
> For a function $f : A \to B$, the set C of elements in B which are images of the elements in A under the function f is called the **image set or range** of the function f. C is always contained in or equal to B. This is written $C \subseteq B$.

This means that the range could be the same as the codomain or it could be a smaller set of this.

> **Key point**
>
> To define a function fully it is necessary to specify:
>
> - The **domain**
> - The **codomain**
> - The rule

Previously in your mathematics studies all of these may not have been defined for the function. This is because it is assumed that the domain and codomain are known. However, there are times when it is not obvious and they need to be defined. This topic will demonstrate this.

Example

Take the function $f(x) = x^2$.

The domain is defined as: x is a member of the set {1,2,3,4,5}.

In shorthand this is $x \in$ { 1,2,3,4,5}

Evaluating each of these values we have the **range**: $f(x) \in \{1, 4, 9, 16, 25\}$

However, the **codomain** is the collection of all the possible output values of $f(x) = x^2$.

So the codomain is $f(x) \in \mathbb{N}$. This is since the domain was originally stated as elements from the Natural numbers.

If the domain was given as: $x \in \mathbb{R}$, then the range would have been $f(x) \in \mathbb{R}$, which would be the same as the codomain.

Sometimes the range of a function is unknown, because it is too complicated or the function is not fully known. In this case the codomain defines the set of numbers that the range lies in so that we can continue with the problem. As can be seen in the example above, the number set of the codomain is linked to the number set of the domian.

Initially the domain had elements belonging to the Natural numbers so the codomain was all Natural numbers.

In the second part the domain changed to Real numbers, so the codomain was all Real numbers.

Function notation

We now have definitions for a function, domain, range and codomain. It is important therefore that when a function is defined it should be followed by its domain.

An example of familiar notation is: $f(x) = x^2$, where $-2 \leqslant x \leqslant 2$.

However, if you were to look at other texts there are various ways of writing this down and you should be familiar with these. The most common form however, is the one above.

- Let f be the function defined by $f(x) = x^2$, where $x \in [-2, 2]$
- $f : [-2, 2] \to \mathbb{R}$
  $f : x \to x^2$
- f is defined by $f : x \to x^2$, where $x \in [-2, 2]$
- f is defined on the set T, where $T = \{x : x \in [-2, 2]\}$ by $f : x \to x^2$
- f is defined by $f(x) = x^2$ with domain $\{ x : -2 \leq x \leq 2\}$

This would give the following graph. This type of function is well known.

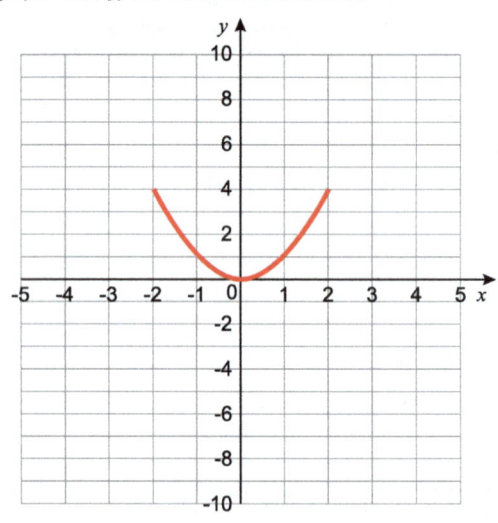

$y = x^2$

A different type of function (with a different notation) is: $f(x) = \begin{cases} 2x : x \leqslant 2 \\ x^2 : x > 2 \end{cases}$

TOPIC 8. CURVE SKETCHING

This function looks like:

$$f(x) = \begin{cases} 2x : x \leqslant 2 \\ x^2 : x > 2 \end{cases}$$

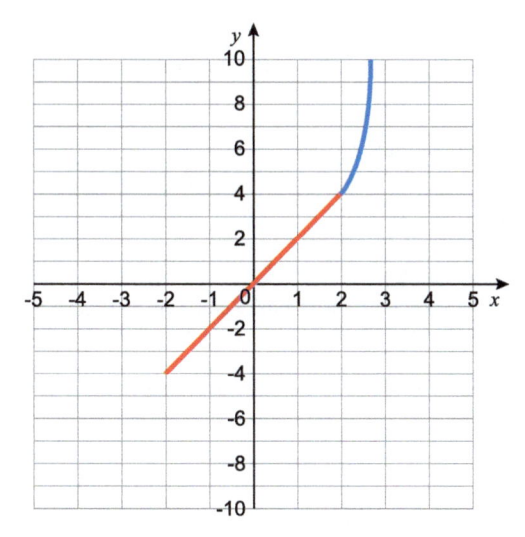

This known as a **discontinuous** or **piecewise function**. It is called this because it is made up of difference functions joined or pieced together.

In the above example, the function of $f(x) = 2x$ is drawn for the values of $x = 2$ and below. After this, for $x > 2$ the graph of x^2 is drawn.

When dealing with functions it is common practice to use the letter defining the function such as f instead of $f(x)$.

The graph of a function $f(x)$ is usually drawn as the curve $y = f(x)$ instead of $f(x)$ when referring to it. This has already been seen in the function notation previously shown.

Examples

1. Problem:

$f(x) = x + 7$ with a domain of {1, 2, 3}.

Solution:

This function has a **codomain** of $\mathbb{N}$ and a **range** of (8, 9, 10).

This shows that the range $\subseteq$ codomain.

....

2. Problem:

As shown earlier the function $f(x) = x^2$ has the graph of $y = x^2$.

Define the function by giving the domain, codomain and rule given in the example above.

© HERIOT-WATT UNIVERSITY

Solution:

It had a domain of [-2, 2] and a codomain of the real numbers. The rule is that every value of x maps to x^2.

Note that the range is actually [0, 4].

x	-2	1	0	1	2
x²	4	1	0	1	4

This is all possible values of y for $x \in [-2, 2]$

Care must be taken to cover all situations. Here, if the end points of the domain are used then $f(x) = 4$ in both cases and the fact that $0^2 = 0$ could be missed.

Notice that the range [0, 4] lies inside the codomain of $\mathbb{R}$ as required.

Finding the range of a function is not always easy.

Examples

1. Problem:
Determine the largest possible domain for the following rule.

$k(z) = \sqrt{\frac{1}{z}}$

Solution:

$\sqrt{\frac{1}{z}}$ is undefined when $z = 0$ since $\frac{1}{0}$ is not defined and also when z is negative.

This means that $z > 0$ (the set of all positive real number excluding zero).
The largest possible domain is therefore $\mathbb{R}^+$ (the set of all positive real numbers).

....

2. Problem:
Using the definition $f : [-2, 2] \to \mathbb{R}$, write the function f defined by:
$f : [0, \pi] \to \mathbb{R}$
$f : x \to \sin x$
using two other methods of notation for a function.

This means that the function f has domain of $[0, \pi]$, codomain of the real numbers and maps every value of x to the corresponding value $\sin x$.

Note again that only part of this codomain of $\mathbb{R}$ will actually be the range. This range however will be totally included in $\mathbb{R}$.

Solution:
When writing the rule of a function. There are two ways of representing the function:
$f(x) = \sin x$ or $f : x \to \sin x$

TOPIC 8. CURVE SKETCHING

And then there are several ways of defining the domain:

$x \in [0, \pi]$

$f : [0, \pi] \to \mathbb{R}$

$x : x \in [0, \pi]$

$0 \leq x \leq \pi$

Any combinations of these would be accepted. For instance:

$f(x) = \sin x$, where $0 \leq x \leq \pi$

$f : x \to \sin x$, where $x \in [0, \pi]$

....

3. Problem:

What is the range of the function f which maps x to $\sin x$ where x has a domain of $[0, \pi]$ and codomain of $\mathbb{R}$?

Solution:

$f(x) = \sin x$ and the domain is $[0, \pi]$

Draw the graph and note that $f(0) = 0$ and $f(90) = 1$

The range is therefore [0, 1]

The following points need to be considered when determining the range of the function.

SHAPE: Think about the shape of a sin graph and its maximum and minimum values. It may be that this is enough to determine the range.

GRAPH VALUES: Consider the values which can be given to x.

For example: 0, $\frac{\pi}{2}$, $\frac{\pi}{4}$, $\frac{\pi}{6}$ and so on. Work out the values of $\sin x$. Sketch the graph if it helps.

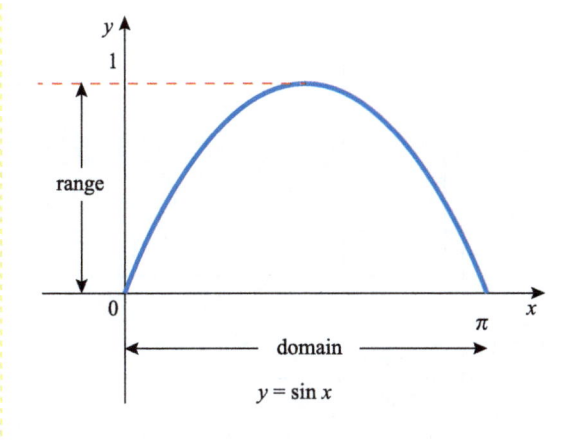

$y = \sin x$

Here the range is [0, 1]

This can also be written as $0 \leq f(x) \leq 1$ or $\{f(x) \in \mathbb{R} : 0 \leqslant f(x) \leqslant 1\}$.

This means that the codomain is the real number line, but the range of this function is between 0 and 1 and includes 0 and 1.

Domain and codomain range

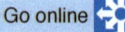

Go online

Q14: Use these domains and ranges to complete the details of the four graphs.

$x \in [0, 2\pi]$	$y \in \mathbb{R}$	$x \neq 3$	$-\infty \leqslant y \leqslant \infty$
$-\infty \leqslant x \leqslant \infty$	$x \geqslant \frac{3}{5}$	$y \in [-1, 1]$	$y \geqslant 0$

Graph of $y = \sqrt{5x - 3}$

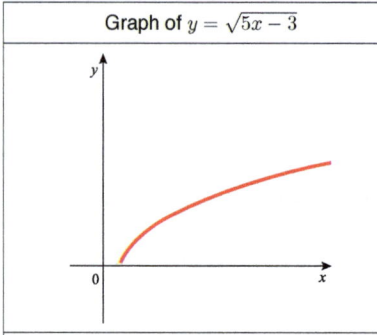

Domain:
Range:

Graph of $y = \frac{1}{x-3}$

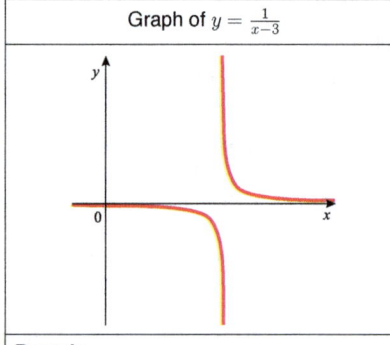

Domain:
Range:

Graph of $y = -\cos x$

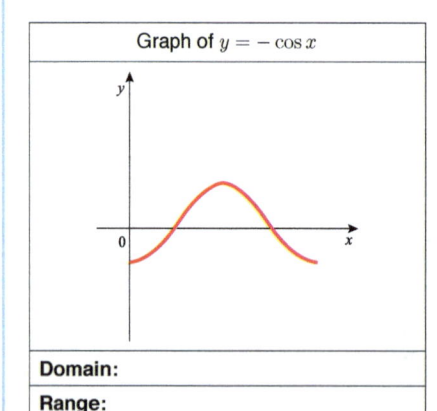

Domain:
Range:

Graph of $y = -3x + 2$

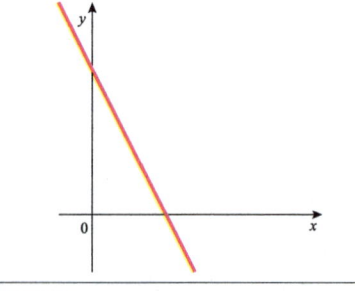

Domain:
Range:

TOPIC 8. CURVE SKETCHING

Function definition exercise   Go online

Q15: What is the largest domain and range of $y = \frac{1}{x+5}$?

Q16: What is the largest domain and range of $f(x) = \frac{1}{x^2+3x+2}$?

Q17: What is the largest domain and range of $y = 2x^2 - 5x + 16$?

Q18: What is the range of the function $y = \tan x$ when the domain is $-45° \leqslant x \leqslant 45°$?

Q19: What is the range of $f(x) = x - 2$ with domain $\{x \in \mathbb{Z} : x > 2\}$?

Q20: What is the range of the function $y = x^2 + 4$ with the domain $\{x \in \mathbb{Z}\}$?

Q21: What is the range of $y = \frac{1}{x+1}$ with the domain $\{x \in \mathbb{N} : x > 4\}$?

Q22: What is the largest possible domain and range of the function $y = \sqrt{x^2 - 1}$?

8.2.2 One-to-one and onto functions

In the definition of a function the elements in set A must map to one and only one element of set B. As seen previously this can be shown in a function diagram.

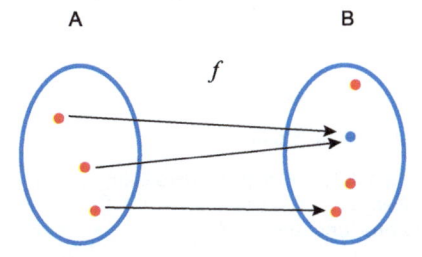

Notice that an element of Set A goes to only one element of Set B, but one element in set B can go to more than one element in Set A. Look at the blue dot.

A **one-to-one** function does not allow this. This means that each value of $f(x)$ in the range is produced by one and only one value of x in the domain. One x goes to one y and one y goes to one x.

A formal definition of this is:

> **Key point**
>
> A function $f : A \rightarrow B$ is a **one-to-one function** if whenever $f(s) = f(t)$, then $s = t$ where $s \in A$ and $t \in B$. The function is said to be in one-to-one correspondence.

Consider the function $y = x + 3$

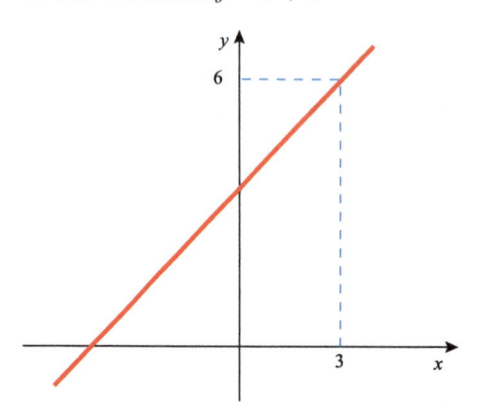

It is clear to see from this graph that when $x = 3$ there is one y value, $y = 6$. Conversely, when $y = 6$, there is only one possible x value, $x = 3$. This is the same for all x and y. This is a **one-to-one function**.

At any point on the graph of a one-to-one function this property will exist.

A one-to-one function can be identified from its graph. Consider these examples.

One-to-one demonstration   Go online

1. Place your ruler over the graph.
2. Move the ruler up & down.
3. The ruler meets the graph no more than once at any point, therefore $f(x) = x + 1$ is a **one-to-one function.**

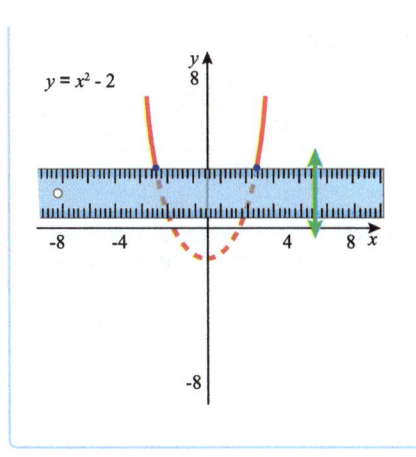

1. Place your ruler over the graph.
2. Move the ruler up & down.
3. The ruler meets the graph at more than one point above the ruler, therefore $f(x) = x^2 - 2$ is **NOT** a **one-to-one function.**

It is possible, however, for more than one element of set A to map to the same element in set B.

This is demonstrated by the example $y = x^2 - 2$ above. When $x = 4$, $y = 14$ and when $x = -4$, $y = 14$. The y-coordinates are the same.

> **Key point**
>
> A function which maps more than one element in the domain to the same element in the range or image set is called a **many-to-one** or a many-one function.
>
> The function is said be in many-to-one correspondence.
>
> It is also common to say that such a function is not one-to-one.

If we take another example, $f(x) = x^2$ we see that $f(3) = 3^2 = 9$ and $f(-3) = (-3)^2 = 9$. So there are two different elements, 3 and -3, in the domain (set A) which map to 9 in the codomain (set B).

The function is not one-to-one. Another way of showing this is a function diagram.

$f(x) = x^2$

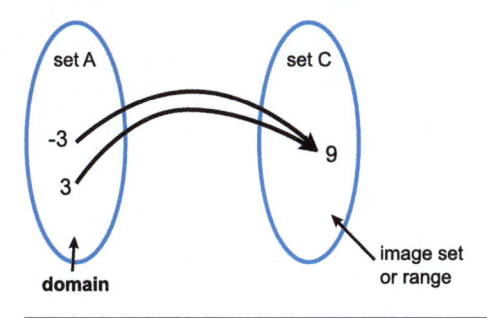

When the range is equal to the codomain the function has a special name.

> **Key point**
>
> An **onto function** is one in which the range is equal to the codomain.

Example

Consider the straight line $y = x + 2$ given above. This is a one-to-one function, but it is also an onto mapping since the range of y is the same as the codomain i.e. the set of real numbers, $\mathbb{R}$.

On the other hand, the graph $y = x^2 + 2$, $x \in \mathbb{R}$ and codomain $y \in \mathbb{R}$ is not an onto mapping. When we look at the values that $y = x^2 + 2$ can take, it gives a range of $\{y \in \mathbb{R} : y \geqslant 2\}$. All the values $y < 2$, which are part of the real numbers is not in the range. Therefore, $y = x^2 + 2$ is not an onto function.

Types of functions Go online

Q23: What type of function is this graph?

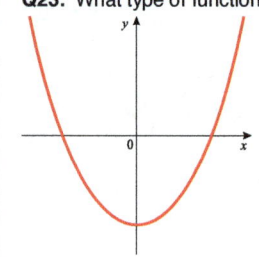

a) Many-to-one
b) One-to-one

Q24: What type of function is this graph?

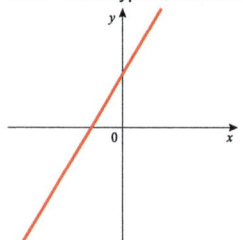

a) Many-to-one
b) One-to-one

TOPIC 8. CURVE SKETCHING

Q25: What type of function is this graph?

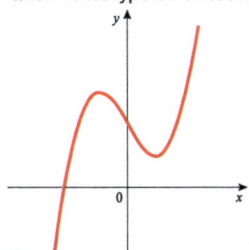

a) Many-to-one
b) One-to-one

Q26: What type of function is this graph?

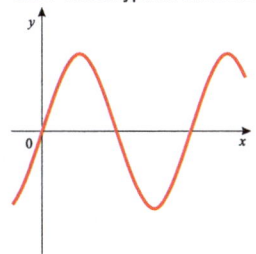

a) Many-to-one
b) One-to-one

Q27: What type of function is this graph?

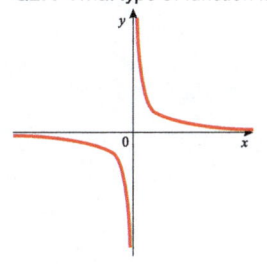

a) Many-to-one
b) One-to-one

Q28: What type of function is this graph?

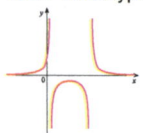

a) Many-to-one
b) One-to-one

One-to-one and onto functions exercise   Go online

Q29: What type of mapping is the function $y = 4x^2 - 1$, where $x \in \mathbb{R}$.

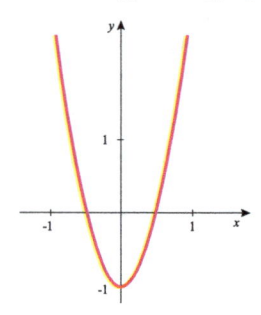

....

Q30: Is the function $f(t) = 5t - t^2$, where $t \in [-4, 2]$ a many-to-one mapping? (make a sketch of the graph)

....

Q31: What type of mapping is $y = \frac{3}{x+5}$, where $-2 \leqslant x \leqslant 5$?

....

TOPIC 8. CURVE SKETCHING

Q32: What type of mapping is $f(x) = \ln|x+1|$, where $x \in \mathbb{N}$?

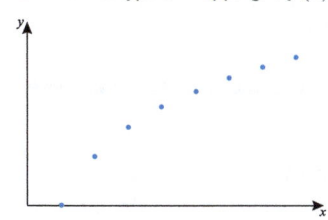

....

Q33: The function $y = \sin x$ where $x \in \mathbb{R}$ is a many-to-one mapping. Write down a possible domain that would change this into a one-to-one mapping.

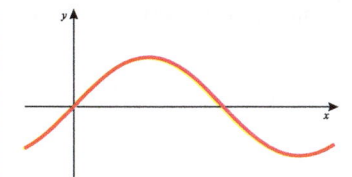

....

Q34: Is $y = x^3 - 3x^2 + 1$ an onto mapping?

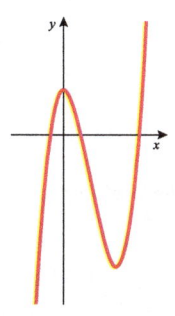

....

Q35: Is the function $y = x^2 - 3x$, where $\{x \in \mathbb{Z} : x \in [-1, 3]\}$, codomain of $\{y \in \mathbb{Z} : [-2, 6]\}$ an onto mapping?

8.2.2.1 One-to-one and onto function proofs

Earlier in this topic the terms **one-to-one** and **onto** were defined.

Recall that when defining odd and even functions $f : A \to B$ is a one-to-one function, if whenever $f(s) = f(t)$ then $s = t$ where $s \in A$ and $t \in A$.

An onto function is one in which the range is equal to the codomain.

When the unit explained odd and even functions it was stated that although the graphs of the functions could show that the property existed, a proof was still needed to be able to state that the function was odd or even.

In the same way it is necessary to show, by proof, that functions are one-to-one and/or onto before inverse functions can be explored further.

Proof 1: One-to-One functions

The technique is to start with two elements in the domain.
Now suppose that $f(s) = f(t)$.
If the function is one-to-one then it is possible to reach the conclusion that $s = t$.

> Prove that the function $f(x) = 3x^3 + 4$ is a one-to-one function.
> Let s and t be elements in the domain of the function.
> Suppose, $f(s) = f(t)$
> Then, $3s^3 + 4 = 3t^3 + 4$
> So, $s^3 = t^3$
> Thus, $s = t$

If $f(s) = f(t)$ then $s = t$ for all elements s, t belonging to the domain of the function f.

The function is a one-to-one function.

Proof 2: Onto functions - Strategy

This is much harder to establish than showing that a function is one-to-one. The proof is an *existence proof*. For functions with formulas the strategy is as follows.

Take a general element t which belongs to the codomain. Solve $t = f(s)$ for s in the domain. This gives a possible domain element. By substituting the value of this element into the function and showing that it is a suitable domain element completes the proof.

TOPIC 8. CURVE SKETCHING

Prove that the function $f: \mathbb{R} \to \mathbb{R}$ where $f(x) = 3x + 4$ is an onto function.
Suppose that t is an element belonging to $\mathbb{R}$.
Let $s = \frac{t-4}{3}$, (To find the value of s, solve the equation $t = 3s + 4$ for s.)
So,

$$f(s) = f\left(\frac{t-4}{3}\right)$$
$$= 3\left(\frac{t-4}{3}\right) + 4$$
$$= t$$

Thus for every element $t \in$ codomain of f there exists an element s in the domain of f such that $f(s) = t$.

The function is an onto function.

If, on the other hand, the task is to show that a function is not one-to-one (or onto) an example which shows this to be untrue is all that is required.

Proving a function to be untrue

Function is not one-to-one

Determine whether the function $f(x) = x^2$ is a one-to-one function.
Take the value of $s = -1$ and $t = 1$. These both give $f(x) = 1$ so here $f(s) = f(t)$ does not mean that $s = t$.
The function is not one-to-one.

Function is not onto

Determine whether the function $f: \mathbb{R}^- \to \mathbb{R}$, where $f(x) = \sqrt[3]{x}$ is onto.
Let $f(s) = 2$. There is no element $s \in \mathbb{R}^-$, the domain of f, such that $f(s) = 2$ as $f(s) = \sqrt[3]{x} \leqslant 0$ for all s.
Hence the function is not onto.

Curve sketching proofs exercise   Go online

Q36: Prove that $f(x) = 2x - 4$ for $x \in \mathbb{R}$ is a one-to-one function.
....

Q37: Prove that the function $g(y) = 2y^5$ for $y \in \mathbb{R}$ is a one-to-one function.
....

Q38: Prove that the function $h(z) = \frac{1}{z+2}$ for $z \in \mathbb{R}$, $z \neq -2$ is a one-to-one function.
....

Q39: Prove that the function $f: \mathbb{R} \to \mathbb{R}$, where $f(x) = 3x$ is an onto function.
....

© HERIOT-WATT UNIVERSITY

Q40: Prove that the function $g : \mathbb{R}^+ \to \mathbb{R}^+$ where $g(y) = y^3$ is an onto function.

....

Q41: Prove that the function $h : \{z \in \mathbb{R}; z \neq 0\} \to \{z \in \mathbb{R}; z \neq 1\}$, where $h(z) = 1 + \frac{1}{z}$ is an onto function.

....

Q42: Show that the function $f(x) = \frac{1}{2x^2}$ is not one-to-one.

....

Q43: Show that the function $f : \mathbb{R}^+ \to \mathbb{R}$ where $f(x) = \sqrt{x}$ is not onto.

8.3 Inverse functions

Consider the function $f(x) = y$ where $x \in A$ and $y \in B$.

This function maps the elements of A to elements of B.

Is it possible to find a function that reverses this?

Such a function will map elements of B back to elements of A.

If this function exists it will be the inverse function of f and is denoted f^{-1}.

Remember that a condition of a function is that it maps one element in the domain (set A) to only one element in the range (set B).

For an inverse of this function to exist, and still be a function, one element of set B will map to the one element in set A from which it came.

This is precisely what a one-to-one function does.

Using function diagrams this differences mentioned above can be seen more clearly.

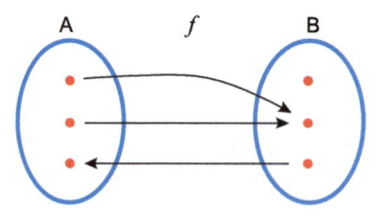

This is a function where one element of A goes to one element of B. The idea of having an inverse is to reverse this and go from B to A.

TOPIC 8. CURVE SKETCHING

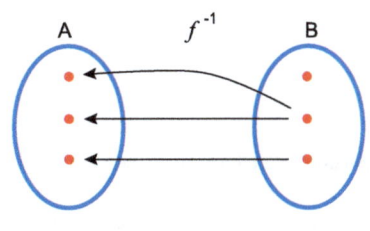

If the arrows were just to be reversed we may think that this is the inverse function f^{-1}. However, a function can only map one element of the domain to one element in the range.

It is clear here that the second element B (now the domain) maps to two different values.

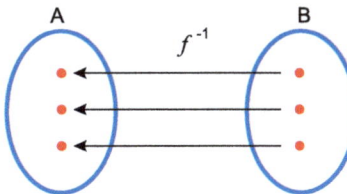

It turns out therefore, for a function to have an inverse it must have a one-to-one mapping. To do this restrictions may need to be put onto the domain and range.

However this only occurs when the image set of f is the whole of the codomain B.

Otherwise the range of the inverse function is not contained in the domain of the original function.

This is precisely the definition of an inverse function.

> **Key point**
>
> Suppose that f is a one-to-one and onto function. For each $y \in B$ (codomain) there is exactly one element $x \in A$ (domain) such that $f(x) = y$.
>
> The **inverse function** is denoted $f^{-1}(y) = x$.

This means that each element in the range of the function f is mapped back to the element from which it came.

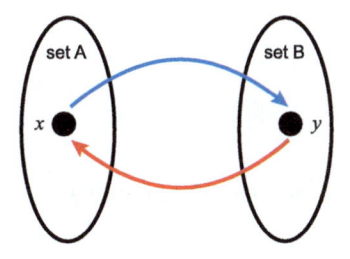

The **domain** of the inverse function is the range of the original function.

The **codomain** of the inverse function is the domain of the original function.

The relationship between functions and their inverses is clearly seen using graphs.

© HERIOT-WATT UNIVERSITY

Two ways of finding an inverse function are:

- Algebraically
- Graphically

To find the rule for an inverse function algebraically, make x the subject of the formula then swap x and y to give the inverse equation.

The effect of interchanging x and y in the equation is the same as interchanging the axes on the graph.

Graphically, this is the same as reflecting the graph of a function in the line $y = x$.

> **Example Problem:**
>
> Sketch the inverse function f^{-1} where $f(x) = 2x$
>
> **Solution:**
>
> First sketch the function f which is the line $y = 2x$.
>
> Reflect this graph in the line $y = x$ to give the inverse function f^{-1}.
>
> By interchanging x and y the equation $y = 2x$ becomes $x = 2y$ and solving for y gives the line $y = \frac{1}{2}x$.
>
> The inverse function is $f^{-1}(x) = \frac{1}{2}x$.
>
> This shows that the element 2 has an image of 4 under f and the element 4 has an image of 2 under f^{-1}

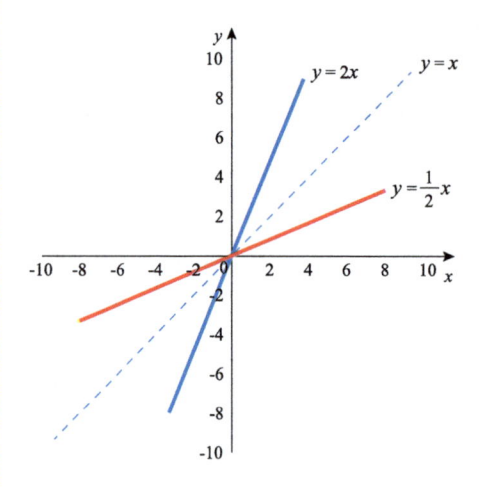

When asked to find and sketch the inverse of a function it is important to check that the function in

TOPIC 8. CURVE SKETCHING

question is actually one-to-one and onto. It could be that the codomain of the original function will have to be restricted in order to find an inverse which actually exists.

This is very important.

> **Key point**
>
> If (a, b) is a point on the function $f(x)$ then (b,a) is a point on the inverse function $f^{-1}(x)$.

Example Problem:

$f(x) = 4x^2 - 3$, where $[0, \infty)$

Let $f(x) = y$ then $y = 4x^2 - 3$. Find the inverse function and draw its graph.

Solution:

The domain of $f(x)$ has been restricted. This is so that it is a one-to-one function and onto so that an inverse may exist. This can be seen from the diagram.

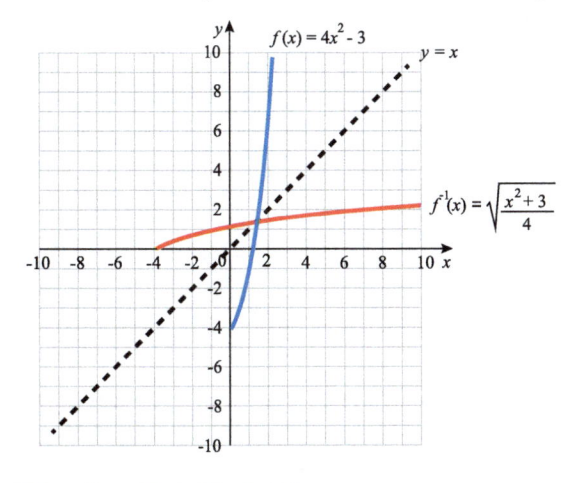

Make x the subject of the formula:

$$y = 4x^2 - 3$$
$$y + 3 = 4x^2$$
$$\frac{y+3}{4} = x^2$$
$$\sqrt{\frac{y+3}{4}} = x$$

Swap x and y which gives, $y = \sqrt{\frac{x+3}{4}}$

Note: Only the +ve square root is used for the inverse function.

The fact that only the positive square root is plotted fits in with the graph that has been drawn.

Inverse functions

Go online

From these examples you can see that reflecting the function in the line $y = x$ gives the inverse function f^{-1}.

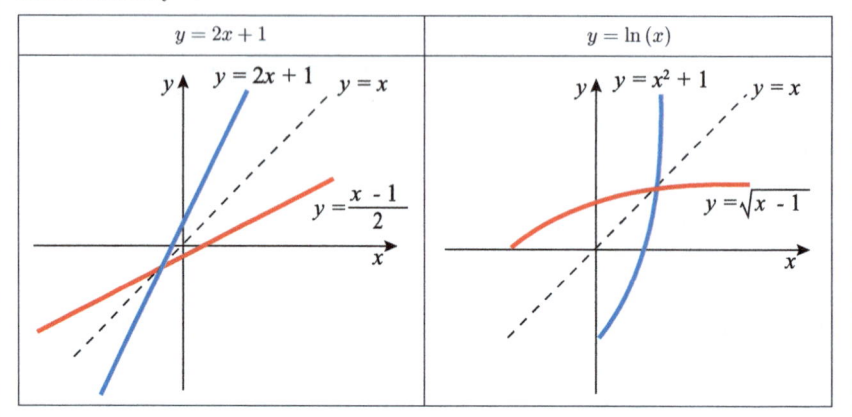

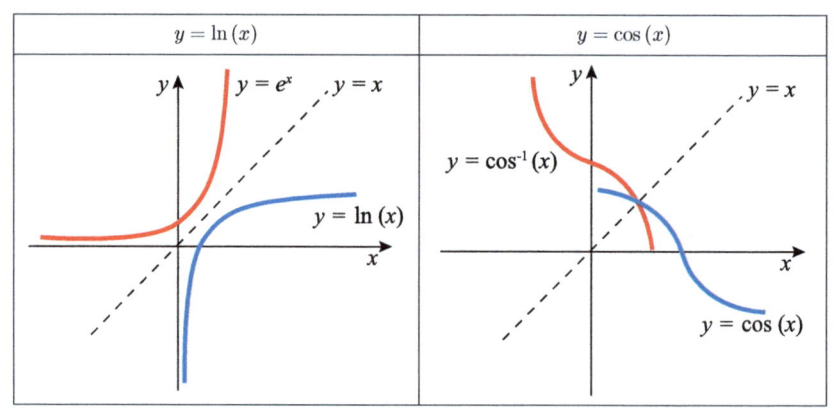

$y = 2x + 1$	$y = \ln(x)$
graph showing $y = 2x+1$, $y = x$, and $y = \frac{x-1}{2}$	graph showing $y = x^2 + 1$, $y = x$, and $y = \sqrt{x-1}$

$y = \ln(x)$	$y = \cos(x)$
graph showing $y = e^x$, $y = x$, and $y = \ln(x)$	graph showing $y = \cos^{-1}(x)$, $y = x$, and $y = \cos(x)$

Q44: Use the graph of $y = \log_{10} x$ for $x \in \mathbb{R}^+$ to sketch the graph of $y = 10^x$. Note that 10^x is the inverse of the graph $y = \log_{10} x$.

....

Q45: Using the graph of $y = 2^x$ sketch the graph of $y = \log_2 x$.

....

Q46: Find and sketch the inverse of the function $f(x) = 8 - 2x$, where $x \in \mathbb{R}$. A graphics calculator can be used to check your understanding of the concept.

....

TOPIC 8. CURVE SKETCHING

Q47: Find and sketch the inverse of the function $g(x) = \frac{1}{2-x}$ where $x < 2 : x < 2$

Q48: Find and sketch the inverse of the function $h(x) = 4x - 6$ where $x \in \mathbb{R}$

Graphs of trigonometric functions

Is it possible to define any type of inverse function for sin (x), cos (x) or tan (x) functions?

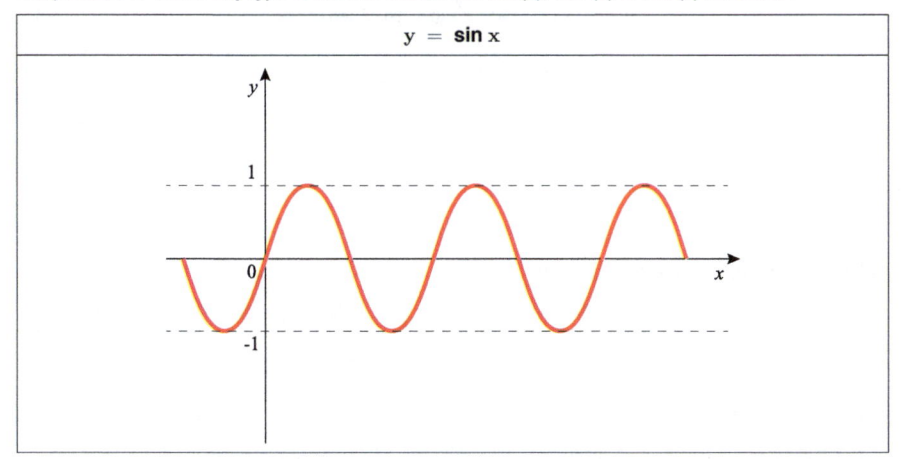

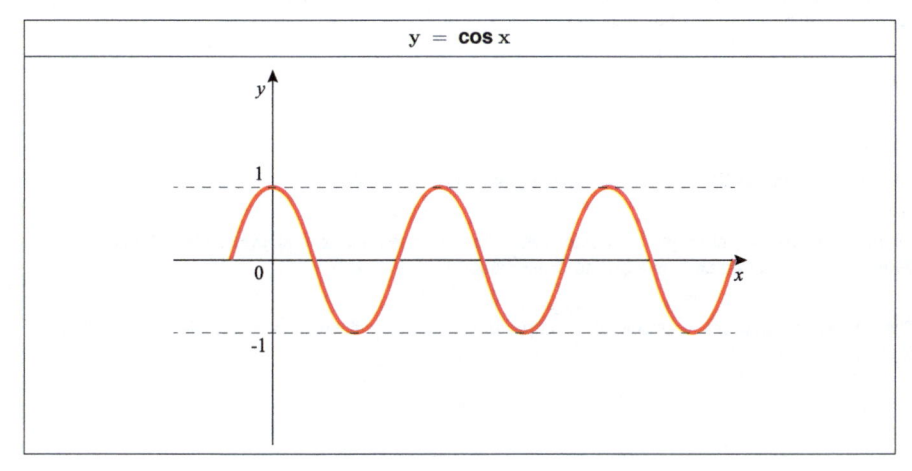

$$y = \sin x$$

$$y = \cos x$$

© HERIOT-WATT UNIVERSITY

y = tan x

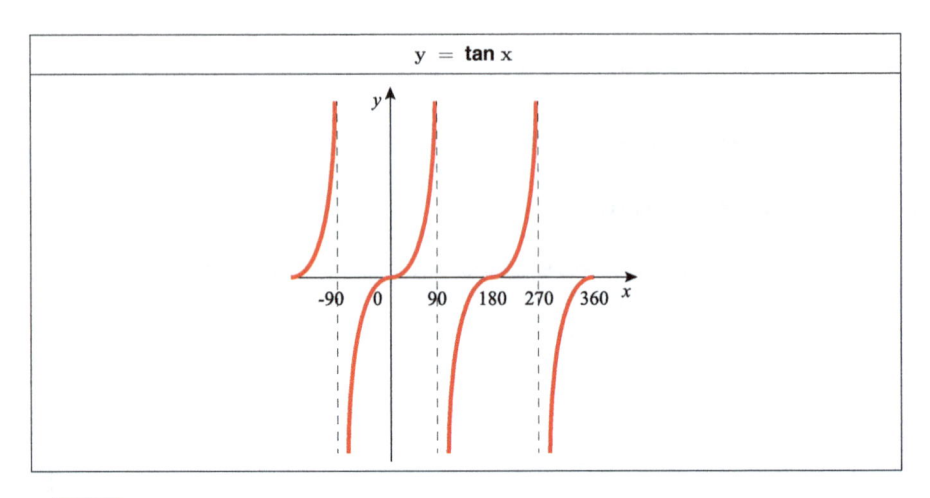

> **Top tip**
>
> Notice the repeating pattern for all three functions.
>
> They are not one-to-one functions. In their current form it does not appear that they have an inverse.

The graph of the function $f : \mathbb{R} \to [-1,1]$ where $f(x) = \sin x$ is a very familiar one.

This graph is cyclical. The wave pattern repeats itself every 360 degrees. Therefore there are many values of x that give the same value of y.

For example $\sin x = 0$, when $x = 0, \pi, 2\pi, \ldots$

The graph however is onto.

The range of the function is [-1,1] which is equal to the codomain given.

By changing the minimum and maximum values for x on the calculator is it possible to find a domain for x such that $\sin x$ is a one-to-one function?

The answer is that there are many different restrictions which will produce a suitable domain. One of these is $x \in \left[\frac{-\pi}{2}, \frac{\pi}{2}\right]$

TOPIC 8. CURVE SKETCHING

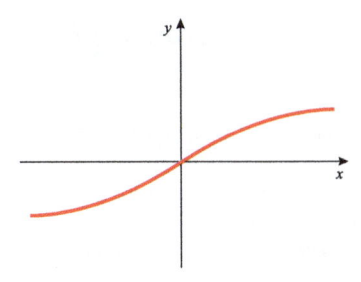

So if the domain is restricted to give a one-to-one, onto sine function then the inverse sine function exists. This is denoted $\sin^{-1} x$ and should not be confused with $\frac{1}{\sin x}$.

$y = \sin^{-1} x$ means 'y is the angle whose sin is the value x'. The functions cos and tan have similarly defined inverses.

The inverses of sine, cosine and tangent functions

The following table demonstrates the sketching of sine, cosine, and tangent graphs and their inverses.

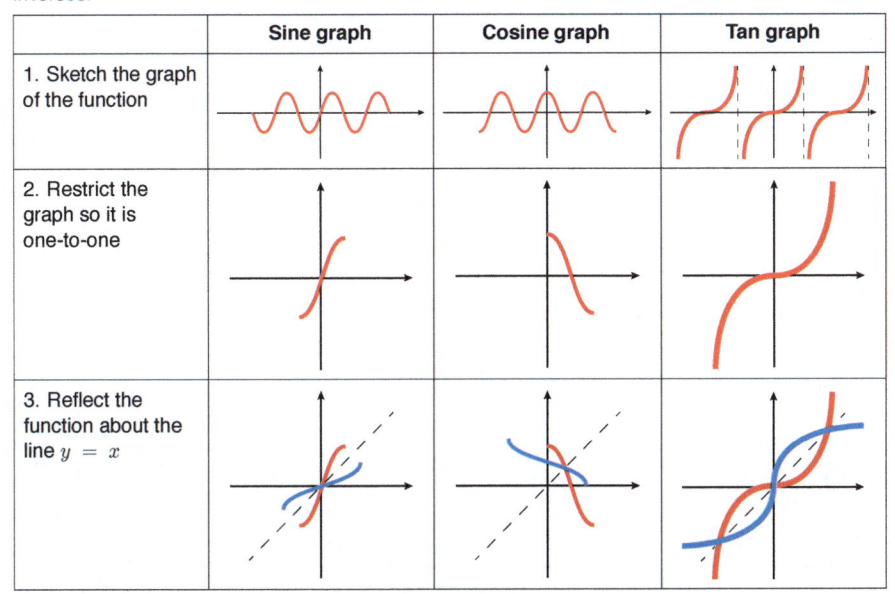

	Sine graph	Cosine graph	Tan graph
1. Sketch the graph of the function			
2. Restrict the graph so it is one-to-one			
3. Reflect the function about the line $y = x$			

Notice that the graph of $y = \tan^{-1} x$ has **horizontal asymptotes** and these are related to the **vertical asymptotes** of tan x. (Remember that an inverse is a reflection in the line $y = x$ which is the same as interchanging the axes.)

© HERIOT-WATT UNIVERSITY

Exploring inverse trigonometric functions Go online

Q49:

a) Show that the sine function when restricted to a domain of $\left[\frac{-\pi}{2}, \frac{\pi}{2}\right]$ and codomain of [-1,1] is both one-to-one and onto.

b) By exploring the cosine function, with or without a graphic calculator, suggest a possible restricted domain on cos x which will make the function a one-to-one function.

c) Explore the graph of $y = \tan x$ and suggest a possible restricted domain to make this a one-to-one function.

d) What is the domain of $y = \sin^{-1} x$ if sin x has a domain of $\left[\frac{-\pi}{2}, \frac{\pi}{2}\right]$?

e) Sketch the graph of $y = \tan^{-1} x$ from the graph of tan x where $x \in \left[\frac{-\pi}{2}, \frac{\pi}{2}\right]$ by using reflection in the line $y = x$

f) Sketch the graph of $y = \cos^{-1} x$ from the graph of cos x where $x \in [0, \pi]$ by using reflection in the line $y = x$

....

Q50: With the aid of a graphics calculator or online graphing tool, determine a general rule for the domain and range of the following functions so that an inverse may exist.

a) $y = k \sin(x)$
b) $y = \cos(bx)$
c) $y = \tan(cx)$

If you do not have a graphics calculator use this online graphing tool to explore:

http://www.mathsisfun.com/data/function-grapher.php

To draw inverse sine etc. type in a sin(x). There is a list of functions that can be drawn.

Construction of different angles Go online

Sketch an equilateral triangle with side lengths of 2 units.

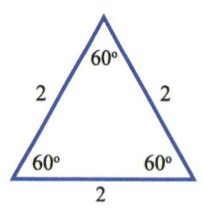

Draw in the perpendicular bisector from the apex to the base, focus on one of the triangles now formed.

The angles of this triangle are 30°, 60° and 90°. The sides of the triangle are 2 units, 1 unit (base halved by bisector) and $\sqrt{3}$ units (by Pythagoras).

© HERIOT-WATT UNIVERSITY

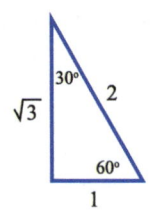

TOPIC 8. CURVE SKETCHING

Read off the exact values for sin, cos and tan for the angles 30° and 60° using the SOH-CAH-TOA ratios.

$$\sin 30° = \frac{O}{H} = \frac{1}{2}$$

$$\cos 30° = \frac{A}{H} = \frac{\sqrt{3}}{2}$$

$$\tan 30° = \frac{O}{A} = \frac{1}{\sqrt{3}}$$

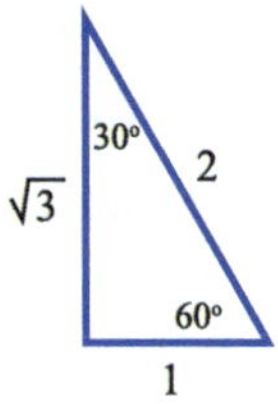

$$\sin 60° = \frac{O}{H} = \frac{\sqrt{3}}{2}$$

$$\cos 60° = \frac{A}{H} = \frac{1}{2}$$

$$\tan 60° = \frac{O}{A} = \sqrt{3}$$

Sketch a square with side length 1 unit.

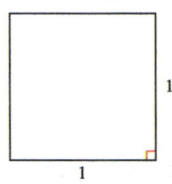

Draw in one of the diagonals, focus on one of the triangles now formed.

The angles of this triangle are 45° (twice) and 90°. The sides of the triangle are 1 unit, 1 unit and $\sqrt{2}$ units (by Pythagoras).

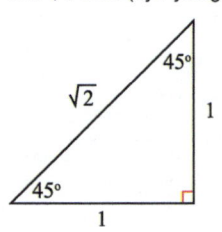

Read off the exact values for sin, cos and tan for the angles 30° and 60° using the SOH-CAH-TOA ratios.

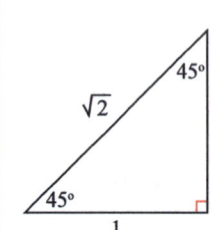

$$\sin 45° = \frac{1}{\sqrt{2}}$$

$$\cos 45° = \frac{1}{\sqrt{2}}$$

$$\tan 45° = 1$$

Inverse functions exercise   Go online

Q51: Copy and complete the blanks in the following table giving exact values throughout. Give angles in degrees.

x	$\sin^{-1} x$	$\tan^{-1} x$	$\cos^{-1} x$
		********	30°
		45°	
	30°	********	
$\frac{1}{\sqrt{2}}$		********	
		60°	

....

Q52: Complete the blanks in the following table giving exact values throughout. Give angles in radians.

x	$\sin^{-1} x$	$\tan^{-1} x$	$\cos^{-1} x$
$\frac{\sqrt{3}}{2}$		********	
		$\frac{\pi}{4}$	
	$\frac{\pi}{4}$		
	********	$\frac{\pi}{3}$	********
	$\frac{\pi}{6}$	********	

....

Q53: The fact that $\ln x$ is the inverse of e^x has already been stated and used in this section. The graph of each clearly shows the reflection in $y = x$

Given the fact that $\ln x$ crosses the x-axis at the point (1, 0), determine where the graph of e^x will cross the y-axis.

....

TOPIC 8. CURVE SKETCHING

Q54: Given the function $f(x) = \frac{1}{a} \ln(x)$, what is the inverse of this function.

Examples of log and exp inverses

$\log_2 x$

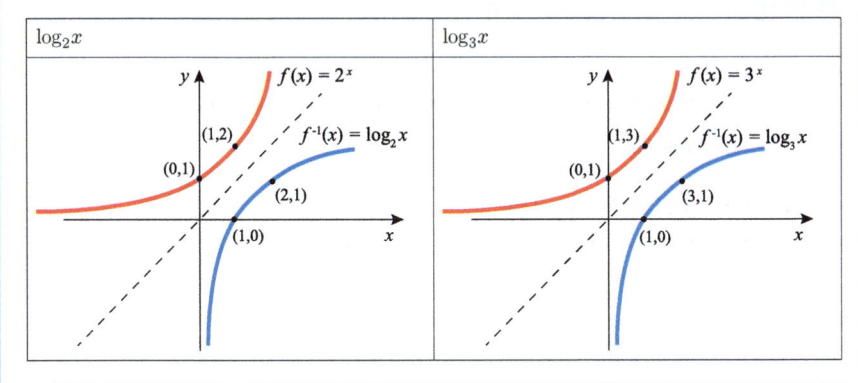

$\log_3 x$

$\log_4 x$

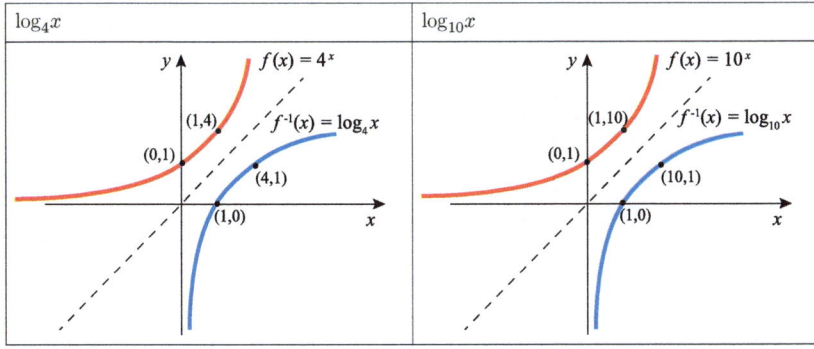

$\log_{10} x$

$\log_e x$

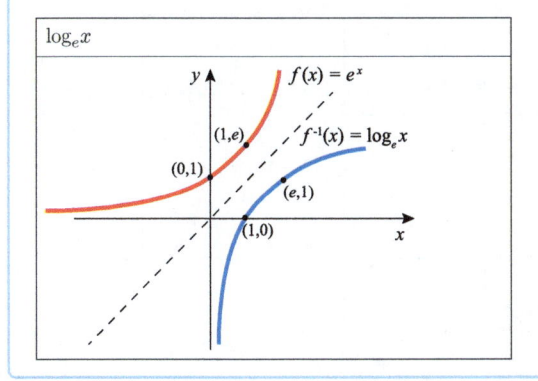

8.4 Odd, even or neither functions

The symmetry of some graphs can make sketching easier.

Take the curve $y = x^2$. This graph is symmetrical about the line $x = 0$.

A simple symmetry involves reflection in the primary axis.

That is:

- $y = -f(x)$ is a reflection of $y = f(x)$ in the x-axis.
- $y = f(-x)$ is a reflection of $y = f(x)$ in the y-axis.

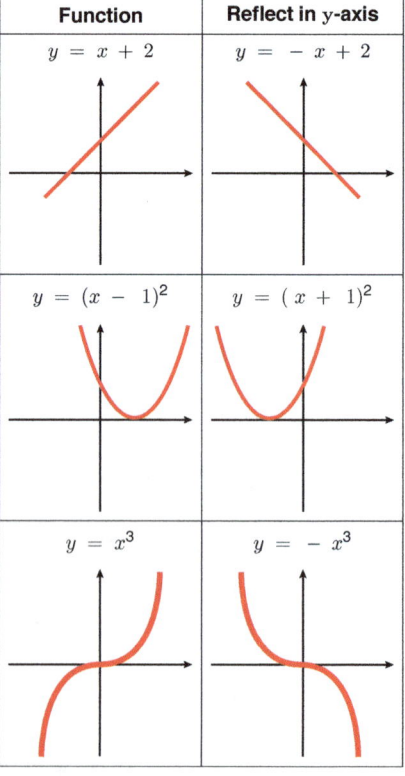

Reflections in the x-axis and in the y-axis

Reflections in the x-axis

$y = f(x) \rightarrow y = -f(x)$

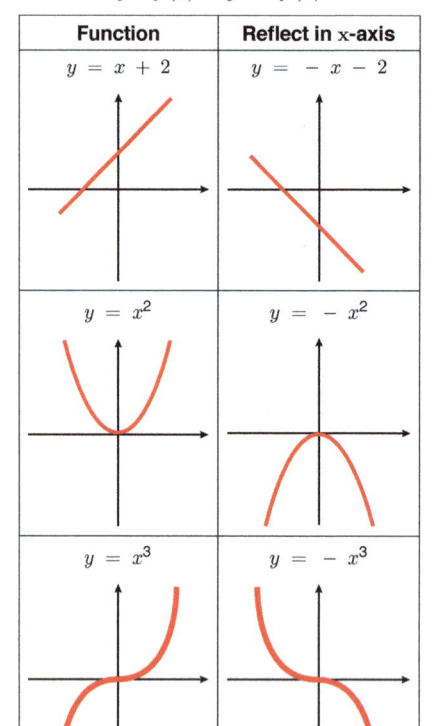

Function	Reflect in x-axis
$y = x + 2$	$y = -x - 2$
$y = x^2$	$y = -x^2$
$y = x^3$	$y = -x^3$

Reflections in the y-axis

$y = f(x) \rightarrow y = f(-x)$

Function	Reflect in y-axis
$y = x + 2$	$y = -x + 2$
$y = (x - 1)^2$	$y = (x + 1)^2$
$y = x^3$	$y = -x^3$

TOPIC 8. CURVE SKETCHING

Reflections in the x-axis

$$y = f(x) \to y = -f(x)$$

Function	Reflect in x-axis
$y = \cos x$	$y = -\cos x$
$y = \log_e x$	$y = -\log_e x$

Reflections in the y-axis

$$y = f(x) \to y = f(-x)$$

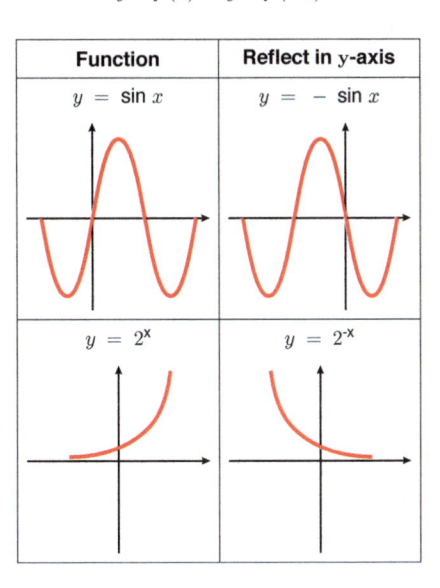

Function	Reflect in y-axis
$y = \sin x$	$y = -\sin x$
$y = 2^x$	$y = 2^{-x}$

Odd, even functions and neither

There are further relationships between the function $f(-x)$ and either $-f(x)$ or $f(x)$ which determines whether a function has further symmetry: odd, even or whether it is neither odd nor even.

> **Key point**
>
> A function is **odd** if $f(-x) = -f(x)$ for every value of x within the domain of the function.
>
> The graph is symmetrical under 180° rotation about the origin.
> It is said to have rotational symmetry of order 2.

> **Key point**
>
> A function is **even** if $f(-x) = f(x)$ for every value of x within the domain of the function.
>
> The graph is symmetrical under reflection in the y-axis.

> **Key point**
>
> If it is not odd or even it is neither odd or even.

© HERIOT-WATT UNIVERSITY

Examples

1. Odd function

A function is **odd** if $f(-x) = -f(x)$ for every value of x within the domain of the function. Consider the following graph $y = \sin(x)$.

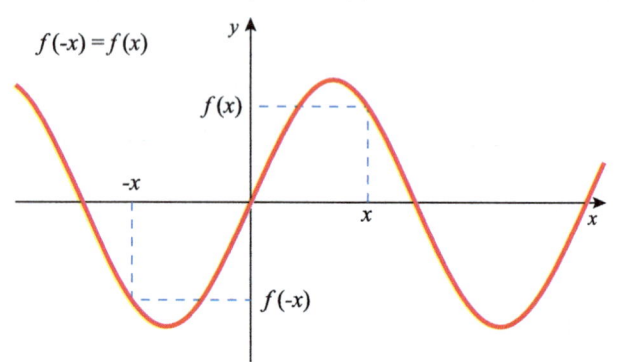

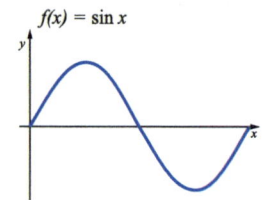

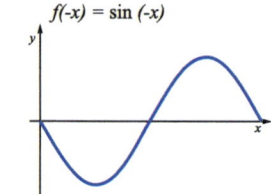

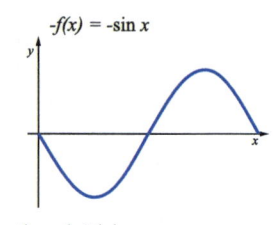

From the graph it is clear to see that $f(-x)$ is equal to the negative of $f(x)$.

We can look at the graph, but we can do this more efficiently and accurately algebraically with more complex functions as will be seen later.

$$f(-x) = \sin(-x)$$
$$= -\sin(x)$$

(We can state this because the graph of the function shows us this.)

and

$$-f(x) = -\sin(x)$$
$$f(-x) = -f(x) \textbf{ so it is odd.}$$

The graph is symmetrical under 180° rotation about the origin. It is said to have rotational symmetry of order 2.

....

2. Even Function

A function is even if $f(-x) = f(x)$ for every value of x within the domain of the function.
Consider the following graph $y = \cos x$.

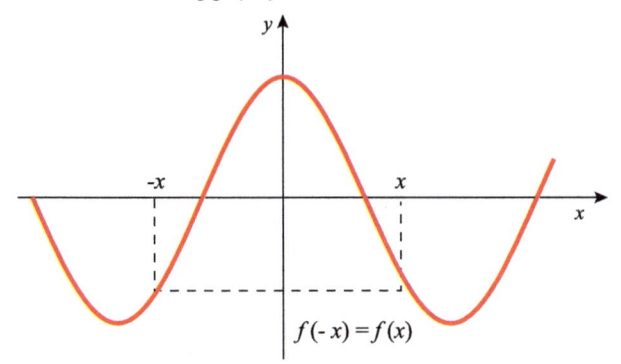

Instead of looking at the graph, we can do this more efficiently and accurately algebraically.

$f(-x) = \cos(-x)$
$\qquad = \cos(x)$

(We can state this because the graph of the function shows us this.)

$f(x) = \cos(x)$

f (−x) = f (x) so it is even.

....

3. Neither odd or even functions

Functions do not have to be odd or even, they can be neither.
Consider the following graph $y = e^x$.

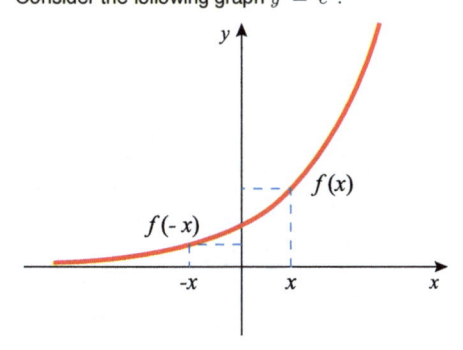

From the graph it is clear the $f(-x)$ does not give the same answer as $f(x)$. It is not even. If we take $-f(x)$ it does not equal $f(-x)$. It is not odd. It is neither odd nor even.

Then consider algebraically,

$f(x) = e^x$

$f(-x) = e^{-x}$

$-f(x) = -e^x$

$f(-x) \neq -f(x)$, **not odd.**

$f(-x) \neq f(x)$, **not even.**

$\therefore$ **neither.**

This property of being odd or even certainly aids the sketching of such a graph but beware, many graphs are neither.

A graph on its own however does not formally prove that a function is odd or even. Certain properties have to be established by algebraic means.

At times it is faster to use the algebraic method than sketch an unknown function.

The following examples will demonstrate the form of such proofs.

Examples

1.

Problem:

Prove that the function $f(x) = x^2 \cos x$ is even.

Solution:

$f(x) = x^2 \cos x$

$f(-x) = (-x)^2 \cos(-x) = x^2 \cos x = f(x)$

Since $f(x) = f(-x)$ the function is even.

....

2.

Problem:

Prove that the function $g(x) = \frac{x}{x^2+1}$ is odd.

Solution:

$g(x) = \frac{x}{x^2+1}$

$g(-x) = \frac{-x}{(-x)^2+1} = \frac{-x}{x^2+1} = -g(x)$

Since $g(-x) = -g(x)$ the function is odd.

TOPIC 8. CURVE SKETCHING 331

> **Top tip**
>
> Note that if a function is a product or quotient of two other functions the following rules apply:
>
> - even × even = even
> - odd × odd = even
> - odd × even = odd

So $g(x) = \frac{x}{x^2+1} = x \bullet \frac{1}{x^2+1}$ = odd × even = odd (as already proved).

Odd, even or neither function exercise Go online

Q55: Prove whether the following graphs are odd, even or neither. If the graph is odd or even, sketch it using the property found.

a) $f(x) = \cos x$
b) $g(x) = \frac{1}{2x-1}$
c) $k(x) = \frac{1}{x}$

....

Q56: For each graph decide whether is:

- Odd
- Even
- Neither

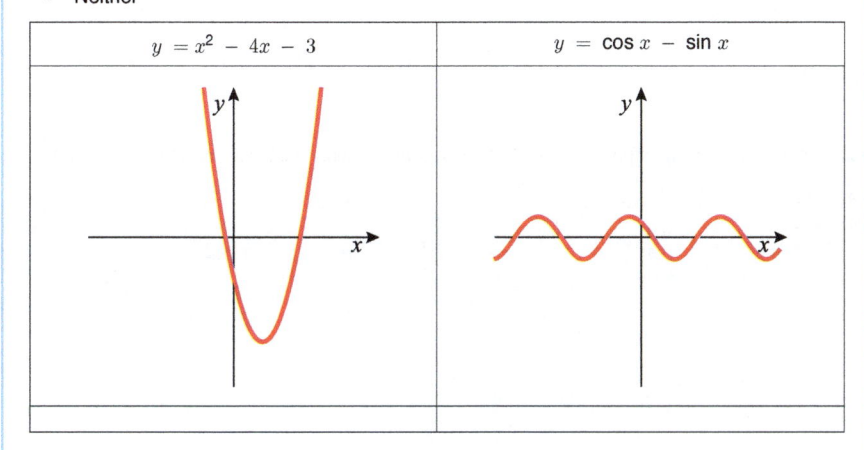

$y = x^2 - 4x - 3$	$y = \cos x - \sin x$

© HERIOT-WATT UNIVERSITY

$y = x^3 \sin x$	$y = \frac{x}{x^2-1}$

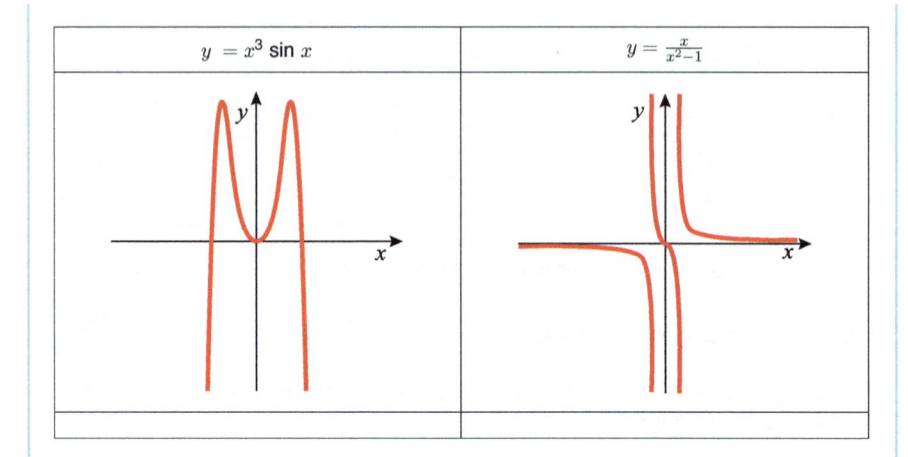

Identify if these functions are odd, even or neither.

Q57: $f(x) = x^2 e^{-x}$

Q58: $f(x) = \frac{4x}{(x-1)^2}$

Q59: $f(\theta) = \frac{\sin \theta}{\theta^2 + 3}$

Q60: $f(x) = \sqrt{5x^2 - 2}$

Q61: If I take an odd function and multiply it by an even function what is the resulting function?

a) Odd
b) Even
c) Neither

8.5 Critical and stationary points

Critical point

A critical point is any point on a curve where the slope of the tangent to the curve is zero (parallel to the x-axis) or where the slope of the tangent to the curve is undefined (parallel to the y-axis).

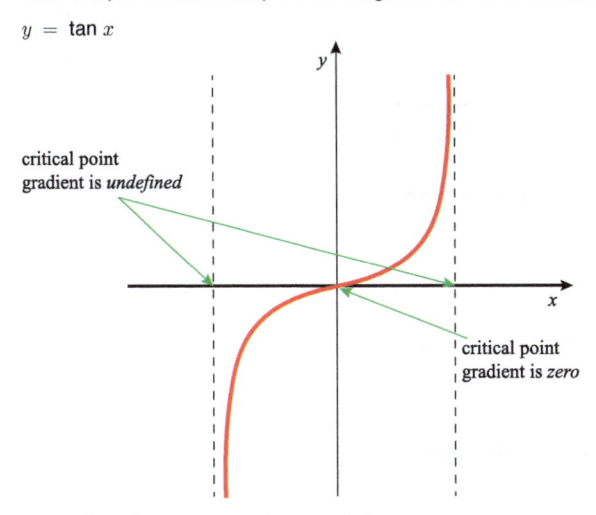

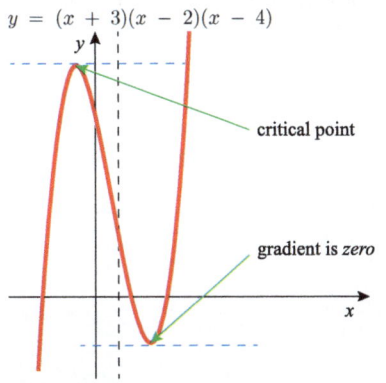

$y = \tan x$

$y = (x + 3)(x - 2)(x - 4)$

Stationary point

A stationary point is any point on a curve where the slope of the tangent to the curve is zero (parallel to the x-axis). The term stationary point can refer to:

- a maximum turning point;
- a minimum turning point;
- a horizontal point of inflection.

$y = x^5 - 6x^3 + 2x^2 - 3$

maximum stationary point

falling point of inflection

minimum stationary point

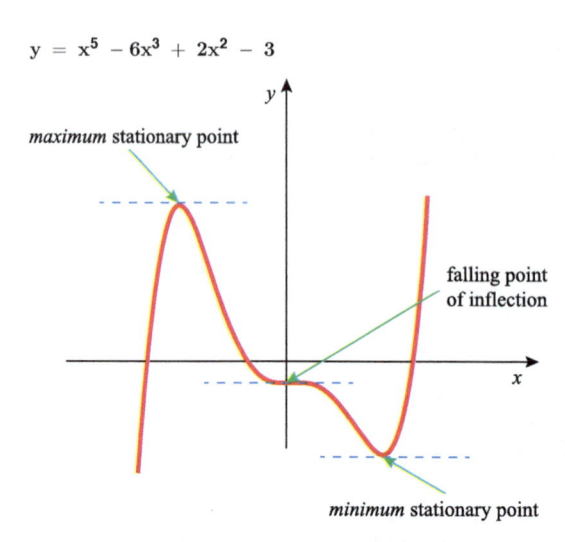

Maximum and minimums

A **local maximum point** occurs when a function has a greater value at that point than at any points close to it. It is not necessarily the greatest value of the function. There can be more than one local maximum turning point.

A **global maximum point** occurs when f is defined over a domain A and the value of the function at this point is greater than or equal to that at any other point within the domain. This includes endpoints if they are contained within the domain.

global maximum point

local maximum point

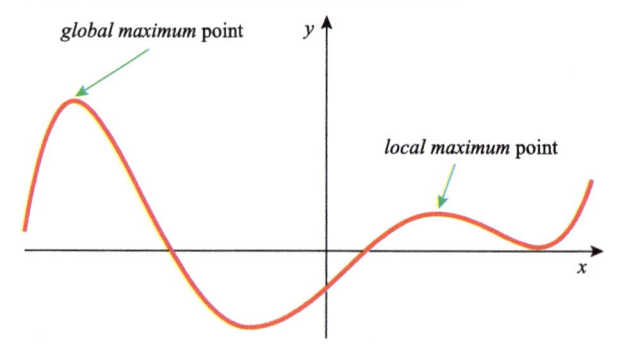

A **local minimum point** occurs when a function has a lesser value at that point than at any points close to it. It is not necessarily the least value of the function. There can be more than one local minimum turning point.

A **global minimum point** occurs when f is defined over a domain A and the value of the function at this point is less than or equal to that at any other point within the domain. This includes endpoints if they are contained within the domain.

TOPIC 8. CURVE SKETCHING

global maximum point

local maximum point

local minimum point

global minimum point

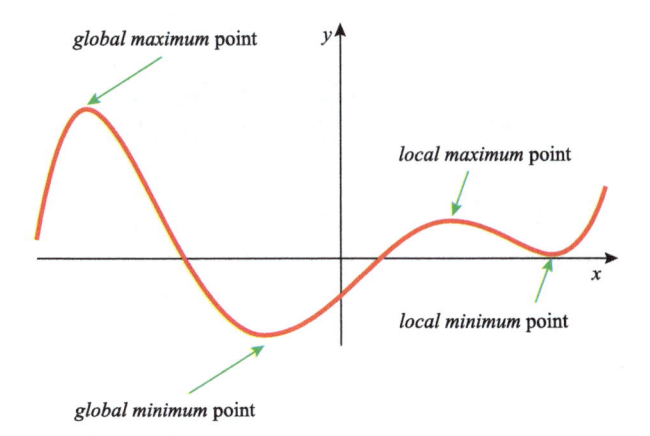

Critical and stationary points exercise   Go online

For each question, identify the global maximum/minimum, local maximum/minimum and points of inflection choosing from the following labels.

Q62:

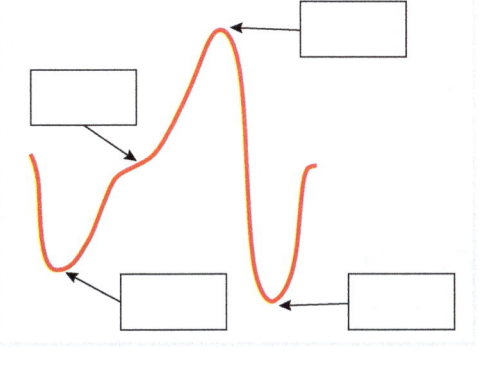

Local maximum	Local minimum	Global maximum	Global minimum	Point of inflection

....

© HERIOT-WATT UNIVERSITY

Q63:

Local maximum	Local minimum	Global maximum	Global minimum	Point of inflection

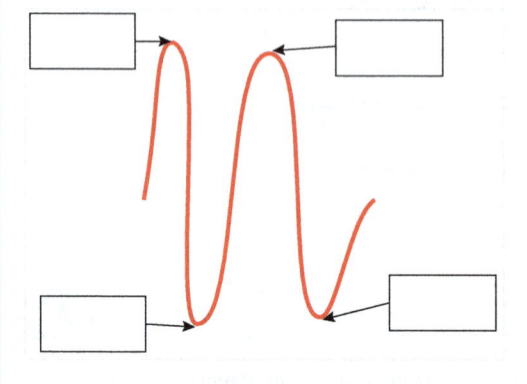

Q64:

Local maximum	Local minimum	Global maximum	Global minimum	Point of inflection

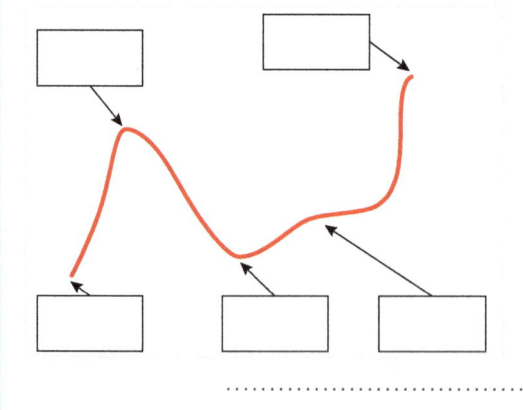

TOPIC 8. CURVE SKETCHING

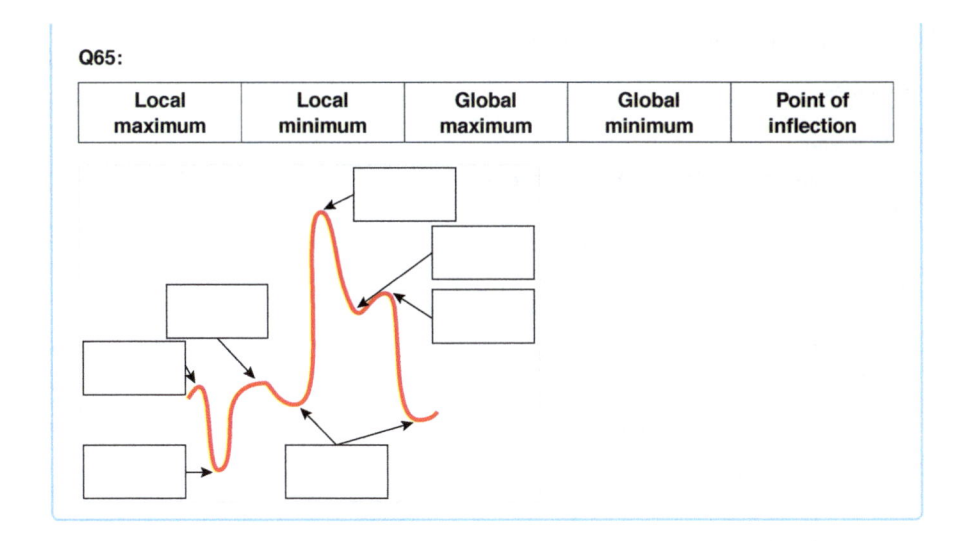

Q65:

Local maximum	Local minimum	Global maximum	Global minimum	Point of inflection

8.6 Derivative tests

Finding stationary points is one of the key features to drawing an "accurate" sketch of a function. When evaluating the derivative at a point x on the curve it tells us how steep the slope is at that point.

By stating that the gradient of the slope has to be zero, i.e. the derivative has to be zero, we can find the x coordinates of these stationary points. To find the corresponding y coordinate, the x value is substituted into the original function.

To determine the nature of the turning point we would draw a nature table using the first derivative. (This is from Higher). However, there is an alternative method to identify the nature. The second derivative can be used. This section will explore these two methods.

8.6.1 The first derivative test

The first derivative test works as follows:

- Calculate values for the first derivative of the function using values on either side of the x coordinate of the turning point. The signs of the values are what is really needed.
- Construct a table of signs from this data to show the nature of the turning point.

x	a^-	a	a^+
$\frac{dy}{dx}$			
slope			

© HERIOT-WATT UNIVERSITY

(Note: the variable allocated to the horizontal axis may be denoted other than x.)

Examples

1. Problem:
Determine the nature of the turning point on the curve $y = 3x^2 - 12x + 4$.
Solution:
Calculate the first derivative of the function $y = 3x^2 - 12x + 4$.
$\frac{dy}{dx} = 6x - 12$
Solve the equation $\frac{dy}{dx} = 0$
This is when $6x - 12 = 0 \Rightarrow x = 2$
So the turning point is at $x = 2$

The y-coordinate of this turning point is $y = 3 \times 2^2 - 12 \times 2 + 4 = -8$

The turning point is at (2, -8)
Now determine the nature of the point (2, -8) using the **first derivative test**.
At this turning point, $x = 2$ look at the value of the derivative on either side of this.
In the table the notation 2⁻ is used for just below $x = 2$ and 2⁺ for just above $x = 2$.
Do not confuse these with - 2 and + 2

	2⁻	2	2⁺
signs of $\frac{dy}{dx}$	-	0	+
slope	↘	→	↗

This clearly shows that the turning point is a minimum.
(If the arrows show as ↗ → ↘ then the turning point will be a maximum.)

....

2. Problem:
Determine the nature of the turning point of the curve $y = \frac{x^2}{x-1}$, $x \neq 1$.
Solution:
$y = \frac{x^2}{x-1}$
First derivative (find using the quotient rule):
$\frac{dy}{dx} = \frac{2x(x-1) - (x^2)1}{(x-1)^2}$
$= \frac{2x^2 - 2x - x^2}{(x-1)^2}$
$= \frac{x^2 - 2x}{(x-1)^2}$
$= \frac{x(x-2)}{(x-1)^2}$

TOPIC 8. CURVE SKETCHING

For stationary points $\frac{dy}{dx} = 0$

$\frac{x(x-2)}{(x-1)^2} = 0 \Rightarrow x(x-2) = 0 \Rightarrow x = 0, x = 2$

Corresponding y coordinate is:

$y = \frac{(0)^2}{0-1} \Rightarrow y = 0$

$y = \frac{(2)^2}{2-1} \Rightarrow y = 4$

Nature table:

x	0^-	0	$\rightarrow$	2	2^+
$\frac{dy}{dx}$	+	0	-	0	+
slope	↗	→	↘	→	↗

There is a maximum turning point at (0,0) and a minimum turning point at (2,4).

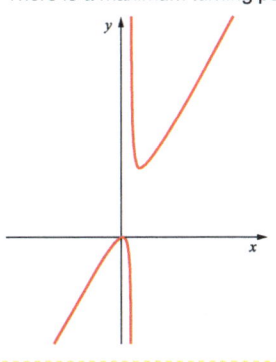

Key point

The **first derivative test** is a means of determining the nature of the turning point by finding the signs of the derivative to the left and to the right of the turning point.

The first derivative test exercise

Q66: Determine the nature of the turning point of the curve $y = -5x^2 + 4x - 6$.

Q67: Determine the nature of the turning point of the curve $y = 2x^3 + 6$.

Q68: Determine the nature of the turning point of the curve $y = \frac{1}{3}x^3 - 4x$.

Q69: Determine the nature of the turning point of the curve $y = 4x^2 + 8x - 3$.

8.6.2 The second derivative test

We were introduced to using implicit differentiation to find the second derivative in the Differentiation topic.

Here we will look at how the second derivative of a function gives an alternative way of determining the nature of a turning point.

In this section we will look at its application. The next section on concavity will explain how this works.

The second derivative test works as follows:

- Once the turning point has been identified, the second derivative $\frac{d^2y}{dx^2}$ of the function is calculated.
- The x value of the stationary point is substituted into the expression for $\frac{d^2y}{dx^2}$.

If the value of $\frac{d^2y}{dx^2}$ is negative (- ve) the turning point will be a maximum.

If the value of $\frac{d^2y}{dx^2}$ is positive (+ ve) the turning point will be a minimum.

If the value of $\frac{d^2y}{dx^2}$ is zero the nature is indeterminate (using this method) and the first derivative test has to be used instead.

$f''(x) < 0$ it is a maximum turning point	$f''(x) = 0$ (possible point of inflection)			$f''(x) > 0$
	x	0^-	0	0^+	
	$\frac{dy}{dx}$	$+$	0	$+$	
	slope	↗	→	↗	

Example

Problem:

Find the coordinates of any turning points on the graph of $y = x^3 + 3x^2 - 4$. Determine the nature of them.

Solution:

$y = x^3 + 3x^2 - 4$

$\frac{dy}{dx} = 3x^2 + 6x$

At turning points $\frac{dy}{dx} = 0 \Rightarrow x = 0$ or $x = -2$

TOPIC 8. CURVE SKETCHING

This means that there are two turning points.

The y coordinate when $x = 0$ is found by substituting the value $x = 0$ into the original equation.

So the turning point is (0, -4)

Similarly when $x = -2$ the turning point is (-2, 0)

Now using the **second derivative test** $\frac{d^2y}{dx^2} = 6x + 6$

At $x = 0$, $\frac{d^2y}{dx^2} = 6$ which is positive so it is a minimum turning point.

At $x = -2$, $\frac{d^2y}{dx^2} = -6$ which is negative so this turning point is a maximum.

There are two turning points; a maximum at (-2, 0) and a minimum at (0, -4)

> **Key point**
>
> The **second derivative test** of a function is a means of determining the nature of a turning point. A positive second derivative gives a minimum turning point and a negative second derivative gives a maximum turning point.

> **Top tip**
>
> The second derivative test is very useful but does not always work because:
>
> - The second derivative may not exist.
> - If the second derivative does exist or is zero, the nature of the point is still undecided.

In these circumstances it is necessary to revert to the first derivative test and make a nature table to determine the nature of the turning point.

These examples will give practice in finding turning points and using the derivative tests to determine their nature.

Example

Problem:

For the curve $y = \frac{x^3}{2} + 5$ find the coordinates of the turning points and determine their nature.

Solution:

$y = \frac{x^3}{2} + 5$

The first derivative: $\frac{dy}{dx} = \frac{3x^2}{2}$

For stationary points $\frac{dy}{dx} = 0$: $\frac{3x^2}{2} = 0 \Rightarrow x = 0$

Corresponding y coordinate:

When $x = 0$ the $y = \frac{0^3}{2} + 5 \Rightarrow y = 5$ coordinate (0,5).

The second derivative: $\frac{d^2y}{dx^2} = 3x$

When $x = 0$ then $\frac{d^2y}{dx^2} = 3(0) \Rightarrow \frac{d^2y}{dx^2} = 0$.

Since $\frac{d^2y}{dx^2} = 0$ then a nature tale is needed to determine the nature of the turning point.

x	0^-	0	0^+
$\frac{dy}{dx}$	+	0	+
slope	↗	→	↗

There is a rising point of inflection at (0,5).

The second derivative test exercise   Go online

Q70: By calculating the second derivative determine the nature of the turning points of the curve $y = 4x^3 + 5x^2$.

....

Q71: Find the nature of the turning points of the curve $y = 2x + \frac{3}{x+2}$, $x \neq -2$.

....

Q72: Find the nature of the turning points of the curve $y = \frac{x-1}{x+2}$, $x \neq -2$.

....

Q73: Find the nature of the turning points of the curve $y = \frac{x^3}{x-1}$, $x \neq 1$.

....

Q74: Find the nature of the turning points of the curve $y = \frac{x^2+1}{x-2}$, $x \neq 2$.

....

Q75: Find the nature of the turning points of the curve $y = \frac{2x^3}{x+5}$, $x \neq -5$.

....

Q76: By calculating the second derivative determine the nature of the turning points of the curve $y = \frac{3}{5}x^5 - 4x^3 + 8x$.

8.7 Concavity

The second derivative of a function can provide information about the shape of the graph of the function without having to draw a nature table to determine if it has turning points.

$\frac{dy}{dx}$ provides an expression that determines the gradient at a particular point on a curve.

Consider,

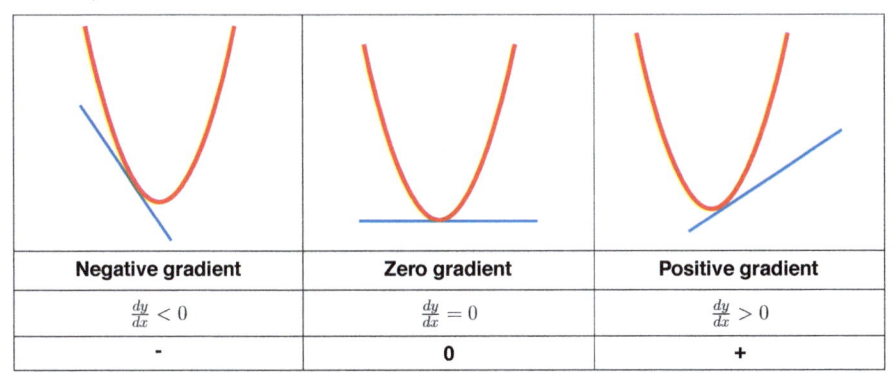

Negative gradient	Zero gradient	Positive gradient
$\frac{dy}{dx} < 0$	$\frac{dy}{dx} = 0$	$\frac{dy}{dx} > 0$
-	0	+

When we calculate $\frac{d^2y}{dx^2}$, we are calculating the rate of change of this first derivative.

We can see in the subsequent diagram that the tangent at this point (blue line) is increasing. In fact all the gradients of this curve range from large negative on the left of the graph, becoming less steep as they approach the turning point, zero at the turning point, becoming steeper and finally to large positive.

$\frac{d^2y}{dx^2}$ is increasing $\Rightarrow \frac{d^2y}{dx^2} > 0$

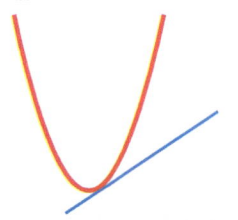

The graph of $y = f(x)$ is **concave upward** in an interval if $f(x)$ is a function which has a second derivative $\frac{d^2y}{dx^2} > 0$ for all x in the open interval.

Here we should be able to see that when $\frac{d^2y}{dx^2} > 0$ then the curve has a minimum turning point at x. Instead of using a nature table, this is an alternative way of determining if a stationary point is a maximum or minimum.

> **Top tip**
>
> Concave upward can be pictured as a 'u' shaped graph. Think of 'u' for **UP**.
>
> Alternatively, remember that it is like a smiley face.

$\frac{dy}{dx}$ provides an expression that determines the gradient at a particular point on a curve.

TOPIC 8. CURVE SKETCHING

Consider,

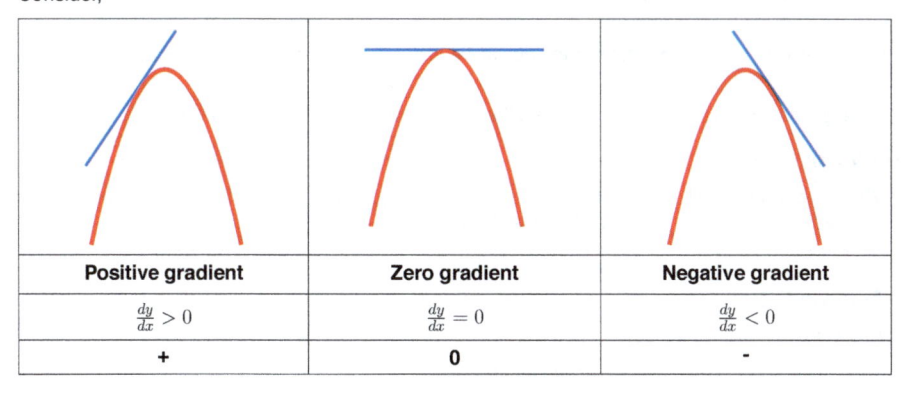

Positive gradient	Zero gradient	Negative gradient
$\frac{dy}{dx} > 0$	$\frac{dy}{dx} = 0$	$\frac{dy}{dx} < 0$
+	0	-

When we calculate $\frac{d^2y}{dx^2}$, we are calculating the rate of change of this first derivative.

We can see that the gradient of the tangent (blue line) is decreasing. Starting from the left hand side of the curve the tangent has a positive gradient then decreases to zero at the maximum turning point and the becomes more negative. The gradient of the tangent is decreasing from large positive values to small negative values. It is getting smaller.

$\frac{d^2y}{dx^2}$ is decreasing $\Rightarrow \frac{d^2y}{dx^2} < 0$

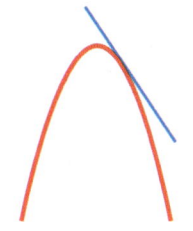

The graph of $y = f(x)$ is concave downward in an interval if $f(x)$ is a function which has a second derivative $\frac{d^2y}{dx^2} < 0$ for all x in the open interval.

Here we should be able to see that when $\frac{d^2y}{dx^2} < 0$ then the curve has a maximum turning point at x.

> **Top tip**
>
> Concave downward can be pictured as a 'n' shaped graph.
>
> Alternatively, you could think of it like a sad face.

Thus:

- concave downwards $\Leftrightarrow \frac{d^2y}{dx^2} < 0$
- concave upwards $\Leftrightarrow \frac{d^2y}{dx^2} > 0$

© HERIOT-WATT UNIVERSITY

The sign ⇔ means "if and only if". That is, the condition works both ways.

For example, if it is known that the second derivative is positive then the curve is concave upwards.

If, however, the curve is concave downwards then the second derivative is negative.

> **Key point**
>
> When the concavity of a curve changes, a point of inflection occurs. In fact this is a sufficient and necessary condition for a point of inflection.

Example

Problem:

Determine the concavity of the following functions:

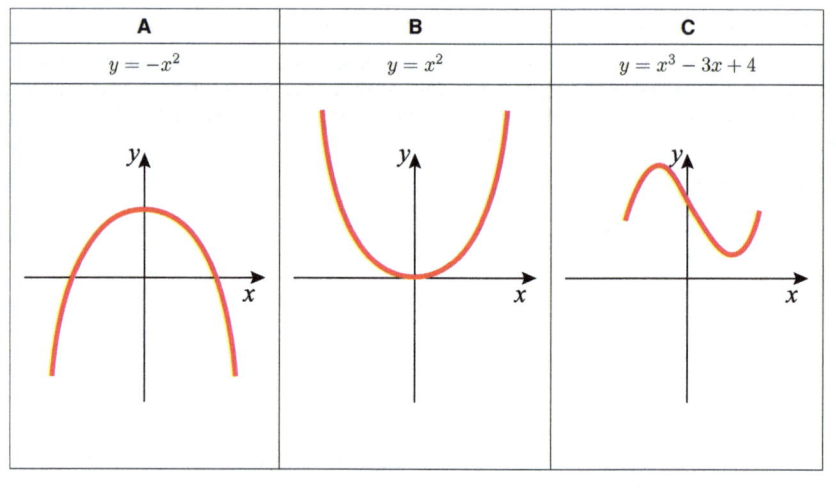

A	B	C
$y = -x^2$	$y = x^2$	$y = x^3 - 3x + 4$

Solution:

A) This is concave downwards. Check: $\frac{d^2y}{dx^2}$ = -2 which is negative. It is an 'n' shape / a sad face.

B) This is concave upwards. Check: $\frac{d^2y}{dx^2}$ = 2 which is positive. It is a 'u' shape / smiley face.

C) This is concave downwards leading to concave upwards.

Although the maximum and minimum turning points are easily identified, the point of inflection when the concavity changes is not identified by the methods shown so far in this section. (It is not a horizontal point of inflection)

Non-horizontal points of inflection

A **non horizontal point of inflection** occurs when $\frac{dy}{dx} \neq 0$ but $\frac{d^2y}{dx^2} = 0$ or $\frac{d^2y}{dx^2}$ does not exist. Note there must be a change of concavity.

Consider: $y = x^3 - 6x^2$

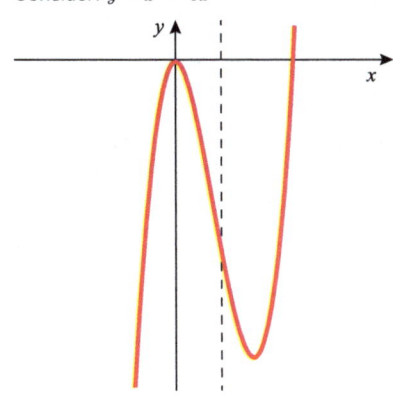

$$\frac{dy}{dx} = 3x^2 - 12x$$

$$\frac{d^2y}{dx^2} = 6x - 12$$

For $\frac{d^2y}{dx^2} = 0$:

$$6x - 12 = 0$$
$$x = 2$$

When $x = 2$, $\frac{dy}{dx} = -12$ i.e. $\frac{dy}{dx} \neq 0$. It is not a stationary point.

Now we check the concavity in the small neighbourhood of $x = 2$.

When $x = 2^-$, then $\frac{d^2y}{dx^2} < 0 \Rightarrow concave\ down$

When $x = 2^+$, then $\frac{d^2y}{dx^2} > 0 \Rightarrow concave\ up$

There is a change in concavity at $x = 2$. It is a non horizontal point of inflection.

It is also worth noting that any point of inflection occurs at a point P on a curve when the tangent to the curve crosses the curve **at that point**.

This is a useful visual check to note the existence of points of inflection.

Examples

1. Problem:

Find the non horizontal point of inflection on the curve $y = 2x^3 - 3x^2$

Solution:

$$\frac{dy}{dx} = 6x^2 - 6x$$

$$\frac{d^2y}{dx^2} = 12x - 6 = 0 \Rightarrow x = \frac{1}{2}$$

When $x = \frac{1}{2}$, $\frac{dy}{dx} = 6x^2 - 6x \neq 0$

At $x = \frac{1}{2}$, $\frac{dy}{dx} \neq 0$ but $\frac{d^2y}{dx^2} = 0$.

We will now evaluate the second derivative just before and after $x = \frac{1}{2}$ to check if there is a change in concavity. We can use the values $x = \frac{1}{4}$ and $x = \frac{3}{4}$ since we need to use values close to $x = \frac{1}{2}$.

When $x = \frac{1}{2}^-$, then $\frac{d^2y}{dx^2} < 0$ $\Rightarrow$ concave down

When $x = \frac{1}{2}^+$, then $\frac{d^2y}{dx^2} > 0$ $\Rightarrow$ concave up

There is a change in concavity at $x = \frac{1}{2}$. It is a point of inflection.

There is a non horizontal point of inflection at $\left(\frac{1}{2}, -\frac{1}{2}\right)$.

....

2. Problem:

Find the non horizontal point of inflection on the curve $y = x^{\frac{1}{3}}$.

Solution:

$$\frac{dy}{dx} = \frac{1}{3}x^{-\frac{2}{3}}$$

$$\frac{d^2y}{dx^2} = -\frac{2}{9}x^{-\frac{5}{3}} \quad \Rightarrow \quad \frac{d^2y}{dx^2} = -\frac{2}{9\sqrt[3]{x^5}}$$

The second derivative is undefined when $x = 0$. This is a possible point of inflection.

When $x = 0^-$, then $\frac{d^2y}{dx^2} > 0$ $\Rightarrow$ concave up

When $x = 0^+$, then $\frac{d^2y}{dx^2} < 0$ $\Rightarrow$ concave down

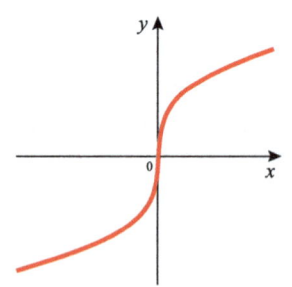

There is a non-horizontal point of inflection at $x = 0$.

TOPIC 8. CURVE SKETCHING

> **Key point**
>
> A **non horizontal point of inflection** occurs when $\frac{dy}{dx} \neq 0$ but $\frac{d^2y}{dx^2} = 0$ or $\frac{d^2y}{dx^2}$ does not exist.

Concavity exercise Go online

For each of the functions below state the coordinates of any non-horizontal points of inflection if they exist.

Q77: $y = x^3 - 4x^2$

Q78: $y = 3x^4 - 2x^2 + 1$

Q79: $y = 5x^2 + 3x - 2$

Q80: $y = -3x + 7$

Q81: $y = x^5 + 3x^4 - 2x$

Q82: $y = \frac{x^3}{6} - x^2 + 2x$

Q83: $y = \frac{5}{x^3} + x$, $x \neq 0$

Q84: $y = 4x^{\frac{1}{2}} + 2x$, $x \geqslant 0$

Q85: $y = 2x^{\frac{3}{5}} + x^2$, $x \geqslant 0$

8.8 Continuity and asymptotic behaviour

Continuity

A function $f(x)$ is said to be continuous if there is no break in the curve of the function for all values of x in the domain of the function.

Formally this is stated as:

> **Key point**
>
> A **continuous function** f (x) is a function where at every point P on the domain $\lim_{x \to P} f(x) = f(P)$

This means that as any point on the curve is approached (from either side), the value of the function tends closer and closer to the value of the function at that point itself.

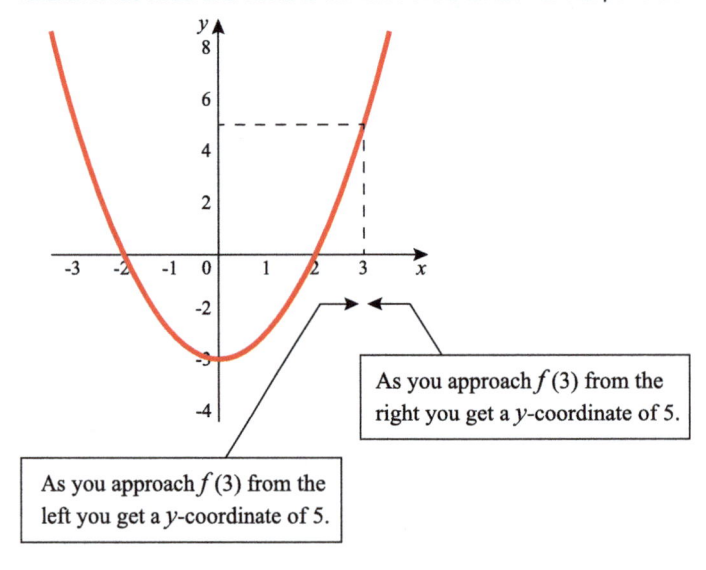

As you approach $f(3)$ from the right you get a y-coordinate of 5.

As you approach $f(3)$ from the left you get a y-coordinate of 5.

If a graph can be drawn without lifting the pencil from the paper, the function of the graph is continuous.

It follows from this that a function $f(x)$ is said to be discontinuous if there is a break in the curve at any value of x within its domain.

Again this can be stated formally.

> **Key point**
>
> A function $f(x)$ is **discontinuous** at a point P if $f(x)$ is not defined at P or if $\lim_{x \to P} f(x) \neq f(P)$

TOPIC 8. CURVE SKETCHING

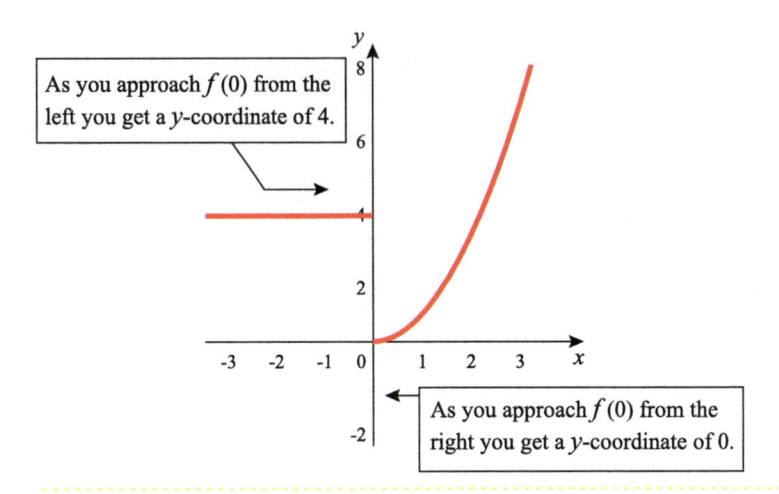

As you approach $f(0)$ from the left you get a y-coordinate of 4.

As you approach $f(0)$ from the right you get a y-coordinate of 0.

Example

Problem:

Examine the following functions

$$f(x) = x^2 \text{ and } g(x) = \begin{cases} x : x \leqslant 1 \\ 3 : x > 1 \end{cases}$$

Solution:

The function $f(x) = x^2$ is continuous.

The function $g(x) = \begin{cases} x : x \leqslant 1 \\ 3 : x > 1 \end{cases}$ is discontinuous.

$y = x^2$

function $g(x)$

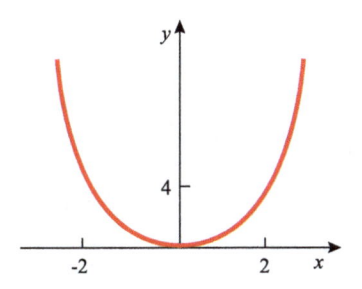

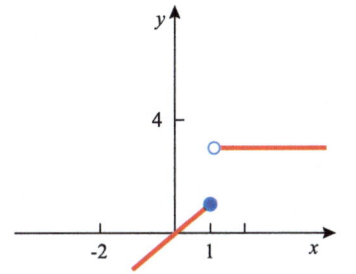

This graph for the function g has a distinct 'jump' in it after $x = 1$.

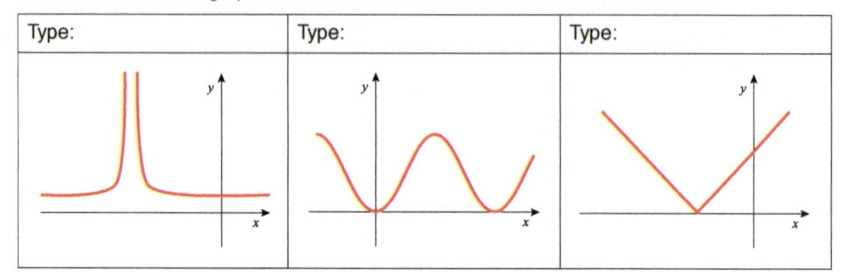

Q86: For each of the graphs decide whether is **continuous** or **discontinuous**.

Type:	Type:	Type:

Type:	Type:	Type:

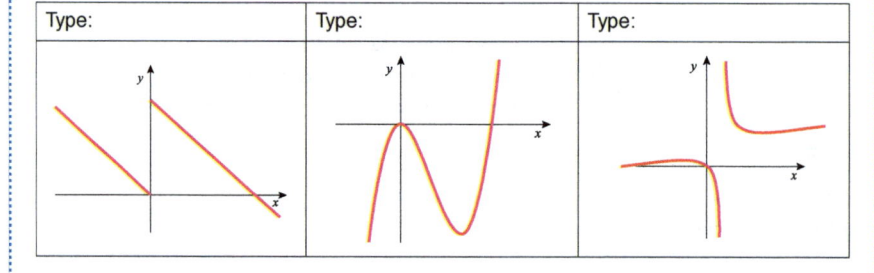

TOPIC 8. CURVE SKETCHING

Asymptotes

As demonstrated in the previous question, some discontinuities occur at asymptotes. (Note, this does not mean that all asymptotes give discontinuities).

An asymptote is a line where the distance between this line and a point P tends to zero as P tends to infinity.

The graph $y = \tan x$ clearly demonstrates this at the lines $x = -90$, $x = 90$ for example.

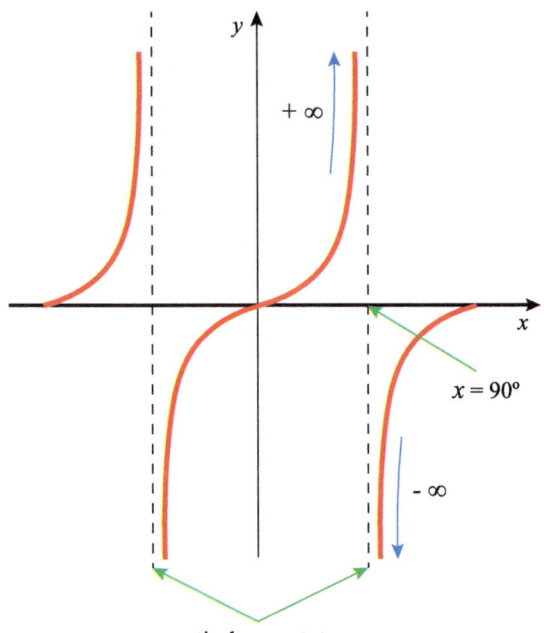

vertical asymptote

When the point $x = 90°$ is approached from the left the value of $\tan x$ approaches ∞ as x approaches $90°$. (The distance between $\tan x$ as x approaches the line $x = 90$ tends towards zero).

Check this by taking values of $\tan x$ for x between $89°$ and $89.9°$ on a calculator. The answers are larger and larger positive values. At $x = 90$ the value of $\tan x$ is undefined.

When the point is approached from the right the value of $\tan x$ approaches $-\infty$ as x approaches $90°$.

Again check this by taking values of $\tan x$ for x from $91°$ down to $90.1°$. The answers in this case are larger and larger negative values.

The line $x = 90°$ is called an asymptote. Asymptotes are also present when sketching functions other than $\tan x$.

© HERIOT-WATT UNIVERSITY

Vertical asymptotes

> **Key point**
>
> For rational functions, a vertical asymptote is a vertical line with equation of the form $x = k$ at which the function in question is undefined.

Consider $y = \frac{x}{x-3}$, where $x \neq 3$.

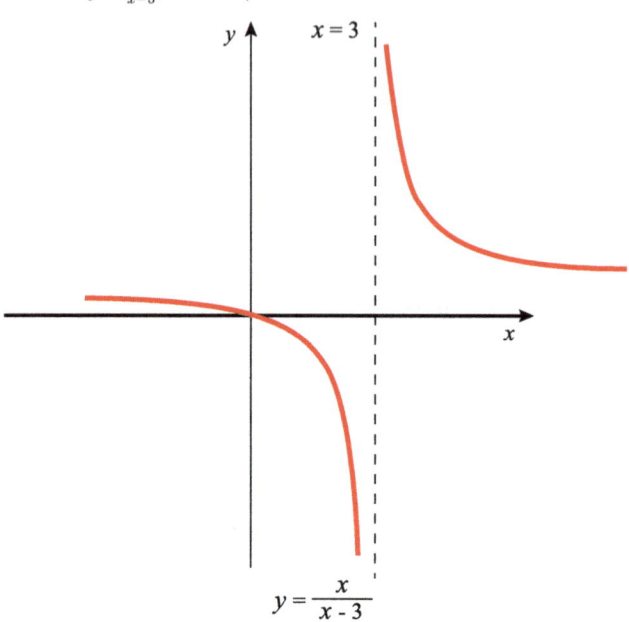

Vertical asymptotes occur when the function is undefined. i.e. **When the denominator is zero.**

$x - 3 = 0$
$\quad x = 3$

The function values either increase rapidly towards $+\infty$ or decrease rapidly towards $-\infty$ as x gets closer and closer to the value 3.

TOPIC 8. CURVE SKETCHING

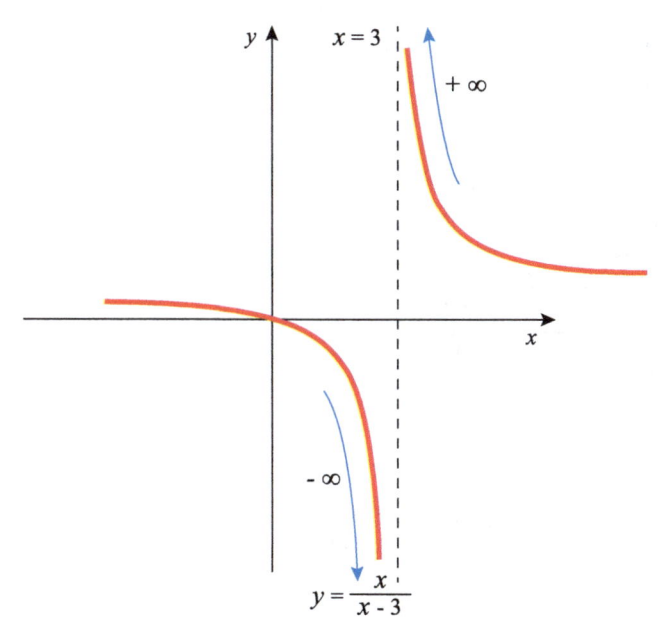

[Graph showing $y = \dfrac{x}{x-3}$ with vertical asymptote at $x = 3$, branch going to $+\infty$ from above and $-\infty$ from below.]

Values of x on either side of the asymptote can be chosen and the function evaluated to see if the function is greatly increasing (towards $+\infty$) or greatly decreasing (towards $-\infty$). This will give enough information to show the direction of the graph near this asymptote.

From below $x = 3$:

x	2	2·5	2·9
$\frac{x}{x-3}$	-2	-5	-29

As x tends to 3 from below y tends to $-\infty$
i.e. as $x \to 3^-$, $y \to -\infty$.

From above $x = 3$:

x	3·1	3·5	4
$\frac{x}{x-3}$	31	7	4

As x tends to 3 from above y tends to $+\infty$
i.e. as $x \to 3^+$, $y \to +\infty$.

Horizontal asymptotes

> **Key point**
>
> For rational functions, a horizontal asymptote is a horizontal line with equation of the form $y = m$ for which the function value gets closer and closer to the value m as x tends towards $+\infty$ and / or $-\infty$.
>
> A horizontal asymptote occurs when the degree of the numerator is less than or equal to the degree of the denominator.

356   TOPIC 8. CURVE SKETCHING

Consider $y = \frac{2x}{x-1}$, where $x \neq 1$.

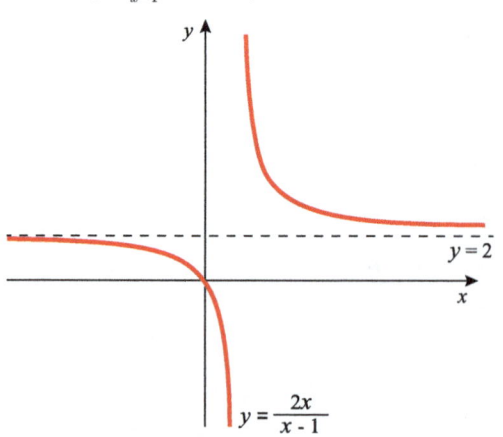

To find the **horizontal asymptote** divide the numerator and denominator by the highest power of x and let $x \to \infty$. The expression that is left is the horizontal asymptote.

$$y = \frac{2x}{x-1}$$
$$y = \frac{2}{1-\frac{1}{x}}$$

So

$$\lim_{x \to \infty} \frac{2}{1-\frac{1}{x}} = 2$$

The horizontal asymptote: $y = 2$.

The function values either increase rapidly towards 2 or decrease rapidly towards 2 as x goes off to $\pm\infty$.

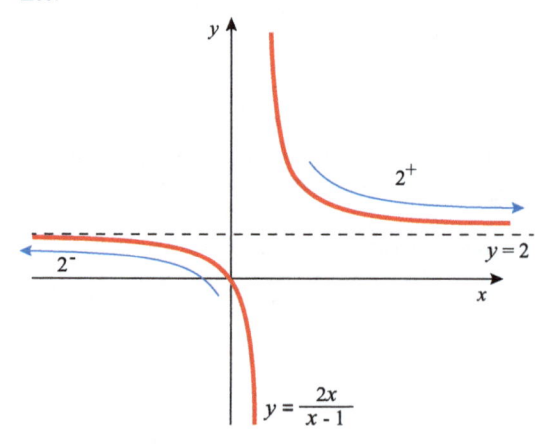

TOPIC 8. CURVE SKETCHING

Large and small values of x can be chosen and evaluated to see if the function is increasing towards 2 from below or decreasing from above. This will give enough information to show the direction of the graph near this asymptote.

From below $y = 2$:

x	-1000	-100	-10
$\frac{2x}{x-1}$	1·998	1·980	1·818

As x tends to $-\infty$, y tends to 2 from below i.e. as $x \to -\infty$, $y \to 2^-$.

From above $y = 2$:

x	10	100	1000
$\frac{2x}{x-1}$	2·222	2·020	2·002

As x tends to $+\infty$, y tends to 2 from above i.e. as $x \to +\infty$, $y \to 2^+$.

Slant asymptotes

> **Key point**
>
> For rational functions, a slant or oblique asymptote is a line, neither horizontal nor vertical, with equation of the form $y = ax + b$ for which the function value gets closer and closer to the line $y = ax + b$ as x tends towards $+\infty$ and / or $-\infty$.
>
> A slant asymptote occurs when the degree of the numerator is greater than the degree of the denominator.

Consider $y = \frac{x^2}{x-2}$, where $x \neq 2$.

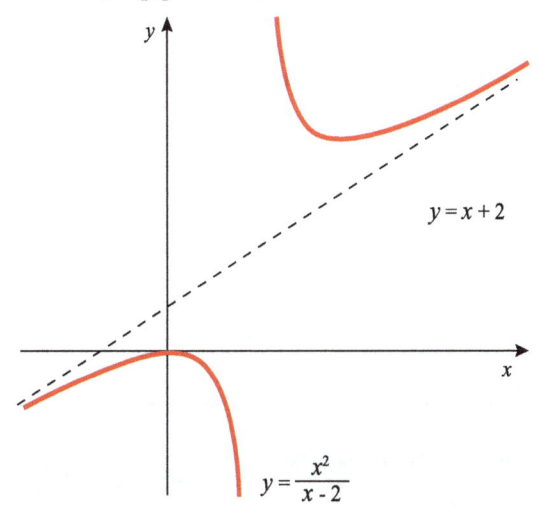

To find the **slant asymptote** divide the numerator by the denominator (using algebraic long division). The **quotient** is the slant asymptote.

$$y = \frac{x^2}{x-2}$$

$$y = x + 2 + \frac{4}{x-2}$$

The slant asymptote: $y = x + 2$

© HERIOT-WATT UNIVERSITY

(Note there will also be a vertical asymptote at $x = 2$.)

The function values either increase rapidly towards $x + 2$ or decrease rapidly towards $x + 2$ as x goes off to $\pm\infty$.

One way of determining the behaviour close to the asymptote is to look at the function instead of picking particular values.

1. Take the case $x \to +\infty$:

$$y = x + 2 + \boxed{\dfrac{4}{x-2}}^{\text{remainder}}$$

The behaviour is driven by $x + 2$ and a small **remainder is added** on since x is large and positive.

When $x \to +\infty$, $\dfrac{4}{x-2} \to 0^+$

Since $\dfrac{4}{x-2}$ tend to zero from above the curve must approach the asymptote $x + 2$ from above.

As $x \to +\infty$, $y \to (x+2)^+$

2. Take the case $x \to -\infty$:

$$y = x + 2 + \boxed{\dfrac{4}{x-2}}^{\text{remainder}}$$

The behaviour is driven by $x+2$ and a small **remainder is substracted** off since x is negative.

When $x \to -\infty$, $\dfrac{4}{x-2} \to 0^-$

Since $\dfrac{4}{x-2}$ tend to zero from below the curve must approach the asymptote $x + 2$ from below.

As $x \to -\infty$, $y \to (x+2)^-$

Examples

1. Problem:

What are the asymptotes for the graph $y = \dfrac{x^2}{x+1}$, where $x \neq -1$?

Solution:

Vertical asymptotes:
Occur when the denominator is equal to zero for rational functions.

For $y = \dfrac{x^2}{x+1}$ the denominator is zero when $x + 1 = 0 \Rightarrow x = -1$.

Vertical asymptote: $x = -1$

As $x \to -1^-$, $y \to -\infty$

As $x \to -1^+$, $y \to +\infty$

TOPIC 8. CURVE SKETCHING

> **Horizontal asymptotes:**
>
> To find the horizontal asymptote, for rational functions, **divide the numerator and denominator by the highest power of** x **and let** $x \to \infty$.
>
> When $x \to \infty$ the expression y may tend to a limit (not to infinity). This limit is the horizontal asymptote of the form $y = b$.

For $y = \frac{x^2}{x+1}$, divide by the highest power, which is x^2: $y = \frac{1}{\frac{1}{x} + \frac{1}{x^2}}$

Take the limit as $x \to \infty$: $\lim\limits_{x \to \infty} \frac{1}{\frac{1}{x} + \frac{1}{x^2}} = \infty$

There is no horizontal asymptote.

> **Slant asymptote:**
>
> A slant asymptote occurs when the degree of the numerator is greater than the degree of the denominator, for rational functions.

For $y = \frac{x^2}{x+1}$ when we divide: $y = x - 1 + \frac{1}{x+1}$

Slant asymptote: $y = x - 1$

As $x \to -\infty, y \to (x-1)^-$

As $x \to +\infty, y \to (x-1)^+$

This is enough information to sketch this graph to show the asymptotes. Note the dashed lines indicate that these are asymptotes. All asymptotes should be shown with dashed lines.

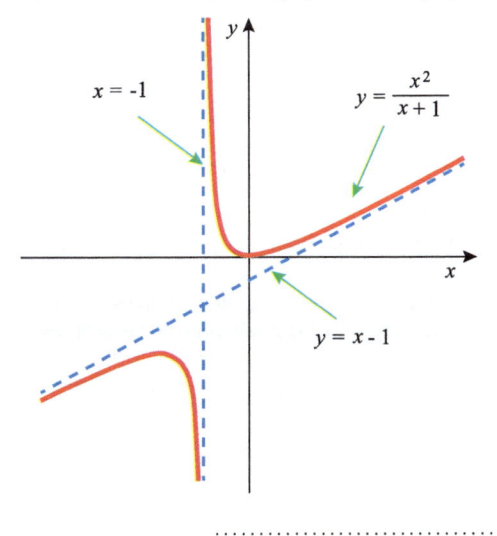

© HERIOT-WATT UNIVERSITY

2. Problem:

What are the asymptotes for the graph $y = \frac{1}{x}$, where $x \neq 0$?

Solution:

Vertical asymptotes:

Occur when the denominator is equal to zero for rational functions.

For $y = \frac{1}{x}$ the denominator is zero when $x = 0$.

Vertical asymptote: $x = 0$

As $x \to 0^-$, $y \to -\infty$ ($\frac{1}{x}$ is negative)

As $x \to 0^+$, $y \to +\infty$ ($\frac{1}{x}$ is positive)

Horizontal asymptotes:

To find the horizontal asymptote, for rational functions, **divide the numerator and denominator by the highest power of** x **and let** $x \to \infty$.

When $x \to \infty$ the expression y may tend to a limit (not to infinity). This limit is the horizontal asymptote of the form $y = b$.

For $y = \frac{1}{x}$, divide by the highest power, which is x: $y = \frac{\frac{1}{x}}{\frac{1}{1}}$

Take the limit as $x \to \infty$: $\lim\limits_{x \to \infty} \frac{\frac{1}{x}}{1} = 0$

Horizontal asymptote: $y = 0$

As $x \to -\infty$, $y \to 0^-$ ($\frac{1}{x}$ is negative)

As $x \to +\infty$, $y \to 0^+$ ($\frac{1}{x}$ is positive)

Slant asymptote:

A slant asymptote occurs when the degree of the numerator is greater than the degree of the denominator, for rational functions.

For $y = \frac{1}{x}$ the degree of the numerator is not bigger than the denominator so there is no slant asymptote.

This is enough information to sketch this graph to show the asymptotes. Note the dashed axes indicate that these are asymptotes. All asymptotes should be shown with dashed lines.

Graph showing $y = \frac{1}{x}$ with vertical asymptote $x = 0$ and horizontal asymptote $y = 0$.

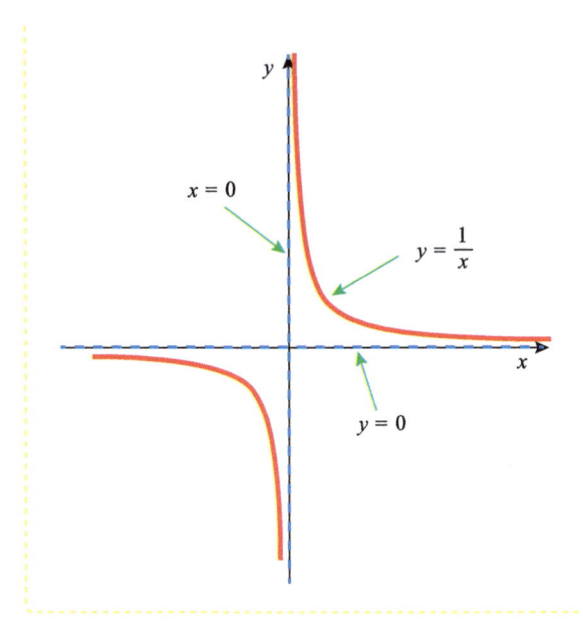

Identifying asymptotes exercise   Go online

> **Key point**
>
> **Vertical asymptotes:**
>
> Occur when the denominator is equal to zero for rational functions.

> **Key point**
>
> **Horizontal asymptotes:**
>
> To find the horizontal asymptote, for rational functions, **divide the numerator and denominator by the highest power of** x **and let** $x \to \infty$.
>
> When $x \to \infty$ the expression y may tend to a limit (not to infinity). This limit is the horizontal asymptote of the form $y = b$.

> **Key point**
>
> **Slant asymptote:**
>
> A slant asymptote occurs when the degree of the numerator is greater than the degree of the denominator, for rational functions.

Complete the tables with different functions. If the function does not have an asymptote then use "none".

Q87:

Function	$y = \frac{5}{x-3}$
Vertical asymptote	
Horizontal asymptote	
Slant asymptote	

Q88:

Function	$y = \frac{3x+1}{x-2}$
Vertical asymptote	
Horizontal asymptote	
Slant asymptote	

Q89:

Function	$y = \frac{2x+3}{x^2-3x+2}$
Vertical asymptote	
Horizontal asymptote	
Slant asymptote	

Q90:

Function	$y = x + 1 - \frac{1}{x}$
Vertical asymptote	
Horizontal asymptote	
Slant asymptote	

Q91:

Function	$y = \frac{x^2+5}{x+1}$
Vertical asymptote	
Horizontal asymptote	
Slant asymptote	

TOPIC 8. CURVE SKETCHING

Q92:

Function	$y = \frac{x+2}{x^2+x-2}$
Vertical asymptote	
Horizontal asymptote	
Slant asymptote	

8.9 Sketching and rational functions

Before considering a formal strategy for sketching graphs, the following example shows some of the techniques already explained.

Example Problem:

Sketch the graph of $y = x - \frac{4}{x}$, where $x \neq 0$.

Solution:

Let $y = f(x)$ then $y = x - \frac{4}{x}$

Symmetry

$$f(-x) = -\left(x - \frac{4}{x}\right)$$
$$= -f(x)$$

The function is odd and has rotational symmetry of order 2.

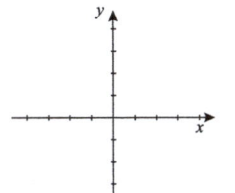

Crossing x-axis

When $y = 0$:

$$x - \frac{4}{x} = 0$$
$$x^2 - 4 = 0$$
$$(x-2)(x+2) = 0$$
$$x = \pm 2$$

$\Rightarrow$ (-2,0) and (2,0)

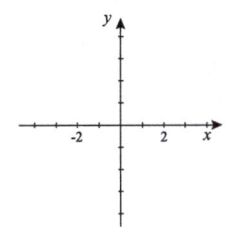

Crossing y-axis

When $x = 0$:

It is not defined since $\frac{4}{0}$ is not defined.

It does not cross the y-axis.

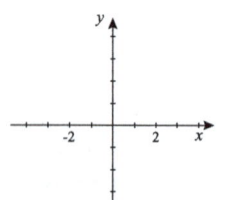

Turning points

$\frac{dy}{dx} = 1 + \frac{4}{x^2}$

$1 + \frac{4}{x^2} \neq 0$ and is always positive. There are no turning points.

$\frac{d^2y}{dx^2} = -\frac{8}{x^3}$

$-\frac{8}{x^3} \neq 0$

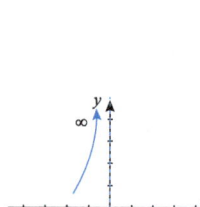

$\frac{d^2y}{dx^2}$ is undefined when $x = 0$. There is a possibility that this is a point of inflection.

However, if we look at the original function $y = x - \frac{4}{x}$, $x \neq 0$. Therefore, it is not a point of inflection.

Vertical asymptotes

When the denominator is zero, i.e. when $x = 0$

As $x \to 0^-$, $y \to \infty$

As $x \to 0^+$, $y \to -\infty$

Horizontal asymptotes

Divide through by the highest power of x in the fraction and take limit as $x \to \infty$.

Divide by x:

$y = x - \frac{\frac{4}{x}}{1}$

So $\lim\limits_{x \to \infty} \left(x - \frac{\frac{4}{x}}{1} \right)$, $y \to x$

So no horizontal asymptote.

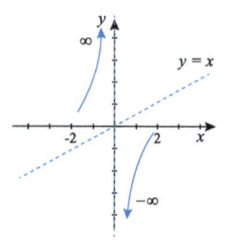

TOPIC 8. CURVE SKETCHING

Slant asymptotes

$y = x - \frac{4}{x}$

As $x \to -\infty$, $y \to x^+$ (x from above)

As $x \to +\infty$, $y \to x^-$ (x from below).

We can already see whether function approach $y = x$ from above or below from previous investigation.

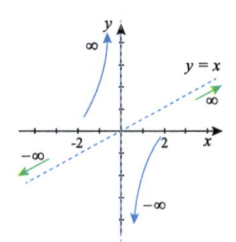

Recall the symmetry to the sketch to check if it seems reasonable.

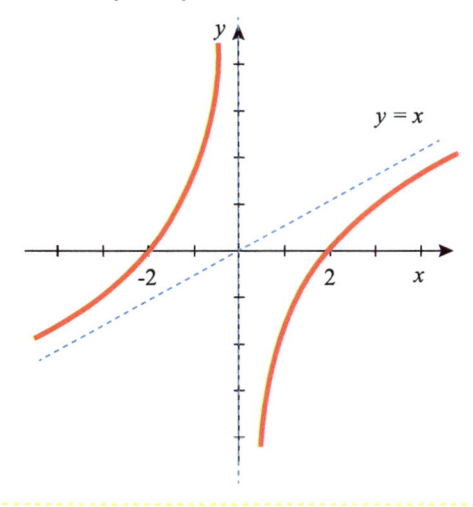

It is possible to identify the relevant features of a graph and develop a strategy for sketching curves by investigating functions in this manner.

Strategy for sketching curves

1. Identify any odd or even symmetry.
2. Identify crossing of the x-axis.
3. Identify crossing of the y-axis.
4. Look for turning points and points of inflection and their nature.
5. Find asymptotes and check the behaviour of x or y as the asymptotes are approached.

With the information obtained from this topic it is possible to sketch a wide variety of functions, including rational functions.

© HERIOT-WATT UNIVERSITY

There are five general types of rational function to consider in this section. Specific examples of each type including a sketch are shown in turn.

| Type 1 | constant / linear | $\dfrac{c}{ex+f}$ | $\dfrac{1}{x-1}, \ x \neq 1$ |

$y = \dfrac{1}{x-1}$

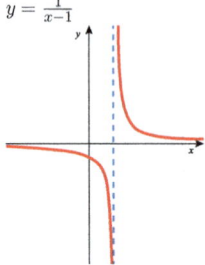

| Type 2 | linear / linear | $\dfrac{bx+c}{ex+f}$ | $\dfrac{2x+3}{4x-1}, \ x \neq \dfrac{1}{4}$ |

$y = \dfrac{2x+3}{4x-1}$

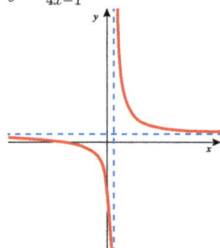

| Type 3 | constant / linear over quadratic | c or $(bx+c)$ over $dx^2 + ex + f$ | 5 or $(2x-3)$ over $x^2 + 3x - 4$, $x \neq -4$ or 1 |

$y = \dfrac{2x-3}{x^2+3x-4}$

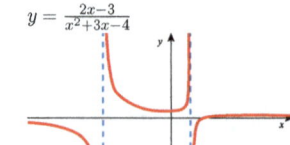

TOPIC 8. CURVE SKETCHING

Type 4	$\dfrac{\text{quadratic}}{\text{quadratic}}$	$\dfrac{ax^2 + bx + c}{dx^2 + ex + f}$	$\dfrac{3x^2 - 4x + 2}{x^2 - 2x + 1},\ x \neq 1$

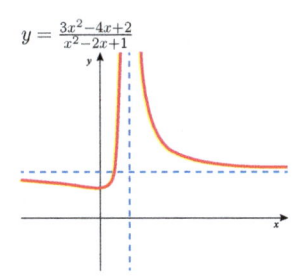

$y = \dfrac{3x^2 - 4x + 2}{x^2 - 2x + 1}$

Type 5	$\dfrac{\text{quadratic}}{\text{linear}}$	$\dfrac{ax^2 + bx + c}{ex + f}$	$\dfrac{x^2 + x - 6}{x - 1},\ x \neq 1$

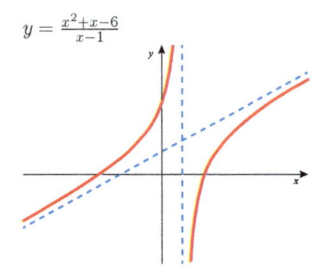

$y = \dfrac{x^2 + x - 6}{x - 1}$

The strategy for sketching these types of rational functions is given in the following examples. The strategy is basically the same for each type.

Make sure that any common factors on the top and bottom are first cancelled before classifying rational functions into the above types.

For example $\dfrac{x^2 - 3x + 2}{x^2 - 4x + 3} = \dfrac{(x-1)(x-2)}{(x-1)(x-3)} = \dfrac{x-2}{x-3}$.

This is a **Type 2** rational function and not as it might first seem, a Type 4.

After each example, simplified techniques are explained which help to reduce the effort when a sketch rather than a detailed graph is required.

© HERIOT-WATT UNIVERSITY

8.10 Type 1 rational function: Constant over linear

Example

Problem:

Sketch the curve of $y = \frac{1}{x+1}$, $x \neq -1$.

Solution:

Let $y = f(x)$ and $y = \frac{1}{x+1}$

Symmetry

$f(-x) = \dfrac{1}{-x+1}$

$-f(x) = -\dfrac{1}{x+1}$

$f(x) = \dfrac{1}{x+1}$

$f(-x) \neq -f(x)$ and $f(-x) \neq f(x)$

$\therefore$ the function is neither even nor odd.

Crosses the x-axis

When $y = 0$: $\frac{1}{x+1} = 0 \Rightarrow 1 = 0$

Which is impossible so does not cross the x-axis.

Crosses the y-axis

When $x = 0$:

$y = \dfrac{1}{0+1}$

$y = 1$

It crosses the y-axis at (0,1).

Turning points

First derivative

$\frac{dy}{dx} = -\frac{1}{(x+1)^2}$

For $\frac{dy}{dx} = 0$: $-\frac{1}{(x+1)^2} = 0 \Rightarrow -1 = 0$

Therefore there are no stationary points.

Second derivative

$\frac{d^2y}{dx^2} = \frac{2}{(x+1)^3}$

For possible points of inflection $\frac{d^2y}{dx^2} = 0$: $\frac{2}{(x+1)^3} = 0 \Rightarrow 2 = 0$

$\frac{d^2y}{dx^2}$ is undefined when $x = -1$. There is a possible point of inflection here since $\frac{dy}{dx}$ does not equal zero. However, when $x = -1$ the original function is undefined.

TOPIC 8. CURVE SKETCHING

Therefore there is no non-horizontal points of inflection.

Asymptotes

Vertical asymptote

When the denominator is zero,

$x + 1 = 0$

$\Rightarrow x = -1$ is a vertical asymptote.

As $x \to -1^-$, $y \to -\infty$

As $x \to -1^+$, $y \to +\infty$

Horizontal asymptote

Divide by the highest power and take the limit as $x \to \infty$.

$$\lim_{x \to \infty} \left(\frac{\frac{1}{x}}{1 + \frac{1}{x}} \right) = 0$$

There is a horizontal asymptote at $y = 0$.

As $x \to -\infty$, $y \to 0^-$

As $x \to +\infty$, $y \to 0^+$

Slant asymptote

When the degree of the numerator is bigger than the denominator.

The degree of the numerator is less than the denominator so there is no slant asymptote.

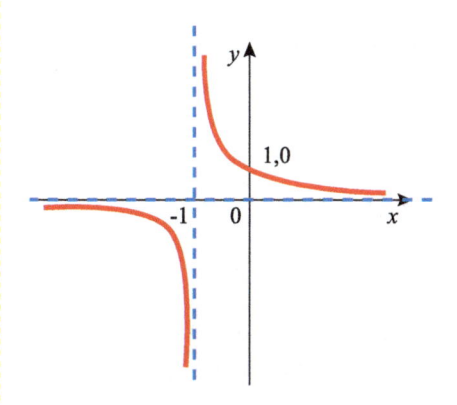

The horizontal axis is drawn with a dashed line to show that it is an asymptote.

Type 1: Shortcuts

The equation takes the form $\frac{c}{ex+f}$.

The vertical asymptote has the equation $x = \frac{-f}{e}$, $e \neq 0$.

All graphs of this type have a horizontal asymptote at $y = 0$

The graph crosses the y-axis at $y = \frac{c}{f}$ if $f \neq 0$.

Apply these shortcuts to the previous graph of $\frac{1}{x+1}$.

$c = 1, e = 1, f = 1$
The vertical asymptote is $x = \frac{-f}{e} = \frac{-1}{1} = -1$.
The horizontal asymptote is at $y = 0$ as it is Type 1.
The graph crosses the y-axis at $y = \frac{1}{1} = 1$.

The results match those found by the formal techniques but of course the turning points may still be needed.

Example

Problem:

What are the asymptotes for the function $y = \frac{2}{x+5}$ and where does it cross the y-axis?

Solution:

$c = 2, e = 1, f = 5$

The vertical asymptote is at $x = \frac{-f}{e} = \frac{-5}{1} = -5$

The horizontal asymptote is at $y = 0$

The graph crosses the y-axis at $\frac{c}{f} = \frac{2}{5}$

TOPIC 8. CURVE SKETCHING

Type 1 rational function exercise — Go online

Q93: A function f is defined by $f(x) = \frac{2}{3x+2}$, $x \neq -\frac{2}{3}$.

a) Identify any odd or even symmetry.
b) Identify where the graph crosses the x-axis and the y-axis.
c) Look for turning points and points of inflection and their nature.
d) Find asymptotes and check the behaviour of x or y as the asymptotes are approached.
e) Sketch the graph of $f(x)$.

....

Q94: The function f is defined by $f(x) = -\frac{4}{x}$, $x \neq 0$ and the diagram shows part of its graph.

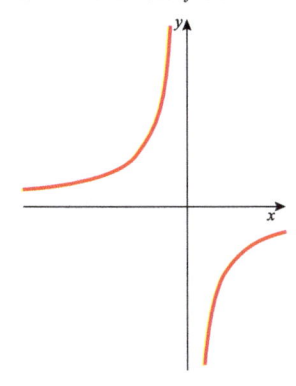

a) Write down the asymptotes of the graph of f.
b) Prove that f has no stationary values.
c) Does the graph of f have any points of inflexion? Justify your answer.
d) Is f an even or odd function?

....

Q95: Sketch the curve of $f(x) = \frac{5}{4x-1}$, $x \neq \frac{1}{4}$.

8.11 Type 2 rational function: Linear over linear

Example

Problem:

Sketch the curve of $y = \frac{2x+3}{4x-1}$, $x \neq \frac{1}{4}$.

Solution:

Let $y = f(x)$ and $y = \frac{2x+3}{4x-1}$

Symmetry

$$f(-x) = \frac{-2x+3}{-4x-1}$$

$$-f(x) = -\frac{2x+3}{4x-1}$$

$$f(x) = \frac{2x+3}{4x-1}$$

$f(-x) \neq -f(x)$ and $f(-x) \neq f(x)$

$\therefore$ the function is neither even nor odd.

Crosses the x-axis

When $y = 0$: $\frac{2x+3}{4x-1} = 0 \Rightarrow 2x+3 = 0 \Rightarrow x = -\frac{3}{2}$

Crosses at $\left(-\frac{3}{2}, 0\right)$.

Crosses the y-axis

When $x = 0$:

$$y = \frac{0+3}{0-1}$$

$$y = -3$$

It crosses the y-axis at (0,-3).

Turning points

First derivative

$\frac{dy}{dx} = -\frac{14}{(4x-1)^2}$

For $\frac{dy}{dx} = 0$: $-\frac{14}{(4x-1)^2} = 0 \Rightarrow 14 = 0$

Therefore there are no turning points.

Second derivative

$\frac{d^2y}{dx^2} = \frac{112}{(4x-1)^3}$

For possible points of inflection $\frac{d^2y}{dx^2} = 0$

$\frac{112}{(4x-1)^3} = 0 \Rightarrow 112 = 0$

$\frac{d^2y}{dx^2}$ is undefined when $x = \frac{1}{4}$. There is a possible point of inflection here since $\frac{dy}{dx} \neq 0$.

However, the original function is undefined for $x = \frac{1}{4}$.

Therefore there are no non-horizontal points of inflection.

Asymptotes

Vertical asymptote

When the denominator is zero:

$4x - 1 = 0$

$\Rightarrow x = \frac{1}{4}$ is a vertical asymptote.

As $x \to \frac{1}{4}^{-}$, $y \to -\infty$

As $x \to \frac{1}{4}^{+}$, $y \to +\infty$

Horizontal asymptote

Divide by the highest power and take the limit as $x \to \infty$.

$$\lim_{x \to \infty} \left(\frac{2 + \frac{3}{x}}{4 - \frac{1}{x}} \right) = 0$$

There is a horizontal asymptote at $y = \frac{1}{2}$.

As $x \to -\infty$, $y \to \frac{1}{2}^{-}$

As $x \to +\infty$, $y \to \frac{1}{2}^{+}$

Slant asymptote

When the degree of the numerator is bigger than the denominator.

The degree of the numerator is less than the denominator so there is no slant asymptote.

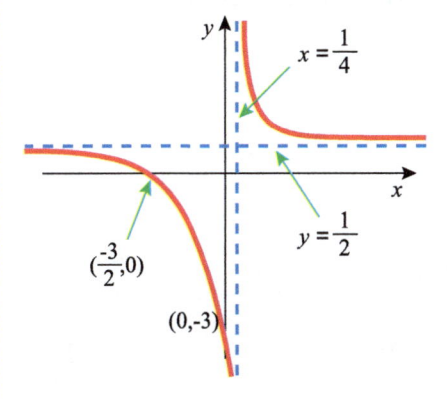

374 TOPIC 8. CURVE SKETCHING

Type 2: Shortcuts

This takes the form $\frac{bx+c}{ex+f}$

The vertical asymptote for a graph of this type will have the equation $x = \frac{-f}{e}$

The horizontal asymptote will have the equation $y = \frac{b}{e}$

The graph will cross the y-axis at the point $(0, \frac{c}{f})$ if $f \neq 0$

The graph will cross the x-axis at the point $(\frac{-c}{b}, 0)$

Apply these shortcuts to the previous graph of $y = \frac{2x+3}{4x-1}$

$b = 2, c = 3, e = 4, f = -1$
The vertical asymptote is $x = \frac{-f}{e} = \frac{1}{4}$
The horizontal asymptote is at $y = \frac{b}{e} = \frac{1}{2}$
The graph crosses the y-axis at $y = \frac{c}{f} = \frac{3}{-1} = -3$
The graph crosses the x-axis at $x = \frac{-c}{b} = \frac{-3}{2}$

The results match those found by the formal techniques but of course the turning points may still be needed.

Example

Problem:

Find the asymptotes and the points where the graph of $y = \frac{2x-3}{3x+4}$ crosses the axes.

Solution:

$b = 2, c = -3, e = 3, f = 4$

The vertical asymptote has equation $x = \frac{-f}{e} = \frac{-4}{3}$

The horizontal asymptote has the equation $y = \frac{b}{e} = \frac{2}{3}$

The graph crosses the y-axis at the point $(0, \frac{c}{f}) = (0, \frac{-3}{4})$

The graph crosses the x-axis at the point $(\frac{-c}{b}, 0) = (\frac{3}{2}, 0)$

TOPIC 8. CURVE SKETCHING

Type 2 rational function exercise

Q96: The diagram shows part of the graph of $y = \frac{3x-2}{2x+1}$, $x \neq -\frac{1}{2}$.

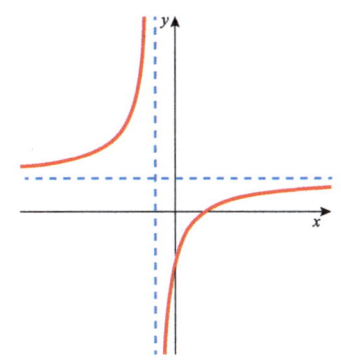

a) Identify any odd or even symmetry.
b) Identify where the graph crosses the x-axis and the y-axis.
c) Look for turning points and points of inflection and their nature.
d) Find asymptotes and check the behaviour of x or y as the asymptotes are approached.
e) Sketch the graph of $f(x)$.

....

Q97: A function f is defined by $f(x) = \frac{2x+1}{2x+3}$, $x \neq -\frac{3}{2}$.

a) Write down an equation for the asymptotes.
b) Show that f does not have any stationary points.
c) Sketch the graph of f.

....

Q98: Sketch the curve of $y = \frac{5x+1}{4x-3}$, $x \neq \frac{3}{4}$.

8.12 Type 3 rational function: Constant or linear over quadratic

Examples

1. Problem:
Sketch the curve of $y = \frac{5}{x^2+3x-4}$.

Solution:
Let $y = f(x)$ and $y = \frac{5}{x^2+3x-4}$

Symmetry

$f(-x) = \frac{5}{x^2 - 3x - 4}$

$-f(x) = -\frac{5}{x^2 + 3x - 4}$

$f(x) = \frac{5}{x^2 + 3x - 4}$

$f(-x) \neq -f(x)$ and $f(-x) \neq f(x)$

∴ the function is neither even nor odd.

Crosses the x-axis

When $y = 0$:

$\frac{5}{x^2+3x-4} = 0 \Rightarrow 5 = 0$

Which is impossible so **does not** cross the x-axis.

Crosses the y-axis

When $x = 0$:

$y = \frac{5}{0+0-4}$

$y = -\frac{5}{4}$

It crosses the y-axis at $(0, -\frac{5}{4})$.

Turning points

First derivative

$\frac{dy}{dx} = -\frac{5(2x+3)}{(x^2+3x-4)^2}$

For $\frac{dy}{dx} = 0$: $-\frac{5(2x+3)}{(x^2+3x-4)^2} = 0 \Rightarrow 5(2x+3) = 0 \Rightarrow x = -\frac{3}{2}$

Therefore there is a turning point at $x = -\frac{3}{2}$.

Second derivative

$\frac{d^2y}{dx^2} = -\frac{10(x^2+3x-4)^2 - (10x+15)2(x^2+3x-4)(2x+3)}{(x^2+3x-4)^4}$

Taking out a common factor of $2(x^2 + 3x - 4)$

$\frac{d^2y}{dx^2} = -\frac{10(x^2+3x-4)\left[(x^2+3x-4) - (2x+3)(2x+3)\right]}{(x^2+3x-4)^4}$

$\frac{d^2y}{dx^2} = -\frac{10(x^2+3x-4)\left[(x^2+3x-4) - (4x^2+12x+9)\right]}{(x^2+3x-4)^4}$

$\frac{d^2y}{dx^2} = -\frac{10(x^2+3x-4)\left[(-3x^2-9x-13)\right]}{(x^2+3x-4)^4}$

$\frac{d^2y}{dx^2} = -\frac{10\left[(-3x^2-9x-13)\right]}{(x^2+3x-4)^3}$

$\frac{d^2y}{dx^2} = \frac{10\left[(3x^2+9x+13)\right]}{(x^2+3x-4)^3}$

TOPIC 8. CURVE SKETCHING

At $x = -\frac{3}{2}$:

$\frac{d^2y}{dx^2} < 0$

Which is a maximum turning point.

If $x = -\frac{3}{2}$ then

$y = \dfrac{5}{\left(-\frac{3}{2}\right)^2 + 3\left(-\frac{3}{2}\right) - 4}$

$y = -\dfrac{4}{5}$

Turning point is at $\left(-\frac{3}{2}, -\frac{4}{5}\right)$.

Non-horizontal point of inflection

$\frac{d^2y}{dx^2} = 0 \Rightarrow 3x^2 + 9x + 13 = 0$

$b^2 - 4ac = 81 - 4(3)(13)$

$ = -75$

Since $b^2 - 4ac < 0$ there are no real roots.

There are no non-horizontal point of inflection.

Asymptotes

Vertical asymptote

When the denominator is zero

$x^2 + 3x - 4 = 0$

$(x+4)(x-1) = 0$

(You may need to use the quadratic formula here.)

$\Rightarrow x = -4$ and $x = 1$ are **vertical asymptotes**.

As $x \to -4^-$, $y \to \infty$

As $x \to -4^+$, $y \to -\infty$

As $x \to 1^-$, $y \to -\infty$

As $x \to 1^+$, $y \to +\infty$

Horizontal asymptote

Divide by the highest power and take the limit as $x \to \infty$.

$\lim\limits_{x \to \infty} \left(\dfrac{\frac{5}{x^2}}{1 + \frac{3}{x} - \frac{4}{x^2}} \right) = 0$

There is a **horizontal asymptote** at $y = 0$.

As $x \to -\infty$, $y \to 0^+$

As $x \to +\infty$, $y \to 0^+$

Slant asymptote

When the degree of the numerator is bigger than the denominator.

The degree of the numerator is less than the denominator so there is no slant asymptote.

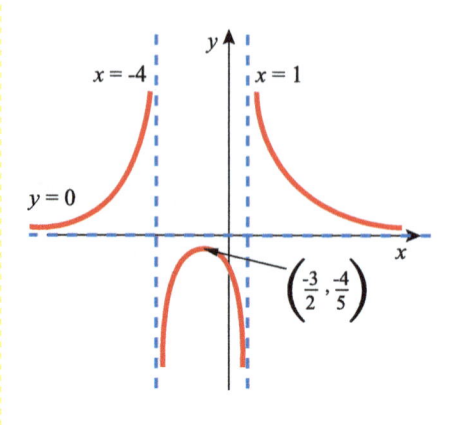

2.

Problem:

Sketch the curve of $y = \frac{x+1}{x^2}$, $x \neq 0$

Solution:

Let $y = f(x)$ and $y = \frac{x+1}{x^2}$.

Symmetry

$$f(-x) = \frac{-x+1}{x^2}$$

$$-f(x) = -\frac{x+1}{x^2}$$

$$f(x) = \frac{x+1}{x^2}$$

$f(-x) \neq -f(x)$ and $f(-x) \neq f(x)$, therefore the function is neither even nor odd.

Crosses the x-axis

When $y = 0$:

$\frac{x+1}{x^2} = 0$

$\Rightarrow x + 1 = 0$

$\Rightarrow x = -1$

It crosses the x-axis at (-1,0).

Crosses the y-axis

When $x = 0$: $y = \frac{0+1}{0}$

Undefined at $x = 0$ so does not cross the y-axis.

TOPIC 8. CURVE SKETCHING

Turning points

First derivative

Using the quotient rule:
$$\frac{dy}{dx} = \frac{1\left(x^2\right) - (x+1)\,2x}{x^4}$$
$$\frac{dy}{dx} = \frac{\left(x^2\right) - \left(2x^2 + 2x\right)}{x^4}$$
$$\frac{dy}{dx} = \frac{\left(-x^2 - 2x\right)}{x^4}$$
Taking out a common factor of $-x$.
$$\frac{dy}{dx} = -\frac{x(x+2)}{x^4}$$
$$\frac{dy}{dx} = -\frac{x+2}{x^3}$$
For $\frac{dy}{dx} = 0$: $\frac{x+2}{x^3} = 0 \;\Rightarrow\; x = -2$
If $x = -2$ then $y = \frac{(-2)+1}{(-2)^2}$
Turning point at $\left(-2, -\frac{1}{4}\right)$.

Second derivative

Turning point
$$\frac{d^2y}{dx^2} = -\frac{1\left(x^3\right) - (x+2)\,3x^2}{x^6}$$
$$\frac{d^2y}{dx^2} = -\frac{x^3 - 3x^3 - 6x^2}{x^6}$$
$$\frac{d^2y}{dx^2} = -\frac{-2x^3 - 6x^2}{x^6}$$
Taking out a factor of $-2x^2$:
$$\frac{d^2y}{dx^2} = -\frac{-2x^2(x+3)}{x^6}$$
$$\frac{d^2y}{dx^2} = \frac{2(x+3)}{x^4}$$
At $x = -2$:
$$\frac{d^2y}{dx^2} > 0$$
Which is a minimum turning point.

Non-horizontal point of inflection

$\frac{d^2y}{dx^2} = 0 \;\Rightarrow\; x + 3 = 0 \;\Rightarrow\; x = -3$

When $x = -3^-$ then $\frac{d^2y}{dx^2} < 0 \;\Rightarrow\;$ *concave down*

When $x = -3^+$ then $\frac{d^2y}{dx^2} > 0 \;\Rightarrow\;$ *concave up*

When $x = -3$, $\frac{dy}{dx} \neq 0$ and $\frac{d^2y}{dx^2} = 0$

There is a non-horizontal point of inflection at $\left(-3, -\frac{2}{9}\right)$.

$\frac{d^2y}{dx^2}$ and $\frac{dy}{dx}$ are undefined when $x = 0$.

© HERIOT-WATT UNIVERSITY

However, the original function is also undefined for $x = 0$.

This is not a point of inflection.

Asymptotes

Vertical asymptote

When the denominator is zero

$x^2 = 0$

$\Rightarrow \quad x = 0$ is a vertical asymptote.

As $x \to 0^-$, $y \to +\infty$

As $x \to 0^+$, $y \to +\infty$

Horizontal asymptote

Divide by the highest power and take the limit as $x \to \infty$.

$$\lim_{x \to \infty} \left(\frac{\frac{1}{x} + \frac{1}{x^2}}{1} \right) = 0$$

There is a horizontal asymptote at $y = 0$.

As $x \to -\infty$, $y \to 0^-$

As $x \to +\infty$, $y \to 0^+$

Slant asymptote

When the degree of the numerator is bigger than the denominator.

The degree of the numerator is the same as the denominator so there is no slant asymptote.

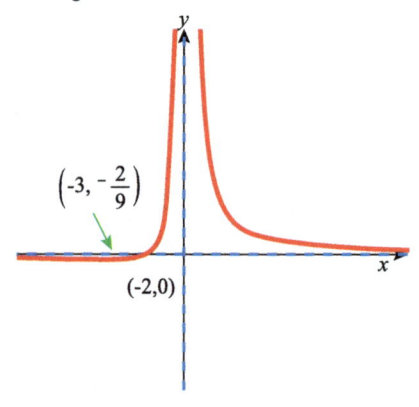

Type 3: Shortcuts

This takes the form $\frac{c}{dx^2+ex+f}$ or $\frac{bx+c}{dx^2+ex+f}$.

The vertical asymptotes exist only if the equation $dx^2 + ex + f = 0$ has real solutions for x. This may not factorise, in which case the quadratic formula will have to be used.

Suppose that there are two roots, namely, $x = w$ and $x = z$ then these equations will be the

TOPIC 8. CURVE SKETCHING

equations of the vertical asymptotes.

If there are no solutions then there are no vertical asymptotes.

The horizontal asymptote will have the equation $y = 0$

The graph crosses the x-axis at the point ($\frac{-c}{b}$, 0) **only when** the numerator is of the form $bx + c$. Otherwise there is no point at which the graph crosses the x-axis.

The graph crosses the y-axis at the point (0, $\frac{c}{f}$) if $f \neq 0$.

Apply these shortcuts to the previous graph of $y = \frac{5}{x^2+3x-4}$

$b = 0$, $c = 5$, $d = 1$, $e = 3$, $f = -4$
The denominator factorises to give $(x + 4)(x - 1)$
The vertical asymptotes are $x = 1$ and $x = -4$
The horizontal asymptote is at $y = 0$
The graph crosses the y-axis at $y = \frac{c}{f} = \frac{-5}{4}$
The graph does not cross the x-axis since $b = 0$

The results match those found by the formal techniques but of course the turning points may still be needed.

Examples

1. Problem:
What are the asymptotes for the function $f(x) = \frac{2x-3}{x^2-3x-4}$, $x \neq 1, -4$, and where does it cross the axes?

Solution:
$b = 2$, $c = -3$, $d = 1$, $e = 3$, $f = -4$

If there are vertical asymptotes then the quadratic on the denominator will have real roots.

This quadratic factorises to $(x + 1)(x - 4)$ giving the solutions when equated to zero of $x = -1$ and $x = 4$.
There are two vertical asymptotes with equations $x = -1$ and $x = 4$

The horizontal asymptote has the equation $y = 0$

The graph crosses the x-axis at the point ($\frac{-c}{b}$, 0) = ($\frac{3}{2}$, 0)

The graph crosses the y-axis at the point (0, $\frac{c}{f}$) = (0, $\frac{-3}{-4}$) = (0, $\frac{3}{4}$)

....

© HERIOT-WATT UNIVERSITY

2. Problem:

What are the asymptotes for the function $f(x) = \frac{-4}{x^2-2x+5}$ and where does it cross the axes?

Solution:

$c = -4, d = 1, e = -2, f = 5$

$x^2 - 2x + 5$ has no real roots so there are no vertical asymptotes.

There is a horizontal asymptote at $y = 0$

The graph does not cross the x-axis since the numerator is not of the form $bx + c$

The graph crosses the y-axis at the point $(0, \frac{c}{f}) = (0, \frac{-4}{5})$.

Type 3 rational function exercise   Go online

Q99: The diagram show the shape of the graph $y = \frac{2}{x^2-x-6}, x \neq -2, 3$.

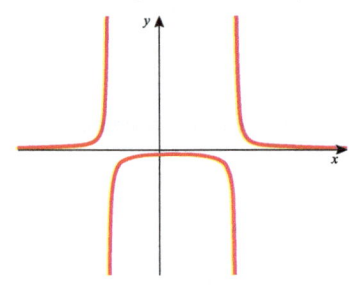

a) Obtain the stationary point of the graph.
b) Show that there are no non-horizontal points of inflection.
c) Write down equations for the asymptotes.
d) Where does the graph cut the y-axis?
e) Is the graph symmetrical?

....

Q100: A function f is defined by $f(x) = \frac{2x}{x^2-1}$.

a) Write down an equation for each of the asymptotes.
b) Find the stationary points and justify their nature.
c) Show that there is a non-horizontal point of inflection.
d) Sketch the curve.

....

Q101: Sketch the curve of $y = \frac{3}{1+2x^2}$.

TOPIC 8. CURVE SKETCHING

8.13 Type 4 rational function: Quadratic over quadratic

Example

Problem:

Sketch the curve of $y = \frac{3x^2-4x+2}{x^2-2x+1}$, $x \neq -1$.

Solution:

Let $y = f(x)$ and $y = \frac{3x^2-4x+2}{x^2-2x+1}$

Symmetry

$$f(-x) = \frac{3x^2+4x+2}{x^2+2x+1}$$

$$-f(x) = -\frac{3x^2-4x+2}{x^2-2x+1}$$

$$f(x) = \frac{3x^2-4x+2}{x^2-2x+1}$$

$f(-x) \neq -f(x)$ and $f(-x) \neq f(x)$

∴ the function is neither even nor odd.

Crosses the x-axis

When $y = 0$: $\frac{3x^2-4x+2}{x^2-2x+1} = 0 \Rightarrow 3x^2 - 4x + 2 = 0$

$b^2 - 4ac < 0 \Rightarrow$ no real roots so **does not** cross the x-axis.

Crosses the y-axis

When $x = 0$:

$$y = \frac{0-0+2}{0-0+1}$$

$y = 2$

It crosses the y-axis at (0,2).

Turning points

First derivative

$$\frac{dy}{dx} = \frac{-2x^2+2x}{(x^2-2x+1)^2}$$

$$= \frac{-2x(x-1)}{\left((x-1)^2\right)^2}$$

$$= \frac{-2x}{(x-1)^3}$$

For $\frac{dy}{dx} = 0$

$$\frac{-2x}{(x-1)^3} = 0$$

$$x = 0$$

© HERIOT-WATT UNIVERSITY

Second derivative

$$\frac{d^2y}{dx^2} = \frac{2(2x+1)}{(x-1)^4}$$

At $x = 0$: $\frac{d^2y}{dx^2} > 0$

Which is a minimum turning point.

If $x = 0$ then

$$y = \frac{3x^2 - 4x + 2}{x^2 - 2x + 1}$$

$y = 2$

Turning point is at (0,2).

Non-horizontal point of inflection

$\frac{d^2y}{dx^2} = 0 \Rightarrow 2x + 1 = 0 \Rightarrow x = -\frac{1}{2}$

Evaluating $x = -\frac{1}{2}$ for $\frac{dy}{dx}$: $\frac{dy}{dx} \neq 0$

When $x \to -\frac{1}{2}^{-}$, then $\frac{d^2y}{dx^2} < 0 \quad \Rightarrow \quad$ concave down

When $x \to -\frac{1}{2}^{+}$, then $\frac{d^2y}{dx^2} > 0 \quad \Rightarrow \quad$ concave up

There is a change in concavity.

There is a non-horizontal point of inflection at $\left(-\frac{1}{2}, \frac{19}{9}\right)$

Asymptotes

Vertical asymptote

When the denominator is zero

$x^2 - 2x + 1 = 0$

$(x-1)(x-1) = 0$

(You may need to use the quadratic formula here.)

$\Rightarrow x = 1$ is a **vertical asymptote**.

As $x \to 1^{-}$, $y \to +\infty$

As $x \to 1^{+}$, $y \to +\infty$

Horizontal asymptote

Divide by the highest power and take the limit as $x \to \infty$.

$$\lim_{x \to \infty} \left(\frac{3 - \frac{4}{x} + \frac{2}{x^2}}{1 - \frac{2}{x} + \frac{1}{x^2}} \right) = 3$$

There is a **horizontal asymptote** at $y = 3$.

As $x \to -\infty$, $y \to 3^{-}$

As $x \to +\infty$, $y \to 3^{+}$

Slant asymptote

When the degree of the numerator is bigger than the denominator.

The degree of the numerator is less than the denominator so there is no slant asymptote.

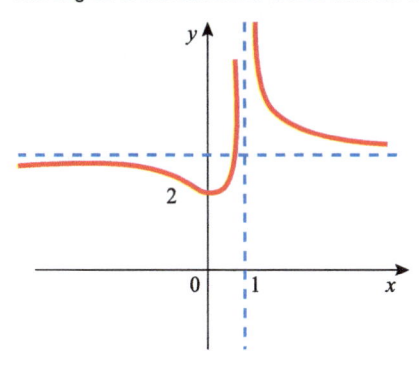

Type 4: Shortcuts

This takes the form $\frac{ax^2+bx+c}{dx^2+ex+f}$

Vertical asymptotes exist only if the equation $dx^2 + ex + f = 0$ has solutions for x.

As for type 3 if they exist then these solutions are the equations of the asymptotes.

The horizontal asymptote has equation $y = \frac{a}{d}$.

The graph crosses the y-axis at the point (0, $\frac{c}{f}$) if $f \neq 0$.

The graph crosses the x-axis at the real roots of the equation $ax^2 + bx + c = 0$ if there are any.

Apply these shortcuts to the previous graph of $y = \frac{3x^2-4x+2}{x^2-2x+1}$.

$a = 3, b = -4, c = 2, d = 1, e = -2, f = 1$
The denominator factorises to give $(x - 1)^2$.
The vertical asymptote is $x = 1$
The horizontal asymptote is at $y = \frac{a}{d} = \frac{3}{1}$.
The graph crosses the y-axis at $y = \frac{c}{f} = \frac{2}{1} = 2$.
The numerator does not have real roots and so the graph does not cross the x-axis.

The results match those found by the formal techniques but of course the turning points may still be needed.

TOPIC 8. CURVE SKETCHING

Example

Problem:

Find the asymptotes for the function $f(x) = \frac{2x^2+3x-2}{4x^2+13x+3}$, $x \neq -3, -\frac{1}{4}$, and the points where it crosses the axes.

Solution:

$a = 2$, $b = 3$, $c = -2$, $d = 4$, $e = 13$, $f = 3$

The denominator factorises to give $(4x + 1)(x + 3)$. Solving this equal to zero gives the asymptote equations.

These vertical asymptote equations are $x = -3$ and $x = \frac{-1}{4}$.

The horizontal asymptote has the equation $y = \frac{a}{d} = \frac{2}{4} = \frac{1}{2}$.

The graph crosses the y-axis at the point (0, $\frac{c}{f}$) = (0, $\frac{-2}{3}$)

The numerator factorises to give $(2x - 1)(x + 2)$

Solving this equal to zero gives $x = -2$ and $x = \frac{1}{2}$

The graph crosses the x-axis at the points (-2, 0) and ($\frac{1}{2}$, 0)

Type 4 rational function exercise

Go online

Q102: The function $f(x)$ is defined by $f(x) = \frac{x^2}{x^2-1}$, $x \neq 1$.

a) Obtain equations for the asymptotes of the graph of $f(x)$.
b) Find the coordinates of the stationary points.
c) Determine if there are any non-horizontal points of inflection.
d) Find the coordinates of the points where the graph of $f(x)$ crosses the x and y-axis.
e) Sketch the graph of $f(x)$.

....

Q103: A curve is defined by $y = \frac{x^2+3x-3}{(x-2)^2}$, $x \neq 2$

a) Write down the equations for its asymptotes.
b) Find the stationary point and justify its nature.
c) Are there any non-horizontal points of inflection? Justify your answer.
d) Sketch the curve.

....

Q104: Sketch the function $f(x)$ defined by $f(x) = \frac{3x(x+1)}{(x+3)^2}$, $x \neq 3$.

8.14 Type 5 rational function: Quadratic over linear

> **Key point**
>
> When the power of the numerator is greater than the denominator apply algebraic division. This will make the first and second derivative easier to calculate and evaluate.

Example

Problem:

Sketch the curve of $y = \frac{x^2+x+2}{x-1}$, $x \neq 1$.

Solution:

This function is an improper function as the numerator has a higher degree than the denominator.

First divide by the denominator to give: $y = x + 2 + \frac{4}{x-1}$

The reason we do this is because it is easier to differentiate and use the derivative tests in this form than the original form given in the question.

Let $y = f(x)$ and $y = x + 2 + \frac{4}{x-1}$

Symmetry

$$f(-x) = -x + 2 + \frac{4}{-x-1}$$

$$-f(x) = -x - 2 - \frac{4}{x-1}$$

$$f(x) = x + 2 + \frac{4}{x-1}$$

$f(-x) \neq -f(x)$ and $f(-x) \neq f(x)$

$\therefore$ the function is neither even nor odd.

Crosses the x-axis

When $y = 0$: $x + 2 + \frac{4}{x-1} = 0 \Rightarrow x^2 + x + 2 = 0$

$b^2 - 4ac < 0 \Rightarrow$ no real roots so **does not** cross the x-axis.

Crosses the y-axis

When $x = 0$:

$$y = 0 + 2 + \frac{4}{0-1}$$

$$y = -2$$

It crosses the y-axis at (0,-2).

Stationary points

First derivative

$\frac{dy}{dx} = 1 - \frac{4}{(x-1)^2}$

For $\frac{dy}{dx} = 0$

$1 - \frac{4}{(x-1)^2} = 0$

$x^2 - 2x - 3 = 0$

When $x = -1$ and $x = 3$.

Second derivative

$\frac{d^2y}{dx^2} = \frac{8}{(x-1)^3}$

At $x = -1$: $\frac{d^2y}{dx^2} < 0$

Which is a maximum turning point at (-1,-1).

$\frac{d^2y}{dx^2} = \frac{8}{(x-1)^3}$

At $x = 3$: $\frac{d^2y}{dx^2} > 0$

Which is a minimum turning point at (3,7).

Non-horizontal point of inflection

$\frac{d^2y}{dx^2} = 0 \Rightarrow 8 = 0$, which has no real roots.

There is no non-horizontal point of inflection.

Asymptotes

Vertical asymptote

When the denominator is zero

$x - 1 = 0 \Rightarrow x = 1$ is a **vertical asymptote**.

As $x \to 1^-$, $y \to -\infty$

As $x \to 1^+$, $y \to +\infty$

Horizontal asymptote

Divide by the highest power and take the limit as $x \to \infty$.

$\lim_{x \to \infty} \left(x + 2 + \frac{\frac{4}{x}}{1 - \frac{1}{x}} \right) = x + 2$

As $x \to \infty$, $y \to x + 2$

So there is not a horizontal asymptote, but a slant asymptote.

Slant asymptote

When the degree of the numerator is bigger than the denominator.

$y = x + 2$

TOPIC 8. CURVE SKETCHING

As $x \to -\infty$, $y \to (x+2)^{-}$
As $x \to +\infty$, $y \to (x+2)^{+}$

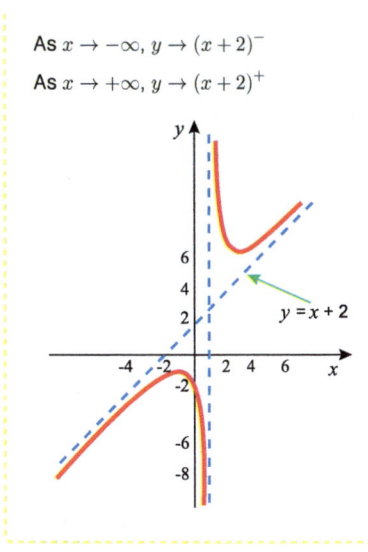

Type 5: Shortcuts

This takes the form $\frac{ax^2+bx+c}{ex+f}$

The vertical asymptote has the equation $x = \frac{-f}{e}$

The slant asymptote is the quotient resulting from the division of the numerator by the denominator.

The graph crosses the y-axis at the point (0, $\frac{c}{f}$) if $f \neq 0$

The graph crosses the x-axis at the real roots of the equation $ax^2 + bx + c = 0$ if there are any.

Otherwise it does not cross the axis.

Apply these shortcuts to the previous graph of $y = \frac{x^2+x+2}{x-1}$
$a = 1$, $b = 1$, $c = 2$, $e = 1$, $f = -1$
The vertical asymptote has the equation $x = \frac{-f}{e} = \frac{1}{1} = 1$
The slant asymptote is at $y = x + 2$
(The quotient found upon dividing the numerator of $x^2 + x + 2$ by the denominator of $x - 1$)
The graph crosses the y-axis at $y = \frac{c}{f} = \frac{2}{-1} = -2$.
The numerator does not have real roots and so the graph does not cross the x-axis.

The results match those found by the formal techniques but of course the turning points may still be needed.

© HERIOT-WATT UNIVERSITY

Example

Problem:

Find the asymptotes for the function $y = \frac{x^2+1}{x}$ and where the graph crosses the axes.

Solution:

$a = 1, b = 0, c = 1, e = 1, f = 0$

The vertical asymptote is $x = 0$ since $f = 0$

The slant asymptote, found by dividing $x^2 + 1$ by x, is $y = x$

The graph does not cross the y-axis since $f = 0$ and the point $\frac{c}{f}$ is undefined.

The graph does not cross the x-axis since $x^2 + 1$ has no real roots.

Type 5 rational function exercise

Q105: The diagram shows the shape of the graph $f(x) = \frac{x^2+2x+5}{x+1}$, $x \neq -1$.

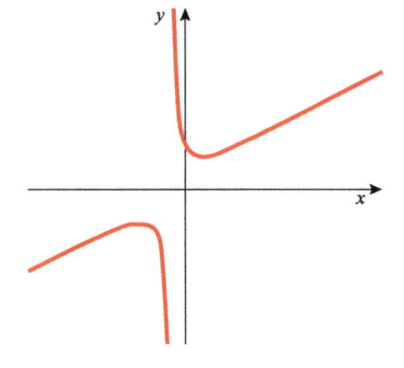

a) Obtain the stationary points of the graph.
b) Find the asymptotes of $f(x)$.
c) Determine there are no non-horizontal points of inflection.

....

Q106: A function f is defined by $f(x) = \frac{x^2-3x-2}{x-3}$, $x \neq 3$.

a) Express $f(x)$ in the form $x + \frac{a}{x-3}$.
b) Write down an equation for each of the asymptotes.

TOPIC 8. CURVE SKETCHING

c) Show that $f(x)$ has two stationary points.
Determine the coordinates and the nature of the stationary points.
d) Sketch the graph of f.

....

Q107: Sketch the function $f(x)$ is defined by $f(x) = \frac{2x^2 - x - 6}{x+2}$, $x \neq -2$

8.15 Summary of shortcuts to sketching rational functions

Using the techniques and shortcuts mentioned, most of the rational functions of the types shown can be sketched easily.

Remember however that the turning points will still have to be calculated and shown on any questions which ask for the important features of a graph.

The following table shows a simplified picture of the various formulae used.

The types explained earlier in the section can be identified by setting any of the coefficients (a, b, c, d, e or f) equal to zero if it is not present.

Remember that there will be instances when the formula is undefined.

In such instances it means that the graph does not possess the property shown.

> **Key point**
>
> **General formula:**
>
> $$\frac{ax^2 + bx + c}{dx^2 + ex + f}$$

	$\frac{a}{ex+f}$	$\frac{bx+c}{ex+f}$	$\frac{c \text{ or } (bx+c)}{dx^2+ex+f}$	$\frac{ax^2+bx+c}{dx^2+ex+f}$	$\frac{ax^2+bx+c}{ex+f}$
Horizontal asymptote	$y = 0$	$y = \frac{b}{e}$	$y = 0$	$y = \frac{a}{d}$	slant
Vertical asymptote	$x = \frac{-f}{e}$	$x = \frac{-f}{e}$	roots of denominator	roots of denominator	$x = \frac{-f}{e}$
crosses x-axis	no	$x = \frac{-c}{b}$	$x = \frac{-c}{b}$	roots of numerator	roots of numerator
crosses y-axis	$y = \frac{a}{f}$	$y = \frac{c}{f}$	$y = \frac{c}{f}$	$y = \frac{c}{f}$	$y = \frac{c}{f}$

© HERIOT-WATT UNIVERSITY

8.16 Graphical relationships between functions

In previous sections simple relationships between functions have been used to aid with the sketching of the graph.

The following two relationships have already been encountered:

1. $y = -f(x)$ is a reflection of $y = f(x)$ in the x-axis.
2. $y = f(-x)$ is a reflection of $y = f(x)$ in the y-axis.

There are other relationships known as transformations, which have already been encountered in Higher. There are four to consider.

If $y = f(x)$ then the following functions can be related to it:

1. $y = f(x) + k$
2. $y = f(x + k)$
3. $y = kf(x)$
4. $y = f(kx)$

Where k is a constant.

Note that more than one transformation may occur in a function.

For example the graph of $y = -2x^3$ is a combination of being scaled vertically by a factor of 2 and a reflection in the x-axis.

Each of the transformation types can be reviewed in the *Looking back: Transformations of graphs* sub-topic at the beginning of this topic.

8.16.1 Modulus function

There is a particular function of x which uses part reflection in the x-axis. This function is called the modulus function.

The modulus of a real number is the value of the number (called its absolute value) regardless of the sign.

So the modulus of 3 (written $|3|$) is 3 and the modulus of -3 (written $|-3|$) is also 3.

This is the basis of the definition of the modulus function.

> **Key point**
>
> **The modulus function**
>
> For $x \in \mathbb{R}$ the modulus function of $f(x)$, denoted by $|f(x)|$ is defined by
>
> $$|f(x)| = f(x) \text{ if } f(x) \geq 0$$
> $$|f(x)| = -f(x) \text{ if } f(x) < 0$$

TOPIC 8. CURVE SKETCHING

This can be made clearer with an example.

> **Examples**
>
> **1.** Let $f(x) = 4$.
> Using the first definition since $f(x) \geqslant 0:$ $\quad |f(x)| = f(x)$
> So we have: $|4| = 4$
>
> ....
>
> **2.** Let $f(x) = -4$.
> Using the second definition since $f(x) < 0:$ $\quad |f(x)| = -f(x)$
> So we have: $|-4| = -(-4)$
> $\quad\quad\quad\quad\quad |-4| = 4$

The result is that no matter if you start with a function where the y-coordinates are positive or negative when the modulus is applied to it all y-coordinates become positive.

> **Key point**
>
> **Rule for graphing** $y = |f(x)|$
>
> To sketch the graph of a modulus function $|f(x)|$, first sketch the graph of the function $y = f(x)$
>
> Take any part of it that lies below the x-axis and reflect it in the x-axis.
>
> The modulus function $y = |f(x)|$ is the combined effect of the positive part of the original function and the new reflected part.

Modulus of a quadratic function   Go online

Example Problem:

Sketch the graph of $f(x) = |x^2 - x - 6|$

Solution:

The graph of $y = x^2 - x - 6$ crosses the x-axis at $x = -2$ and $x = 3$.
It has a minimum turning point at $\left(\frac{1}{2}, \frac{-25}{4}\right)$, reflect the curve between $x = -2$ and $x = 3$.
Keep the remainder of the curve $y = f(x)$

© HERIOT-WATT UNIVERSITY

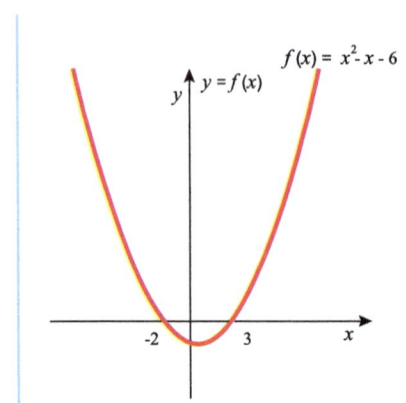

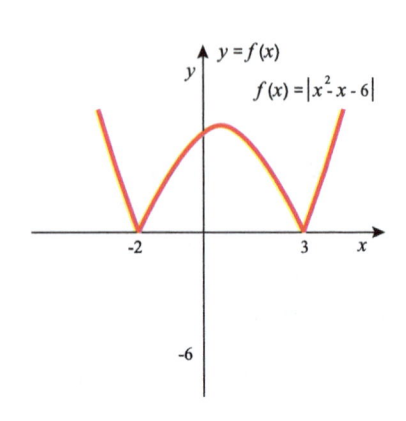

394 TOPIC 8. CURVE SKETCHING

Modulus function exercise

Q108: If $f(x) = x + 3$, which graph represents $|f(x)|$?

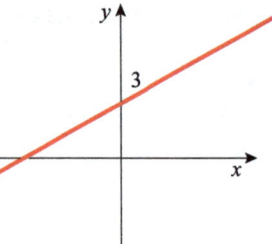

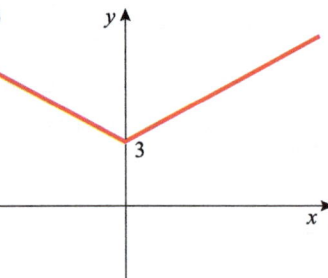

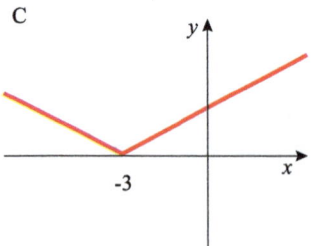

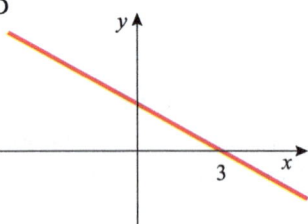

A

B

C

D

a) A
b) B
c) C
d) D

TOPIC 8. CURVE SKETCHING

Q109: The diagram shows part of the function of $f(x)$. Which graph represents $\left|f^{-1}(x-1)\right|$?
Note that the points of intersection with the axes are shown.

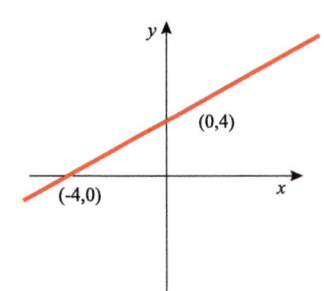

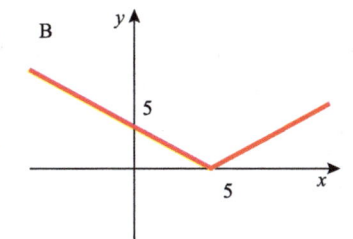

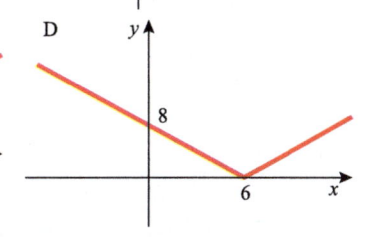

Q110: Sketch the graph of $y = \left|2x^2 - x - 10\right|$.

Q111: Sketch the graph of $y = |\tan x|$ for $x \in [-270°, 270°]$.

Q112: The diagram shows part of the function of $f(x)$. Sketch the graph of $|f(-x)| + 3$. Show the points of intersection with the axes.

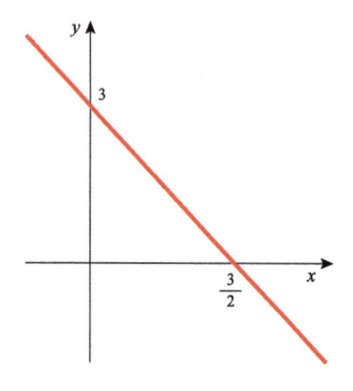

...

Q113: The diagram shows part of the function of $f(x)$. Sketch the graph of $|3f(x+2)|$. Show the points of intersection with the axes.

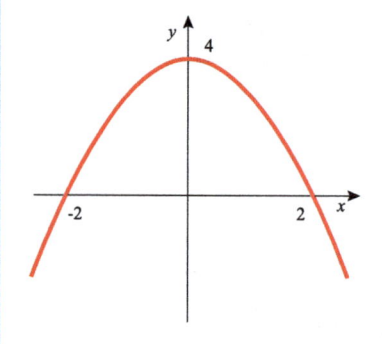

...

TOPIC 8. CURVE SKETCHING

Q114: The diagram show the shape of the graph $y = \frac{x}{2+x^2}$.

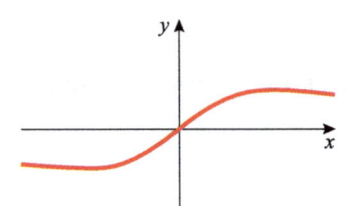

Sketch the graph of $y = \left|\frac{x}{2+x^2}\right|$.

....

Q115: The diagram show the shape of the graph $y = \frac{x^2}{x-1}$.

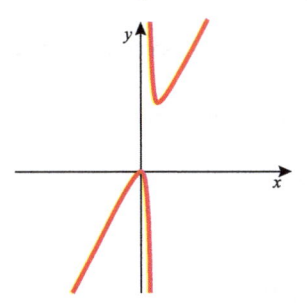

Sketch the graph of $y = \left|\frac{x^2}{x-1}\right| - 2$.

8.17 Learning points

Curve sketching
Function definition

- A function f from set A to set B is a rule which assigns to each element in A, exactly one element in B;
- Set A is the domain on the function;
- Set B is the codomain;
- Elements in set B which are the image of elements of A under the function f is called the image set or range of the function. This new set is always contained inside or equal to the codomain.

One-to-one and onto functions

- **One-to-one:** A function which maps one element in the domain to one element in the range or image set and vice versa.
- **Many-to-one:** A function which maps more than one element in the domain to the same element in the range or image set.
- **Onto function:** A function in which the range is equal to the codomain.

Inverse functions

- For a function to have an inverse it must be one-to-one and onto;
- To draw an inverse function reflect the original in the line $y = x$;
- To derive the equation of an inverse function:
  - make x the subject of the formula;
  - swap the x and y.
-

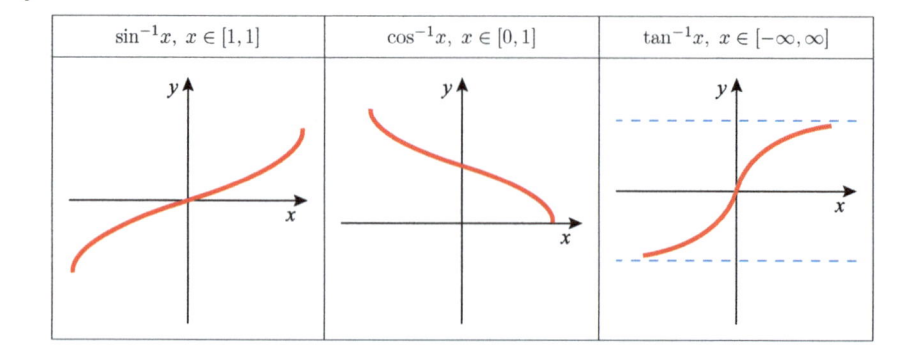

$\sin^{-1} x,\ x \in [1, 1]$	$\cos^{-1} x,\ x \in [0, 1]$	$\tan^{-1} x,\ x \in [-\infty, \infty]$

Odd and even functions

- A function is **odd** if $f(-x) = -f(x)$ for every value of x within the domain of the function. The graph will have rotational symmetry of order 2.
- A function is **even** if $f(-x) = f(x)$ for every value of x within the domain of the function. The graph is symmetrical under reflection in the y-axis.
- Functions do not have to be odd or even, they can be neither.

Critical and stationary points

- A **critical point** is any point on a curve where the slope of the tangent to the curve is zero or undefined.
- **Stationary points** is the name given to critical points where the slope of the tangent to the curve is zero.
- A **local maximum point** occurs when a function has a greater value at that point than at any points close to it.
- A **global maximum point** occurs when f is defined over a domain A and the value of the function at this point is greater than or equal to that at any other point within the domain.
- A **local minimum point** occurs when a function has a lesser value at that point than at any points close to it.
- A **global minimum point** occurs when f is defined over a domain A and the value of the function at this point is less than or equal to that at any other point within the domain.

Derivative tests

- If $\frac{dy}{dx} > 0$ the gradient is positive.
- If $\frac{dy}{dx} < 0$ the gradient is negative.
- The following patterns give the different types of turning points.

First derivative **Second derivative**

Minimum turning point

x	$\rightarrow$	x	$\rightarrow$
$\frac{dy}{dx}$	$-$	0	$+$
	$\searrow$	$\rightarrow$	$\nearrow$

$$\frac{d^2y}{dx^2} > 0$$

Maximum turning point

x	$\rightarrow$	x	$\rightarrow$
$\frac{dy}{dx}$	$+$	0	$-$
	$\nearrow$	$\rightarrow$	$\searrow$

$$\frac{d^2y}{dx^2} < 0$$

- **Rising and falling points of inflection**

x	$\rightarrow$	x	$\rightarrow$
$\frac{dy}{dx}$	$-$	0	$-$
	↘	$\rightarrow$	↘

Points of inflection can only be determined by making a nature table.

x	$\rightarrow$	x	$\rightarrow$
$\frac{dy}{dx}$	$+$	0	$+$
	↗	$\rightarrow$	↗

- If $\frac{d^2y}{dx^2} = 0$ the nature of the turning point is indeterminate. We would need to apply the first derivative and draw a nature table.

Concavity

- The graph of $y = f(x)$ is **concave downward** in an interval if $f(x)$ is a function which has a second derivative $\frac{d^2y}{dx^2} < 0$ for all x in the interval.

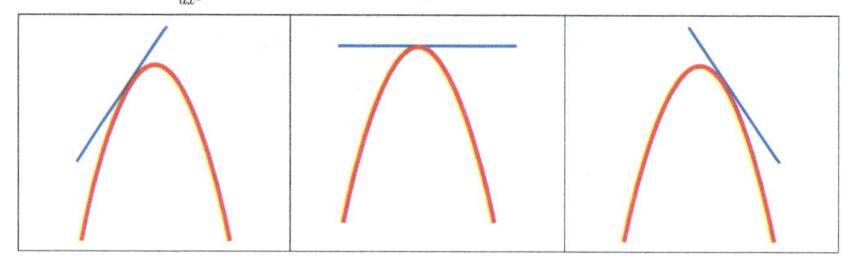

- The graph of $y = f(x)$ is **concave upward** in an interval if $f(x)$ is a function which has a second derivative $\frac{d^2y}{dx^2} > 0$ for all x in the interval.

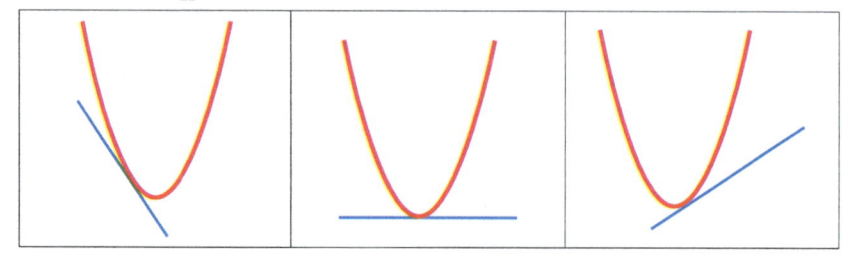

TOPIC 8. CURVE SKETCHING

- A non **horizontal point of inflection** occurs when $\frac{dy}{dx} \neq 0$ but $\frac{d^2y}{dx^2} = 0$ or $\frac{d^2y}{dx^2}$ is undefined and there is a change in concavity about that point.

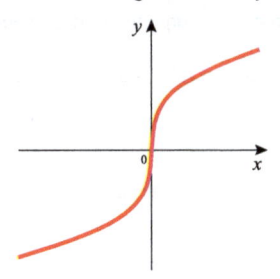

Continuity and asymptotic behaviour

- A continuous function $f(x)$ is a function where at every point P on the domain $\lim_{x \to P} f(x) = P$. (i.e. the curve can be drawn without lifting the pen from the paper.)
- A function $f(x)$ is discontinuous at a point P if $f(x)$ is not defined at P or if $\lim_{x \to P} f(x) \neq P$. (i.e. the pen has to be lifted off of the paper to draw the curve.)
- For rational functions **vertical asymptotes** occur when the denominator is equal to zero.
- For rational functions **horizontal asymptotes** can be found by dividing the numerator and denominator by the highest power of x and letting $x \to \infty$. The expression that is left is the horizontal asymptote.
- For rational functions slant asymptotes occur when the degree of the numerator is greater than the degree of the denominator. The quotient that is left after dividing the numerator by the denominator is the slant asymptote.

Sketching and rational functions

- When sketching a graph of the function $y = \frac{ax^2+bx+c}{dx^2+ex+f}$ you need to check:
    - Symmetry: is the graph odd, even or neither.
    - Cuts the x-axis: solve for when $y = 0$.
    - Cuts the y-axis: solve for when $x = 0$.
    - Turning Points: find $\frac{dy}{dx}$ and solve for zero then find the associated y coordinates.
    - Nature: find $\frac{d^2y}{dx^2}$.
        - If $\frac{d^2y}{dx^2} > 0$ it is a minimum.
        - If $\frac{d^2y}{dx^2} < 0$ it is a maximum.
    - Non-horizontal points of inflection: may occur when $\frac{d^2y}{dx^2} = 0$ or is undefined but $\frac{dy}{dx} \neq 0$ then there must be a change of concavity.
    - Vertical Asymptotes: solve for when the denominator is zero.

© HERIOT-WATT UNIVERSITY

- Horizontal Asymptotes: divide through by the highest degree of x and take the limit as $x \to \infty$.
- Slant Asymptote: exists when the degree of the numerator is bigger than the degree on the denominator. It is the quotient after algebraic division.
- Check the behaviour of the function as $x \to \pm\infty$ and x gets close to the asymptotes.

Graphical relationships between functions

- For $x \in \mathbb{R}$ the modulus function of $f(x)$, is denoted by $|f(x)|$ and is defined by:

$$|f(x)| = \begin{cases} f(x), & \text{if } f(x) \geqslant 0 \\ -f(x), & \text{if } f(x) < 0 \end{cases}$$

- To draw $|f(x)|$ reflect any part that is below the x-axis in the x-axis.

8.18 Extended Information

Leibniz

Gottfried von Leibniz is attributed with first using the term 'function' but in those days (1694) the term was used to denote the slope of the curve.
He was a very famous mathematician and is best known for his work on calculus. He also developed the binary system of arithmetic.

Euler

In 1749 Leonhard Euler defined a function in terms of two related quantities, which is more in keeping with the modern definition. Euler made a considerable contribution to analysis and his name will appear in other sections of the course. He made contributions in the fields of geometry, calculus and number theory.

Fourier

Joseph Fourier modified Euler's definition by noting that the domain of a function was important. He was an oustanding teacher and developed the theory of heat.

Dirichlet

Lejeune Dirichlet introduced the concept of a correspondence relationship. This is similar to the definition of an onto function. He was interested in algebraic number theory and was also considered to be the founder of the theory of Fourier series (and not Fourier as one might expect).

8.19 End of topic test

End of topic 8 test — Go online

Q116: What is the domain of $\sqrt{3x-2}$?

Q117: What is the domain of $\frac{5x}{\sqrt{2x+1}}$?

Q118: What is the domain of $\frac{6}{5x-2}$?

Q119: If the domain of $\tan x$ is $x \in \left[-\frac{\pi}{4}, \frac{\pi}{4}\right]$ what is the range?

Q120: If the domain of $3x^2 + x - 2$ are all the odd integers between -4 and 3 what is the range?

Give in ascending order.

Q121: Which of the following gives the biggest restricted domain to allow the function $(x-3)^2$ to be one-to-one?

Q122: Choose a codomain that makes $g(x)$ an onto function.
$g(x) = x^2 - 5$, where x is a positive odd integer in the interval [1,7].

Q123: What gives the biggest restricted domain to allow the function $\sin \theta$ to be one-to-one?

Q124: What gives the biggest restricted domain to allow the function $x^2 + 4$ to be one-to-one?

Q125: What gives the biggest restricted domain to allow the function $\frac{1}{x+1}$, $x \neq -1$ to be one-to-one?

Q126: What codomain makes $g(x)$ an onto function?
$g(x) = (x+1)^2$, where x is a negative even integer in the interval [-10,-5].

Q127: Write down the inverse function of $y = 3x + 5$.

Q128: For the function $f(\theta) = \cos\theta$ to have an inverse it must be one-to-one and onto. Choose from the options an appropriate domain for $f(\theta) = \cos\theta$ so that it has an inverse.
a) $\left[-\frac{\pi}{2}, \frac{\pi}{2}\right]$
b) $[0, 2\pi]$
c) $[0, \pi]$
d) $\left[-\frac{3\pi}{2}, -\frac{\pi}{2}\right]$

....

Q129: The function $f(\theta) = \sin\theta$ has an inverse.
Choose from the options an appropriate domain for $f(\theta) = \sin^{-1}\theta$.
a) $\left[-\frac{\pi}{2}, \frac{\pi}{2}\right]$
b) $[0, \pi]$
c) [-1,0]
d) [-1,1]

....

Q130: The function $f(\theta) = \tan\theta$ has an inverse.
Choose from the options an appropriate range for $f(\theta) = \tan^{-1}\theta$.
a) $\left[-\frac{\pi}{2}, \frac{\pi}{2}\right]$
b) $[-\pi, \pi]$
c) $[-\infty, \infty]$
d) $[-2\pi, 2\pi]$

....

Q131: For the graph shown, choose the correct inverse function from the four options.

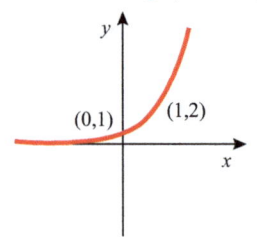

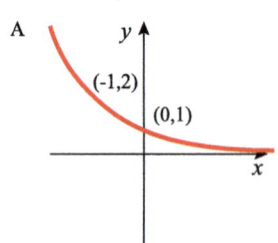

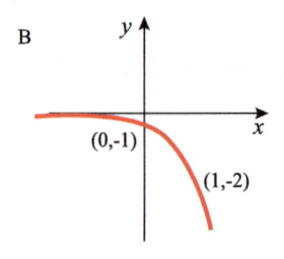

TOPIC 8. CURVE SKETCHING

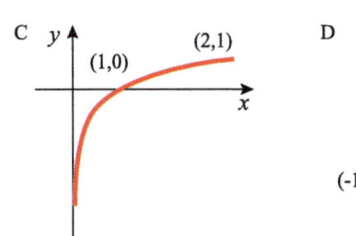

C — graph through (1,0) and (2,1)

D — graph through (0,-1) and (-1,-2)

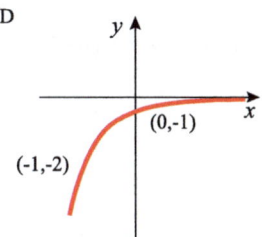

Q132: For the graph shown, choose the correct inverse function from the four options.

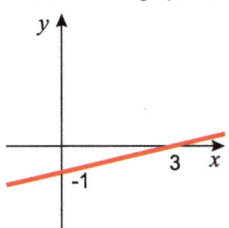

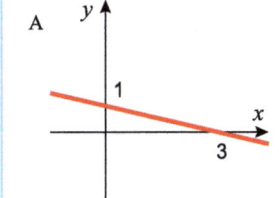

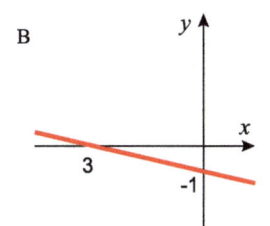

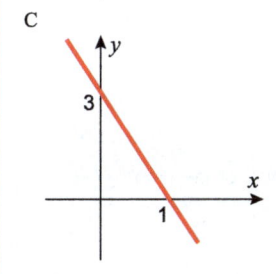

(graph through (-1, 0) and (3, ...) — line with positive slope)

A — line with negative slope, y-intercept 1, x-intercept 3

B — line with negative slope, x-intercept 3, passing through (0,-1) area

C — line with steep negative slope, y-intercept 3, x-intercept 1

D — line with steep positive slope, y-intercept 3, x-intercept -1

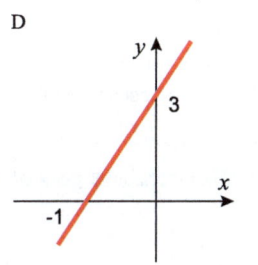

Q133: Is function $f(x) = x^3 \cos x$ odd, even or neither?

a) Odd
b) Even
c) Neither

Q134: Is function $f(x) = \frac{1}{x}\sin x$ odd, even or neither?

a) Odd
b) Even
c) Neither

Q135: Is function $f(x) = x^2 e^x$ odd, even or neither?

a) Odd
b) Even
c) Neither

Q136: Is function $f(x) = \frac{x^2+3}{x}$ odd, even or neither?

a) Odd
b) Even
c) Neither

Q137: Is function $f(x) = \frac{x^3-2x+1}{x^2+3x+2}$ odd, even or neither?

a) Odd
b) Even
c) Neither

Q138: For the graph of $f(x) = \frac{3x^2+2x+1}{x-3}$, $x \neq 3$

a) Give the equation of the vertical asymptote.
b) Determine the equation of the non-vertical asymptote.

Q139: Find the coordinates of the non-horizontal point of inflection of the graph $f(x) = 2x^3 + 12x^2 + 9x + 15$ if it exists.

Q140: Show that there is a non-horizontal point of inflection on the graph of $f(x) = 2\cos\left(x + \frac{\pi}{2}\right)$ at $x = 0$.

Q141: Show that there is a non-horizontal point of inflection on the graph of $f(x) = 3\sin\left(x - \frac{\pi}{4}\right)$ at $x = \frac{\pi}{4}$.

TOPIC 8. CURVE SKETCHING

Q142: Let $f(x)$ be a cubic function with a minimum turning point (-3,-4) and maximum turning point at (1,5).

Choose the graph of $g(x) = |3 - f(x)|$

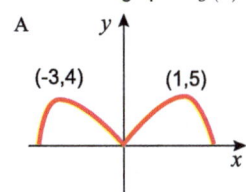

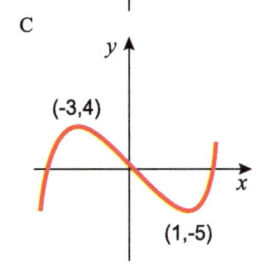

A (-3,4) (1,5)

B (-3,7) (1,2)

C (-3,4) (1,-5)

D (-3,1) (1,-8)

....

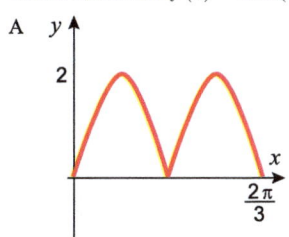

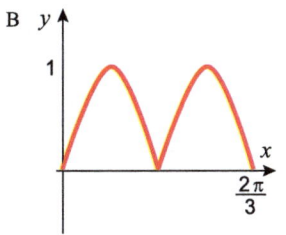

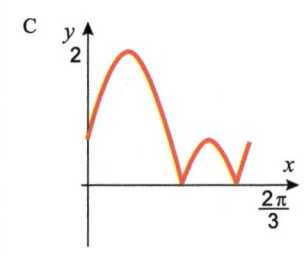

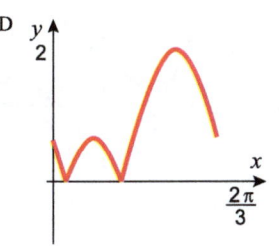

Q143: Given that $f(x) = 2\sin(3x)$, choose the correct graph of $|f(x) + 1|$ for $0 \leqslant x \leqslant \frac{2\pi}{3}$.

A

B

C

D

....

© HERIOT-WATT UNIVERSITY

Q144: Given that $f(x) = 4x - 1$, choose the correct graph of $|f(x-2)|$.

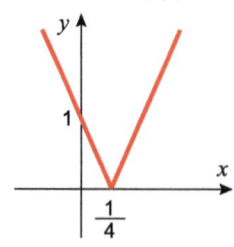

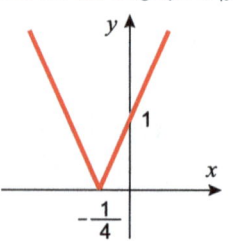

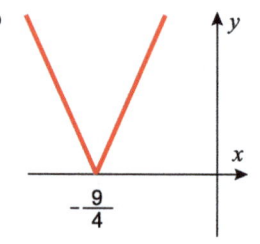

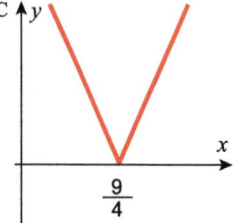

Q145: Given that $f(x) = x^2 - 4$, choose the correct graph of $|-0.5f(x)|$.

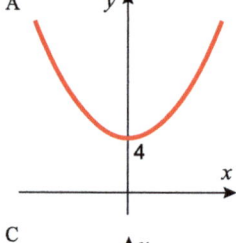

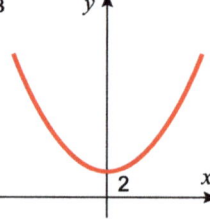

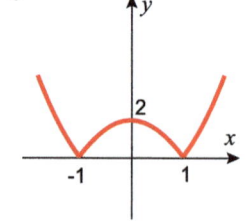

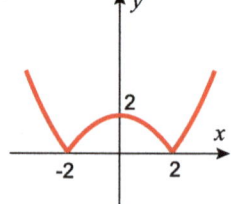

Q146: For the graph shown choose the correct option for the function $|f^{-1}(x)| - 2$.

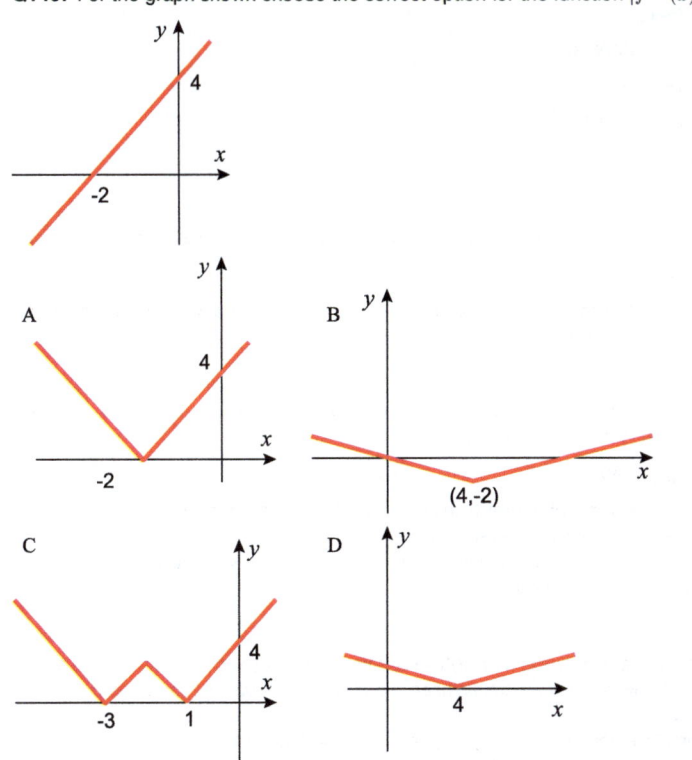

A

B

(4,-2)

C

D

Glossary

adding complex numbers
> to add two complex numbers:
> - add the real parts;
> - add the imaginary parts.
>
> e.g. $(a + ib) + (c + id) = (a + c) + i(b + d)$

alternating sequence
> an alternating sequence is any sequence which has alternate positive and negative terms.

argument of a complex number
> the angle, θ, between the positive x-axis and the line representing the complex number on an Argand diagram, denoted $\arg(z)$

arithmetic sequence
> an arithmetic sequence is one which takes the form $a, a + d, a + 2d, a + 3d, \ldots$ where a is the first term and d is the common difference.

asymptotes of rational functions
> for rational functions, a **vertical asymptote** is a vertical line with equation of the form $x = k$ at which the function in question is undefined.
> The function values either increase rapidly towards $+\infty$ or decrease rapidly towards $-\infty$ as x gets closer and closer to the value k
>
> For rational functions, a **horizontal asymptote** is a horizontal line with equation of the form $y = m$ for which the function value gets closer and closer to the value m as x tends towards $+\infty$ and / or $-\infty$
> A horizontal asymptote occurs when the degree of the numerator is less than or equal to the degree of the denominator.
>
> For rational functions, a **slant or oblique asymptote** is a line, neither horizontal nor vertical, with equation of the form $y = ax + b$ for which the function value gets closer and closer to the line $y = ax + b$ as x tends towards $+\infty$ and / or towards $-\infty$
> A slant asymptote occurs when the degree of the numerator is greater than the degree of the denominator.

binomial coefficient formula 1
> The **Binomial coefficient formula** is $\binom{n}{r} = \frac{n!}{r!(n-r)!}$

binomial coefficient rule 1
> The **first binomial coefficient rule** is $\binom{n}{r} = \binom{n}{n-r}$

binomial coefficient rule 2
> The **second binomial coefficient rule** is $\binom{n}{r-1} + \binom{n}{r} = \binom{n+1}{r}$

© HERIOT-WATT UNIVERSITY

GLOSSARY

binomial theorem

The **Binomial Theorem** states that if $x, y \in \mathbb{R}$ and $n \in \mathbb{N}$ then $(x+y)^n$

$$= \binom{n}{r}x^n + \binom{n}{r}x^{n-1}y + \binom{n}{r}x^{n-2}y^2 + \cdots + \binom{n}{r}x^{n-r}y^r + \cdots + \binom{n}{r}y^n$$

codomain

for a function $f : A \to B$, B is called the **codomain** of the function f.

common difference

the common difference in an arithmetic sequence is the difference between any two consecutive terms in the sequence.

common ratio

the common ratio in a geometric sequence is the ratio $r = \frac{u_{n+1}}{u_n}$ of two consecutive terms.

complex number

a number of the form $a + bi$ (which may also be written as $a + ib$) where a and b are real numbers and $i = \sqrt{-1}$

conjugate of a complex number

denoted as $\overline{z}$ (or sometimes z^*) and defined by $\overline{z} = a - ib$

conjugate roots property

states that if $P(x)$ is a polynomial with real coefficients and $z = \alpha$ is a solution of $P(x) = 0$, then $z = \overline{\alpha}$ is also a solution of $P(x) = 0$

continuous function

a **continuous function** f (x) is a function where at every point P on the domain

$$\lim_{x \to P} f(x) = f(P)$$

convergent sequence

an infinite sequence $\{u_n\}$ for which $\lim_{n \to \infty} u_n = k$ is called a convergent sequence with limit k.

convergent series

a convergent series is one for which the limit of partial sums exists. This limit is called the **sum** and is denoted by S_∞ or $\sum_{n=1}^{\infty} u_n$a

de Moivre's theorem

states that if $z = r\cos\theta + ir\sin\theta$, then $z^n = r^n(cos(n\theta) + i\sin(n\theta))$ for all $n \in \mathbb{N}$

de Moivre's theorem for fractional powers

$(r(\cos\theta + i\sin\theta))^{\frac{p}{q}} = r^{\frac{p}{q}}\left(\cos\left(\frac{p}{q}\theta\right) + i\sin\left(\frac{p}{q}\theta\right)\right)$

discontinuous function

a function $f(x)$ is **discontinuous** at a point P if $f(x)$ is not defined at P or if

$$\lim_{x \to P} f(x) \neq f(P)$$

divergent series

a divergent series is one which is not convergent.
For example 1 + 2 + 3 + 4 + 5 ... is a divergent series. The sum of this series will continue increasing the more terms we add. It will not tend towards a limit.

dividing complex numbers

to divide two complex numbers:

- find the conjugate of the denominator;
- multiply the complex fraction, both top and bottom, by this conjugate to give an integer on the denominator;
- express the answer in the form $a + bi$

e.g.
$$\frac{a+bi}{c+di} = \frac{(a+bi)(c-di)}{(c+di)(c-di)}$$
$$= \frac{(ac+bd)+i(bc-ad)}{(c^2+d^2)}$$
$$= \frac{(ac+bd)}{(c^2+d^2)} + \frac{i(bc-ad)}{(c^2+d^2)}$$

dividing two complex numbers in polar form

to divide two complex numbers in polar form:

- divide the moduli;
- subtract the arguments.

e.g. $\frac{r_1(\cos\theta + i\sin\theta)}{r_2(\cos\varphi + i\sin\varphi)} = \frac{r_1}{r_2}(\cos(\theta - \varphi) + i\sin(\theta - \varphi))$

domain

For a function $f : A \to B$, A is called the **domain** of the function f.

even function

a function is **even** if $f(-x) = f(x)$ for every value of x within the domain of the function. The graph is symmetrical under reflection in the y-axis.

Fibonacci

Fibonacci was born in 1170 and died in 1250 in Italy. Fibonacci is actually a nickname, his real name being Leonardo Pisano; he also sometimes called himself Bigollo, which may mean either a traveller or a good-for-nothing

Fibonacci sequence

the Fibonacci sequence is defined by a second order recurrence relation of the form $u_{n+2} = u_{n+1} + u_n$.
An example is: 1, 1, 2, 3, 5, 8, 13, ...

finite sequence

a finite sequence is one which has a last term.

first derivative test

the **first derivative test** is a means of determining the nature of the turning point by finding the signs of the derivative to the left and to the right of the turning point.

function f

a **function** f from set A to set B is a rule which assigns to each element in A exactly one element in B. This is often written as $f : A \to B$.

fundamental theorem of algebra

let $P(z) = a_n z^n + a_{n-1} z^{n-1} + \ldots + a_1 z + a_0$ be a polynomial of degree n (with real or complex coefficients), then the fundamental theorem of algebra states that:

- $P(z) = 0$ has n solutions $\alpha_1, \ldots, \alpha_n$ in the complex numbers;
- $P(z) = (z - \alpha_1)(z - \alpha_2)\ldots(z - \alpha_n)$

general term of (x + y)ⁿ

The **general term** of $(x + y)^n$ is given by $\binom{n}{r} x^{n-r} y^r$

geometric sequence

a geometric sequence is one which has the form $a, ar, ar^2, ar^3, \ldots$ where a is the first term and r is the common ratio.

image set or range

for a function $f : A \to B$, the set C of elements in B which are images of the elements in A under the function f is called the **image set or range** of the function f. C is always contained in or equal to B. This is written $C \subseteq B$.

infinite sequence

an infinite sequence is one which continues indefinitely.

inverse function

suppose that f is a one-to-one and onto function. For each $y \in B$ (codomain) there is exactly one element $x \in A$ (domain) such that $f(x) = y$.
The **inverse function** is denoted $f^{-1}(y) = x$.

Maclaurin's theorem

Maclaurin's theorem states that

$$f(x) = \sum_{r=0}^{\infty} f^{(r)}(0)\frac{x^r}{r!}$$
$$= f(0) + f^{(1)}(0)\frac{x}{1!} + f^{(2)}(0)\frac{x^2}{2!} + f^{(3)}(0)\frac{x^3}{3!} + \ldots + f^{(n)}(0)\frac{x^n}{n!} + \ldots$$

many-to-one

a function which maps more than one element in the domain to the same element in the range or image set is called a **many-to-one** or a many-one function.

The function is said be in many-to-one correspondence.

It is also common to say that such a function is not one-to-one.

modulus function

for $x \in \mathbb{R}$ the modulus function of $f(x)$, denoted by $|f(x)|$ is defined by

$$|f(x)| = f(x) \text{ if } f(x) \geq 0$$
$$|f(x)| = -f(x) \text{ if } f(x) < 0$$

modulus *r* of a complex number

written as $|z|$ and defined by $|z| = \sqrt{a^2 + b^2}$

multiplying complex numbers

to multiply complex numbers, use the technique for multiplying out two brackets,

e.g. $(a + ib)(c + id) = (ac - bd) + i(bc + ad)$

multiplying two complex numbers in polar form

to multiply two complex numbers in polar form:

- multiply the moduli;
- add the arguments.

e.g. $r_1(\cos\theta + i\sin\theta) \times r_2(\cos\phi + i\sin\phi) = r_1 r_2(\cos(\theta + \phi) + i\sin(\theta + \phi))$

n factorial

n! (called n factorial) is the product of the integers $n, n-1, n-2, \ldots, 2, 1$, i.e.
$n! = n \times (n-1) \times (n-2) \times \ldots \times 2 \times 1 \text{ for } n \in \mathbb{N}$

non horizontal point of inflection

a **non horizontal point of inflection** occurs when $\frac{dy}{dx} \neq 0$ but $\frac{d^2y}{dx^2} = 0$ or is undefined.

null sequence

a convergent sequence which converges to the limit 0 is called a null sequence.

odd function

a function is **odd** if $f(-x) = -f(x)$ for every value of x within the domain of the function. The graph is symmetrical under 180° rotation about the origin.

It is said to have rotational symmetry of order 2.

GLOSSARY

one-to-one

a function $f : A \to B$ is a **one-to-one function** if whenever $f(s) = f(t)$, then $s = t$ where $s \in A$ and $t \in B$. The function is said to be in one-to-one correspondence.

onto

an **onto function** is one in which the range is equal to the codomain.

partial sum

the partial sum is the sum of the terms from 1 to n, where $n \in \mathbb{N}$. It is denoted by S_n and represented as $S_n = \sum_{r=1}^{n} u_r$.

polar form of a complex number

$z = r(\cos\theta + i\sin\theta)$ where r is the modulus and θ is the argument

power series

a power series is an expression of the form

$$\sum_{n=0}^{\infty} a_n x^n = a_0 + a_1 x + a_2 x^2 + a_3 x^3 + \ldots + a_n x^n + \ldots$$

, where $a_0, a_1, a_2, a_3, \ldots, a_n, \ldots$ are constants and x is a variable. It is called a power series as it is made up of a sequence of powers of x with coefficients $a_0, a_1, a_2, a_3, \ldots, a_n, \ldots$

principal value of an argument

the value which lies between $-\pi$ and π

recurrence relation

a sequence in which each term is a function of the previous term or terms

roots of unity

the nth roots of unity are those numbers which satisfy the equation $z^n = 1$

rule for graphing y = |f(x)|

to sketch the graph of a modulus function $|f(x)|$, first sketch the graph of the function $y = f(x)$ Take any part of it that lies below the x-axis and reflect it in the x-axis.
The modulus function $y = |f(x)|$ is the combined effect of the positive part of the original function and the new reflected part.

second derivative test

the **second derivative test** of a function is a means of determining the nature of a turning point. A positive second derivative gives a minimum turning point and a negative second derivative gives a maximum turning point.

sequence

a series of terms with a definite pattern; can be defined by a rule or a formula for the nth term

series

a series is the sum of the terms in an infinite sequence.

set of complex numbers

$\mathbb{C} = \{a + bi : a, b \in \mathbb{R}\}$

© HERIOT-WATT UNIVERSITY

standard number sets

the **standard number sets** are:

- $\mathbb{N} = \{1, 2, 3, 4, 5, ...\}$ the set of natural numbers.
- $\mathbb{W} = \{0, 1, 2, 3, 4, 5, ...\}$ the set of whole numbers.
- $\mathbb{Z} = \{..., -3, -2, -1, 0, 1, 2, 3, ...\}$ the set of integers.
- $\mathbb{Q}$ = the set of all numbers which can be written as fractions, called the set of rational numbers.
- $\mathbb{R}$ = the set of rational and irrational numbers, called the set of real numbers.

stationary points

stationary points are points on a curve where the gradient of the tangent to the curve is zero; at these points $f'(x) = 0$.

subtracting complex numbers

to subtract two complex numbers:

- subtract the real parts;
- subtract the imaginary parts.

e.g. $(a + ib) - (c + id) = (a - c) + i(b - d)$

sum to n terms of a geometric series

the sum to n terms of a geometric series (the nth partial sum) is given by $S_n = \frac{a(1-r^n)}{1-r}$ where a is the first term of the sequence, r ($\neq 1$) represents the common ratio and $n \in \mathbb{N}$.

term

each number in a sequence is called a term or an element. The nth term (or general term) is often denoted by u_n.

triangle inequality

if z and w are complex numbers, then $|z + w| \leq |z| + |w|$

triangular number sequence

the triangular number sequence comprises the natural numbers which can be drawn as dots in a triangular shape.

Hints for activities

Topic 8: Curve sketching

Types of functions

Hint 1:

One-to-one function

A function which maps one element in the domain to one element in the range or image set.

Many-to-one function

A function which maps more than one element in the domain to the same element in the range or image set.

Answers to questions and activities

Topic 5: Binomial theorem

Revision exercise (page 7)

Q1:

$$\begin{aligned}(2y-1)(5y+3)^2 &= (2y-1)(5y+3)(5y+3) \\ &= (2y-1)(25y^2+30y+9) \\ &= 2y(25y^2+30y+9) - (25y^2+30y+9) \\ &= (50y^3+60y^2+18y) - (25y^2+30y+9) \\ &= 50y^3+35y^2-12y-9\end{aligned}$$

Q2:

$$\begin{aligned}(3y-4)^3 &= (3y-4)(3y-4)(3y-4) \\ &= (3y-4)(9y^2-24y+16) \\ &= 3y(9y^2-24y+16) - 4(9y^2-24y+16) \\ &= (27y^3-72y^2+48y) - (36y^2-96y+64) \\ &= 27y^3-108y^2+144y-64\end{aligned}$$

Q3:

Hints:

- The sum of the products of the inner and outer terms gives $1g + 3g = 4g$

Answer: $3g^2 + 4g + 1 = (3g+1)(g+1)$

Q4:

Hints:

- The sum of the products of the inner and outer terms gives $-2h + 5h = 3h$

Answer: $5h^2 + 3h - 2 = (5h-2)(h+1)$

Q5:

Hints:

- The simple common factor is 2 giving $2(j^2+j-6)$ and $j^2+j-6 = (j+3)(j-2)$

Answer: $2j^2 + 2j - 12 = 2(j+3)(j-2)$

Q6:

Hints:

- The simple common factor is 2 giving $2(2k^2-k-1)$ and $2k^2-k-1 = (2k+1)(k-1)$
- Remember that the sum of the product of the inner and outer terms gives $1k + (-2k) = -k$

Answer: $4k^2 - 2k - 2 = 2(2k+1)(k-1)$

ANSWERS: UNIT 1 TOPIC 5

Q7: Find the lowest common multiple of $\frac{4}{5} + \frac{3}{7}$, which is 35:

$$\frac{4}{5} + \frac{3}{7} = \frac{4}{5} \times \frac{7}{7} + \frac{3}{7} \times \frac{5}{5}$$
$$= \frac{28}{35} + \frac{15}{35}$$
$$= \frac{43}{35}$$
$$= 1\frac{8}{35}$$

Q8: Find the lowest common multiple of $\frac{7}{8} - \frac{2}{3}$, which is 24:

$$\frac{7}{8} - \frac{2}{3} = \frac{7}{8} \times \frac{3}{3} - \frac{2}{3} \times \frac{8}{8}$$
$$= \frac{21}{24} - \frac{16}{24}$$
$$= \frac{5}{24}$$

Q9: $\frac{1}{5} \times \frac{4}{7} \times \frac{2}{3} = \frac{8}{105}$

Q10:

$$\frac{5}{6} \times \frac{3}{8} \times \frac{2}{5} = \frac{30}{240}$$
$$= \frac{1}{8}$$

Q11:

```
        56
28 ) 1568
     140
     ───
     168
     168
     ───
       0
```

Therefore, $1568 \div 28 = 56$

Q12:

```
        85
32 ) 2720
     256
     ───
     160
     160
     ───
       0
```

Therefore, $2720 \div 32 = 85$

Answers from page 8.

Q13: 6

Q14: 24

© HERIOT-WATT UNIVERSITY

Factorials calculator activity (page 9)

Expected answer

It is dependent upon the calculator that you have that will determine which factorial you can put into your calculator without an error being displayed.

Some calculators can only give up to $69!$ so when $70!$ is entered, an error is displayed.

However, we have to consider the accuracy of these answers.

$15!$ when calculated by hand gives 1307674368000, which is the same on the calculator.

$16!$ by hand gives 20922789888000, but on the calculator gives 20922789890000. There is a very subtle difference in the fourth last digit, but since the screen of the calculator is limited, the numbers are starting to be rounded.

This illustrates the need to be aware of the limitations of a calculator.

Factorials exercise (page 10)

Q15: $8 \times 7!$

Q16: $12 \times 11!$

Q17:
$8! = 8 \times 7 \times 6 \times 5 \times 4 \times 3 \times 2 \times 1$
$= 8 \times 7 \times 6 \times 5!$
$= 336 \times 5!$

Q18:
$10! = 10 \times 9 \times 8 \times 7 \times 6 \times 5 \times 4 \times 3 \times 2 \times 1$
$= 10 \times 9 \times 8!$
$= 90 \times 8!$

Q19:
$11! = 11 \times 10 \times 9!$
$= 110 \times 9!$

Q20:
$6! = 6 \times 5 \times 4 \times 3!$
$= 120 \times 3!$

ANSWERS: UNIT 1 TOPIC 5

Binomial coefficients: ^nC_r exercise (page 12)

Q21:

$$\binom{5}{2} = \frac{5!}{2!3!}$$
$$= \frac{5 \times 4 \times 3!}{2 \times 1 \times 3!}$$
$$= \frac{5 \times 4}{2 \times 1}$$
$$= 10$$

Q22:

$$\binom{6}{4} = \frac{6!}{4!2!}$$
$$= \frac{6 \times 5 \times 4!}{2 \times 1 \times 4!}$$
$$= \frac{6 \times 5}{2 \times 1}$$
$$= 15$$

Q23:

$$\binom{5}{3} = \frac{5!}{3!2!}$$
$$= \frac{5 \times 4 \times 3!}{2 \times 1 \times 3!}$$
$$= \frac{5 \times 4}{2 \times 1}$$
$$= 10$$

Q24:

$$^7C_4 = \binom{7}{4}$$
$$= \frac{7!}{4!3!}$$
$$= \frac{7 \times 6 \times 5 \times 4!}{3 \times 2 \times 1 \times 4!}$$
$$= \frac{7 \times 6 \times 5}{3 \times 2 \times 1}$$
$$= 35$$

© HERIOT-WATT UNIVERSITY

First binomial coefficient rule exercise (page 14)

Q25:
$$\binom{n}{r} = \binom{n}{n-r}$$
$$\binom{7}{4} = \binom{7}{7-4}$$
$$= \binom{7}{3}$$

Q26:
$$\binom{n}{r} = \binom{n}{n-r}$$
$$\binom{21}{17} = \binom{21}{21-17}$$
$$= \binom{21}{4}$$

Second binomial coefficient rule exercise (page 15)

Q27:
$$\binom{n}{r-1} + \binom{n}{r} = \binom{n+1}{r}$$
$$\binom{8}{6} + \binom{8}{7} = \binom{9}{7}$$

Q28:
$$\binom{n}{r-1} + \binom{n}{r} = \binom{n+1}{r}$$
$$\binom{14}{11} + \binom{14}{12} = \binom{15}{12}$$

ANSWERS: UNIT 1 TOPIC 5

Finding n given the value of a binomial coefficient and r exercise (page 16)

Q29:
$$\binom{n}{2} = 10$$
$$\frac{n!}{2!(n-2)!} = 10$$
$$\frac{n \times (n-1)}{2} = 10$$
$$n^2 - n = 20$$
$$n^2 - n - 20 = 0$$
$$(n-5)(n+4) = 0$$
$$n = 5 \text{ or } n = -4$$
Since $n \in \mathbb{N}$ then $n = 5$

Q30:
$$\binom{n}{2} = 21$$
$$\frac{n!}{2!(n-2)!} = 21$$
$$\frac{n \times (n-1)}{2} = 21$$
$$n^2 - n = 42$$
$$n^2 - n - 42 = 0$$
$$(n-7)(n+6) = 0$$
$$n = 7 \text{ or } n = -6$$
Since $n \in \mathbb{N}$ then $n = 7$

Q31:
$$\binom{n}{2} = 36$$
$$\frac{n!}{2!(n-2)!} = 36$$
$$\frac{n \times (n-1)}{2} = 36$$
$$n^2 - n = 72$$
$$n^2 - n - 72 = 0$$
$$(n-9)(n+8) = 0$$
$$n = 9 \text{ or } n = -8$$
Since $n \in \mathbb{N}$ then $n = 9$

© HERIOT-WATT UNIVERSITY

Binomial coefficients exercise (page 16)

Q32: 1

Q33: 5040

Q34: $100 \times 99!$

Q35: $1 \times 0!$

Q36:

$$^6C_2 = \binom{6}{2}$$
$$= \frac{6!}{2!4!}$$
$$= \frac{6 \times 5 \times 4!}{2 \times 1 \times 4!}$$
$$= \frac{6 \times 5}{2 \times 1}$$
$$= 15$$

Q37:

$$\binom{9}{5} = \frac{9!}{5!4!}$$
$$= \frac{9 \times 8 \times 7 \times 6 \times 5!}{4 \times 3 \times 2 \times 1 \times 5!}$$
$$= \frac{9 \times 8 \times 7 \times 6}{4 \times 3 \times 2 \times 1}$$
$$= 126$$

Q38:

$$\binom{7}{4} = \frac{7!}{4!3!}$$
$$= \frac{7 \times 6 \times 5 \times 4!}{3 \times 2 \times 1 \times 4!}$$
$$= \frac{7 \times 6 \times 5}{3 \times 2 \times 1}$$
$$= 35$$

Q39: $\binom{13}{1}$

Q40: $\binom{6}{2}$

Q41: $\binom{8}{5}$

ANSWERS: UNIT 1 TOPIC 5

Q42: $\binom{13}{10}$

Pascal's triangle (page 18)

Q43:

Row 0					1				
Row 1				1		1			
Row 2				1	2	1			
Row 3			1	3	3	1			
Row 4		1	4	6	4	1			
Row 5	1	5	10	10	5	1			
Row 6	1	6	15	20	15	6	1		
Row 7	1	7	21	35	35	21	7	1	

Q44:

Row 0				1		
Row 1			1		1	
Row 2		1		2		1
Row 3	1		3		3	1

Q45:

Binomial coefficients

Row 4: $\binom{4}{0}$ $\binom{4}{1}$ $\binom{4}{2}$ $\binom{4}{3}$ $\binom{4}{4}$

Row 5: $\binom{5}{0}$ $\binom{5}{1}$ $\binom{5}{2}$ $\binom{5}{3}$ $\binom{5}{4}$ $\binom{5}{5}$

Row 6: $\binom{6}{0}$ $\binom{6}{1}$ $\binom{6}{2}$ $\binom{6}{3}$ $\binom{6}{4}$ $\binom{6}{5}$ $\binom{6}{6}$

Row 7: $\binom{7}{0}$ $\binom{7}{1}$ $\binom{7}{2}$ $\binom{7}{3}$ $\binom{7}{4}$ $\binom{7}{5}$ $\binom{7}{6}$ $\binom{7}{7}$

© HERIOT-WATT UNIVERSITY

	Values of binomial coefficients							
Row 4		1	4	6	4	1		
Row 5		1	5	10	10	5	1	
Row 6	1	6	15	20	15	6	1	
Row 7	1	7	21	35	35	21	7	1

Q46: When the binomial coefficients are evaluated, they produce Pascal's triangle.

Q47: $\binom{5}{3}$

Q48: $\binom{7}{2}$

Q49: $\binom{10}{6}$

Q50: $\binom{12}{4}$

Pascal's triangle exercise (page 20)

Q51:

Using the rule: $\binom{n}{r-1} + \binom{n}{r} = \binom{n+1}{r}$

$$\binom{3}{1} + \binom{3}{2} = \binom{4}{2}$$

Q52:

Using the rule: $\binom{n}{r-1} + \binom{n}{r} = \binom{n+1}{r}$

$$\binom{5}{3} + \binom{5}{2} = \binom{6}{3}$$

ANSWERS: UNIT 1 TOPIC 5

Q53:

Using the rule: $\binom{n}{r-1} + \binom{n}{r} = \binom{n+1}{r}$

so $\binom{n+1}{r} - \binom{n}{r} = \binom{n}{r-1}$

$\binom{6}{2} - \binom{5}{2} = \binom{5}{1}$

Q54:

Using the rule: $\binom{n}{r-1} + \binom{n}{r} = \binom{n+1}{r}$

so $\binom{n+1}{r} - \binom{n}{r} = \binom{n}{r-1}$

$\binom{7}{4} - \binom{6}{3} = \binom{6}{4}$

Finding coefficients exercise (page 29)

Q55:

The general term in this problem is:

$$\binom{7}{r} x^{7-r} \left(\frac{-3}{x}\right)^r = \binom{7}{r} x^{7-r}(-3)^r x^{-r}$$

$$= \binom{7}{r} x^{7-2r}(-3)^r$$

We need to use the rules of indices to simplify the power of x.

The coefficient is then $\binom{7}{r}(-3)^r$

For $7 - 2r = 5$, we require $r = 1$

The coefficient is:

$$\binom{7}{1}(-3)^1 = 7 \times -3$$
$$= -21$$

Q56:

The general term in this problem is: $\binom{3}{r}(1)^{3-r}(2x^2)^r = \binom{3}{r}(1)^{3-r}(2)^r x^{2r}$

The coefficient is then $\binom{3}{r}2^r$

For $2r = 4$, we require $r = 2$
The coefficient is:

$\binom{3}{2}2^2 = 3 \times 4$

$= 12$

Q57:

The general term in this problem is: $\binom{8}{r}(3x)^{8-r}(-2y)^r = \binom{8}{r}3^{8-r}x^{8-r}(-2)^r y^r$

The coefficient is then $\binom{8}{r}3^{8-r}(-2)^r$

For $8 - r = 2$, we require $r = 6$
The coefficient is:

$\binom{8}{6}3^{8-6}(-2)^6 = 28 \times 9 \times 64$

$= 16128$

Binomial applications exercise (page 31)

Q58:

$(1 + (-0 \cdot 5))^5 = \sum_{r=0}^{5}\binom{5}{r}1^{5-r}(-0 \cdot 5)^r$

$= \binom{5}{0}1^5 + \binom{5}{1}1^4(-0 \cdot 5) + \binom{5}{2}1^3(-0 \cdot 5)^2 + \binom{5}{3}1^2(-0 \cdot 5)^3 + \binom{5}{4}1(-0 \cdot 5)^4 +$

$\binom{5}{5}(-0 \cdot 5)^5$

$= 1 \times 1 + (5 \times 1 \times (-0 \cdot 5)) + (10 \times 1 \times 0 \cdot 25) + (10 \times 1 \times (-0 \cdot 125)) +$
$(5 \times 1 \times 0 \cdot 0625) + (1 \times (-0 \cdot 03125))$

$= 1 - 2 \cdot 5 + 2 \cdot 5 - 1 \cdot 25 + 0 \cdot 3125 - 0 \cdot 03125$

$= 0 \cdot 03125$

Q59:

$$(2+0\cdot 7)^3 = \sum_{r=0}^{3} \binom{3}{r} 2^{3-r}(0\cdot 7)^r$$

$$= \binom{3}{0} 2^3 + \binom{3}{1} 2^2 (0\cdot 7) + \binom{3}{2} 2(0\cdot 7)^2 + \binom{3}{3} (0\cdot 7)^3$$

$$= 8 + (3\times 4\times 0\cdot 7) + (3\times 2\times 0\cdot 49) + (1\times 0\cdot 343)$$

$$= 8 + 8\cdot 4 + 2\cdot 94 + 0\cdot 343$$

$$= 19\cdot 683$$

End of topic 5 test (page 37)

Q60:
a)
$$(2x-5)^4 = \sum_{r=0}^{4} \binom{4}{r} (2x)^{4-r}(-5)^r$$

b)
$\binom{4}{0}, \binom{4}{1}, \binom{4}{2}, \binom{4}{3}, \binom{4}{4}$, which is equivalent to 1, 4, 6, 4, 1

c)
$$\binom{4}{0}(2x)^4 + \binom{4}{1}(2x)^3(-5) + \binom{4}{2}(2x)^2(-5)^2 + \binom{4}{3}(2x)^1(-5)^3 + \binom{4}{4}(-5)^4$$

$$= 16x^4 + 4\times 8x^3(-5) + 6\times 4x^2(-5)^2 + 4\times 2x(-5)^3 + (-5)^4$$

$$= 16x^4 - 160x^3 + 600x^2 - 1000x + 625$$

Q61:
a)
$$(5u-3v)^5 = \sum_{r=0}^{5} \binom{5}{r} (5u)^{5-r}(-3v)^r$$

b)
$\binom{5}{0}, \binom{5}{1}, \binom{5}{2}, \binom{5}{3}, \binom{5}{4}, \binom{5}{5}$, which is equivalent to 1, 5, 10, 10, 5, 1

c)
$$\binom{5}{0}(5u)^5 + \binom{5}{1}(5u)^4(-3v) + \binom{5}{2}(5u)^3(-3v)^2 + \binom{5}{3}(5u)^2(-3v)^3 +$$
$$\binom{5}{4}(5u)(-3v)^4 + \binom{5}{5}(-3v)^5$$

$$= 3125u^5 + 5\times 625u^4(-3)v + 10\times 125u^3\times 9\times v^2 + 10\times 25u(-27)v^3$$
$$+ 5\times 5u\times 81v^4 - 243v^5$$

$$= 3125u^5 - 9375u^4v + 11250u^3v^2 - 6750uv^3 + 2025uv^4 - 243v^5$$

Q62:

a)

$$(y^2+7)^5 = \sum_{r=0}^{5} \binom{5}{r} (y^2)^{5-r}(7)^r$$

b)

$\binom{5}{0}, \binom{5}{1}, \binom{5}{2}, \binom{5}{3}, \binom{5}{4}, \binom{5}{5}$, which is equivalent to 1, 5, 10, 10, 5, 1

c)

$$\binom{5}{0}y^{10} + \binom{5}{1}y^8 7 + \binom{5}{2}y^6 7^2 + \binom{5}{3}y^4 7^3 + \binom{5}{4}y^2 7^4 + \binom{5}{5}7^5$$
$$= y^{10} + 5 \times y^8 7 + 10 \times y^6 7^2 + 10 \times y^4 7^3 + 5 \times y^2 7^4 + 7^5$$
$$= y^{10} + 35y^8 + 490y^6 + 3430y^4 + 12005y^2 + 16807$$

Q63:

a)

$$\left(k - \frac{5}{k^2}\right)^4 = \sum_{r=0}^{4} \binom{4}{r}(k)^{4-r}\left(-\frac{5}{k^2}\right)^r$$
$$= \sum_{r=0}^{4} \binom{4}{r}k^{4-3r}(-5)^r$$

b)

$\binom{4}{0}, \binom{4}{1}, \binom{4}{2}, \binom{4}{3}, \binom{4}{4}$, which is equivalent to 1, 4, 6, 4, 1

c)

$$\binom{4}{0}k^4 + \binom{4}{1}k^1(-5) + \binom{4}{2}k^{-2}(-5)^2 + \binom{4}{3}k^{-5}(-5)^3 + \binom{4}{4}k^{-8}(-5)^4$$
$$= k^4 + 4k(-5) + 6\frac{1}{k^2} \times 25 + 4\frac{1}{k^5}(-125) + 625$$
$$= k^4 - 20k + \frac{150}{k^2} - \frac{500}{k^5} + \frac{625}{k^8}$$

Q64:

a)

Steps:

- What is the general term for $\binom{?}{?}(2x)^?\left(\frac{-3}{x^2}\right)^?$? $\binom{9}{r}(2x)^{9-r}\left(\frac{-3}{x^2}\right)^r$

- What is the general term for $\binom{?}{?}(2)^?(x)^?(-3)^?$?

Answer: $\binom{9}{r} 2^{9-r} x^{9-3r} (-3)^r$

b)

Steps:

- When does the power of x equal 0?
  For a term to be independent of x, then $9 - 3r = 0$, i.e. $r = 3$
- Substitute $r = 3$ into $\binom{9}{r} 2^{9-r} x^{9-3r} (-3)^r$

Answer:

$$\binom{9}{3} 2^6 x^0 (-3)^3 = 84 \times 64 \times 1 \times (-27)$$
$$= -145152$$

Q65:

a)

Steps:

- What is the general term for $\binom{?}{?} (3x^2)^? \left(\frac{-2}{x}\right)^?$? $\binom{10}{r} (3x^2)^{10-r} \left(\frac{-2}{x}\right)^r$
- What is the general term for $\binom{?}{?} (3)^? (x)^? (-2)^?$?

Answer: $\binom{10}{r} 3^{10-r} x^{20-3r} (-2)^r$

b)

Steps:

- When does the power of x equal 14?
  For the term x^{14}, then $20 - 3r = 14$, i.e. $r = 2$
- Substitute $r = 2$ into $\binom{10}{r} 3^{10-r} x^{20-3r} (-2)^r$

Answer:

$$\binom{10}{2} 3^8 x^{14} (-2)^2 = 45 \times 6561 \times x^{14} \times 4$$
$$= 1180980 x^{14}$$

Q66:

a)
Hints:

- Use Pascal's triangle for the coefficients or the binomial expansion which is:
$$\sum_{r=0}^{n}\binom{n}{r}(1)^{n-r}(x)^r$$

Answer:

$$(1+x)^5 = \binom{5}{0} + \binom{5}{1}x + \binom{5}{2}x^2 + \binom{5}{3}x^3 + \binom{5}{4}x^4 + \binom{5}{5}x^5$$
$$= 1 + 5x + 10x^2 + 10x^3 + 5x^4 + x^5$$

b)
Steps:

- What is the expression for $0 \cdot 9$ in terms of $(1+x)^5$? $(1 + (-0 \cdot 1))^5$

Answer:

$$(1+(-0 \cdot 1))^5 = \binom{5}{0} + \binom{5}{1}(-0 \cdot 1) + \binom{5}{2}(-0 \cdot 1)^2 + \binom{5}{3}(-0 \cdot 1)^3 +$$
$$\binom{5}{4}(-0 \cdot 1)^4 + \binom{5}{5}(-0 \cdot 1)^5$$
$$= 1 + 5(-0 \cdot 1) + 10(0 \cdot 01) + 10(-0 \cdot 001) + 5(0 \cdot 0001) + (-0 \cdot 00001)$$
$$= 1 - 0 \cdot 5 + 0 \cdot 1 - 0 \cdot 01 + 0 \cdot 0005 - 0 \cdot 00001$$
$$= 0 \cdot 59049$$

Q67:

a)
Hints:

- Use Pascal's triangle for the coefficients or the binomial expansion which is:
$$\sum_{r=0}^{n}\binom{n}{r}(1)^{n-r}(x)^r$$

Answer:

$$(1+x)^4 = \binom{4}{0} + \binom{4}{1}x + \binom{4}{2}x^2 + \binom{4}{3}x^3 + \binom{4}{4}x^4$$
$$= 1 + 4x + 6x^2 + 4x^3 + x^4$$

b)

$$(1+0 \cdot 2)^4 = \binom{4}{0} + \binom{4}{1}0 \cdot 2 + \binom{4}{2}0 \cdot 2^2 + \binom{4}{3}0 \cdot 2^3 + \binom{4}{4}0 \cdot 2^4$$
$$= +4(0 \cdot 2) + 6(0 \cdot 04) + 4(0 \cdot 008) + (0 \cdot 0016)$$
$$= 1 + 0 \cdot 8 + 0 \cdot 24 + 0 \cdot 032 + 0 \cdot 0016$$
$$= 2 \cdot 0736$$

Topic 6: Complex numbers
Multiplying out brackets exercise (page 44)

Q1:
$$(3x+1)(2x+5) = 3x(2x+5) + 1(2x+5)$$
$$= 6x^2 + 15x + 2x + 5$$
$$= 6x^2 + 17x + 5$$

Q2:
$$(3x+2)(4x-1) = 3x(4x-1) + 2(4x-1)$$
$$= 12x^2 - 3x + 8x - 2$$
$$= 12x^2 + 5x - 2$$

Q3:
$$(4-3x)(x+2) = 4(x+2) - 3x(x+2)$$
$$= 4x + 8 - 3x^2 - 6x$$
$$= -3x^2 - 2x + 8$$

Q4:
$$(2-3x)(1-x) = 2(1-x) - 3x(1-x)$$
$$= 2 - 2x - 3x + 3x^2$$
$$= 3x^2 - 5x + 2$$

Q5:
$$(b+2)(b^2-4b+3) = b(b^2-4b+3) + 2(b^2-4b+3)$$
$$= b^3 - 4b^2 + 3b + 2b^2 - 8b + 6$$
$$= b^3 - 2b^2 - 5b + 6$$

Q6:
$$(g-3)(2g^2+5g-7) = g(2g^2+5g-7) - 3(2g^2+5g-7)$$
$$= 2g^3 + 5g^2 - 7g - 6g^2 - 15g + 21$$
$$= 2g^3 - g^2 - 22g + 21$$

Factorising a trinomial exercise (page 46)

Q7: Given $ax^2 + bx + c$ factorises to $(dx + e)(fx + g)$, what factors would satisfy:
1. $e \times g = c$
2. $d \times f = a$
3. $dg + ef = b$

$6x^2 + 17x + 5 = (3x + 1)(2x + 5)$

Q8: Given $ax^2 + bx + c$ factorises to $(dx + e)(fx + g)$, what factors would satisfy:
1. $e \times g = c$
2. $d \times f = a$
3. $dg + ef = b$

$7 + 40x - 12x^2 = (7 - 2x)(1 + 6x)$

Q9: Given $ax^2 + bx + c$ factorises to $(dx + e)(fx + g)$, what factors would satisfy:
1. $e \times g = c$
2. $d \times f = a$
3. $dg + ef = b$

$5x^2 - 16x + 3 = (5x - 1)(x - 3)$

Q10: Given $ax^2 + bx + c$ factorises to $(dx + e)(fx + g)$, what factors would satisfy:
1. $e \times g = c$
2. $d \times f = a$
3. $dg + ef = b$

$25 - 4x^2 = (5 - 2x)(2x + 5)$

Q11: Given $ax^2 + bx + c$ factorises to $(dx + e)(fx + g)$, what factors would satisfy:
1. $e \times g = c$
2. $d \times f = a$
3. $dg + ef = b$

$3 - 14x + 8x^2 = (1 - 4x)(3 - 2x)$

Arithmetic of surds exercise (page 49)

Q12:
$$\sqrt{50} = \sqrt{25} \times \sqrt{2}$$
$$= 5\sqrt{2}$$

Q13: Simplify each surd by finding the biggest square number that goes into each and use the rule $\sqrt{ab} = \sqrt{a} \times \sqrt{b}$.

$$4\sqrt{20} + 2\sqrt{45} = 4\sqrt{4 \times 5} + 2\sqrt{9 \times 5}$$
$$= 4\sqrt{4}\sqrt{5} + 2\sqrt{9}\sqrt{5}$$
$$= 4 \times 2\sqrt{5} + 2 \times 3\sqrt{5}$$
$$= 8\sqrt{5} + 6\sqrt{5}$$
$$= 14\sqrt{5}$$

ANSWERS: UNIT 1 TOPIC 6

Q14: Simplify each surd by finding the biggest square number that goes into each and use the rule $\sqrt{ab} = \sqrt{a} \times \sqrt{b}$.

$$\begin{aligned} 3\sqrt{8} - 2\sqrt{18} &= 3\sqrt{4 \times 2} - 2\sqrt{9 \times 2} \\ &= 3\sqrt{4}\sqrt{2} - 2\sqrt{9}\sqrt{2} \\ &= 3 \times 2\sqrt{2} - 2 \times 3\sqrt{2} \\ &= 6\sqrt{2} - 6\sqrt{2} \\ &= 0 \end{aligned}$$

Q15: Multiply the surds under the one root and then simplify.

$$\begin{aligned} 5\sqrt{6} \times \sqrt{3} &= 5\sqrt{6 \times 3} \\ &= 5\sqrt{18} \\ &= 5 \times \sqrt{2 \times 9} \\ &= 5 \times 3 \times \sqrt{2} \\ &= 15\sqrt{2} \end{aligned}$$

Q16: Multiply out the brackets then apply the surd rules.

$$\begin{aligned} \left(2\sqrt{2}+3\right)\left(5-2\sqrt{3}\right) &= 2\sqrt{2}\left(5-2\sqrt{3}\right)+3\left(5-2\sqrt{3}\right) \\ &= 10\sqrt{2} - 4\sqrt{2}\sqrt{3} + 15 - 6\sqrt{3} \\ &= 10\sqrt{2} - 4\sqrt{6} + 15 - 6\sqrt{3} \end{aligned}$$

Q17: Multiply the fraction by the conjugate of the denominator, e.g. $\frac{(\sqrt{a}-b)}{(c+\sqrt{d})} \times \frac{(c-\sqrt{d})}{(c-\sqrt{d})}$

$$\begin{aligned} \frac{4\sqrt{3}-2}{1+\sqrt{3}} &= \frac{(4\sqrt{3}-2)}{(1+\sqrt{3})} \times \frac{(1-\sqrt{3})}{(1-\sqrt{3})} \\ &= \frac{4\sqrt{3} \times -4\sqrt{3}\sqrt{3} - 2 + 2\sqrt{3}}{1 - \sqrt{3} + \sqrt{3} - 3} \\ &= \frac{6\sqrt{3} - 4 \times 3 - 2}{-2} \\ &= \frac{6\sqrt{3} - 14}{-2} \\ &= -3\sqrt{3} + 7 \\ &= 7 - 3\sqrt{3} \end{aligned}$$

Q18: Multiply the fraction by the conjugate of the denominator, e.g. $\frac{(\sqrt{a}-b)}{(c+\sqrt{d})} \times \frac{(c-\sqrt{d})}{(c-\sqrt{d})}$

$$\begin{aligned} \frac{2\sqrt{5}+3}{4-\sqrt{2}} &= \frac{(2\sqrt{5}+3)}{(4-\sqrt{2})} \times \frac{(4+\sqrt{2})}{(4+\sqrt{2})} \\ &= \frac{8\sqrt{5} + 2\sqrt{10} + 12 + 3\sqrt{2}}{16 - 2} \\ &= \frac{8\sqrt{5} + 2\sqrt{10} + 12 + 3\sqrt{2}}{14} \end{aligned}$$

© HERIOT-WATT UNIVERSITY

Q19:
$$2\sqrt{27} + 5\sqrt{12} = 2\sqrt{9 \times 3} + 5\sqrt{4 \times 3}$$
$$= 2 \times 3\sqrt{3} + 5 \times 2\sqrt{3}$$
$$= 16\sqrt{3}$$

Q20:
$$\sqrt{18} - 3\sqrt{32} = \sqrt{9 \times 2} - 3\sqrt{16 \times 2}$$
$$= 3\sqrt{2} - 3 \times 4\sqrt{2}$$
$$= -9\sqrt{2}$$

Q21:
$$2\sqrt{45} - 5\sqrt{80} + \sqrt{2} = 2\sqrt{9 \times 5} - 5\sqrt{16 \times 5} + \sqrt{2}$$
$$= 2 \times 3\sqrt{5} - 5 \times 4\sqrt{5} + \sqrt{2}$$
$$= \sqrt{2} - 14\sqrt{5}$$

Q22:
$$3\sqrt{28} + 4\sqrt{63} = 3\sqrt{4 \times 7} + 4\sqrt{9 \times 7}$$
$$= 3 \times 2\sqrt{7} + 4 \times 3\sqrt{7}$$
$$= 18\sqrt{7}$$

Q23:
$$2\sqrt{12} \times \sqrt{3} = 2\sqrt{4 \times 3} \times \sqrt{3}$$
$$= 2 \times 2\sqrt{3} \times \sqrt{3}$$
$$= 12$$

Q24:
$$\sqrt{10} \times 7\sqrt{5} = 7\sqrt{10 \times 5}$$
$$= 7\sqrt{50}$$
$$= 7\sqrt{25 \times 2}$$
$$= 35\sqrt{2}$$

Q25:
$$\left(3\sqrt{2} - 2\right)\left(5 + \sqrt{3}\right) = 3\sqrt{2}\left(5 + \sqrt{3}\right) - 2\left(5 + \sqrt{3}\right)$$
$$= 15\sqrt{2} + 3\sqrt{2}\sqrt{3} - 10 - 2\sqrt{3}$$
$$= 15\sqrt{2} + 3\sqrt{6} - 10 - 2\sqrt{3}$$

Q26:
$$\left(7 - 3\sqrt{5}\right)\left(4 - 2\sqrt{5}\right) = 7\left(4 - 2\sqrt{5}\right) - 3\sqrt{5}\left(4 - 2\sqrt{5}\right)$$
$$= 28 - 14\sqrt{5} - 12\sqrt{5} + 6\sqrt{5}\sqrt{5}$$
$$= 58 - 26\sqrt{5}$$

ANSWERS: UNIT 1 TOPIC 6

Q27:

$$\frac{5\sqrt{2}+6}{1-\sqrt{5}} = \frac{(5\sqrt{2}+6)}{(1-\sqrt{5})} \times \frac{(1+\sqrt{5})}{(1+\sqrt{5})}$$

$$= \frac{5\sqrt{2}+5\sqrt{2}\sqrt{5}+6+6\sqrt{5}}{1-5}$$

$$= -\frac{5\sqrt{2}+5\sqrt{10}+6+6\sqrt{5}}{4}$$

Q28:

$$\frac{2\sqrt{2}-3}{4-\sqrt{2}} = \frac{(2\sqrt{2}-3)}{(4-\sqrt{2})} \times \frac{(4+\sqrt{2})}{(4+\sqrt{2})}$$

$$= \frac{8\sqrt{2}+2\sqrt{2}\sqrt{2}-12-3\sqrt{2}}{16-2}$$

$$= \frac{5\sqrt{2}-8}{14}$$

Trigonometric identities exercise (page 52)

Q29:

Use the expansion $\sin(A - B) = \sin A \cos B - \cos A \sin B$ to expand the expression first and then evaluate.

$$\sin(x - \pi) = \sin x \cos \pi - \cos x \sin \pi$$
$$= -\sin x$$

Since $\cos \pi = -1$ and $\sin \pi = 0$

Q30:

Use the expansion $\cos(A + B) = \cos A \cos B - \sin A \sin B$ to expand the expression first and then evaluate.

$$\cos\left(\frac{\pi}{2} + x\right) = \cos\left(\frac{\pi}{2}\right)\cos x - \sin\left(\frac{\pi}{2}\right)\sin x$$
$$= -\sin x$$

Since $\cos\left(\frac{\pi}{2}\right) = 0$ and $\sin\left(\frac{\pi}{2}\right) = 1$

Q31:

Use the identity $\cos^2 A + \sin^2 A = 1$

Rearrange to give an expression for $\sin^2 \theta$ in terms of cosine:

$\sin^2 \theta = 1 - \cos^2 \theta$

Now replace $\sin^2 \theta$ in the original expression:

$$\cos\theta \sin^2\theta = \cos\theta\left(1 - \cos^2\theta\right)$$
$$= \cos\theta - \cos^3\theta$$

© HERIOT-WATT UNIVERSITY

Q32:
Use the identity $\cos^2 A + \sin^2 A = 1$
Rearrange to give an expression for $\cos^2 \theta$ in terms of sine:
$\cos^2 \theta = 1 - \sin^2 \theta$
Now rewrite the original expression and replace $\cos^2 \theta$:
$$\begin{aligned}\cos^4 \theta &= \left(\cos^2 \theta\right)^2 \\ &= \left(1 - \sin^2 \theta\right)^2 \\ &= 1 - 2\sin^2 \theta + \sin^4 \theta\end{aligned}$$

Q33:
$$\begin{aligned}\cos(\theta - \pi) &= \cos \theta \cos \pi + \sin \theta \sin \pi \\ &= -\cos x\end{aligned}$$
Since $\cos \pi = -1$ and $\sin \pi = 0$

Q34:
$$\begin{aligned}\sin\left(x + \frac{3\pi}{2}\right) &= \sin x \cos\left(\frac{3\pi}{2}\right) + \cos x \sin\left(\frac{3\pi}{2}\right) \\ &= -\cos x\end{aligned}$$
Since $\cos\left(\frac{3\pi}{2}\right) = 0$ and $\sin\left(\frac{3\pi}{2}\right) = -1$

Q35:
Use the identity $\cos^2 A + \sin^2 A = 1$
Rearrange to give an expression for $\cos^2 \theta$ in terms of sine: $\cos^2 \theta = 1 - \sin^2 \theta$
Now replace $\cos^2 \theta$ in the original expression:
$$\begin{aligned}\cos^2 \theta \sin^3 \theta &= \left(1 - \sin^2 \theta\right)\sin^3 \theta \\ &= \sin^3 \theta - \sin^5 \theta\end{aligned}$$

Q36:
Use the identity $\cos^2 A + \sin^2 A = 1$
Rearrange to give an expression for $\sin^2 \theta$ in terms of cosine: $\sin^2 \theta = 1 - \cos^2 \theta$
Now replace $\sin^2 \theta$ in the original expression:
$$\begin{aligned}\sin^4 x \cos^2 x &= \left(1 - \cos^2 x\right)^2 \cos^2 x \\ &= \left(1 - 2\cos^2 x + \cos^4 x\right) \cos^2 x \\ &= \cos^2 x - 2\cos^4 x + \cos^6 x\end{aligned}$$

Revision exercise (page 52)

Q37: $(3x - 1)(x - 4) = 0$
The two roots are $x = \frac{1}{3}$ and $x = 4$

Q38: $2x^2 - x - 6$

Q39:

$$(2x-y)^4 = \binom{4}{0}(2x)^4 + \binom{4}{1}(2x)^3(-y) + \binom{4}{2}(2x)^2(-y)^2 + \binom{4}{3}(2x)(-y)^3 + \binom{4}{4}(-y)^4$$
$$= 1 \times 16x^4 + 4 \times 8x^3 \times -y + 6 \times 4x^2 \times y^2 + 4 \times 2x \times -y^3 + 1 \times y^4$$
$$= 16x^4 - 32x^3y + 24x^2y^2 - 8xy^3 + y^4$$

Q40:

$\cos(a + \pi) = \cos a \cos \pi - \sin a \sin \pi$
But $\sin \pi = 0$ and $\cos \pi = -1$
Hence $\cos(a + \pi) = -\cos a$

Q41:

$$\frac{(3\sqrt{48} - \sqrt{27})}{(2\sqrt{12} + \sqrt{75})} = \frac{(3\sqrt{16 \times 3} - \sqrt{9 \times 3})}{(2\sqrt{4 \times 3} + \sqrt{25 \times 3})}$$
$$= \frac{(12\sqrt{3} - 3\sqrt{3})}{(4\sqrt{3} + 5\sqrt{3})}$$
$$= \frac{9\sqrt{3}}{9\sqrt{3}}$$
$$= 1$$

Square roots of a negative number and quadratic equations exercise (page 56)

Q42:

$-16 = 16 \times (-1)$
Since $i^2 = -1$ then
$-16 = 16i^2$
So $\sqrt{-16} = \sqrt{16i^2}$
$= \pm 4i$

Q43:

$-64 = 64 \times (-1)$
Since $i^2 = -1$ then
$-64 = 64i^2$
So $\sqrt{-64} = \sqrt{64i^2}$
$= \pm 8i$

Q44:

$-8 = 8 \times (-1)$
Since $i^2 = -1$ then
$-8 = 8i^2$
So $\sqrt{-8} = \sqrt{8i^2}$
$= \pm 2i\sqrt{2}$

© HERIOT-WATT UNIVERSITY

Q45:

For $3x^2 - 2x + 4 = 0$
$a = 3, b = -2, c = 4$
$x = \dfrac{-b \pm \sqrt{b^2 - 4ac}}{2a}$
$= \dfrac{2 \pm \sqrt{(-2)^2 - 4(3)(4)}}{2(3)}$
$= \dfrac{2 \pm \sqrt{-44}}{6}$
$= \dfrac{2 \pm \sqrt{44i^2}}{6}$
$= \dfrac{2 \pm 2i\sqrt{11}}{6}$
$= \dfrac{1 \pm i\sqrt{11}}{3}$

Q46:

For $x^2 - 2x + 5 = 0$
$a = 1, b = -2, c = 5$
$x = \dfrac{-b \pm \sqrt{b^2 - 4ac}}{2a}$
$= \dfrac{2 \pm \sqrt{(-2)^2 - 4(1)(5)}}{2(1)}$
$= \dfrac{2 \pm \sqrt{-16}}{2}$
$= \dfrac{2 \pm \sqrt{16i^2}}{2}$
$= \dfrac{2 \pm 4i}{2}$
$= 1 \pm 2i$

Q47:

For $x^2 + 2x + 6 = 0$
$a = 1, b = 2, c = 6$
$x = \dfrac{-b \pm \sqrt{b^2 - 4ac}}{2a}$
$= \dfrac{-2 \pm \sqrt{(2)^2 - 4(1)(6)}}{2(1)}$
$= \dfrac{-2 \pm \sqrt{-20}}{2}$
$= \dfrac{-2 \pm \sqrt{20i^2}}{2}$
$= \dfrac{-2 \pm 2i\sqrt{5}}{2}$
$= -1 \pm i\sqrt{5}$

Q48:
For $x^2 + x + 1 = 0$
$a = 1, b = 1, c = 1$
$$x = \frac{-b \pm \sqrt{b^2 - 4ac}}{2a}$$
$$= \frac{-1 \pm \sqrt{(1)^2 - 4(1)(1)}}{2(1)}$$
$$= \frac{-1 \pm \sqrt{-3}}{2}$$
$$= \frac{-1 \pm \sqrt{3i^2}}{2}$$
$$= \frac{-1 \pm i\sqrt{3}}{2}$$

Identifying the real and imaginary parts of complex numbers exercise (page 57)

Q49: Re$(-5 + 3i) = -5$ and Im$(-5 + 3i) = 3$

Q50: Re$(-6i - 2) = -2$ and Im$(-6i - 2) = -6$

Complex numbers and the complex plane exercise (page 59)

Q51: $1 + 3i$

Q52: $-2 - i$

Q53: $1 - 3i$

Multiplying imaginary numbers exercise (page 64)

Q54:
$i^2 = \left(\sqrt{-1}\right)^2$
$= -1$

Q55: $i^2 = -1$ so:
$i^3 = i^2 \times i$
$= -1 \times i$
$= -i$

Q56: $i^2 = -1$ so:
$i^4 = i^2 \times i^2$
$= (-1) \times (-1)$
$= 1$

Q57: $i^4 = 1$ so:
$i^5 = i^4 \times i$
$= 1 \times i$
$= i$

Q58: $i^5 = i$ so:
$i^6 = i^5 \times i$
$= i \times i$
$= -1$

Q59: $i^2 = -1$ so:
$-i^2 = (-1) \times i^2$
$= (-1) \times (-1)$
$= 1$

Q60: $i^3 = -i$ so:
$-i^3 = (-1) \times i^3$
$= (-1) \times (-i)$
$= i$

Q61: $i^4 = 1$ so:
$-i^4 = (-1) \times i^4$
$= (-1) \times 1$
$= -1$

Q62: $i^5 = i$ so:
$-i^5 = (-1) \times i^5$
$= (-1) \times i$
$= -i$

Q63: $i^6 = -1$ so:
$-i^6 = (-1) \times i^6$
$= (-1) \times (-1)$
$= 1$

Finding the square roots of a complex number exercise (page 67)

Q64:
Let the square roots of $8 + 6i$ be $a + ib$:
$8 + 6i = (a + ib)^2$
$ = a^2 + 2iab + i^2b^2$
$ = a^2 - b^2 + 2iab$

Equate the real and imaginary parts:
$$a^2 - b^2 = 8 \quad (1)$$
$$2ab = 6 \quad (2)$$

Rearranging (2): $$b = \frac{3}{a} \quad (3)$$

Substituting (3) into (1): $$a^2 - \left(\frac{3}{a}\right)^2 = 8$$

Simplifying: $$a^2 - \frac{9}{a^2} = 8$$

Multiplying through by a^2: $a^4 - 8a^2 - 9 = 0$

Factorising and solving: $(a^2 - 9)(a^2 + 1) = 0$
$$a^2 = 9 \text{ or } a^2 = -1$$
$a \in \mathbb{R}$ so $a^2 = -1$ is impossible
so $a = \pm 3$

Substituting into (2):
When $a = 3$, then $b = \frac{3}{3} \Rightarrow b = 1$ and when $a = -3$, then $b = \frac{3}{-3} \Rightarrow b = -1$
The square roots of $8 + 6i$ are $3 + i$ and $-3 - i$

Q65:
Let the square roots of $6 - 8i$ be $a + ib$:
$6 - 8i = (a + ib)^2$
$= a^2 + 2iab + i^2b^2$
$= a^2 - b^2 + 2iab$

Equate real and imaginary parts:
$$a^2 - b^2 = 6 \quad (1)$$
$$2ab = 8 \quad (2)$$

Rearranging (2): $$b = \frac{4}{a} \quad (3)$$

Substituting (3) into (1): $$a^2 - \left(\frac{4}{a}\right)^2 = 6$$

Simplifying: $$a^2 - \frac{16}{a^2} = 6$$

Multiplying through by a^2: $a^4 - 6a^2 - 16 = 0$

Factorising and solving: $(a^2 - 8)(a^2 + 2) = 0$
$$a^2 = 8 \text{ or } a^2 = -2$$
$a \in \mathbb{R}$ so $a^2 = -2$ is impossible
so $a = \pm 2\sqrt{2}$

Substituting into (2):
When $a = 2\sqrt{2}$, then $b = \frac{4}{2\sqrt{2}} \Rightarrow b = \sqrt{2}$ and when $a = -2\sqrt{2}$, then $b = \frac{4}{-2\sqrt{2}} \Rightarrow b = -\sqrt{2}$
The square roots of $6 - 8i$ are $2\sqrt{2} + i\sqrt{2}$ and $-2\sqrt{2} - i\sqrt{2}$

© HERIOT-WATT UNIVERSITY

Q66:
Let the square roots of $-8 + 15i$ be $a + ib$:
$$-8 + 15i = (a + ib)^2$$
$$= a^2 + 2iab + i^2b^2$$
$$= a^2 - b^2 + 2iab$$
Equate real and imaginary parts:
$$a^2 - b^2 = -8 \quad (1)$$
$$2ab = 15 \quad (2)$$

Rearranging (2): $\quad b = \dfrac{15}{2a} \quad (3)$

Substituting (3) into (1): $\quad a^2 - \left(\dfrac{15}{2a}\right)^2 = -8$

Simplifying: $\quad a^2 - \dfrac{225}{4a^2} = -8$

Multiplying through by a^2: $\quad 4a^4 + 32a^2 - 225 = 0$

Factorising and solving: $\quad (2a^2 - 9)(2a^2 + 25) = 0$
$$a^2 = \dfrac{9}{2} \text{ or } a^2 = -\dfrac{25}{2}$$
$$a \in \mathbb{R} \text{ so } a^2 = -\dfrac{25}{2} \text{ is impossible}$$
$$\text{so } a = \pm\dfrac{3}{\sqrt{2}} \Rightarrow a = \pm\dfrac{3\sqrt{2}}{2}$$

Substituting into (2):

When $a = \dfrac{3\sqrt{2}}{2}$, then $b = \dfrac{15}{2\left(\frac{3\sqrt{2}}{2}\right)} \Rightarrow b = \dfrac{5\sqrt{2}}{2}$ and when $a = -\dfrac{3\sqrt{2}}{2}$, then $b = \dfrac{15}{2\left(-\frac{3\sqrt{2}}{2}\right)} \Rightarrow b = -\dfrac{5\sqrt{2}}{2}$

The square roots of $-8 + 15i$ are $\dfrac{3\sqrt{2}}{2} + \dfrac{5i\sqrt{2}}{2}$ and $-\dfrac{3\sqrt{2}}{2} - \dfrac{5i\sqrt{2}}{2}$

Q67:
Let the square roots of $-20 - 21i$ be $a + ib$:
$$-20 - 21i = (a + ib)^2$$
$$= a^2 + 2iab + i^2b^2$$
$$= a^2 - b^2 + 2iab$$
Equate real and imaginary parts:
$$a^2 - b^2 = -20 \quad (1)$$
$$2ab = -21 \quad (2)$$

Rearranging (2): $\quad b = -\dfrac{21}{2a} \quad (3)$

Substituting (3) into (1): $\quad a^2 - \left(\dfrac{21}{2a}\right)^2 = -20$

Simplifying: $\quad a^2 - \dfrac{441}{4a^2} = -20$

ANSWERS: UNIT 1 TOPIC 6 445

Multiplying through by a^2: $4a^4 + 80a^2 - 441 = 0$

Factorising and solving: $(2a^2 - 9)(2a^2 + 49) = 0$

$$a^2 = \frac{9}{2} \text{ or } a^2 = -\frac{49}{2}$$

$$a \in \mathbb{R} \text{ so } a^2 = -\frac{49}{2} \text{ is impossible}$$

$$\text{so } a = \pm \frac{3}{\sqrt{2}} \Rightarrow a = \pm \frac{3\sqrt{2}}{2}$$

Substituting into (2):

When $a = \frac{3\sqrt{2}}{2}$, then $b = -\frac{21}{2\left(\frac{3\sqrt{2}}{2}\right)} \Rightarrow b = -\frac{7\sqrt{2}}{2}$ and when $a = -\frac{3\sqrt{2}}{2}$, then $b = -\frac{21}{2\left(-\frac{3\sqrt{2}}{2}\right)} \Rightarrow b = \frac{7\sqrt{2}}{2}$

The square roots of $-20 - 21i$ are $\frac{3\sqrt{2}}{2} - \frac{7i\sqrt{2}}{2}$ and $-\frac{3\sqrt{2}}{2} + \frac{7i\sqrt{2}}{2}$

Multiplication by i exercise (page 70)

Q68:

$(3 - 2i)i = 3i - 2i^2$
$= 2 + 3i$

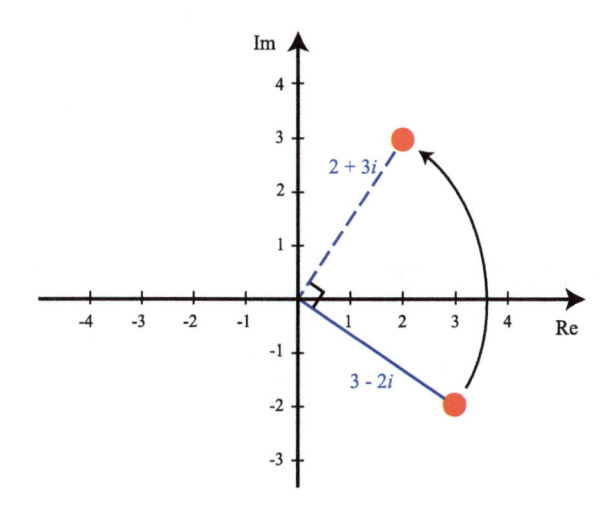

© HERIOT-WATT UNIVERSITY

Q69:

$$(-1-3i)\,i = -i - 3i^2$$
$$= 3 - i$$

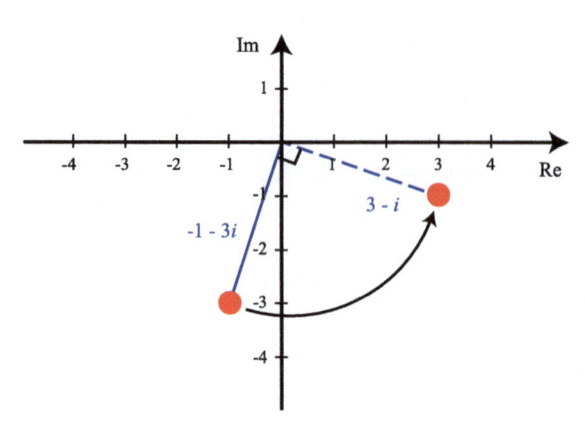

Q70:

$$(-3+2i)\,i = -3i + 2i^2$$
$$= -2 - 3i$$

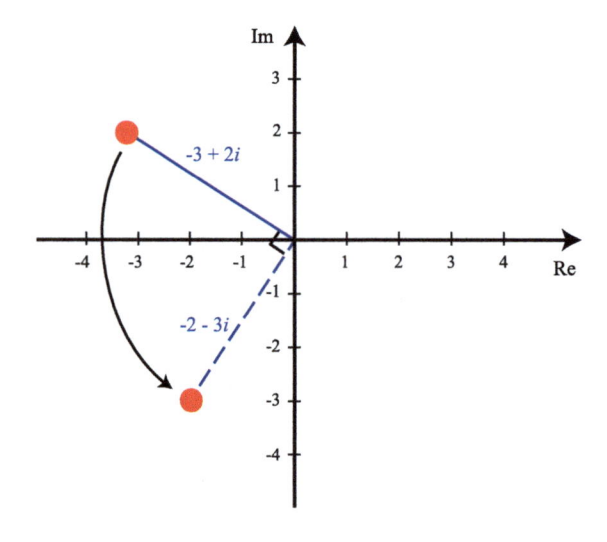

Q71:

$$(-3+2i)(-i) = 3i - 2i^2$$
$$= 2 + 3i$$

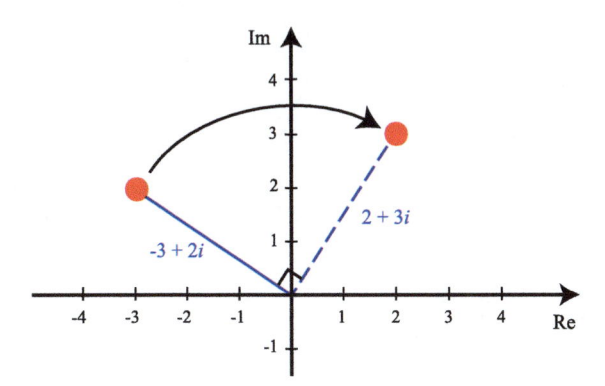

Q72:

$$(1-4i)(-i) = -i + 4i^2$$
$$= -4 - i$$
$$(-4-i)(-i) = 4i + i^2$$
$$= -1 + 4i$$
$$(-1+4i)(-i) = i - 4i^2$$
$$= 4 + i$$
$$(4+i)(-i) = -4i - i^2$$
$$= 1 - 4i$$

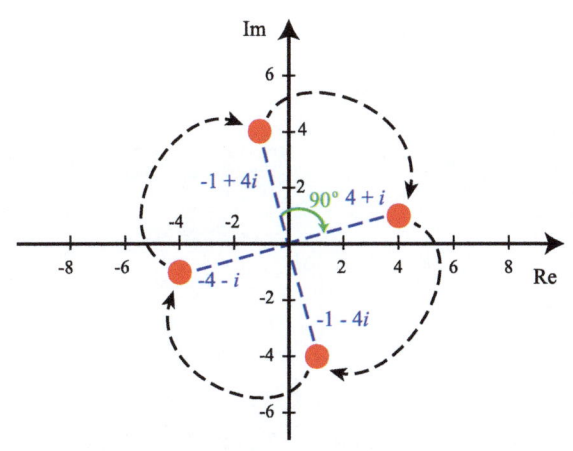

Multiplying by $-i$ is the same as a clockwise rotation of $90°$ or $\frac{\pi}{2}$

Conjugates of complex numbers exercise (page 72)

Q73:
The conjugate is $\bar{z} = 4 - 7i$

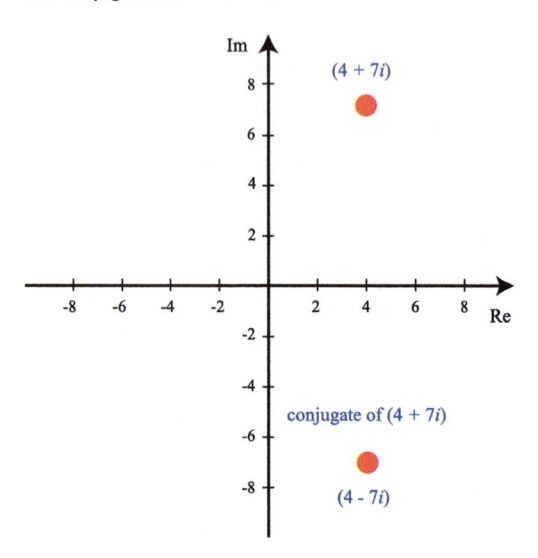

[Argand diagram showing (4 + 7i) in the upper half plane and its conjugate (4 - 7i) in the lower half plane]

Q74: $5 - i$

Q75: $-3 + 5i$

Q76: $-4i$

Q77: 5

Q78: The conjugate is $\bar{z} = 3 - 2i$
$z\bar{z} = (3 + 2i)(3 - 2i)$
$= 9 - 6i + 6i - 4i^2$
$= 13$

Q79: $z = -4 + 5i$
$3iz = 3i(-4 + 5i)$
$= -12i + 15i^2$
$= -15 - 12i$

Q80: The conjugate is $\bar{z} = 2 + i$
$i(\bar{z} + 1) = i(2 + i + 1)$
$= i(3 + i)$
$= -1 + 3i$

ANSWERS: UNIT 1 TOPIC 6

Division by i exercise (page 75)

Q81:

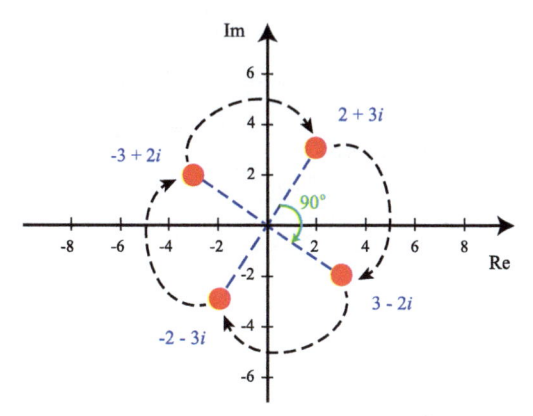

When dividing by i, the complex number is rotated $90°$ in a clockwise direction.

Cartesian form exercise (page 80)

Q82:
Given the modulus r and argument θ, we know that $a = r\cos\theta$ and $b = r\sin\theta$.
Substitute for r and θ and evaluate.

$$a = 2\cos\left(\frac{\pi}{3}\right)$$
$$= 2 \times \frac{1}{2}$$
$$= 1$$
$$b = 2\sin\left(\frac{\pi}{3}\right)$$
$$= 2 \times \frac{\sqrt{3}}{2}$$
$$= \sqrt{3}$$

So $z = 1 + i\sqrt{3}$

Q83:
Given the modulus r and argument θ, we know that $a = r\cos\theta$ and $b = r\sin\theta$.
Substitute for r and θ and evaluate.

$$a = 3\cos(30°)$$
$$= 3 \times \frac{\sqrt{3}}{2}$$
$$= \frac{3\sqrt{3}}{2}$$

© HERIOT-WATT UNIVERSITY

$b = 3\sin(30°)$
$= 3 \times \dfrac{1}{2}$
$= \dfrac{3}{2}$
So $z = \dfrac{3\sqrt{3}}{2} + \dfrac{3}{2}i$

Q84:
For the complex number $z = a + ib$, $a = r\cos\theta$ and $b = r\sin\theta$ so:
$z = 4\cos\left(\dfrac{\pi}{4}\right) + 4i\sin\left(\dfrac{\pi}{4}\right)$
$= 4\dfrac{1}{\sqrt{2}} + 4i\dfrac{1}{\sqrt{2}}$
$= 4\dfrac{1}{\sqrt{2}} \times \dfrac{\sqrt{2}}{\sqrt{2}} + 4i\dfrac{1}{\sqrt{2}} \times \dfrac{\sqrt{2}}{\sqrt{2}}$
$= 2\sqrt{2} + 2i\sqrt{2}$

Q85:
For the complex number $z = a + ib$, $a = r\cos\theta$ and $b = r\sin\theta$ so:
$z = 3\cos(240°) + i3\sin(240°)$
$= -\dfrac{3}{2} - \dfrac{3\sqrt{3}}{2}i$

Q86:
For the complex number $z = a + ib$, $a = r\cos\theta$ and $b = r\sin\theta$ so:
$z = 2\cos\left(\dfrac{3\pi}{2}\right) + 2i\sin\left(\dfrac{3\pi}{2}\right)$
$= 2 \times 0 + 2i \times -1$
$= -2i$

Q87:
For the complex number $z = a + ib$, $a = r\cos\theta$ and $b = r\sin\theta$ so:
$z = 5\cos(150°) + 5i\sin(150°)$
$= -\dfrac{5\sqrt{3}}{2} + \dfrac{5}{2}i$

Q88:
$2\left(\cos\left(\dfrac{7\pi}{6}\right) + i\sin\left(\dfrac{7\pi}{6}\right)\right)$ is equivalent to $-2\cos\left(\dfrac{\pi}{6}\right) - 2i\sin\left(\dfrac{\pi}{6}\right)$ so:
$z = -2\dfrac{\sqrt{3}}{2} - 2i\dfrac{1}{2}$
$= -\sqrt{3} - i$

Q89:
$3(\cos(25°) + i\sin(25°))$ is equivalent to $3\cos(25°) + 3i\sin(25°)$
Using a calculator: $z = 2 \cdot 72 + 1 \cdot 27i$

Principal argument exercise (page 85)

Q90:

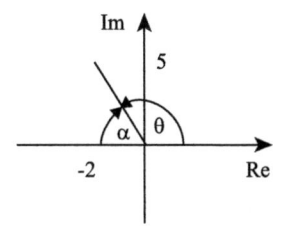

First we need to find α:

$\alpha = \tan^{-1}\left|\dfrac{5}{-2}\right|$

$= 68°$

z is in the second quadrant so $\theta = 180 - \alpha$ and the argument is positive.

The principal argument is:

$\theta = 180° - 68°$

$= 112°$

Q91:

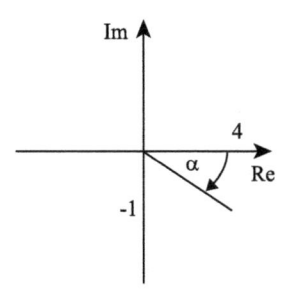

First we need to find α:

$\alpha = \tan^{-1}\left|\dfrac{-1}{4}\right|$

$= 14°$

z is in the fourth quadrant so $\theta = -\alpha$ and the argument is negative.
The principal argument is $\theta = -14°$

Q92:

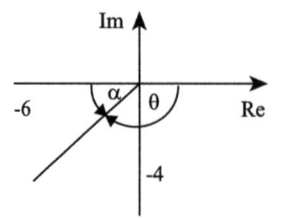

First we need to find α:

$\alpha = \tan^{-1}\left|\dfrac{-6}{-4}\right|$

$= 56°$

z is in the third quadrant so $\theta = -180° + \alpha$ and the argument is negative.
The principal argument is $\theta = -124°$

Q93:

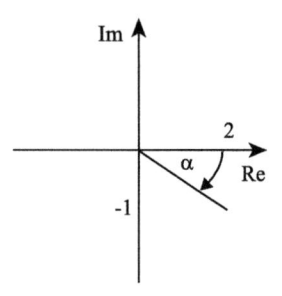

First we need to find α:

$\alpha = \tan^{-1}\left|\dfrac{-1}{2}\right|$

$= 0\cdot 46$ rad

z is in the fourth quadrant so $\theta = -\alpha$ and the argument is negative.
The principal argument is $\theta = -0\cdot 46\ rad$

Q94:

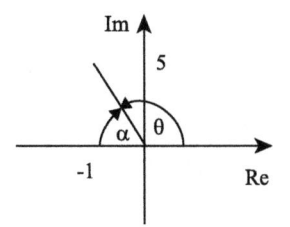

First we need to find α:

$\alpha = \tan^{-1}\left|\dfrac{5}{-1}\right|$

$= 1 \cdot 37$ rad

z is in the second quadrant so $\theta = \pi - \alpha$ and the argument is positive.

The principal argument is:

$\theta = \pi - 1 \cdot 37$

$= 1 \cdot 77$ rad

Q95:

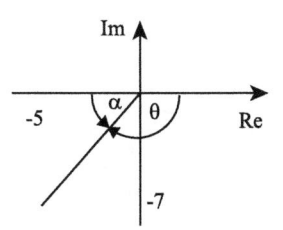

First we need to find α:

$\alpha = \tan^{-1}\left|\dfrac{-7}{-5}\right|$

$= 0 \cdot 95$ rad

z is in the third quadrant so $\theta = -\pi + \alpha$ and the argument is negative.

The principal argument is:

$\theta = -\pi + 0 \cdot 95$

$= -2 \cdot 19$ rad

© HERIOT-WATT UNIVERSITY

Strategy to find modulus and argument exercise (page 90)

Q96:

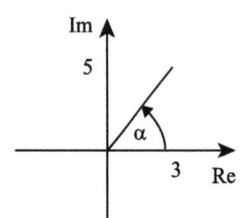

Im
5
α
3 Re

Modulus:
$r = \sqrt{a^2 + b^2}$
$= \sqrt{3^2 + 5^2}$
$= \sqrt{34}$

Argument:
$\alpha = \tan^{-1}\left|\dfrac{5}{3}\right|$
$\alpha = 59°$ or $\alpha = 1 \cdot 03 \; rad$ (2 d.p.)

Since z is in the first quadrant, $\arg(z) = \alpha$
So $\arg(z) = 59°$ or $1 \cdot 03$ radians

Q97:

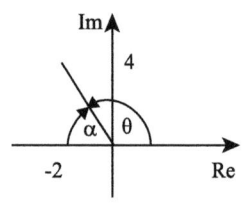

Im
4
α θ
-2   Re

Modulus:
$r = \sqrt{a^2 + b^2}$
$= \sqrt{4^2 + (-2)^2}$
$= 2\sqrt{5}$

Argument:
$\alpha = \tan^{-1}\left|\dfrac{4}{-2}\right|$
$\alpha = 63 \cdot 4°$ or $\alpha = 1 \cdot 11 \; rad$ (2 d.p.)

Since z is in the second quadrant, $\arg(z) = 180° - \alpha$
So $\arg(z) = 116 \cdot 6°$ or $2 \cdot 03$ radians

Q98:

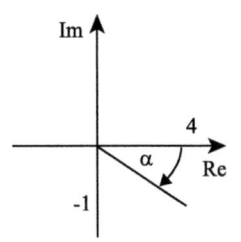

Modulus:
$r = \sqrt{a^2 + b^2}$
$= \sqrt{4^2 + (-1)^2}$
$= \sqrt{17}$

Argument:
$\alpha = \tan^{-1}\left|\frac{-1}{4}\right|$
$\alpha = 14°$ or $\alpha = 0.24 \, rad$ **(2 d.p.)**

Since z is in the fourth quadrant, $\arg(z) = -\alpha$
So $\arg(z) = -14°$ or $-0 \cdot 24$ radians

Q99:

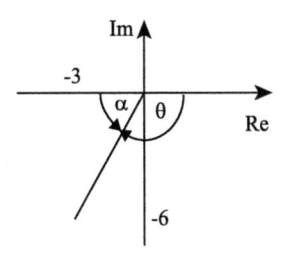

Modulus:
$r = \sqrt{a^2 + b^2}$
$= \sqrt{(-3)^2 + (-6)^2}$
$= 3\sqrt{5}$

Argument:
$\alpha = \tan^{-1}\left|\frac{-6}{-3}\right|$
$\alpha = 63 \cdot 4°$ or $\alpha = 1 \cdot 11 \, rad$ **(2 d.p.)**

Since z is in the third quadrant, $\arg(z) = -180° + \alpha$
So $\arg(z) = -116 \cdot 6°$ or $-2 \cdot 03$ radians

Q100:

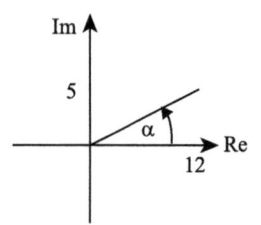

[Diagram: Argand diagram showing point in first quadrant with real part 12, imaginary part 5, angle α from Re axis]

Modulus:
$r = \sqrt{a^2 + b^2}$
$= \sqrt{(12)^2 + (5)^2}$
$= 13$

Argument:
$\alpha = \tan^{-1}\left|\dfrac{5}{12}\right|$
$\alpha = 22 \cdot 6°$ **or** $\alpha = 0 \cdot 39\ rad$ **(1 d.p.)**

Since z is in the first quadrant, $\arg(z) = \alpha$
So $\arg(z) = 22 \cdot 6°$ or $0 \cdot 39$ radians

Q101:

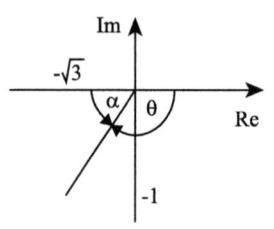

[Diagram: Argand diagram showing point with real part $-\sqrt{3}$, imaginary part -1, with angles α and θ]

Modulus:
$r = \sqrt{a^2 + b^2}$
$= \sqrt{(-1)^2 + \left(-\sqrt{3}\right)^2}$
$= 2$

Argument:
$\alpha = \tan^{-1}\left|\dfrac{-1}{-\sqrt{3}}\right|$
$\alpha = 30°$ **or** $\alpha = \dfrac{\pi}{6}\ rad$

Since z is in the first quadrant, $\arg(z) = -180° + \alpha$
So $\arg(z) = -150°$ or $-\frac{5\pi}{6}$ radians

Geometric interpretations: Circle representations exercise (page 104)

Q102:

Remember that $|z| = \sqrt{x^2 + y^2}$, where $z = x + iy$

Substituting $z = x + iy$ in $|z - 3| > 2$: $|x + iy - 3| > 2$

Grouping the real and imaginary parts together: $|(x - 3) + iy| > 2$

Since we do not want any square roots we will square both sides: $|(x - 3) + iy|^2 > 2^2$

Expanding the left hand side, we use the definition $|z|^2 = x^2 + y^2$: $(x - 3)^2 + y^2 > 2^2$

This is a circle with centre $(3, 0)$ and a radius of 2.

$>$ means that we are interested in the region outside the circle.

Argand diagram: $|z - 3| > 2$

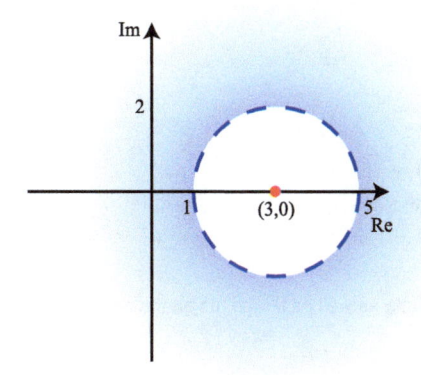

Q103:

Remember that $|z| = \sqrt{x^2 + y^2}$, where $z = x + iy$

Substituting $z = x + iy$ in $|z + 1| = 3$: $|x + iy + 1| = 3$

Grouping the real and imaginary parts together: $|(x + 1) + iy| = 3$

Since we do not want any square roots we will square both sides: $|(x + 1) + iy|^2 = 3^2$

Expanding the left hand side, we use the definition $|z|^2 = x^2 + y^2$: $(x + 1)^2 + y^2 = 3^2$

This is a circle with centre $(-1, 0)$ and a radius of 3.

$=$ means that we are interested in the points on the circle.

Argand diagram: $|z + 1| = 3$

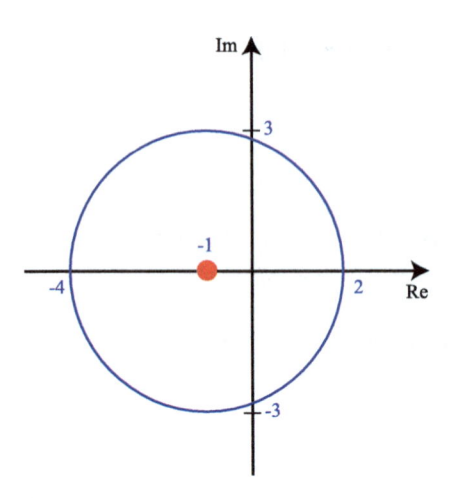

Q104:

Remember that $|z| = \sqrt{x^2 + y^2}$, where $z = x + iy$
Substituting $z = x + iy$ in $|z - 2 + 3i| \leq 4$: $|x + iy - 2 + 3i| \leq 4$
Grouping the real and imaginary parts together: $|(x - 2) + i(y + 3)| \leq 4$
Since we do not want any square roots we will square both sides: $|(x - 2) + i(y + 3)|^2 \leq 4^2$
Expanding the left hand side, we use the definition $|z|^2 = x^2 + y^2$: $(x - 2)^2 + (y + 3)^2 \leq 4^2$
This is a circle with centre $(2, -3)$ and a radius of 4.
$\leq$ means that we are interested in the region inside the circle as well as the points on the circle.
Argand diagram: $|z - 2 + 3i| \leq 4$

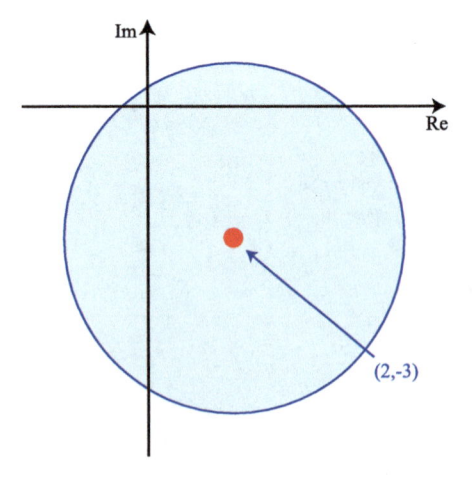

Q105:

Remember that $|z| = \sqrt{x^2 + y^2}$, where $z = x + iy$

Substituting $z = x + iy$ in $|2z - 3i| > 2$: $|2(x + iy) - 3i| > 2$

Grouping the real and imaginary parts together: $|2x + i(2y - 3)| > 2$

Since we do not want any square roots we will square both sides: $|2x + i(2y - 3)|^2 > 2^2$

Expanding the left hand side, we use the definition $|z|^2 = x^2 + y^2$: $(2x)^2 + (2y - 3)^2 > 2^2$

The general equation of a circle is: $(x - a)^2 + (y - b)^2 = r^2$

To find the centre we equate $x - a = 0$ and $y - b = 0$:

$2x = 0 \Rightarrow x = 0$

$2y - 3 = 0 \Rightarrow y = \frac{3}{2}$

This is a circle with centre $(0, \frac{3}{2})$ and a radius of 2.

$>$ means that we are interested in the region outside the circle.

Argand diagram: $|2z - 3i| > 2$

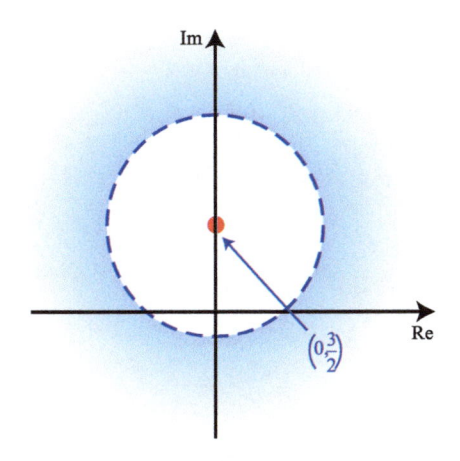

Q106: Remember that $|z| = \sqrt{x^2 + y^2}$, where $z = x + iy$

Substituting $z = x + iy$ in $|z - 3i| > 5$: $|x + iy - 3i| > 5$

Grouping the real and imaginary parts together: $|x + i(y - 3)| > 5$

Since we do not want any square roots we will square both sides: $|x + i(y - 3)|^2 > 5^2$

Expanding the left hand side, we use the definition $|z|^2 = x^2 + y^2$: $x^2 + (y - 3)^2 > 5^2$

  a) Centre: $(0, 3)$
  b) Radius: 5
  c) Shaded region: Outside

Argand diagram: $|z - 3i| > 5$

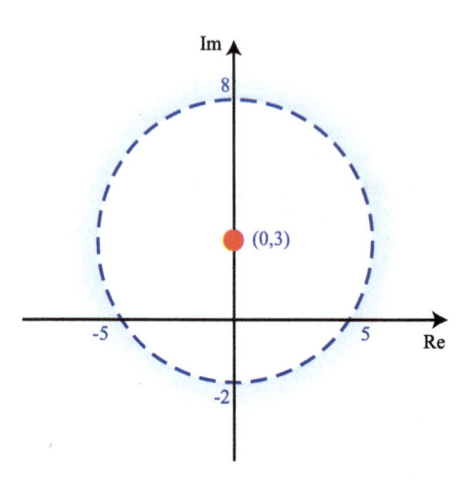

Q107:

Remember that $|z| = \sqrt{x^2 + y^2}$, where $z = x + iy$

Substituting $z = x + iy$ in $|z - 5 + 2i| \leq 4$: $|x + iy - 5 + 2i| \leq 4$

Grouping the real and imaginary parts together: $|(x - 5) + i(y + 2)| \leq 4$

Since we do not want any square roots we will square both sides: $|(x - 5) + i(y + 2)|^2 \leq 4^2$

Expanding the left hand side, we use the definition $|z|^2 = x^2 + y^2$: $(x - 5)^2 + (y + 2)^2 \leq 4^2$

a) Centre: $(5, -2)$
b) Radius: 4
c) Shaded region: On and inside the circle

Argand diagram: $|z - 5 + 2i| \leq 4$

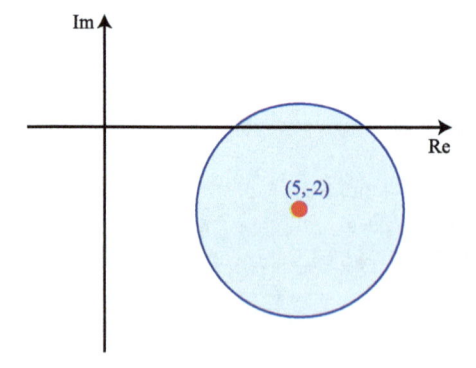

ANSWERS: UNIT 1 TOPIC 6

Q108:

Remember that $|z| = \sqrt{x^2 + y^2}$, where $z = x + iy$
Substituting $z = x + iy$ in $|3z - 2 - 5i| \geq 2$: $|3(x + iy) - 2 - 5i| \geq 2$
Grouping the real and imaginary parts together: $|(3x - 2) + i(3y - 5)| \geq 2$
Since we do not want any square roots we will square both sides: $|(3x - 2) + i(3y - 5)|^2 \geq 2^2$
Expanding the left hand side, we use the definition $|z|^2 = x^2 + y^2$: $(3x - 2)^2 + (3y - 5)^2 \geq 2^2$
To find the centre we equate $3x - 2 = 0$ and $3y - 5 = 0$:
$3x = 2 \Rightarrow x = \frac{2}{3}$
$3y - 5 = 0 \Rightarrow y = \frac{5}{3}$

a) Centre: $\left(\frac{2}{3}, \frac{5}{3}\right)$
b) Radius: 2
c) Shaded region: On and outside the circle

Argand diagram: $|3z - 2 - 5i| \geq 2$

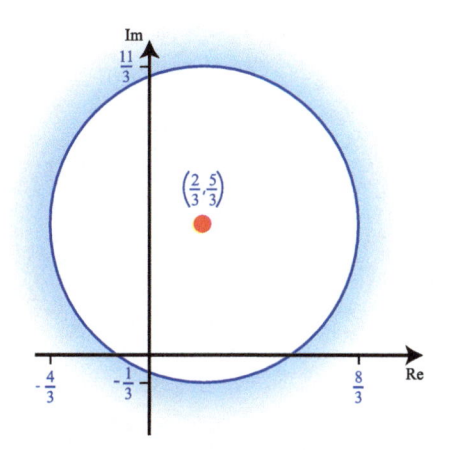

Geometric interpretations: Straight line representation exercise (page 108)

Q109:

Substitute $z = x + iy$ in $|z - 1| = |z - 2|$: $|x + iy - 1| = |x + iy - 2|$
Grouping the real and imaginary parts together: $|(x - 1) + iy| = |(x - 2) + iy|$
Taking the modulus: $(x - 1)^2 + y^2 = (x - 2)^2 + y^2$
This gives:
$x^2 - 2x + 1 + y^2 = x^2 - 4x + 4 + y^2$
$\qquad 2x = 3$
$\qquad x = \frac{3}{2}$

© HERIOT-WATT UNIVERSITY

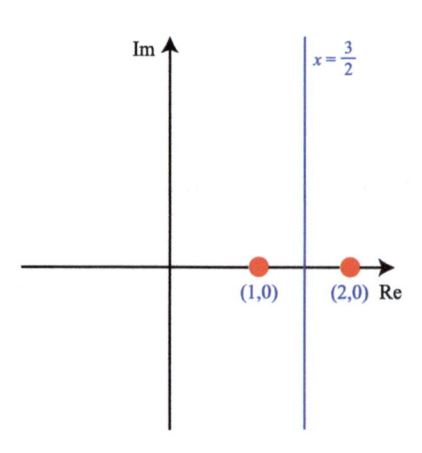

Q110:

Substitute $z = x + iy$ in $|z + i| = |z - 3|$: $|x + iy + i| = |x + iy - 3|$
Grouping the real and imaginary parts together: $|x + i(y + 1)| = |(x - 3) + iy|$
Taking the modulus: $x^2 + (y + 1)^2 = (x - 3)^2 + y^2$
This gives:
$$x^2 + y^2 + 2y + 1 = x^2 - 6x + 9 + y^2$$
$$2y = -6x + 8$$
$$y = -3x + 4$$

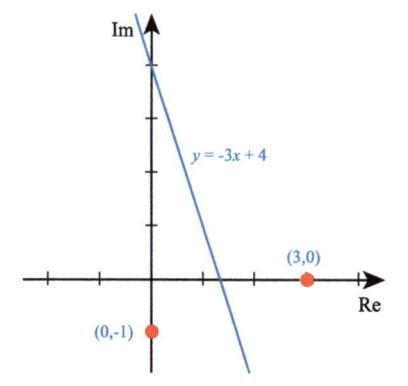

Q111:

Substitute $z = x + iy$ in $|z + 1 - 2i| = |z - 2 + 3i|$: $|x + iy + 1 - 2i| = |x + iy - 2 + 3i|$
Grouping the real and imaginary parts together: $|(x + 1) + i(y - 2)| = |(x - 2) + i(y + 3)|$
Taking the modulus: $(x + 1)^2 + (y - 2)^2 = (x - 2)^2 + (y + 3)^2$
This gives:
$$x^2 + 2x + 1 + y^2 - 4y + 4 = x^2 - 4x + 4 + y^2 + 6y + 9$$
$$2x - 4y + 5 = -4x + 6y + 13$$
$$-10y = -6x + 8$$
$$5y = 3x - 4$$

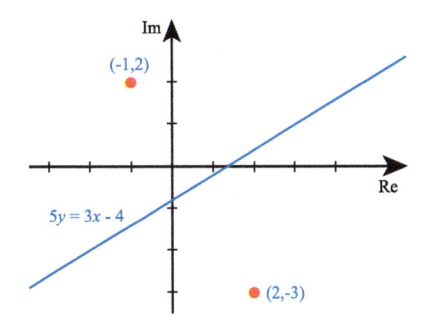

Q112:

Substitute $z = x + iy$ in $|z + 1| = |z - 2|$: $|x + iy + 1| = |x + iy - 2|$
Grouping the real and imaginary parts together: $|(x + 1) + iy| = |(x - 2) + iy|$
Taking the modulus: $(x + 1)^2 + y^2 = (x - 2)^2 + y^2$
This gives:
$$x^2 + 2x + 1 + y^2 = x^2 - 4x + 4 + y^2$$
$$2x + 1 = -4x + 4$$
$$6x = 3$$
$$x = \frac{1}{2}$$

Q113:

Substitute $z = x + iy$ in $|z - 3i| = |z + i|$: $|x + iy - 3i| = |x + iy + i|$
Grouping the real and imaginary parts together: $|x + i(y - 3)| = |x + i(y + 1)|$
Taking the modulus: $x^2 + (y - 3)^2 = x^2 + (y + 1)^2$
This gives:
$$x^2 + y^2 - 6y + 9 = x^2 + y^2 + 2y + 1$$
$$-6y + 9 = 2y + 1$$
$$-8y = -8$$
$$y = 1$$

Q114:

Substitute $z = x + iy$ in $|z - 1 + 2i| = |z + 5 - i|$: $|x + iy - 1 + 2i| = |x + iy + 5 - i|$
Grouping the real and imaginary parts together: $|(x - 1) + i(y + 2)| = |(x + 5) + i(y - 1)|$
Taking the modulus: $(x - 1)^2 + (y + 2)^2 = (x + 5)^2 + (y - 1)^2$
This gives:
$x^2 - 2x + 1 + y^2 + 4y + 4 = x^2 + 10x + 25 + y^2 - 2y + 1$
$\quad -2x + 4y + 5 = 10x - 2y + 26$
$\quad -12x + 6y - 21 = 0$
$\quad -4x + 2y - 7 = 0$

Q115:

Substitute $z = x + iy$ in $|z + 2i| = |z + 3|$: $|x + iy + 2i| = |x + iy + 3|$
Grouping the real and imaginary parts together: $|x + i(y + 2)| = |(x + 3) + iy|$
Taking the modulus: $x^2 + (y + 2)^2 = (x + 3)^2 + y^2$
This gives:
$x^2 + y^2 + 4y + 4 = x^2 + 6x + 9 + y^2$
$\quad 4y + 4 = 6x + 9$
$\quad 4y = 6x + 5$

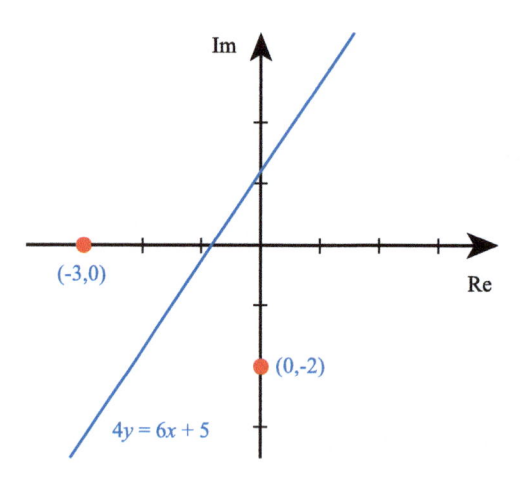

Solving complex equations exercise (page 118)

Q116:

Substitute $z = 3 - 3i$ into $z^4 - 8z^3 + 32z^2 - 48z + 36$ and if that equals zero, it is a root.
Use the binomial theorem to expand the brackets:

ANSWERS: UNIT 1 TOPIC 6

$$(3-3i)^4 = \binom{4}{0}3^4 + \binom{4}{1}3^3(-3i) + \binom{4}{2}3^2(-3i)^2 + \binom{4}{3}3(-3i)^3 + \binom{4}{4}(-3i)^4$$
$$= 1 \times 81 + 4 \times 27 \times (-3i) + 6 \times 9 \times 9i^2 + 4 \times 3 \times -27i^3 + 1 \times 81i^4$$
$$= 81 - 324i - 486 + 324i + 81$$
$$= -324$$

$$(3-3i)^3 = \binom{3}{0}3^3 + \binom{3}{1}3^2(-3i) + \binom{3}{2}3(-3i)^2 + \binom{3}{3}(-3i)^3$$
$$= 1 \times 27 + 3 \times 9 \times (-3i) + 3 \times 3 \times 9i^2 + 1 \times -27i^3$$
$$= 27 - 81i - 81 + 27i$$
$$= -54 - 54i$$

$$(3-3i)^2$$
$$(3-3i)^2 = 9 - 18i + 9i^2$$
$$= -18i$$

So:
$$z^4 - 8z^3 + 32z^2 - 48z + 36 = -324 - 8(-54 - 54i) + 32(-18i) - 48(3-3i) + 36$$
$$= -324 + 432 + 432i - 576i - 144 + 144i + 36$$
$$= 0$$

So $(z - (3 - 3i))$ and $(z - (3 + 3i))$ are factors, giving the additional quadratic factor:
$$(z - (3-3i))(z - (3+3i)) = z^2 - 3z - 3zi - 3z + 9 + 9i + 3iz - 9i - 9i^2$$
$$= z^2 - 6z + 18$$

Long division gives:

```
                    z² - 2z  + 2
z² - 6z + 18 ) z⁴ - 8z³ + 32z² - 48z + 36
               -z⁴ + 6z³ - 18z²
               ─────────────────
                0  - 2z³ + 14z² - 48z + 36
                    2z³ - 12z² + 36z
                    ─────────────────
                    0  + 2z² - 12z + 36
                       - 2z² + 12z - 36
                       ─────────────────
                                      0
```

So $z^4 - 8z^3 + 32z^2 - 48z + 36 = (z^2 - 6z + 18)(z^2 - 2z + 2)$.

Using the quadratic formula to solve $z^2 - 2z + 2$:
$$z = \frac{-b \pm \sqrt{b^2 - 4ac}}{2a}$$
$$= \frac{2 \pm \sqrt{(-2)^2 - 4(1)(2)}}{2(1)}$$
$$= \frac{2 \pm \sqrt{-4}}{2}$$
$$= \frac{2 \pm 2i}{2}$$

© HERIOT-WATT UNIVERSITY

The quadratic equation has solutions $z = 1 \pm i$.
Therefore, the four roots of $z^4 - 8z^3 + 32z^2 - 48z + 36 = 0$ are:

- $z = 3 - 3i$
- $z = 3 + 3i$
- $z = 1 - i$
- $z = 1 + i$

Q117:

Substitute $x = 2$ into $x^3 - 6x^2 + 13x - 10$ and if that equals zero, it is a root.

$$x^3 - 6x^2 + 13x - 10 = 2^3 - 6(2)^2 + 13(2) - 10$$
$$= 8 - 24 + 26 - 10$$
$$= 0$$

So $(x - 2)$ is a factor.

Using long division:

```
           x² - 4x + 5
x - 2  ) x³ - 6x² + 13x - 10
         -x³ + 2x²
         ─────────
          0 - 4x² + 13x - 10
              4x² - 8x
              ────────
              0 + 5x - 10
                  -5x + 10
                  ────────
                  0
```

So $x^3 - 6x^2 + 13x - 10 = (x - 2)(x^2 - 4x + 5)$.

Using the quadratic formula to solve $x^2 - 4x + 5$:

$$x = \frac{-b \pm \sqrt{b^2 - 4ac}}{2a}$$

$$= \frac{4 \pm \sqrt{(-4)^2 - 4(1)(5)}}{2(1)}$$

$$= \frac{4 \pm \sqrt{-4}}{2}$$

$$= \frac{4 \pm 2i}{2}$$

The quadratic equation has solutions $x = 2 \pm i$.
Therefore, the three roots of $x^3 - 6x^2 + 13x - 10 = 0$ are:

- $x = 2$
- $x = 2 + i$
- $x = 2 - i$

Q118:

Substitute $z = -1$ into $z^3 - 5z^2 + 7z + 13$ and if that equals zero, it is a root.

$$z^3 - 5z^2 + 7z + 13 = (-1)^3 - 5(-1)^2 + 7(-1) + 13$$
$$= -1 - 5 - 7 + 13$$
$$= 0$$

So $(z + 1)$ is a factor.

Using long division:

```
              z² - 6z + 13
       ┌─────────────────────
z + 1  │ z³ - 5z²  + 7z  + 13
         -z³ -  z²
         ─────────
          0  - 6z² + 7z  + 13
               6z² + 6z
               ─────────
                0  + 13z + 13
                    -13z - 13
                    ─────────
                         0
```

So $z^3 - 5z^2 + 7z + 13 = (z + 1)(z^2 - 6z + 13)$.

Using the quadratic formula to solve $z^2 - 6z + 13$:

$$z = \frac{-b \pm \sqrt{b^2 - 4ac}}{2a}$$

$$= \frac{6 \pm \sqrt{(-6)^2 - 4(1)(13)}}{2(1)}$$

$$= \frac{6 \pm \sqrt{-16}}{2}$$

The quadratic equation has solutions $z = 3 \pm 2i$.

Therefore, the three roots of $z^3 - 5z^2 + 7z + 13 = 0$ are:

- $z = -1$
- $z = 3 + 2i$
- $z = 3 - 2i$

Q119:

If $x - 2i$ is a factor, then $x - 2i = 0$ and $x = 2i$.

Substitute $2i$ into $x^4 + 2x^3 + x^2 + 8x - 12 = 0$ and if that equals zero, it is a root and thus $x - 2i$ is a factor.

$$x^4 + 2x^3 + x^2 + 8x - 12 = (2i)^4 + 2(2i)^3 + (2i)^2 + 8(2i) - 12$$
$$= 16 - 16i - 4 + 16i - 12$$
$$= 0$$

Since roots come in complex conjugate pairs, we have factors $(x - 2i)$ and $(x + 2i)$, which gives us the additional factor:

$$(x - 2i)(x + 2i) = x^2 + 2xi - 2ix - 4i^2$$
$$= x^2 + 4$$

Using long division:

$$\begin{array}{r}
x^2 + 2x - 3 \\
x^2 + 4 \overline{\smash{\big)}\, x^4 + 2x^3 + x^2 + 8x - 12}\\
\underline{-x^4 - 4x^2 }\\
2x^3 - 3x^2 + 8x - 12\\
\underline{-2x^3 - 8x }\\
-3x^2 - 12\\
\underline{3x^2 + 12}\\
0
\end{array}$$

So $x^4 + 2x^3 + x^2 + 8x - 12 = (x^2 + 4)(x^2 + 2x - 3)$.
Factorising: $(x^2 + 2x - 3) = (x - 1)(x + 3)$
Therefore, the four roots of $x^4 + 2x^3 + x^2 + 8x - 12 = 0$ are:
- $x = 2i$
- $x = -2i$
- $x = 1$
- $x = -3$

Q120:
Substitute $x = 1 - 3i$ into $x^4 - 2x^3 + 26x^2 - 32x + 160 = 0$ and if that equals zero, it is a root.

Use the binomial theorem to expand the brackets:

$$(1 - 3i)^4 = \binom{4}{0} 1^4 + \binom{4}{1} 1^3 (-3i) + \binom{4}{2} 1^2 (-3i)^2 + \binom{4}{3} 1 (-3i)^3 + \binom{4}{4} (-3i)^4$$
$$= 1 \times 1 + 4 \times 1 \times (-3i) + 6 \times 1 \times 9i^2 + 4 \times 1 \times -27i^3 + 1 \times 81i^4$$
$$= 1 - 12i - 54 + 108i + 81$$
$$= 28 + 96i$$

$$(1 - 3i)^3 = \binom{3}{0} 1^3 + \binom{3}{1} 1^2 (-3i) + \binom{3}{2} 1 (-3i)^2 + \binom{3}{3} (-3i)^3$$
$$= 1 \times 1 + 3 \times 1 \times (-3i) + 3 \times 1 \times 9i^2 + 1 \times -27i^3$$
$$= 1 - 9i - 27 + 27i$$
$$= -26 + 18i$$

$$(1 - 3i)^2 = 1 - 6i - 9$$
$$= -8 - 6i$$

So:
$$x^4 - 2x^3 + 26x^2 - 32z + 160 = 28 + 96i - 2(-26 + 18i) + 26(-8 - 6i) - 32(1 - 3i) + 160$$
$$= 28 + 96i + 52 - 36i - 208 - 156i - 32 + 96i + 160$$
$$= 0$$

ANSWERS: UNIT 1 TOPIC 6

Since roots come in complex conjugate pairs, we have factors: $(x - (1 - 3i))$ and $(x - (1 + 3i))$ which gives us the additional factor:

$$(x - (1 - 3i))(x - (1 + 3i)) = x^2 - x - 3ix - x + 3ix + 1 - 9i^2$$
$$= x^2 - 2x + 10$$

Long division gives:

```
                  x²       + 16
x² - 2x + 10 | x⁴ - 2x³ + 26x² - 32x + 160
               -x⁴ + 2x³ - 10x²
               ─────────────────────────
                0 + 0 + 16x² - 32x + 160
                        -16x² + 32x - 160
                        ─────────────────
                                       0
```

So $x^4 - 2x^3 + 26x^2 - 32x + 160 = (x^2 - 2x + 10)(x^2 + 16)$.

Solving $x^2 + 16$ we have the roots $x = 4i$ and $x = -4i$.

Therefore, the four roots of $x^4 - 2x^3 + 26x^2 - 32x + 160 = 0$ are:

- $x = 1 - 3i$
- $x = 1 + 3i$
- $x = 4i$
- $x = -4i$

Q121:

Substitute $z = 4 + 3i$ into $z^4 - 12z^3 + 70z^2 - 204z + 325 = 0$ and if that equals zero, it is a root.

Use the binomial theorem to expand the brackets:

$$(4+3i)^4 = \binom{4}{0}4^4 + \binom{4}{1}4^3(3i) + \binom{4}{2}4^2(3i)^2 + \binom{4}{3}4(3i)^3 + \binom{4}{4}(3i)^4$$
$$= 1 \times 256 + 4 \times 64 \times 3i + 6 \times 16 \times 9i^2 + 4 \times 4 \times 27i^3 + 1 \times 81i^4$$
$$= 256 + 768i - 864 - 432i + 81$$
$$= -527 + 336i$$

$$(4+3i)^3 = \binom{3}{0}4^3 + \binom{3}{1}4^2(3i) + \binom{3}{2}4(3i)^2 + \binom{3}{3}(3i)^3$$
$$= 1 \times 64 + 3 \times 16 \times 3i + 3 \times 4 \times 9i^2 + 1 \times 27i^3$$
$$= 64 + 144i - 108 - 27i$$
$$= -44 + 117i$$

$$(4+3i)^2 = 16 + 24i - 9$$
$$= 7 + 24i$$

So:
$$z^4 - 12z^3 + 70z^2 - 204z + 325 = -527 + 336i - 12(-44 + 117i) + 70(7 + 24i) - 204(4 + 3i) + 325$$
$$= -527 + 336i + 528 - 1404i + 490 + 1680i - 816 - 612i + 325$$
$$= 0$$

© HERIOT-WATT UNIVERSITY

Since roots come in complex conjugate pairs, we have factors: $(z - (4 + 3i))$ and $(z - (4 - 3i))$ which gives us the additional factor:

$$(z - (4 + 3i))(z - (4 - 3i)) = z^2 - 4z + 3iz - 4z - 3iz + 16 + 9$$
$$= z^2 - 8z + 25$$

Long division gives:

```
                    z² -  4z  + 13
           ┌─────────────────────────────
z² - 8z + 25 │ z⁴ - 12z³ + 70z² - 204z  + 325
             │ -z⁴ +  8z³ - 25z²
             │ ─────────────────────
             │  0  - 4z³ + 45z² - 204z  + 325
             │       4z³ - 32z² +100z
             │       ──────────────────
             │        0 + 13z² - 104z + 325
             │           -13z² + 104z - 325
             │           ──────────────────
             │                          0
```

So $z^4 - 12z^3 + 70z^2 - 204z + 325 = (z^2 - 8z + 25)(z^2 - 4z + 13)$.
Using the quadratic formula to solve $z^2 - 4z + 13$:

$$z = \frac{-b \pm \sqrt{b^2 - 4ac}}{2a}$$

$$= \frac{4 \pm \sqrt{(-4)^2 - 4(1)(13)}}{2(1)}$$

$$= \frac{4 \pm \sqrt{-36}}{2}$$

The quadratic equation has solutions $z = 2 \pm 3i$.
Therefore, the four roots of $z^4 - 12z^3 + 70z^2 - 204z + 325 = 0$ are:

- $z = 4 + 3i$
- $z = 4 - 3i$
- $z = 2 + 3i$
- $z = 2 - 3i$

Q122:

Substitute $z = 2 + 5i$ into $z^4 - 10z^3 + 63z^2 - 214z + 290 = 0$ and if that equals zero, it is a root.
Use the binomial theorem to expand the brackets:

$$(2+5i)^4 = \binom{4}{0} 2^4 + \binom{4}{1} 2^3 (5i) + \binom{4}{2} 2^2 (5i)^2 + \binom{4}{3} 2(5i)^3 + \binom{4}{4} (5i)^4$$

$$= 1 \times 16 + 4 \times 8 \times 5i + 6 \times 4 \times 25i^2 + 4 \times 2 \times 125i^3 + 1 \times 625i^4$$

$$= 16 + 160i - 600 - 1000i + 625$$

$$= 41 - 840i$$

ANSWERS: UNIT 1 TOPIC 6

$$(2+5i)^3 = \binom{3}{0} 2^3 + \binom{3}{1} 2^2 (5i) + \binom{3}{2} 2(5i)^2 + \binom{3}{3}(5i)^3$$
$$= 1 \times 8 + 3 \times 4 \times 5i + 3 \times 2 \times 25i^2 + 1 \times 125i^3$$
$$= 8 + 60i - 150 - 125i$$
$$= -142 - 65i$$

$$(2+5i)^2 = 4 + 20i - 25$$
$$= -21 + 20i$$

So:
$$z^4 - 10z^3 + 63z^2 - 214z + 290 = 41 - 840i - 10(-142 - 65i) + 63(-21 + 20i) - 214(2 + 5i) + 290$$
$$= 41 - 840i + 1420 + 650i - 1323 + 1260i - 428 - 1070i + 290$$
$$= 0$$

Since roots come in complex conjugate pairs, we have factors: $(z - (2 - 5i))$ and $(z - (2 + 5i))$ which gives us the additional factor:

$$(z - (2 - 5i))(z - (2 + 5i)) = z^2 - 2z - 5zi - 2z + 5iz + 4 + 25$$
$$= z^2 - 4z + 29$$

Long division gives:

```
                    z² -  6z + 10
         ┌─────────────────────────────────
z² - 4z + 29 │ z⁴ - 10z³ + 63z² - 214z + 290
               -z⁴ +  4z³ -  29z²
               ─────────────────────────────
                  0 -  6z³ + 34z² - 214z + 290
                       6z³ - 24z² + 174z
                       ──────────────────────
                          0 + 10z² -  40z + 290
                             -10z² +  40z - 290
                             ──────────────────
                                              0
```

So $z^4 - 10z^3 + 63z^2 - 214z + 290 = (z^2 - 4z + 29)(z^2 - 6z + 10)$.

Using the quadratic formula to solve $z^2 - 6z + 10$:

$$z = \frac{-b \pm \sqrt{b^2 - 4ac}}{2a}$$
$$= \frac{6 \pm \sqrt{(-6)^2 - 4(1)(10)}}{2(1)}$$
$$= \frac{6 \pm \sqrt{-4}}{2}$$

The quadratic equation has solutions $z = 3 \pm i$.

Therefore, the four roots of $z^4 - 10z^3 + 63z^2 - 214z + 290 = 0$ are:

- $z = 2 - 5i$
- $z = 2 + 5i$
- $z = 3 - i$
- $z = 3 + i$

© HERIOT-WATT UNIVERSITY

Multiplication using polar form exercise (page 120)

Q123:

Modulus:

$|zw| = r_1 r_2$
$= 2 \times 5$
$= 10$

Argument:

$arg(zw) = \theta + \varphi$
$= \dfrac{\pi}{3} + \dfrac{\pi}{2}$
$= \dfrac{5\pi}{6}$

So:

$zw = 10 \left(\cos \left(\dfrac{5\pi}{6} \right) + i \sin \left(\dfrac{5\pi}{6} \right) \right)$
$= -5\sqrt{3} + 5i$

Q124:

We need to rewrite z in the form $\cos \theta + i \sin \theta$.
Remember that: $\cos \theta - i \sin \theta = \cos(-\theta) + i \sin(-\theta)$ so:
$z = \cos \left(-\dfrac{\pi}{3} \right) + i \sin \left(-\dfrac{\pi}{3} \right)$

Modulus:

$|zw| = r_1 r_2$
$= 1 \times 4$
$= 4$

Argument:

$arg(zw) = \theta + \varphi$
$= -\dfrac{\pi}{3} + \dfrac{2\pi}{3}$
$= \dfrac{\pi}{3}$

So:

$zw = 4 \left(\cos \left(\dfrac{\pi}{3} \right) + i \sin \left(\dfrac{\pi}{3} \right) \right)$
$zw = 2 + i2\sqrt{3}$

Q125:

We need to rewrite z and w in the form $\cos \theta + i \sin \theta$.
Remember that $\cos \theta - i \sin \theta = \cos(-\theta) + i \sin(-\theta)$ so:
$z = 8 \left(\cos \left(-\dfrac{\pi}{2} \right) + i \sin \left(-\dfrac{\pi}{2} \right) \right)$ and $w = 6 \left(\cos \left(-\dfrac{4\pi}{3} \right) + i \sin \left(-\dfrac{4\pi}{3} \right) \right)$

Modulus:

$|zw| = r_1 r_2$
$= 8 \times 6$
$= 48$

Argument:

$arg(zw) = \theta + \varphi$
$= -\dfrac{\pi}{2} - \dfrac{4\pi}{3}$
$= -\dfrac{11\pi}{6}$
$= \dfrac{\pi}{6}$

So:

$zw = 48 \left(\cos \left(\dfrac{\pi}{6} \right) + i \sin \left(\dfrac{\pi}{6} \right) \right)$
$= 24\sqrt{3} + 24i$

Q126:

Modulus:

$|z^3| = |z^2 z|$

So:

$|z^3| = r^3$
$= 3^3$
$= 27$

Argument:

$arg(z^3) = \theta + \theta + \theta$
$= 3\theta$
$= \dfrac{3\pi}{4}$

So:

$zw = 27\left(\cos\left(\dfrac{3\pi}{4}\right) + i\sin\left(\dfrac{3\pi}{4}\right)\right)$
$= -\dfrac{27\sqrt{2}}{2} + i\dfrac{27\sqrt{2}}{2}$

Division using polar form exercise (page 122)

Q127:

Modulus:

$\left|\dfrac{z}{w}\right| = \dfrac{r_1}{r_2}$
$= \dfrac{3}{6}$
$= \dfrac{1}{2}$

Argument:

$arg\left(\dfrac{z}{w}\right) = \theta - \varphi$
$= 75° - 15°$
$= 60°$

So:

$\dfrac{z}{w} = \dfrac{1}{2}(\cos(60°) + i\sin(60°))$
$= \dfrac{1}{4} + i\dfrac{\sqrt{3}}{4}$

Q128:

Modulus:

$\left|\dfrac{z}{w}\right| = \dfrac{r_1}{r_2}$
$= \dfrac{10}{2}$
$= 5$

Argument:

$arg\left(\dfrac{z}{w}\right) = \theta - \varphi$
$= \dfrac{\pi}{2} - \dfrac{\pi}{3}$
$= \dfrac{\pi}{6}$

So:

$\dfrac{z}{w} = 5\left(\cos\left(\dfrac{\pi}{6}\right) + i\sin\left(\dfrac{\pi}{6}\right)\right)$
$= \dfrac{5\sqrt{3}}{2} + \dfrac{5}{2}i$

© HERIOT-WATT UNIVERSITY

Q129:
We need to rewrite z and w in the form $\cos \theta + i \sin \theta$.
Remember that $\cos \theta - i \sin \theta = \cos(-\theta) + i \sin(-\theta)$ so:
$z = 7 \left(\cos \left(-\frac{\pi}{7} \right) + i \sin \left(-\frac{\pi}{7} \right) \right)$ and $w = 2 \left(\cos \left(-\frac{2\pi}{7} \right) + i \sin \left(-\frac{2\pi}{7} \right) \right)$

Modulus:
$$\left| \frac{z}{w} \right| = \frac{r_1}{r_2}$$
$$= \frac{7}{2}$$

Argument:
$$arg \left(\frac{z}{w} \right) = \theta - \varphi$$
$$= \left(-\frac{\pi}{7} \right) - \left(-\frac{2\pi}{7} \right)$$
$$= \frac{\pi}{7}$$

So:
$$\frac{z}{w} = \frac{7}{2} \left(\cos \left(\frac{\pi}{7} \right) + i \sin \left(\frac{\pi}{7} \right) \right)$$
$$= 3 \cdot 2 + 1 \cdot 5i \quad \text{(1 d.p.)}$$

De Moivre's theorem exercise 1 (page 128)

Q130:
If $z = r(cos\ \theta + i \sin \theta)$, then $z^n = r^n(cos(n\theta) + i\ \sin(n\theta))$ for all $n \in \mathbb{N}$.
Thus, using de Moivre's theorem with $r = 2$ and $\theta = \frac{3\pi}{4}$:

$$\left(2 \left(\cos \left(\frac{3\pi}{4} \right) + i \sin \left(\frac{3\pi}{4} \right) \right) \right)^4 = 2^4 \left(\cos \left(4 \times \frac{3\pi}{4} \right) + i \sin \left(4 \times \frac{3\pi}{4} \right) \right)$$
$$= 16 \left(\cos (3\pi) + i \sin (3\pi) \right)$$
$$= 16 \left(\cos \pi + i \sin \pi \right)$$
$$= -16$$

(The principal argument of 3π is π.)

Q131:
If $z = r(cos\ \theta + i \sin \theta)$, then $z^n = r^n(cos(n\theta) + i\ \sin(n\theta))$ for all $n \in \mathbb{N}$.
Thus, using de Moivre's theorem with $r = 4$ and $\theta = \frac{\pi}{3}$:

$$\left(4 \left(\cos \frac{\pi}{3} + i \sin \frac{\pi}{3} \right) \right)^3 = 4^3 \left(\cos \left(3 \times \frac{\pi}{3} \right) + i \sin \left(3 \times \frac{\pi}{3} \right) \right)$$
$$= 64 \left(\cos \pi + i \sin \pi \right)$$
$$= -64$$

Q132:
If $z = r(cos\ \theta + i \sin \theta)$, then $z^n = r^n(cos(n\theta) + i\ \sin(n\theta))$ for all $n \in \mathbb{N}$.

ANSWERS: UNIT 1 TOPIC 6

Thus, using de Moivre's theorem with $r = 2$ and $\theta = \frac{3\pi}{4}$:

$$\left(2\left(\cos\left(\frac{3\pi}{4}\right) + i\sin\left(\frac{3\pi}{4}\right)\right)\right)^{10} = 2^{10}\left(\cos\left(10 \times \frac{3\pi}{4}\right) + i\sin\left(10 \times \frac{3\pi}{4}\right)\right)$$

$$= 1024\left(\cos\left(\frac{30\pi}{4}\right)\pi + i\sin\left(\frac{30\pi}{4}\right)\right)$$

$$= 1024\left(\cos\left(-\frac{\pi}{2}\right)\pi + i\sin\left(-\frac{\pi}{2}\right)\right)$$

$$= -1024$$

(The principal argument of $\frac{30\pi}{4}$ is $-\frac{\pi}{2}$.)

Q133:

If $z = r(cos\,\theta + i\,sin\,\theta)$, then $z^n = r^n(cos(n\theta) + i\,sin(n\theta))$ for all $n \in \mathbb{N}$.
Thus, using de Moivre's theorem with $r = 3$ and $\theta = 30°$:

$$(3\,(\cos(30°) + i\sin(30°)))^8 = 3^8\,(\cos(8 \times 30°) + i\sin(8 \times 30°))$$

$$= 6561\,(\cos(240°) + i\sin(240°))$$

$$= 6561\,(\cos(-120°) + i\sin(-120°))$$

$$= -3280 \cdot 5 - 5682 \cdot 0i$$

(The principal argument of $240°$ is $-120°$.)

Q134:

If $z = r(cos\,\theta + i\,sin\,\theta)$, then $z^n = r^n(cos(n\theta) + i\,sin(n\theta))$ for all $n \in \mathbb{N}$.
Thus, using de Moivre's theorem with $r = 6$ and $\theta = 42°$:

$$(6\,(\cos(42°) + i\sin(42°)))^4 = 6^4\,(\cos(4 \times 42°) + i\sin(4 \times 42°))$$

$$= 1296\,(\cos(168°) + i\sin(168°))$$

$$= -1267 \cdot 7 + 269 \cdot 5i$$

Q135:

First express $-1 - i$ in polar form.

Modulus:

$$|z| = \sqrt{(-1)^2 + (-1)^2}$$

$$= \sqrt{2}$$

Argument:

As z is in the third quadrant, $\arg(z) = -\pi + \alpha$ where $\tan\alpha = \left|\frac{-1}{-1}\right|$, i.e. $\alpha = \frac{\pi}{4}$.
So $\arg(z) = -\frac{3\pi}{4}$ and, therefore, $z = \sqrt{2}\left(\cos\left(-\frac{3\pi}{4}\right) + i\sin\left(-\frac{3\pi}{4}\right)\right)$.

Using de Moivre's theorem:

$$z^5 = \left(\sqrt{2}\right)^5\left(\cos\left(5 \times -\frac{3\pi}{4}\right) + i\sin\left(5 \times -\frac{3\pi}{4}\right)\right)$$

$$= 4\sqrt{2}\left(\cos\left(-\frac{15\pi}{4}\right) + i\sin\left(-\frac{15\pi}{4}\right)\right)$$

$$= 4\sqrt{2}\left(\cos\left(\frac{\pi}{4}\right) + i\sin\left(\frac{\pi}{4}\right)\right)$$

$$= 4 + 4i$$

(The principal argument of $\frac{-15\pi}{4}$ is $\frac{\pi}{4}$.)

© HERIOT-WATT UNIVERSITY

Q136:
First express $-4 + 5i$ in polar form.
Modulus:
$$|z| = \sqrt{(-4)^2 + (5)^2}$$
$$= \sqrt{41}$$
Argument:
As z is in the second quadrant, $\arg(z) = \pi - \alpha$ where $\tan \alpha = \left|\frac{5}{-4}\right|$, i.e. $\alpha = 0 \cdot 896$ rad (3 d.p.)
So $\arg(z) = 2 \cdot 246$ (3 d.p.) and, therefore, $z = \sqrt{41}\left(\cos\left(2 \cdot 246\right) + i\sin\left(2 \cdot 246\right)\right)$
Using de Moivre's theorem:
$$z^7 = \left(\sqrt{41}\right)^7 \left(\cos\left(7 \times 2 \cdot 246\right) + i\sin\left(7 \times 2 \cdot 246\right)\right)$$
$$= 68921\sqrt{41}\left(\cos\left(15 \cdot 722\right) + i\sin\left(15 \cdot 722\right)\right)$$
$$= 68921\sqrt{41}\left(\cos\left(-3 \cdot 128\right) + i\sin\left(-3 \cdot 128\right)\right)$$
$$= -441266 \cdot 3 - 6194 \cdot 3i$$
(The principal argument of $15 \cdot 722$ radians is $-3 \cdot 128$ radians.)

Q137:
First express $\sqrt{3} - 2i$ in polar form.
Modulus:
$$|z| = \sqrt{\left(\sqrt{3}\right)^2 + (-2)^2}$$
$$= \sqrt{7}$$
Argument:
As z is in the fourth quadrant, $\arg(z) = -\alpha$ where $\tan \alpha = \left|\frac{-2}{\sqrt{3}}\right|$, i.e. $\alpha = 49 \cdot 107°$ (3 d.p.)
So $\arg(z) = -49 \cdot 107$ and, therefore, $z = \sqrt{7}\left(\cos\left(-49 \cdot 107°\right) + i\sin\left(-49 \cdot 107°\right)\right)$
Using de Moivre's theorem:
$$z^3 = \sqrt{7}^3 \left(\cos\left(3 \times -49 \cdot 107°\right) + i\sin\left(3 \times -49 \cdot 107°\right)\right)$$
$$= 7\sqrt{7}\left(\cos\left(-147 \cdot 321°\right) + i\sin\left(-147 \cdot 321°\right)\right)$$
$$= 15 \cdot 6 - 10 \cdot 0i$$

Q138:
First express $\sqrt{3} + i$ in polar form:
Modulus:
$$|z| = \sqrt{\left(\sqrt{3}\right)^2 + (1)^2}$$
$$= 2$$
Argument:
As z is in the first quadrant, $\arg(z) = \alpha$ where $\tan \alpha = \left|\frac{1}{\sqrt{3}}\right|$, i.e. $\alpha = 30°$
So $\arg(z) = 30°$ and, therefore, $z = 2(cos(30°) + i\sin(30°))$.

ANSWERS: UNIT 1 TOPIC 6

Using de Moivre's theorem:
$$z^4 = 2^4\left(\cos\left(4\times 30°\right) + i\sin\left(4\times 30°\right)\right)$$
$$= 16\left(\cos\left(120°\right) + i\sin\left(120°\right)\right)$$
$$= -8 + i8\sqrt{3}$$

Q139:
Change $3 - 3i$ into polar form.

Modulus:
$$|z| = \sqrt{3^2 + (-3)^2}$$
$$= \sqrt{18}$$
$$= 3\sqrt{2}$$

Argument:
As z is in the fourth quadrant, $\arg(z) = -\alpha$ where $\tan\alpha = \left|\frac{-3}{3}\right|$, i.e. $\alpha = 45°$

So $\arg(z) = -45°$ and, therefore, $z = 3\sqrt{2}\left(\cos\left(-45°\right) + i\sin\left(-45°\right)\right)$

Using de Moivre's theorem:
$$z^4 = \left(3\sqrt{2}\right)^4\left(\cos\left(4\times -45°\right) + i\sin\left(4\times -45°\right)\right)$$
$$= 324\left(\cos\left(-180°\right) + i\sin\left(-180°\right)\right)$$
$$= 324\left(\cos\left(180°\right) + i\sin\left(180°\right)\right)$$

Change $\sqrt{3} + i$ into polar form.

Modulus:
$$|z| = \sqrt{\left(\sqrt{3}\right)^2 + (1)^2}$$
$$= 2$$

Argument:
As z is in the first quadrant, $\arg(z) = \alpha$ where $\tan\alpha = \left|\frac{1}{\sqrt{3}}\right|$, i.e. $\alpha = 30°$

So $\arg(z) = 30°$ and, therefore, $z = 2(\cos(30°) + i\sin(30°))$.

Using de Moivre's theorem:
$$z^3 = 2^3\left(\cos\left(3\times 30°\right) + i\sin\left(3\times 30°\right)\right)$$
$$= 8\left(\cos\left(90°\right) + i\sin\left(90°\right)\right)$$

Now do the division:
$$\frac{324\left(\cos\left(180°\right) + i\sin\left(180°\right)\right)}{8\left(\cos\left(90°\right) + i\sin\left(90°\right)\right)} = \frac{324}{8}\left(\cos\left(180° - 90°\right) + i\sin\left(180° - 90°\right)\right)$$
$$= \frac{324}{8}\left(\cos 90° + i\sin 90°\right)$$
$$= \frac{81}{2}i$$

Q140:

Change $\sqrt{3} + i$ into polar form.

Modulus:
$$|z| = \sqrt{\left(\sqrt{3}\right)^2 + (1)^2}$$
$$= 2$$

Argument:

As z is in the first quadrant, $\arg(z) = \alpha$ where $\tan\alpha = \left|\frac{1}{\sqrt{3}}\right|$, i.e. $\alpha = 30°$.

So $\arg(z) = 30°$ and, therefore, $z = 2(\cos 30° + i\sin 30°)$.

Using De Moivre's theorem:
$$z^4 = 2^4 \left(\cos\left(4 \times 30°\right) + i\sin\left(4 \times 30°\right)\right)$$
$$= 16\left(\cos\left(120°\right) + i\sin\left(120°\right)\right)$$

Change $1 - i$ into polar form.

Modulus:
$$|z| = \sqrt{(1)^2 + (-1)^2}$$
$$= \sqrt{2}$$

Argument:

As z is in the fourth quadrant, $\arg(z) = -\alpha$, where $\tan\alpha = \left|\frac{-1}{1}\right|$, i.e. $\alpha = 45°$.

So $\arg(z) = -45°$ and, therefore, $z = \sqrt{2}\left(\cos\left(-45°\right) + i\sin\left(-45°\right)\right)$.

Using De Moivre's theorem:
$$z^3 = \left(\sqrt{2}\right)^3 \left(\cos\left(3 \times -45°\right) + i\sin\left(3 \times -45°\right)\right)$$
$$= 2\sqrt{2}\left(\cos\left(-135°\right) + i\sin\left(-135°\right)\right)$$

Now do the division:
$$\frac{16\left(\cos\left(120°\right) + i\sin\left(120°\right)\right)}{2\sqrt{2}\left(\cos\left(-135°\right) + i\sin\left(-135°\right)\right)} = \frac{16}{2\sqrt{2}}\left(\cos\left(120° - \left(-135°\right)\right) + i\sin\left(120° - \left(-135°\right)\right)\right)$$
$$= \frac{8}{\sqrt{2}}\left(\cos 255° + i\sin 255°\right)$$
$$= 4\sqrt{2}\left(\cos\left(-105°\right) + i\sin\left(-105°\right)\right)$$
$$= -1 \cdot 5 - 5.5i$$

Q141:

Change $-\sqrt{2} + i\sqrt{2}$ into polar form.

Modulus:
$$|z| = \sqrt{\left(-\sqrt{2}\right)^2 + \left(\sqrt{2}\right)^2}$$
$$= 2$$

Argument:

As z is in the second quadrant, $\arg(z) = \pi - \alpha$ where $\tan\alpha = \left|\frac{\sqrt{2}}{-\sqrt{2}}\right|$, i.e. $\alpha = \frac{\pi}{4}$.

So $\arg(z) = \frac{3\pi}{4}$ and, therefore, $z = 2\left(\cos\left(\frac{3\pi}{4}\right) + i\sin\left(\frac{3\pi}{4}\right)\right)$.

ANSWERS: UNIT 1 TOPIC 6

Using de Moivre's theorem:
$$z^5 = 2^5 \left(\cos \left(5 \times \frac{3\pi}{4} \right) + i \sin \left(5 \times \frac{3\pi}{4} \right) \right)$$
$$= 32 \left(\cos \left(\frac{15\pi}{4} \right) + i \sin \left(\frac{15\pi}{4} \right) \right)$$

Change $-\sqrt{3} - i$ into polar form.
Modulus:
$$|z| = \sqrt{\left(-\sqrt{3}\right)^2 + (-1)^2}$$
$$= 2$$

Argument:
As z is in the third quadrant, $\arg(z) = -\pi + \alpha$ where $\tan \alpha = \left| \frac{-1}{-\sqrt{3}} \right|$, i.e. $\alpha = \frac{\pi}{6}$.
So $\arg(z) = -\frac{5\pi}{6}$ and, therefore, $z = 2 \left(\cos \left(-\frac{5\pi}{6} \right) + i \sin \left(-\frac{5\pi}{6} \right) \right)$.

Using de Moivre's theorem:
$$z^4 = 2^4 \left(\cos \left(4 \times -\frac{5\pi}{6} \right) + i \sin \left(4 \times -\frac{5\pi}{6} \right) \right)$$
$$= 16 \left(\cos \left(-\frac{20\pi}{6} \right) + i \sin \left(-\frac{20\pi}{6} \right) \right)$$

Now do the division:
$$\frac{\left(-\sqrt{2} + i\sqrt{2}\right)^5}{\left(-\sqrt{3} - i\right)^4} = \frac{32 \left(\cos \left(\frac{15\pi}{4} \right) + i \sin \left(\frac{15\pi}{4} \right) \right)}{16 \left(\cos \left(-\frac{20\pi}{6} \right) + i \sin \left(-\frac{20\pi}{6} \right) \right)}$$
$$= 2 \frac{\left(\cos \left(\frac{15\pi}{4} \right) + i \sin \left(\frac{15\pi}{4} \right) \right)}{\left(\cos \left(-\frac{10\pi}{3} \right) + i \sin \left(-\frac{10}{3} \right) \right)}$$
$$= 2 \left(\cos \left(\frac{15\pi}{4} - \left(-\frac{10\pi}{3} \right) \right) + i \sin \left(\frac{15\pi}{4} - \left(-\frac{10\pi}{3} \right) \right) \right)$$
$$= 2 \left(\cos \left(\frac{85\pi}{12} \right) + i \sin \left(\frac{85\pi}{12} \right) \right)$$
$$= 2 \left(\cos \left(-\frac{11\pi}{12} \right) + i \sin \left(-\frac{11\pi}{12} \right) \right)$$
$$= -1 \cdot 9 - 0 \cdot 5i$$

De Moivre's theorem exercise 2 (page 130)

Q142:
Using de Moivre's theorem, $z^n = r^n(cos(n\theta) + i \sin(n\theta))$ for all $n \in \mathbb{Z}$.
We have:
$$(4 (\cos 125° + i \sin 125°))^{-4} = 4^{-4} (\cos (-4 \times 125°) + i \sin (-4 \times 125°))$$
$$= \frac{1}{256} (\cos (-500°) + i \sin (-500°))$$
$$= \frac{1}{256} (\cos (-140°) + i \sin (-140°)) \quad \text{principal argument}$$
$$= -0 \cdot 003 - 0 \cdot 003i$$

© HERIOT-WATT UNIVERSITY

Q143:

$$\left(5\left(\cos\left(\frac{7\pi}{4}\right)+i\sin\left(\frac{7\pi}{4}\right)\right)\right)^{-3} = 5^{-3}\left(\cos\left(-3\times\frac{7\pi}{4}\right)+i\sin\left(-3\times\frac{7\pi}{4}\right)\right)$$

$$= \frac{1}{125}\left(\cos\left(-\frac{21\pi}{4}\right)+i\sin\left(-\frac{21\pi}{4}\right)\right)$$

$$= \frac{1}{125}\left(\cos\left(\frac{3\pi}{4}\right)+i\sin\left(\frac{3\pi}{4}\right)\right) \quad \text{principal argument}$$

$$= -\frac{\sqrt{2}}{250}+i\frac{\sqrt{2}}{250}$$

Q144:
First convert $\sqrt{3}-i$ into polar form.
Modulus:
$$|z| = \sqrt{\left(\sqrt{3}\right)^2+(-1)^2}$$
$$= 2$$
Argument:
As z is in the fourth quadrant, $\arg(z) = -\alpha$ where $\tan\alpha = \left|\frac{-1}{\sqrt{3}}\right|$, i.e. $\alpha = \frac{\pi}{6}$.
So $\arg(z) = -\frac{\pi}{6}$ and, therefore, $z = 2\left(\cos\left(-\frac{\pi}{6}\right)+i\sin\left(-\frac{\pi}{6}\right)\right)$.
Using de Moivre's theorem:

$$\left(2\left(\cos\left(-\frac{\pi}{6}\right)+i\sin\left(-\frac{\pi}{6}\right)\right)\right)^{-6} = 2^{-6}\left(\cos\left(-6\times-\frac{\pi}{6}\right)+i\sin\left(-6\times-\frac{\pi}{6}\right)\right)$$

$$= \frac{1}{64}\left(\cos\left(\pi\right)+i\sin\left(\pi\right)\right)$$

$$= \frac{1}{64}\left(\cos\left(\pi\right)+i\sin\left(\pi\right)\right) \quad \text{principal argument}$$

$$= -\frac{1}{64}$$

Q145:
First convert $1+i\sqrt{2}$ into polar form.
Modulus:
$$|z| = \sqrt{\left(\sqrt{2}\right)^2+(1)^2}$$
$$= \sqrt{3}$$
Argument:
As z is in the first quadrant, $\arg(z) = \alpha$, where $\tan\alpha = \left|\frac{\sqrt{2}}{1}\right|$, i.e. $\alpha = 0\cdot 955$ rad (3 d.p.).
So $\arg(z) = 0\cdot 955$ and, therefore, $z = \sqrt{3}\left(\cos\left(6\cdot 3\right)+i\sin\left(6\cdot 3\right)\right)$.

ANSWERS: UNIT 1 TOPIC 6

Using de Moivre's Theorem:

$$\left(\sqrt{3}\left(\cos\left(0\cdot 955\right) + i\sin\left(0\cdot 955\right)\right)\right)^{-7} = \sqrt{3}^{-7}\left(\cos\left(-7\times 0\cdot 955\right) + i\sin\left(-7\times 0\cdot 955\right)\right)$$

$$= \frac{\sqrt{3}}{81}\left(\cos\left(-6\cdot 685\right) + i\sin\left(-6\cdot 685\right)\right)$$

$$= \frac{\sqrt{3}}{81}\left(\cos\left(-0\cdot 402\right) + i\sin\left(-0\cdot 402\right)\right) \quad \text{principal argument}$$

$$= \frac{\sqrt{3}}{81}\left(\cos\left(0\cdot 402\right) - i\sin\left(0\cdot 402\right)\right)$$

$$= 0\cdot 020 - 0\cdot 008i$$

De Moivre's theorem and multiple angle formulae exercise (page 132)

Q146:

Using de Moivre's theorem: $\cos(3\theta) + i\sin(3\theta) = (cos\ \theta + i\sin\ \theta)^3$

Expanding this using binomial theorem:

$(cos\ \theta + i\sin\ \theta)^3 = \cos^3\theta + 3i\cos^2\theta\sin\theta - 3\cos\theta\sin^2\theta - i\sin^3\theta$

Equating imaginary parts: $\sin(3\theta) = 3\cos^2\theta\sin\theta - \sin^3\theta$

Replace $\cos^2\theta$ with $1 - \sin^2\theta$ and simplify:

$\sin(3\theta) = 3\left(1 - \sin^2\theta\right)\sin\theta - \sin^3\theta$

$= 3\sin\theta - 4\sin^3\theta$

Rearrange for $\sin^3\theta$:

$\sin^3\theta = \frac{3\sin\theta - \sin(3\theta)}{4}$

Q147:

Using de Moivre's theorem: $\cos(5\theta) + i\sin(5\theta) = (cos\ \theta + i\sin\ \theta)^5$

Expanding this using binomial theorem:

$(cos\ \theta + i\sin\ \theta)^5 = \cos^5\theta + 5i\cos^4\theta\sin\theta - 10\cos^3\theta\sin^2\theta - 10i\cos^2\theta\sin^3\theta + 5\cos\theta\sin^4\theta + i\sin^5\theta$

Equating imaginary parts: $\cos(5\theta) = \cos^5\theta - 10\cos^3\theta\sin^2\theta + 5\cos\theta\sin^4\theta$

Replace $\sin^2\theta$ with $1 - \cos^2\theta$ and simplify:

$\cos(5\theta) = \cos^5\theta - 10\cos^3\theta\left(1 - \cos^2\theta\right) + 5\cos\theta\left(1 - \cos^2\theta\right)^2$

$= \cos^5\theta - 10\cos^3\theta + 10\cos^5\theta + 5\cos\theta - 10\cos^3\theta + 5\cos^5\theta$

$= 16\cos^5\theta - 20\cos^3\theta + 5\cos\theta$

Q148:

Deal with the numerator first.

Using de Moivre's theorem: $\cos(5\theta) + i\sin(5\theta) = (cos\ \theta + i\sin\ \theta)^5$

Expanding this using the binomial theorem:

$(cos\ \theta + i\sin\ \theta)^5 = \cos^5\theta + 5i\cos^4\theta\sin\theta - 10\cos^3\theta\sin^2\theta - 10i\cos^2\theta\sin^3\theta + 5\cos\theta\sin^4\theta + i\sin^5\theta$

Equating the imaginary parts: $\sin(5\theta) = 5\cos^4\theta\sin\theta - 10\cos^2\theta\sin^3\theta + \sin^5\theta$

© HERIOT-WATT UNIVERSITY

Substituting into the original expression:
$$\frac{\sin(5\theta)}{\sin\theta} = \frac{5\cos^4\theta\sin\theta - 10\cos^2\theta\sin^3\theta + \sin^5\theta}{\sin\theta}$$
$$= 5\cos^4\theta - 10\cos^2\theta\sin^2\theta + \sin^4\theta$$

Replace $\cos^2\theta$ with $1 - \sin^2\theta$ and simplify:
$$\frac{\sin(5\theta)}{\sin\theta} = 5(1-\sin^2\theta)^2 - 10(1-\sin^2\theta)\sin^2\theta + \sin^4\theta$$
$$= 5 - 10\sin^2\theta + 5\sin^4\theta - 10\sin^2\theta + 10\sin^4\theta + \sin^4\theta$$
$$= 5 - 20\sin^2\theta + 16\sin^4\theta$$

Q149:

We know that $\tan(4\theta) = \frac{\sin(4\theta)}{\cos(4\theta)}$

We need to find expressions for $\sin(4\theta)$ and $\cos(4\theta)$ in terms of powers of cosine and sine.

So: $\cos(4\theta) + i\sin(4\theta) = (\cos\theta + i\sin\theta)^4$

Expanding this using binomial theorem:

$(\cos\theta + i\sin\theta)^4 = \cos^4\theta + 4i\cos^3\theta\sin\theta - 6\cos^2\theta\sin^2\theta - 4i\cos\theta\sin^3\theta + \sin^4\theta$

Equating real and imaginary parts:
$$\cos(4\theta) = \cos^4\theta - 6\cos^2\theta\sin^2\theta + \sin^4\theta$$
$$= 4\cos^3\theta\sin\theta - 4\cos\theta\sin^3\theta$$

Putting this into the expression for $\tan(4\theta)$:
$$\tan(4\theta) = \frac{4\cos^3\theta\sin\theta - 4\cos\theta\sin^3\theta}{\cos^4\theta - 6\cos^2\theta\sin^2\theta + \sin^4\theta}$$

We divide through by $\cos^4\theta$ because we only want expressions in terms of $\tan\theta$ and the highest power of sine and cosine is 4.

$$\tan(4\theta) = \frac{4\tan\theta - 4\tan^3\theta}{1 - 6\tan^2\theta + \tan^4\theta}$$

Q150:

Using de Moivre's theorem: $\cos(3\theta) + i\sin(3\theta) = (\cos\theta + i\sin\theta)^3$

Expanding this using binomial theorem:

$(\cos\theta + i\sin\theta)^3 = \cos^3\theta + 3i\cos^2\theta\sin\theta - 3\cos\theta\sin^2\theta - i\sin^3\theta$

Equating real parts: $\cos(3\theta) = \cos^3\theta - 3\cos\theta\sin^2\theta$

Now simplify:
$$\frac{\cos(3\theta)}{\cos^2\theta} = \frac{\cos^3\theta - 3\cos\theta\sin^2\theta}{\cos^2\theta}$$
$$= \cos\theta - \frac{3\sin^2\theta}{\cos\theta}$$
$$= \cos\theta - \frac{3(1-\cos^2\theta)}{\cos\theta}$$
$$= \cos\theta - \frac{3}{\cos\theta} + 3\cos\theta$$
$$= 4\cos\theta - 3\sec\theta$$

De Moivre's theorem with integers exercise (page 135)

Q151:

Let $z = \cos\theta + i\sin\theta$

We will use the form: $2i\sin(n\theta) = z^n - \frac{1}{z^n}$

In this case n = 1 since we wish to find an expression for $(\sin(1 \times \theta))^7$ (rewritten from $\sin^7\theta$) so we have: $(2i\sin\theta)^7 = \left(z - \frac{1}{z}\right)^7$

Using the binomial theorem to expand the RHS:

$-128i\sin^7\theta = z^7 - 7z^6\frac{1}{z} + 21z^5\frac{1}{z^2} - 35z^4\frac{1}{z^3} + 35z^3\frac{1}{z^4} - 21z^2\frac{1}{z^5} + 7z\frac{1}{z^6} - \frac{1}{z^7}$

(Remember that $i^7 = -i$)

Simplifying: $-128i\sin^7\theta = z^7 - 7z^5 + 21z^3 - 35z + \frac{35}{z} - \frac{21}{z^3} + \frac{7}{z^5} - \frac{1}{z^7}$

Grouping terms together into the form $z^n - \frac{1}{z^n}$, we can then replace them with $2i\sin(n\theta)$:

$-128i\sin^7\theta = \left(z^7 - \frac{1}{z^7}\right) - 7\left(z^5 - \frac{1}{z^5}\right) + 21\left(z^3 - \frac{1}{z^3}\right) - 35\left(z - \frac{1}{z}\right)$

$= (2i\sin(7\theta)) - 7(2i\sin(5\theta)) + 21(2i\sin(3\theta)) - 35(2i\sin\theta)$

$= 2i\sin(7\theta) - 14i\sin(5\theta) + 42i\sin(3\theta) - 70i\sin\theta$

We have: $\sin^7\theta = -\frac{1}{64}(\sin(7\theta) - 7\sin(5\theta) + 21\sin(3\theta) - 35\sin\theta)$

Q152:

Let $z = \cos\theta + i\sin\theta$

We will use the form: $2\cos(n\theta) = z^n + \frac{1}{z^n}$

In this case n = 1 since we wish to find an expression for $(\cos(1 \times \theta))^6$ (rewritten from $\cos^6\theta$) so we have: $(2\cos\theta)^6 = \left(z + \frac{1}{z}\right)^6$

Using the binomial theorem to expand the RHS:

$64\cos^6\theta = z^6 + 6z^5\frac{1}{z} + 15z^4\frac{1}{z^2} + 20z^3\frac{1}{z^3} + 15z^2\frac{1}{z^4} + 6z\frac{1}{z^5} + \frac{1}{z^6}$

Simplifying: $64\cos^6\theta = z^6 + 6z^4 + 15z^2 + 20 + 15\frac{1}{z^2} + 6\frac{1}{z^4} + \frac{1}{z^6}$

Grouping terms together into the form $z^n + \frac{1}{z^n}$, we can then replace them with $2\cos(n\theta)$:

$64\cos^6\theta = \left(z^6 + \frac{1}{z^6}\right) + 6\left(z^4 + \frac{1}{z^4}\right) + 15\left(z^2 + \frac{1}{z^2}\right) + 20$

$= (2\cos(6\theta)) + 6(2\cos(4\theta)) + 15(2\cos(2\theta)) + 20$

$= 2\cos(6\theta) + 12\cos(4\theta) + 30\cos(2\theta) + 20$

We have: $\cos^6\theta = \frac{1}{32}(\cos(6\theta) + 6\cos(4\theta) + 15\cos(2\theta) + 10)$

Q153:

Let $z = \cos\theta + i\sin\theta$

We will use the form: $2i\sin(n\theta) = z^n - \frac{1}{z^n}$

In this case n = 1 since we wish to find an expression for $(\sin(1 \times \theta))^6$ (rewritten from $\sin^6\theta$) so we have: $(2i\sin\theta)^6 = \left(z - \frac{1}{z}\right)^6$

Using the binomial theorem to expand the RHS:

$-64\sin^6\theta = z^6 - 6z^5\frac{1}{z} + 15z^4\frac{1}{z^2} - 20z^3\frac{1}{z^3} + 15z^2\frac{1}{z^4} - 6z\frac{1}{z^5} + \frac{1}{z^6}$

(Remember that $i^6 = -1$)

Simplifying: $-64\sin^6\theta = z^6 - 6z^4 + 15z^2 - 20 + 15\frac{1}{z^2} - 6\frac{1}{z^4} + \frac{1}{z^6}$

Grouping terms together into the form $z^n + \frac{1}{z^n}$, we can then replace them with $2\cos(n\theta)$:

$$-64\sin^6\theta = \left(z^6 + \frac{1}{z^6}\right) - 6\left(z^4 + \frac{1}{z^4}\right) + 15\left(z^2 + \frac{1}{z^2}\right) - 20$$
$$= (2\cos(6\theta)) - 6(2\cos(4\theta)) + 15(2\cos(2\theta)) - 20$$
$$= 2\cos(6\theta) - 12\cos(4\theta) + 30\cos(2\theta) - 20$$

We have: $\sin^6\theta = -\frac{1}{32}(\cos(6\theta) - 6\cos(4\theta) + 15\cos(2\theta) - 10)$

Q154:

Using: $2\cos(n\theta) = z^n + \frac{1}{z^n}$

We have:

$$(2\cos\theta)^2 = \left(z + \frac{1}{z}\right)^2$$
$$4\cos^2\theta = z^2 + 2 + \frac{1}{z^2}$$
$$\cos^2\theta = \frac{1}{4}\left(z^2 + 2 + \frac{1}{z^2}\right)$$
$$= \frac{1}{4}\left(z^2 + \frac{1}{z^2} + 2\right)$$

And using: $2i\sin(n\theta) = z^n - \frac{1}{z^n}$

We have:

$$2i\sin\theta = \left(z - \frac{1}{z}\right)$$
$$\sin\theta = \frac{1}{2i}\left(z - \frac{1}{z}\right)$$

Multiplying these together:

$$\cos^2\theta \sin\theta = \frac{1}{4}\left(z^2 + \frac{1}{z^2} + 2\right)\frac{1}{2i}\left(z - \frac{1}{z}\right)$$
$$= \frac{1}{8i}\left(z^3 - z + \frac{1}{z} - \frac{1}{z^3} + 2z - \frac{2}{z}\right)$$
$$= \frac{1}{8i}\left(z^3 - \frac{1}{z^3} + z - \frac{1}{z}\right)$$

De Moivre's theorem and fractional powers exercise (page 137)

Q155:

By de Moivre's theorem:

$$\left(27\left(\cos\left(\frac{\pi}{2}\right) + i\sin\left(\frac{\pi}{2}\right)\right)\right)^{\frac{1}{3}} = 27^{\frac{1}{3}}\left(\cos\left(\frac{\pi}{2}\right) + i\sin\left(\frac{\pi}{2}\right)\right)^{\frac{1}{3}}$$
$$= 3\left(\cos\left(\frac{1}{3} \times \frac{\pi}{2}\right) + i\sin\left(\frac{1}{3} \times \frac{\pi}{2}\right)\right)$$
$$= 3\left(\cos\left(\frac{\pi}{6}\right) + i\sin\left(\frac{\pi}{6}\right)\right)$$

Q156:
By de Moivre's theorem:
$$\left(8\left(\cos\left(\frac{\pi}{4}\right)+i\sin\left(\frac{\pi}{4}\right)\right)\right)^{\frac{2}{3}} = 8^{\frac{2}{3}}\left(\cos\left(\frac{\pi}{4}\right)+i\sin\left(\frac{\pi}{4}\right)\right)^{\frac{2}{3}}$$
$$= 4\left(\cos\left(\frac{2}{3}\times\frac{\pi}{4}\right)+i\sin\left(\frac{2}{3}\times\frac{\pi}{4}\right)\right)$$
$$= 4\left(\cos\left(\frac{\pi}{6}\right)+i\sin\left(\frac{\pi}{6}\right)\right)$$

Q157:
By de Moivre's theorem:
$$(81\left(\cos(36°)+i\sin(36°)\right))^{\frac{1}{4}} = 81^{\frac{1}{4}}(\cos(36°)+i\sin(36°))^{\frac{1}{4}}$$
$$= 3\left(\cos\left(\frac{1}{4}\times 36°\right)+i\sin\left(\frac{1}{4}\times 36°\right)\right)$$
$$= 3\left(\cos(9°)+i\sin(9°)\right)$$

Q158:
By de Moivre's theorem:
$$\left(32\left(\cos\left(\frac{\pi}{3}\right)+i\sin\left(\frac{\pi}{3}\right)\right)\right)^{\frac{3}{5}} = 32^{\frac{3}{5}}\left(\cos\left(\frac{\pi}{3}\right)+i\sin\left(\frac{\pi}{3}\right)\right)^{\frac{3}{5}}$$
$$= 8\left(\cos\left(\frac{3}{5}\times\frac{\pi}{3}\right)+i\sin\left(\frac{3}{5}\times\frac{\pi}{3}\right)\right)$$
$$= 8\left(\cos\left(\frac{\pi}{5}\right)+i\sin\left(\frac{\pi}{5}\right)\right)$$

Q159:
By de Moivre's theorem:
$$\left(64\left(\cos\left(\frac{3\pi}{2}\right)+i\sin\left(\frac{3\pi}{2}\right)\right)\right)^{\frac{4}{3}} = 64^{\frac{4}{3}}\left(\cos\left(\frac{3\pi}{2}\right)+i\sin\left(\frac{3\pi}{2}\right)\right)^{\frac{4}{3}}$$
$$= 256\left(\cos\left(\frac{4}{3}\times\frac{3\pi}{2}\right)+i\sin\left(\frac{4}{3}\times\frac{3\pi}{2}\right)\right)$$
$$= 256\left(\cos(2\pi)+i\sin(2\pi)\right)$$

The argument is not within the range of the principal argument, namely $(-\pi, \pi)$.

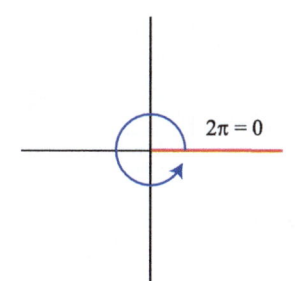

$$\left(64\left(\cos\frac{3\pi}{2}+i\sin\frac{3\pi}{2}\right)\right)^{\frac{4}{3}} = 256\left(\cos 0+i\sin 0\right)$$
$$= 256$$

Q160:

By de Moivre's theorem:

$$\left(16\left(\cos\left(\frac{5\pi}{3}\right)+i\sin\left(\frac{5\pi}{3}\right)\right)\right)^{\frac{3}{2}} = 16^{\frac{3}{2}}\left(\cos\left(\frac{5\pi}{3}\right)+i\sin\left(\frac{5\pi}{3}\right)\right)^{\frac{3}{2}}$$

$$= 64\left(\cos\left(\frac{3}{2}\times\frac{5\pi}{3}\right)+i\sin\left(\frac{3}{2}\times\frac{5\pi}{3}\right)\right)$$

$$= 64\left(\cos\left(\frac{5\pi}{2}\right)+i\sin\left(\frac{5\pi}{2}\right)\right)$$

The argument is not within the range of the principal argument, namely $(-\pi, \pi)$.

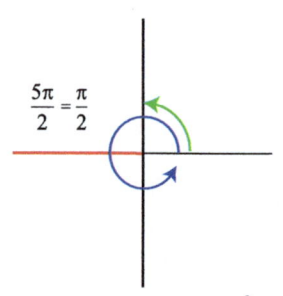

$$\frac{5\pi}{2} = \frac{\pi}{2}$$

$\left(16\left(\cos\left(\frac{5\pi}{3}\right)+i\sin\left(\frac{5\pi}{3}\right)\right)\right)^{\frac{3}{2}} = 64\left(\cos\left(\frac{\pi}{2}\right)+i\sin\left(\frac{\pi}{2}\right)\right)$

Q161:

By de Moivre's theorem:

$$(343\left(\cos\left(243°\right)+i\sin\left(243°\right)\right))^{-\frac{1}{3}} = 343^{-\frac{1}{3}}(\cos\left(243°\right)+i\sin\left(243°\right))^{-\frac{1}{3}}$$

$$= 64\left(\cos\left(-\frac{1}{3}\times 243°\right)+i\sin\left(-\frac{1}{3}\times 243°\right)\right)$$

$$= 64(\cos\left(-81°\right)+i\sin\left(-81°\right))$$

Since cosine is an even function and sine is an odd function, this can be simplified to:

$(343\left(\cos\left(243°\right)+i\sin\left(243°\right)\right))^{-\frac{1}{3}} = 64(\cos\left(81°\right)-i\sin\left(81°\right))$

Q162:

By de Moivre's theorem:

$$(25\left(\cos\left(184°\right)+i\sin\left(184°\right)\right))^{-\frac{5}{2}} = 25^{-\frac{5}{2}}(\cos\left(184°\right)+i\sin\left(184°\right))^{-\frac{5}{2}}$$

$$= \frac{1}{3125}\left(\cos\left(-\frac{5}{2}\times 184°\right)+i\sin\left(-\frac{5}{2}\times 184°\right)\right)$$

$$= \frac{1}{3125}(\cos\left(-460°\right)+i\sin\left(-460°\right))$$

The argument is not within the range of the principal argument, namely $(-180°, 180°)$.

ANSWERS: UNIT 1 TOPIC 6

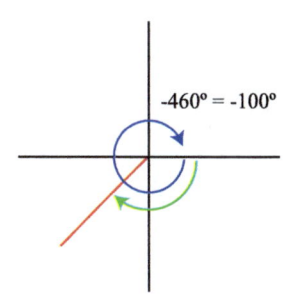

The principal argument for this is:
$$(25\left(\cos\left(184°\right)+i\sin\left(184°\right)\right))^{-\frac{5}{2}} = \frac{1}{3125}\left(\cos\left(-100°\right)+i\sin\left(-100°\right)\right)$$
$$= \frac{1}{3125}\left(\cos\left(100°\right)-i\sin\left(100°\right)\right)$$

De Moivre's theorem and n^{th} roots exercise (page 141)

Q163:
This expression is in polar form. Use $2\pi = 360°$ and $4\pi = 720°$
$z = 64(cos(30°) + i\sin(30°))$
which can also be written as: $z = 64(cos(30+360)° + i\sin(30+360)°)$
and: $z = 64(cos(30+720)° + i\sin(30+720)°)$
Therefore, the cube roots of z are:
$z_1 = 64^{\frac{1}{3}}(\cos\left(30°\right)+i\sin\left(30°\right))^{\frac{1}{3}}$
$= 4\left(\cos\left(10°\right)+i\sin\left(10°\right)\right)$
$z_2 = 64^{\frac{1}{3}}(\cos\left(390°\right)+i\sin\left(390°\right))^{\frac{1}{3}}$
$= 4\left(\cos\left(130°\right)+i\sin\left(130°\right)\right)$
$z_3 = 64^{\frac{1}{3}}(\cos\left(750°\right)+i\sin\left(750°\right))^{\frac{1}{3}}$
$= 4\left(\cos\left(250°\right)+i\sin\left(250°\right)\right)$ convert to principal argument
$= 4\left(\cos\left(-110°\right)+i\sin\left(-110°\right)\right)$
$= 4\left(\cos\left(110°\right)-i\sin\left(110°\right)\right)$

Q164:
The fourth roots of $81\left(\cos\left(\frac{\pi}{2}\right)+i\sin\left(\frac{\pi}{2}\right)\right)$ are:
$z_1 = 81^{\frac{1}{4}}\left(\cos\left(\frac{\pi}{2}\right)+i\sin\left(\frac{\pi}{2}\right)\right)^{\frac{1}{4}} \Rightarrow z_1 = 3\left(\cos\left(\frac{\pi}{8}\right)+i\sin\left(\frac{\pi}{8}\right)\right)$
$z_2 = 81^{\frac{1}{4}}\left(\cos\left(\frac{5\pi}{2}\right)+i\sin\left(\frac{5\pi}{2}\right)\right)^{\frac{1}{4}} \Rightarrow z_2 = 3\left(\cos\left(\frac{5\pi}{8}\right)+i\sin\left(\frac{5\pi}{8}\right)\right)$
$z_3 = 81^{\frac{1}{4}}\left(\cos\left(\frac{9\pi}{2}\right)+i\sin\left(\frac{9\pi}{2}\right)\right)^{\frac{1}{4}} \Rightarrow z_3 = 3\left(\cos\left(\frac{9\pi}{8}\right)+i\sin\left(\frac{9\pi}{8}\right)\right)$
$\Rightarrow z_3 = 3\left(\cos\left(\frac{7\pi}{8}\right)-i\sin\left(\frac{7\pi}{8}\right)\right)$

© HERIOT-WATT UNIVERSITY

$$z_4 = 81^{\frac{1}{4}}\left(\cos\left(\frac{13\pi}{2}\right) + i\sin\left(\frac{13\pi}{2}\right)\right)^{\frac{1}{4}} \Rightarrow z_4 = 3\left(\cos\left(\frac{13\pi}{8}\right) + i\sin\left(\frac{13\pi}{8}\right)\right)$$
$$\Rightarrow z_4 = 3\left(\cos\left(\frac{3\pi}{8}\right) - i\sin\left(\frac{3\pi}{8}\right)\right)$$

Q165: Modulus:

$$\left|\sqrt{3} + i\right| = \sqrt{\left(\sqrt{3}\right)^2 + 1^2}$$
$$= 2$$

Argument:

Since $\sqrt{3} + i$ is in the first quadrant, $\arg(\sqrt{3} + i) = \alpha$ where $\tan\alpha = \left|\frac{1}{\sqrt{3}}\right|$ so:

$\arg\left(\sqrt{3} + i\right) = \frac{\pi}{6}$

The six roots are:

$z_1 = 2^{\frac{1}{6}}\left(\cos\left(\frac{\pi}{6}\right) + i\sin\left(\frac{\pi}{6}\right)\right)^{\frac{1}{6}} \Rightarrow z_1 = \sqrt[6]{2}\left(\cos\left(\frac{\pi}{36}\right) + i\sin\left(\frac{\pi}{36}\right)\right)$

$z_2 = 2^{\frac{1}{6}}\left(\cos\left(\frac{13\pi}{6}\right) + i\sin\left(\frac{13\pi}{6}\right)\right)^{\frac{1}{6}} \Rightarrow z_2 = \sqrt[6]{2}\left(\cos\left(\frac{13\pi}{36}\right) + i\sin\left(\frac{13\pi}{36}\right)\right)$

$z_3 = 2^{\frac{1}{6}}\left(\cos\left(\frac{25\pi}{6}\right) + i\sin\left(\frac{25\pi}{6}\right)\right)^{\frac{1}{6}} \Rightarrow z_3 = \sqrt[6]{2}\left(\cos\left(\frac{25\pi}{36}\right) + i\sin\left(\frac{25\pi}{36}\right)\right)$

$z_4 = 2^{\frac{1}{6}}\left(\cos\left(\frac{37\pi}{6}\right) + i\sin\left(\frac{37\pi}{6}\right)\right)^{\frac{1}{6}} \Rightarrow z_4 = \sqrt[6]{2}\left(\cos\left(\frac{37\pi}{36}\right) + i\sin\left(\frac{37\pi}{36}\right)\right)$
$$= \sqrt[6]{2}\left(\cos\left(\frac{35\pi}{36}\right) - i\sin\left(\frac{35\pi}{36}\right)\right)$$

$z_5 = 2^{\frac{1}{6}}\left(\cos\left(\frac{49\pi}{6}\right) + i\sin\left(\frac{49\pi}{6}\right)\right)^{\frac{1}{6}} \Rightarrow z_5 = \sqrt[6]{2}\left(\cos\left(\frac{49\pi}{36}\right) + i\sin\left(\frac{49\pi}{36}\right)\right)$
$$= \sqrt[6]{2}\left(\cos\left(\frac{23\pi}{36}\right) - i\sin\left(\frac{23\pi}{36}\right)\right)$$

$z_6 = 2^{\frac{1}{6}}\left(\cos\left(\frac{61\pi}{6}\right) + i\sin\left(\frac{61\pi}{6}\right)\right)^{\frac{1}{6}} \Rightarrow z_6 = \sqrt[6]{2}\left(\cos\left(\frac{61\pi}{36}\right) + i\sin\left(\frac{61\pi}{36}\right)\right)$
$$= \sqrt[6]{2}\left(\cos\left(\frac{11\pi}{36}\right) - i\sin\left(\frac{11\pi}{36}\right)\right)$$

De Moivre's theorem and roots of unity exercise (page 146)

Q166:

The fourth roots of unity are the solutions to $z^4 = 1$

$z = \left(\cos\left(0 + 2k\pi\right) + i\sin\left(0 + 2k\pi\right)\right)^{\frac{1}{4}}$
$= \cos\left(\frac{2k\pi}{4}\right) + i\sin\left(\frac{2k\pi}{4}\right)$
$= \cos\left(\frac{k\pi}{2}\right) + i\sin\left(\frac{k\pi}{2}\right)$

$k = 0: \quad z_0 = \cos(0) + i\sin(0) \quad \Rightarrow z_0 = 1$

ANSWERS: UNIT 1 TOPIC 6

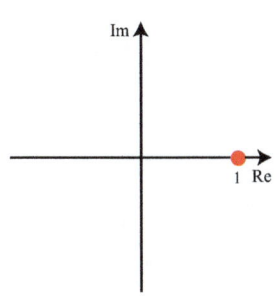

$k = 1:$ $z_1 = \cos\left(\dfrac{\pi}{2}\right) + i\sin\left(\dfrac{\pi}{2}\right)$ $\Rightarrow z_1 = i$

$k = 2:$ $z_2 = \cos(\pi) + i\sin(\pi)$ $\Rightarrow z_2 = -1$

$k = 3:$ $z_3 = \cos\left(\dfrac{3\pi}{2}\right) + i\sin\left(\dfrac{3\pi}{2}\right)$ $\Rightarrow z_3 = -i$

Roots: $z = 1,\ i,\ -1,\ -i$

On an Argand diagram:

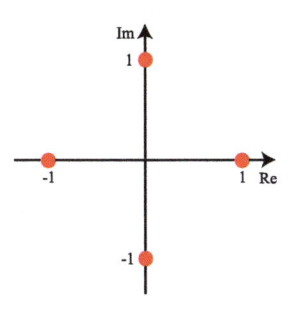

Notice the equal spacing between the roots.

Q167:

Rearrange to give $z^6 = 1$

Given that $1 = \cos(2\pi) + i\sin(2\pi)$:

$z^6 = \cos(0 + 2k\pi) + i\sin(0 + 2k\pi)$

$z = (\cos(2k\pi) + i\sin(2k\pi))^{\frac{1}{6}}$

$ = \cos\left(\dfrac{2k\pi}{6}\right) + i\sin\left(\dfrac{2k\pi}{6}\right)$

$ = \cos\left(\dfrac{k\pi}{3}\right) + i\sin\left(\dfrac{k\pi}{3}\right)$

Evaluate for $k = 0, 1, 2, 3, 4, 5$:

$k = 0:$ $z_0 = \cos 0 + i\sin 0$
$ = 1$

$k = 1:$ $z_1 = \cos\left(\frac{\pi}{3}\right) + i\sin\left(\frac{\pi}{3}\right)$
$= \frac{1}{2} + i\frac{\sqrt{3}}{2}$

$k = 2:$ $z_2 = \cos\left(\frac{2\pi}{3}\right) + i\sin\left(\frac{2\pi}{3}\right)$
$= -\frac{1}{2} + i\frac{\sqrt{3}}{2}$

$k = 3:$ $z_3 = \cos\pi + i\sin\pi$
$= -1$

$k = 4:$ $z_4 = \cos\left(\frac{4\pi}{3}\right) + i\sin\left(\frac{4\pi}{3}\right)$
$= -\frac{1}{2} - i\frac{\sqrt{3}}{2}$

$k = 5:$ $z_5 = \cos\left(\frac{5\pi}{3}\right) + i\sin\left(\frac{5\pi}{3}\right)$
$= \frac{1}{2} - i\frac{\sqrt{3}}{2}$

Roots: $z = 1, \frac{1}{2} + i\frac{\sqrt{3}}{2}, -\frac{1}{2} + i\frac{\sqrt{3}}{2}, -1, -\frac{1}{2} - i\frac{\sqrt{3}}{2}, \frac{1}{2} - i\frac{\sqrt{3}}{2}$
On an Argand diagram:

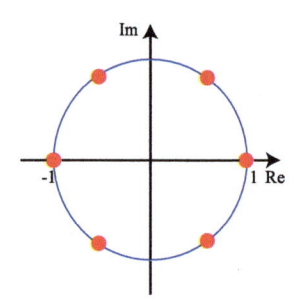

Notice the equal spacing between the roots.

Q168:
$z^4 = i$
$= \cos\left(\frac{\pi}{2} + 2k\pi\right) + i\sin\left(\frac{\pi}{2} + 2k\pi\right)$
$z = \left(\cos\left(\frac{\pi}{2} + 2k\pi\right) + i\sin\left(\frac{\pi}{2} + 2k\pi\right)\right)^{\frac{1}{4}}$
$= \cos\left(\frac{(4k+1)\pi}{8}\right) + i\sin\left(\frac{(4k+1)\pi}{8}\right)$

Evaluate for $k = 0, 1, 2, 3$:
$k = 0:$ $z_0 = \cos\left(\frac{\pi}{8}\right) + i\sin\left(\frac{\pi}{8}\right)$
$= 0.92 + 0.38i$

$k = 1:$ $z_1 = \cos\left(\dfrac{5\pi}{8}\right) + i\sin\left(\dfrac{5\pi}{8}\right)$
$= -0 \cdot 38 + 0 \cdot 92i$

$k = 2:$ $z_2 = \cos\left(\dfrac{9\pi}{8}\right) + i\sin\left(\dfrac{9\pi}{8}\right)$
$= -0 \cdot 92 + i - 0 \cdot 38i$

$k = 3:$ $z_3 = \cos\left(\dfrac{13\pi}{8}\right) + i\sin\left(\dfrac{13\pi}{8}\right)$
$= 0 \cdot 38 - 0 \cdot 92i$

Roots: $z = 0 \cdot 92 + 0 \cdot 38i,\ -0 \cdot 38 + 0 \cdot 92i,\ -0 \cdot 92 + -0 \cdot 38i,\ 0 \cdot 38 - 0 \cdot 92i$
(Answers to 2 d.p.)
Notice that the roots come in conjugate pairs.
On an Argand diagram:

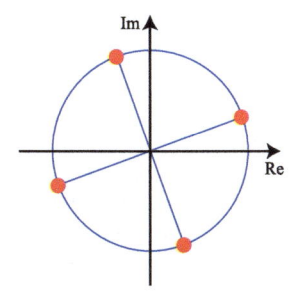

This has the same spacing of its roots as in the first question, but has been rotated through $\dfrac{\pi}{8}$ radians.

Q169:
We would expect the solutions for $z^4 = 16$ to start on the real axis at $z = 2$ and be spaced apart by a rotation of $\dfrac{\pi}{2}$ radians. The solutions of $z^4 = -16$ have the same magnitude, but would start at rotation of $\dfrac{\pi}{4}$ from the first solution of $z^4 = 16$.

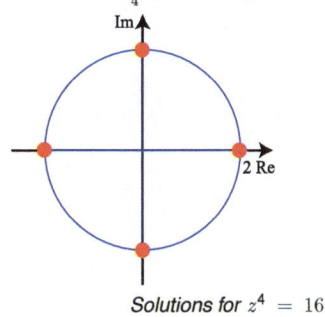

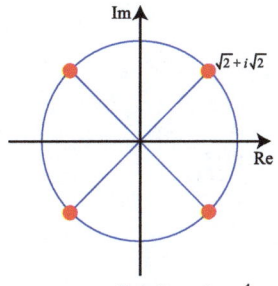

Solutions for $z^4 = 16$ Solutions for $z^4 = -16$

For $z^4 = 16$, on an Argand diagram:

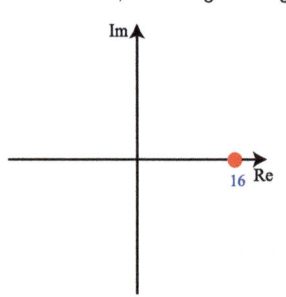

So:
$$z^4 = 16\left(\cos\left(0 + 2k\pi\right) + i\sin\left(0 + 2k\pi\right)\right)$$
$$z = 16^{\frac{1}{4}}\left(\cos\left(\frac{2k\pi}{4}\right) + i\sin\left(\frac{2k\pi}{4}\right)\right)$$
$$= 2\left(\cos\left(\frac{k\pi}{2}\right) + i\sin\left(\frac{k\pi}{2}\right)\right)$$

So:
$k = 0:\quad z_0 = 2\left(\cos\left(0\right) + i\sin\left(0\right)\right)$
$\qquad\qquad = 2$

Every other root will be spaced in $\frac{\pi}{2}$ intervals.

$k = 1:\quad z_1 = 2\left(\cos\left(\frac{\pi}{2}\right) + i\sin\left(\frac{\pi}{2}\right)\right)$
$\qquad\qquad = i$

$k = 2:\quad z_2 = 2\left(\cos\pi + i\sin\pi\right)$
$\qquad\qquad = -1$

$k = 3:\quad z_3 = 2\left(\cos\left(\frac{3\pi}{2}\right) + i\sin\left(\frac{3\pi}{2}\right)\right)$
$\qquad\qquad = -i$

So we have:

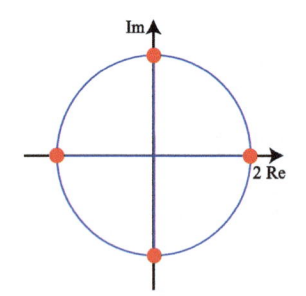

For $z^4 = -16$, on an Argand diagram:

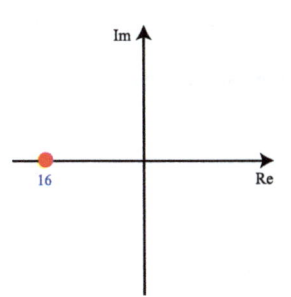

So:
$$z^4 = 16\left(\cos\left(\pi + 2k\pi\right) + i\sin\left(\pi + 2k\pi\right)\right)$$
$$z = 16^{\frac{1}{4}}\left(\cos\left(\frac{(2k+1)\pi}{4}\right) + i\sin\left(\frac{(2k+1)\pi}{4}\right)\right)$$
$$= 2\left(\cos\left(\frac{(2k+1)\pi}{4}\right) + i\sin\left(\frac{(2k+1)\pi}{4}\right)\right)$$

So:
$k = 0:\quad z_0 = 2\left(\cos\left(\frac{\pi}{4}\right) + i\sin\left(\frac{\pi}{4}\right)\right)$
$\qquad\qquad\quad = \sqrt{2} + i\sqrt{2}$

Every other root will be spaced in $\frac{\pi}{2}$ intervals.

$k = 1:\quad z_1 = 2\left(\cos\left(\frac{3\pi}{4}\right) + i\sin\left(\frac{3\pi}{4}\right)\right)$
$\qquad\qquad\quad = -\sqrt{2} + i\sqrt{2}$

$k = 2:\quad z_2 = 2\left(\cos\left(\frac{5\pi}{4}\right) + i\sin\left(\frac{5\pi}{4}\right)\right)$
$\qquad\qquad\quad = -\sqrt{2} - i\sqrt{2}$

$k = 3:\quad z_2 = 2\left(\cos\left(\frac{7\pi}{4}\right) + i\sin\left(\frac{7\pi}{4}\right)\right)$
$\qquad\qquad\quad = \sqrt{2} - i\sqrt{2}$

So we have:

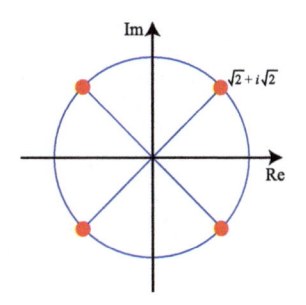

Q170:

From the first question, the solutions of $z^4 = 1$ are $z = 1, i, -1, -i$

Adding these solutions together: $1 + i + (-1) + (-i) = 0$

This can be explained by looking at the symmetry of the solutions on an Argand diagram:

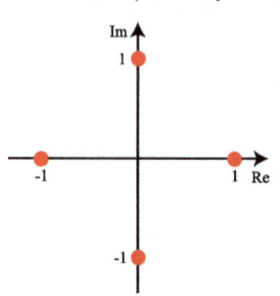

The real parts are symmetrical and the imaginary parts are symmetrical but are the negative of the other. Therefore, the sum of the roots will equal 0.

Q171:

Change into polar form first.

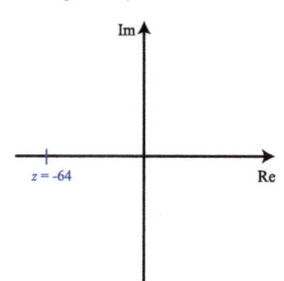

Modulus:

$|z| = \sqrt{(-64)^2 + 0^2}$
$= 64$

Argument:

$\tan \theta = -\dfrac{0}{64}$

$arg(z) = \pi$

We have: $z^3 = 64 \left(\cos(\pi + 2k\pi) + i \sin(\pi + 2k\pi) \right)$

Evaluate for $k = 0, 1, 2$:

$k = 0: \quad z^3 = 64 \left(\cos(\pi) + i \sin(\pi) \right)$

$k = 1: \quad z^3 = 64 \left(\cos(\pi + 2\pi) + i \sin(\pi + 2\pi) \right)$
$\qquad\qquad = 64 \left(\cos(3\pi) + i \sin(3\pi) \right)$

$k = 2: \quad z^3 = 64 \left(\cos(\pi + 4\pi) + i \sin(\pi + 4\pi) \right)$
$\qquad\qquad = 64 \left(\cos(5\pi) + i \sin(5\pi) \right)$

ANSWERS: UNIT 1 TOPIC 6

Using de Moivre's theorem, the roots are:

$k = 0:\quad z = 4\left(\cos\left(\frac{\pi}{3}\right) + i\sin\left(\frac{\pi}{3}\right)\right)$

$k = 1:\quad z = 4\left(\cos\left(\pi\right) + i\sin\left(\pi\right)\right)$

$k = 2:\quad z = 4\left(\cos\left(\frac{5\pi}{3}\right) + i\sin\left(\frac{5\pi}{3}\right)\right)$

$\quad\quad\quad = 4\left(\cos\left(\frac{\pi}{3}\right) - i\sin\left(\frac{\pi}{3}\right)\right)$

Evaluating the roots:

$k = 0:\quad z = 2 + 3 \cdot 46i$
$k = 1:\quad z = -4$
$k = 2:\quad z = 2 - 3 \cdot 46i$

Conjugate properties exercise (page 152)

Q172:

If $z = 4 - 5i$, then the conjugate is $\bar{z} = 4 + 5i$

Therefore, $\bar{\bar{z}} = 4 - 5i \therefore \bar{\bar{z}} = z$

Q173:

$z = -4 + \frac{1}{2}i$ so the real part of z is $\text{Re}(z) = -4$

The conjugate is $\bar{z} = -4 - \frac{1}{2}i$

The real part of $\bar{z}$ is $\text{Re}(\bar{z}) = -4$

So $\text{Re}(z) = -4 = \text{Re}(\bar{z})$

Q174:

If $z = -\frac{\sqrt{3}}{2} + 10i$, then $\text{Re}(z) = -\frac{\sqrt{3}}{2}$ and $\bar{z} = -\frac{\sqrt{3}}{2} - 10i$

$\bar{z} + z = \left(-\frac{\sqrt{3}}{2} - 10i\right) + \left(-\frac{\sqrt{3}}{2} + 10i\right)$

$\quad\quad = 2\left(-\frac{\sqrt{3}}{2}\right)$

$\quad\quad = -\sqrt{3}$

$\quad\quad = 2\,\text{Re}(z)$

Q175:

If $z = 3 + 7i$, then $\bar{z} = 3 - 7i$

$|z| = \sqrt{3^2 + 7^2}$
$\quad = \sqrt{58}$

and

$|\bar{z}| = \sqrt{3^2 + (-7)^2}$
$\quad = \sqrt{58}$

Therefore, $|\bar{z}| = |z|$

© HERIOT-WATT UNIVERSITY

Q176:

If $z = 6 - 2i$, then $\bar{z} = 6 + 2i$
$z\bar{z} = (6 - 2i)(6 + 2i)$
$= 36 - 4i^2$
$= 36 + 4$
$= 40$

and
$|z|^2 = \left(\sqrt{6^2 + 2^2}\right)^2$
$= 6^2 + 2^2$
$= 40$

Therefore, $z\bar{z} = |z|^2$

Q177:

If $z = 5 - 2i$, then $\bar{z} = 5 + 2i$
If $w = 3 + 7i$, then $\bar{w} = 3 - 7i$

So
$\bar{z} + \bar{w} = 5 + 2i + 3 - 7i$
$= 8 - 5i$

and
$z + w = 5 - 2i + 3 + 7i$
$= 8 + 5i$
$\overline{(z + w)} = 8 - 5i$

Therefore, $\overline{z + w} = \bar{z} + \bar{w}$

Q178:

If $z = 6 + i$, then $\bar{z} = 6 - i$
If $w = 2 - 3i$, then $\bar{w} = 2 + 3i$
$\bar{z}\,\bar{w} = (6 - i)(2 + 3i)$
$= 12 + 16i - 3i^2$
$= 15 + 16i$

and
$zw = (6 + i)(2 - 3i)$
$= 12 - 16i - 3i^2$
$= 15 - 16i$
$\overline{zw} = 15 + 16i$

Therefore, $\bar{z}\,\bar{w} = \overline{zw}$

Q179:

If $z = 9 - 5i$, then $\bar{z} = 9 + 5i$
If $w = 1 - 4i$, then $\bar{w} = 1 + 4i$

ANSWERS: UNIT 1 TOPIC 6   497

$$\overline{zw} = \overline{(9-5i)(1-4i)}$$
$$= \overline{-11-41i}$$
$$= -11+41i$$

so

$$|\overline{zw}| = \sqrt{(-11)^2 + (41)^2}$$
$$= \sqrt{1802}$$

$$|z| = \sqrt{(9)^2 + (-5)^2}$$
$$= \sqrt{106}$$

and

$$|w| = \sqrt{(1)^2 + (-4)^2}$$
$$= \sqrt{17}$$

so

$$|z||w| = \sqrt{106} \times \sqrt{17}$$
$$= \sqrt{1802}$$

Therefore, $|\overline{zw}| = |z||w|$

Q180:

Substitute $x + iy$ for z and expand to convert to the form $a + ib$:

$$x + iy + 2i(x - iy) = -5 + 2i$$
$$x + iy + 2ix + 2y = -5 + 2i$$
$$x + 2y + i(y + 2x) = -5 + 2i$$

Equate the real and the imaginary parts:

$$x + 2y = -5 \quad (1)$$
$$y + 2x = 2 \quad (2)$$
$$2 \times (1) - (2) \quad 3y = -12$$
$$y = -4$$

Substituting $y = -4$ into (2) we have $x = 3$

So $z = 3 - 4i$

End of topic 6 test (page 163)

Q181:

Hints:

- Substitute the complex numbers correctly into the expression, i.e. $2(3 - i) - (m - 3i) + 3(3 + 2i)$

Answer:

$$2u - w + 3v = 2(3 - i) - (m - 3i) + 3(3 + 2i)$$
$$= 6 - 2i - m + 3i + 9 + 6i$$
$$= 15 - m + 7i$$

© HERIOT-WATT UNIVERSITY

Q182:

Hints:

- Substitute the complex numbers correctly into the expression, i.e. $3i(2 + 5i) + 2(6 + 2i) - 4(-1 - 2i)$

Answer:

$$3iu + 2\overline{w} - 4v = 3i(2 + 5i) + 2(6 + 2i) - 4(-1 - 2i)$$
$$= 6i + 15i^2 + 12 + 4i + 4 + 8i$$
$$= 6i - 15 + 12 + 4i + 4 + 8i$$
$$= 1 + 18i$$

Q183:

Steps:

- What do you multiply the numerator and denominator by to simplify? $2 + 3i$

Answer:

$$\frac{u}{v} = \frac{(5+i)}{(2-3i)} \times \frac{(2+3i)}{(2+3i)}$$
$$= \frac{10 + 15i + 2i + 3i^2}{4 - 9i^2}$$
$$= \frac{7 + 17i}{13}$$

Q184:

Steps:

- What four terms result from expanding $4 - 5i$ multiplied by $1 + 2i$? $4 + 8i - 5i - 10i^2$

Answer:

$$uv = (4 - 5i)(1 + 2i)$$
$$= 4 + 8i - 5i - 10i^2$$
$$= 4 + 8i - 5i + 10$$
$$= 14 + 3i$$

Q185:

Steps:

- Let the square root be $a + bi$, then what is $(a + bi)^2$ once it is expanded and simplified as a complex number? $a^2 + 2abi - b^2$
- Equate the real term of $16 - 30i$ with the real terms in the expansion of $(a + bi)^2$. Finish the equation $16 = ?$ $16 = a^2 - b^2$
- Equate the imaginary term of $16 - 30i$ with the imaginary term in the expansion of $(a + bi)^2$. Finish the equation $-30i = ?$ $-30i = 2abi$
- Solve the simultaneous equations to find the values for a and b.

Answer:

Let the square root of $16 - 30i$ be $a + ib$
$$16 - 30i = (a + ib)^2$$
$$= a^2 + 2iab + i^2b^2$$
$$= a^2 - b^2 + 2iab$$

Equate the real and imaginary parts:
$$a^2 - b^2 = 16 \quad (1)$$
$$2ab = -30 \quad (2)$$

Rearranging (2): $\quad b = -\dfrac{15}{a} \quad (3)$

Substituting (3) into (1): $\quad a^2 - \left(-\dfrac{15}{a}\right)^2 = 16$

Simplifying: $\quad a^2 - \dfrac{225}{a^2} = 16$

Multiplying through by a^2: $\quad a^4 - 16a^2 - 225 = 0$

Factorising and solving: $\quad (a^2 + 9)(a^2 - 25) = 0$
$$a^2 = -9 \text{ or } a^2 = 25$$
$$a \in \mathbb{R} \text{ so } a^2 = -9 \text{ is impossible}$$
$$\text{so } a = \pm 5$$

Substituting into (2):

When $a = 5$, then $b = -\dfrac{15}{5} \Rightarrow b = -3$; when $a = -5$, then $b = -\dfrac{15}{-5} \Rightarrow b = 3$.

The square root of $16 - 30i$ where the real part is greater than the imaginary part is $5 - 3i$.

Q186:

Steps:

- What is the conjugate needed for the division? $2 + 5i$
- After multiplying the numerator by the conjugate, what is the new numerator?
$$(3 + 4i)(2 + 5i) = 6 + 15i + 8i + 20i^2$$
$$= -14 + 23i$$
- After multiplying the denominator by the conjugate, what is the new denominator?
$$(2 - 5i)(2 + 5i) = 4 - 25i^2$$
$$= 29$$

Answer:
$$\dfrac{3 + 4i}{2 - 5i} = \dfrac{3 + 4i}{2 - 5i} \times \dfrac{2 + 5i}{2 + 5i}$$
$$= \dfrac{6 + 15i + 8i + 20i^2}{4 - 25i^2}$$
$$= -\dfrac{14}{29} + \dfrac{23}{29}i$$

Q187:
Hints:

- The modulus is given by the formula $\sqrt{a^2 + b^2}$
- The argument of a complex number $a + bi$ depends on both the quadrant in which the complex number can be plotted and on the value of $\tan^{-1}\left|\frac{b}{a}\right|$

Steps:

- What is the modulus and principal value of the argument of z? Answer in radians.
  Modulus:
  $$|z| = \sqrt{(-3)^2 + \left(3\sqrt{3}\right)^2}$$
  Argument:
  $$\alpha = \tan^{-1}\left|-\frac{3\sqrt{3}}{3}\right|$$
  The principal argument can be determined using the quadrant diagram:

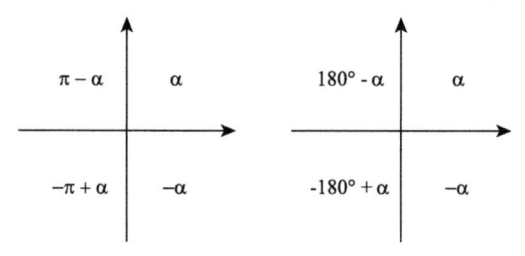

$\pi - \alpha$	α		$180° - \alpha$	α
$-\pi + \alpha$	$-\alpha$		$-180° + \alpha$	$-\alpha$

Answer:

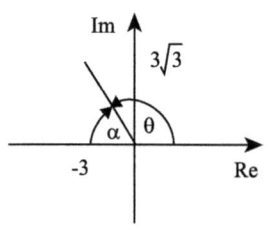

Modulus:
$$|z| = \sqrt{(-3)^2 + \left(3\sqrt{3}\right)^2}$$
$$= 6$$

Argument:
$$\alpha = \tan^{-1}\left|-\frac{3\sqrt{3}}{3}\right|$$
$$= \frac{\pi}{3}$$

z is in the 2nd quadrant so $\theta = \pi - \alpha$ and the argument is positive.
The principal argument is:
$$\theta = \pi - \frac{\pi}{3}$$
$$= \frac{2\pi}{3}$$

ANSWERS: UNIT 1 TOPIC 6   501

Since $r\cos\theta + ir\sin\theta$ is equivalent to $z = a + ib$, then we can rewrite $z = -3 + 3\sqrt{3}i$ in polar form as:
$z = 6\left(\cos\left(\frac{2\pi}{3}\right) + i\sin\left(\frac{2\pi}{3}\right)\right)$

Q188:

Hints:

- The modulus is given by the formula $\sqrt{a^2 + b^2}$
- The argument of a complex number $a + bi$ depends on both the quadrant in which the complex number can be plotted and on the value of $\tan^{-1}\left|\frac{b}{a}\right|$

Steps:

- What is the modulus and principal value of the argument of z? Answer in radians.
  Modulus:
  $|z| = \sqrt{\left(2\sqrt{3}\right)^2 + (-2)^2}$
  Argument:
  $\alpha = \tan^{-1}\left|-\frac{2}{2\sqrt{3}}\right|$
  The principal argument can be determined using the quadrant diagram:

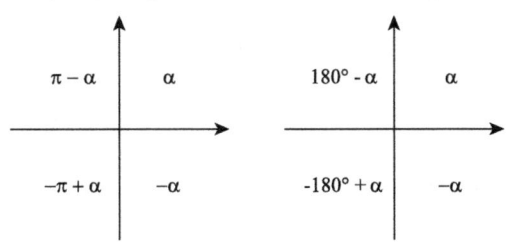

Answer:

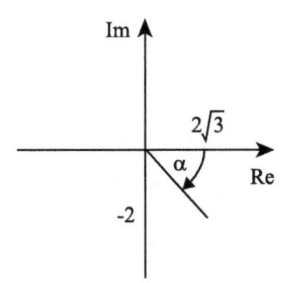

Modulus: Argument:
$|z| = \sqrt{\left(2\sqrt{3}\right)^2 + (-2)^2}$ $\alpha = \tan^{-1}\left|-\frac{2}{2\sqrt{3}}\right|$
$= 4$ $= \frac{\pi}{6}$

z is in the 4th quadrant so $\theta = -\alpha$ and the argument is negative.

© HERIOT-WATT UNIVERSITY

The principal argument is:

$\theta = -\frac{\pi}{6}$

Since $r\cos\theta + ir\sin\theta$ is equivalent to $z = a + ib$, then we can rewrite $z = 2\sqrt{3} - 2i$ in polar form as:

$z = 4\left(\cos\left(-\frac{\pi}{6}\right) + i\sin\left(-\frac{\pi}{6}\right)\right)$ or $z = 4\left(\cos\left(\frac{\pi}{6}\right) - i\sin\left(\frac{\pi}{6}\right)\right)$

Q189:

Hints:

- The modulus is given by the formula $\sqrt{a^2 + b^2}$
- The argument of a complex number $a + bi$ depends on both the quadrant in which the complex number can be plotted and on the value of $\tan^{-1}\left|\frac{b}{a}\right|$

Steps:

- What is the modulus and principal value of the argument of z? Answer in radians.
  Modulus:
  $|z| = \sqrt{(-5)^2 + (-3)^2}$
  Argument:
  $\alpha = \tan^{-1}\left|\frac{3}{5}\right|$
  The principal argument can be determined using the quadrant diagram:

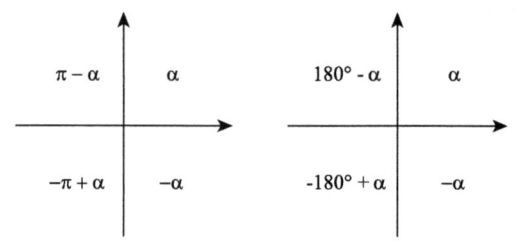

Answer:

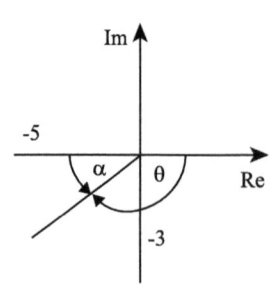

Modulus:

$|z| = \sqrt{(-5)^2 + (-3)^2}$
$= \sqrt{34}$

Argument:

$\alpha = \tan^{-1}\left|\frac{-3}{-5}\right|$
$= 0 \cdot 540$

… ANSWERS: UNIT 1 TOPIC 6

z is in the 3rd quadrant so $\theta = -\pi + \alpha$ and the argument is negative.
The principal argument is:
$$\theta = -\pi + 0 \cdot 540$$
$$= -2 \cdot 602$$
Since $r\cos\theta + ir\sin\theta$ is equivalent to $z = a + ib$, then we can rewrite $z = -5 - 3i$ in polar form as:
$$z = \sqrt{34}\left(\cos\left(-2\cdot 602\right) + i\sin\left(-2\cdot 602\right)\right) \text{ or } z = \sqrt{34}(\cos(2\cdot 602) - i\sin(2\cdot 602))$$

Q190:

Hints:

- The modulus is given by the formula $\sqrt{a^2 + b^2}$
- The argument of a complex number $a + bi$ depends on both the quadrant in which the complex number can be plotted and on the value of $\tan^{-1}\left|\frac{b}{a}\right|$

Steps:

- What is the modulus and principal value of the argument of z? Answer in radians.
  Modulus:
  $$|z| = \sqrt{(0)^2 + (-8)^2}$$
  Argument:
  $z = -8i$ lies on the negative imaginary axis so $\alpha = \frac{\pi}{2}$
  The principal argument can be determined using the quadrant diagram:

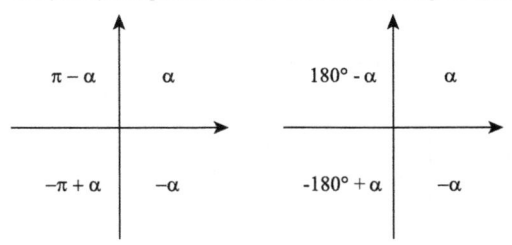

$\pi - \alpha$	α	180° - α	α
$-\pi + \alpha$	$-\alpha$	-180° + α	$-\alpha$

Answer:

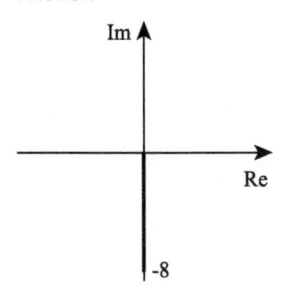

Im ↑

Re →

-8

Modulus:
$$|z| = \sqrt{(0)^2 + (-8)^2}$$
$$= 8$$
Argument:
$z = -8i$ lies on the negative imaginary axis so $\alpha = \frac{\pi}{2}$
The principal argument is:
$\theta = -\frac{\pi}{2}$
Since $r \cos \theta + ir \sin \theta$ is equivalent to $z = a + ib$, then we can rewrite $z = -8i$ in polar form as:
$z = 8\left(\cos\left(-\frac{\pi}{2}\right) + i\sin\left(-\frac{\pi}{2}\right)\right)$ or $z = 8\left(\cos\left(\frac{\pi}{2}\right) - i\sin\left(\frac{\pi}{2}\right)\right)$

Q191:

Hints:

- The modulus is given by the formula $\sqrt{a^2 + b^2}$
- The argument of a complex number $a + bi$ depends on both the quadrant in which the complex number can be plotted and on the value of $\tan^{-1}\left|\frac{b}{a}\right|$

Steps:

- What is the modulus and principal value of the argument of z? Answer in radians.
  Modulus:
  $$|z| = \sqrt{(9)^2 + (0)^2}$$
  Argument: $z = 9$ lies on the positive real axis so $\alpha = 0$
  The principal argument can be determined using the quadrant diagram:

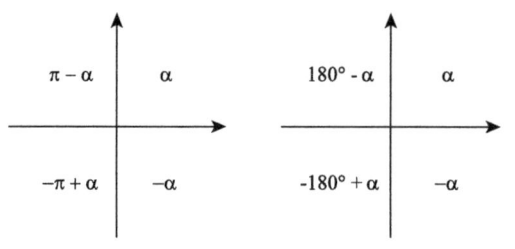

$\pi - \alpha$	α		$180° - \alpha$	α
$-\pi + \alpha$	$-\alpha$		$-180° + \alpha$	$-\alpha$

Answer:

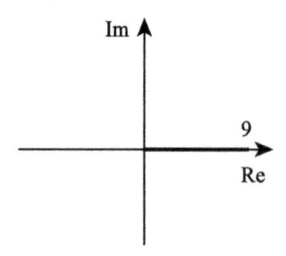

Im ↑

9 → Re

Modulus:
$$|z| = \sqrt{(9)^2 + (0)^2}$$
$$= 9$$
Argument:
$z = 9$ lies on the positive real axis so $\alpha = 0$
The principal argument is: $\theta = 0$
Since $r\cos\theta + ir\sin\theta$ is equivalent to $z = a + ib$, then we can rewrite $z = 9$ in polar form as:
$z = 9(\cos 0 + i\sin 0)$

Q192:
Hints:

- The modulus is the length of the line from the origin given by the formula $\sqrt{a^2 + b^2}$
- Arguments, $\theta \geq 0$, are drawn in an anti-clockwise direction from the positive direction of the real axis.

Steps:

- Plot z on an Argand diagram.

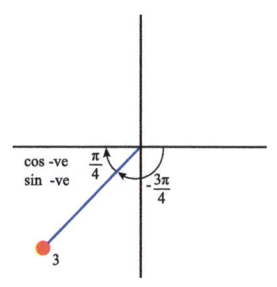

- What is the modulus and the argument? The modulus is 3 and the argument is $\frac{-3\pi}{4}$
- Given the Cartesian form $a + bi$, what does a equate to in terms of part of the expression in $\sin\theta$ and $\cos\theta$? $a = 3\cos\left(-\frac{3\pi}{4}\right)$
- Given the Cartesian form $a + bi$, what does b equate to in terms of part of the expression in $\sin\theta$ and $\cos\theta$? $b = 3\sin\left(-\frac{3\pi}{4}\right)$

Answer:
The polar form is given by: $z = r(cos\theta + isin\theta)$ where $r = 3$ and $\theta = -\frac{3\pi}{4}$
Since $\theta = -\frac{3\pi}{4}$ is not an acute angle, we must work out the equivalent acute angle. It is easiest to see this on a diagram. The principle argument is always measured from 0 anti-clockwise or clockwise. The equivalent acute angle is the smallest angle measured from z to the horizontal axis.

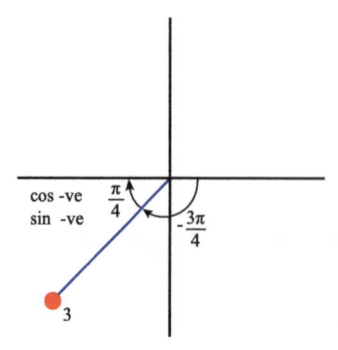

So we have:

$$z = 3\left(\cos\left(-\frac{3\pi}{4}\right) + i\sin\left(-\frac{3\pi}{4}\right)\right)$$
$$= 3\left(-\cos\left(\frac{\pi}{4}\right) - i\sin\left(\frac{\pi}{4}\right)\right)$$
$$= 3\left(-\frac{1}{\sqrt{2}} - i\frac{1}{\sqrt{2}}\right)$$
$$= -\frac{3}{\sqrt{2}} - \frac{3}{\sqrt{2}}i$$

Q193:

Hints:

- The modulus is the length of the line from the origin given by the formula $\sqrt{a^2 + b^2}$
- Arguments, $\theta \geq 0$, are drawn in an anti-clockwise direction from the positive direction of the real axis.

Steps:

- Plot z on the Argand diagram.

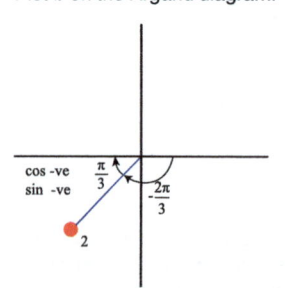

- What is the modulus and the argument? The modulus is 2 and the argument is $-\frac{2\pi}{3}$
- Given the Cartesian form $a + bi$, what does a equate to in terms of part of the expression in $\sin\theta$ and $\cos\theta$? $a = 2\cos\left(-\frac{2\pi}{3}\right)$
- Given the Cartesian form $a + bi$, what does b equate to in terms of part of the expression in $\sin\theta$ and $\cos\theta$? $b = 2\sin\left(-\frac{2\pi}{3}\right)$

Answer:

The polar form is given by:

$z = r(cos\theta + isin\theta)$ where $r = 2$ and $\theta = \frac{4\pi}{3}$

Since $\theta = \frac{4\pi}{3}$ is not an acute angle, we must work out the equivalent acute angle. It is easiest to see this on a diagram. The principle argument is always measured from 0 anti-clockwise or clockwise. The equivalent acute angle is the smallest angle measured from z to the horizontal axis. Therefore, $\theta = -\frac{2\pi}{3}$ is the principle argument and $\frac{\pi}{3}$ is the equivalent acute angle.

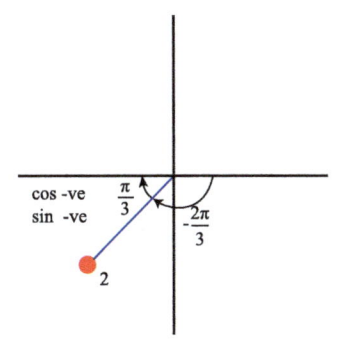

So we have:
$$z = 2\left(\cos\left(\frac{4\pi}{3}\right) + i\sin\left(\frac{4\pi}{3}\right)\right)$$
$$= 2\left(-\cos\left(\frac{\pi}{3}\right) - i\sin\left(\frac{\pi}{3}\right)\right)$$
$$= 2\left(-\frac{1}{2} - i\frac{\sqrt{3}}{2}\right)$$
$$= -1 - \sqrt{3}i$$

Q194:
Steps:

- What is the modulus of the expression $(5(cos\ y + i\sin y))^3$? 5
- What is the modulus of the answer? 125
- What is the argument of the expression $(5(cos\ y + i\sin y))^3$? y
- What is the argument of the answer? $3y$

Answer:

If $z = r(\cos\theta + i\sin\theta)$, then $z^n = r^n(\cos(n\theta) + i\sin(n\theta))$ for all $n \in \mathbb{N}$.
Thus, using de Moivre's theorem with $r = 5$ and $\theta = y$:
$$(5(\cos y + i\sin y))^3 = 5^3(\cos y + i\sin y)^3$$
$$= 125(\cos(3y) + i\sin(3y))$$

Q195:

Steps:

- What is the modulus of the expression $(3(\cos x + i \sin x))$? 3
- What is the modulus of the answer? $\frac{1}{81}$
- What is the argument of the expression $(3(\cos x + i \sin x))$? x
- What is the argument of the answer? $-4x$

Answer:

If $z = r(\cos \theta + i \sin \theta)$, then $z^n = r^n(\cos(n\theta) + i \sin(n\theta))$ for all $n \in \mathbb{N}$.
Thus, using de Moivre's theorem with $r = 3$ and $\theta = x$:

$$(3(\cos x + i \sin x))^{-4} = 3^{-4}(\cos x + i \sin x)^{-4}$$
$$= \frac{1}{81}(\cos(-4x) + i \sin(-4x))$$
$$= \frac{1}{81}(\cos(4x) - i \sin(4x))$$

Q196:

Steps:

- What is the modulus of the expression $(8(\cos x + i \sin x))^{\frac{2}{3}}$? 8
- What is the modulus of the answer? 4
- What is the argument of the expression $(8(\cos x + i \sin x))^{\frac{2}{3}}$? x
- What is the argument of the answer? $\frac{2}{3}x$

Answer:

If $z = r(\cos \theta + i \sin \theta)$, then $z^n = r^n(\cos(n\theta) + i \sin(n\theta))$ for all $n \in \mathbb{N}$.
Thus, using de Moivre's theorem with $r = 8$ and $\theta = x$:

$$(8(\cos x + i \sin x))^{\frac{2}{3}} = 8^{\frac{2}{3}}(\cos x + i \sin x)^{\frac{2}{3}}$$
$$= \left(\sqrt[3]{8}\right)^2 \left(\cos\left(\frac{2}{3}x\right) + i \sin\left(\frac{2}{3}x\right)\right)$$
$$= 4\left(\cos\left(\frac{2}{3}x\right) + i \sin\left(\frac{2}{3}x\right)\right)$$

Q197:

Hints:

- The modulus is given by the formula $\sqrt{a^2 + b^2}$
- Arguments, $\theta \geq 0$, are drawn in an anti-clockwise direction from the positive direction of the real axis.

Steps:

- What is the modulus of $-3 - 2i$?
$$|z| = \sqrt{(-3)^2 + (-2)^2}$$
$$= \sqrt{13}$$

- What is the principal argument (in radians) of $-3 - 2i$?

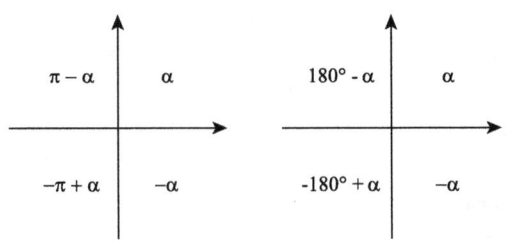

The principal argument is in the third quadrant so:
$arg(z) = -\pi + 0 \cdot 588$
$= -2 \cdot 554$

- What is the correct formula for z^n? $r^n(cos(n\theta) + i\sin(n\theta))$

Answer:

$z = -3 - 2i$ but we need it in the form $z(cos\,\theta + i\sin\theta)$ to use de Moivre's theorem.
Consider its representation on an Argand diagram:

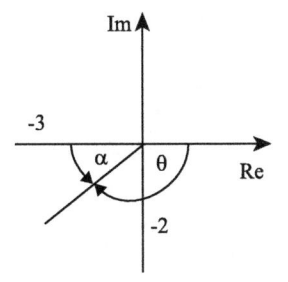

Modulus:
$|z| = \sqrt{(-3)^2 + (-2)^2}$
$= \sqrt{13}$

Argument:
As the argument is in the third quadrant, $arg(z) = -\pi + \alpha$, where $\tan\alpha = \left|\frac{-2}{-3}\right|$, i.e. $\alpha = 0 \cdot 588$
Therefore, $\theta = -2 \cdot 554$
Using de Moivre's theorem with $r = \sqrt{13}$ and $\theta = -2 \cdot 554$
$z^6 = \left(\sqrt{13}\right)^6 (\cos(-2 \cdot 554) + i\sin(-2 \cdot 554))^6$
$= 2197 \times (\cos(-15 \cdot 324) + i\sin(-15 \cdot 324))$
The angle $-15 \cdot 324$ is an anti-clockwise rotation. It is equivalent to $-4\pi - 2 \cdot 758$
$z^6 = 2197 \times (\cos(-2 \cdot 758) + i\sin(-2 \cdot 758))$
$= 2197 \times (\cos(2 \cdot 758) - i\sin(2 \cdot 758))$

© HERIOT-WATT UNIVERSITY

Q198:
Steps:

- Given $z = x + iy$, what is $z - 5$ in Cartesian form? $(x - 5) + iy$

Answer:

Remember that $|z| = \sqrt{x^2 + y^2}$, where $z = x + iy$
Substituting $z = x + iy$ in $|z - 5| = 8$: $|x + iy - 5| = 8$
Grouping the real and imaginary parts together: $|(x - 5) + iy| = 8$
Squaring both sides: $(x - 5)^2 + y^2 = 8^2$
This is a circle with centre $(5, 0)$ and radius 8

Q199:
Steps:

- Given $z = x + iy$, what is $z + 3i$ in Cartesian form? $x + i(y + 3)$

Answer:

Remember that $|z| = \sqrt{x^2 + y^2}$, where $z = x + iy$
Substituting $z = x + iy$ in $|z + 3i| = 4$: $|x + iy + 3i| = 4$
Grouping the real and imaginary parts together: $|x + i(y + 3)| = 4$
Squaring both sides: $x^2 + (y + 3)^2 = 4^2$
This is a circle with centre $(0, -3)$ and radius 4

Q200:
Steps:

- Given $z = x + iy$, what is $z - 4 + 2i$ in Cartesian form? $(x - 4) + i(y + 2)$

Answer:

Remember that $|z| = \sqrt{x^2 + y^2}$, where $z = x + iy$
Substituting $z = x + iy$ in $|z - 4 + 2i| = 6$: $|x + iy - 4 + 2i| = 6$
Grouping the real and imaginary parts together: $|(x - 4) + i(y + 2)| = 6$
Squaring both sides: $(x - 4)^2 + (y + 2)^2 = 6^2$
This is a circle with centre $(4, -2)$ and radius 6

Q201:
Hints:

- The modulus is given by the formula $\sqrt{a^2 + b^2}$

Steps:

- Given $z = x + iy$, what is $|z + 2i|$ in Cartesian form? $x + i(y + 2)$
- Given $z = x + iy$, what is $|z - 3i|$ in Cartesian form? $x + i(y - 3)$

Answer:

Let $z = x + iy$

$|z + 2i| = |z - 3i|$

Replace z with $x + iy$: $|x + iy + 2i| = |x + iy - 3i|$

Group the real and imaginary parts together: $|x + i(y + 2)| = |x + i(y - 3)|$

Square both sides:

$|x + i(y+2)|^2 = |x + i(y-3)|^2$

$x^2 + (y+2)^2 = x^2 + (y-3)^2$

Expand out and simplify:

$x^2 + y^2 + 4y + 4 = x^2 + y^2 - 6y + 9$

$\qquad\quad 4y + 4 = -6y + 9$

$\qquad\quad\quad 10y = 5$

$\qquad\quad\quad\quad y = \dfrac{1}{2}$

Q202:

Hints:

- The modulus is given by the formula $\sqrt{a^2 + b^2}$.

Steps:

- Given $z = x + iy$, what is $z + 2i$ in Cartesian form? $x + i(y + 2)$
- Given $z = x + iy$, what is $z - 5$ in Cartesian form? $(x - 5) + iy$

Answer:

Let $z = x + iy$

$|z + 2i| = |z - 5|$

Replace z with $x + iy$: $|x + iy + 2i| = |x + iy - 5|$

Group the real and imaginary parts together: $|x + i(y + 2)| = |(x - 5) + iy|$

Square both sides:

$|x + i(y+2)|^2 = |(x-5) + iy|^2$

$x^2 + (y+2)^2 = (x-5)^2 + y^2$

Expand out and simplify:

$x^2 + y^2 + 4y + 4 = x^2 - 10x + 25 + y^2$

$\qquad\quad 4y + 4 = -10x + 25$

$\qquad\quad\quad 4y = -10x + 21$

$\qquad\quad\quad\quad y = -\dfrac{5}{2}x + \dfrac{21}{4}$

© HERIOT-WATT UNIVERSITY

Q203:

Hints:

- The modulus is given by the formula $\sqrt{a^2 + b^2}$.

Steps:

- Given $z = x + iy$, what is $z - 4$ in Cartesian form? $(x - 4) + iy$
- Given $z = x + iy$, what is $z - 7$ in Cartesian form? $(x - 7) + iy$

Answer:

Let $z = x + iy$

$|z - 4| = |z - 7|$

Replace z with $x + iy$: $|x + iy - 4| = |x + iy - 7|$

Group the real and imaginary parts together: $|(x - 4) + iy| = |(x - 7) + iy|$

Square both sides:

$|(x - 4) + iy|^2 = |(x - 7) + iy|^2$

$(x - 4)^2 + y^2 = (x - 7)^2 + y^2$

Expand out and simplify:

$x^2 - 8x + 16 + y^2 = x^2 - 14x + 49 + y^2$

$-8x + 16 = -14x + 49$

$6x = 33$

$x = \dfrac{11}{2}$

Q204:

Steps:

- From the root $-3 + i$, what is the first factor of the quadratic in the form $(x - k)$? $x - (-3 + i)$
- From the root $2 - 5i$, what is the first factor of the quadratic in the form $(x - k)$? $x - (2 - 5i)$
- By multiplying the two factors obtained in the previous steps and rearranging, what is the quadratic obtained in the form $x^2 + bx + c$? Hence determine b and c.

Answer:

If $-3 + i$ is a root of $x^2 + bx + c$, then a factor is $x - (-3 + i) = x + 3 - i$

If $2 - 5i$ is a root of $x^2 + bx + c$, then a factor is $x - (2 - 5i) = x - 2 + 5i$

Multiplying the two factors:

$(x + 3 - i)(x - 2 + 5i) = x^2 - 2x + 5ix + 3x - 6 + 15i - ix + 2i + 5$

$= x^2 + x + 4ix + 17i - 1$

$= x^2 + (1 + 4i)x + 17i - 1$

Hence, $b = 1 + 4i$ and $c = 17i - 1$

ANSWERS: UNIT 1 TOPIC 6

Q205:
Hints:
- If the complex number $a + bi$ is a root, then its conjugate $a - bi$ is also a root.

Steps:
- The cubic has a root of $2 + 3i$. What is another? $2 - 3i$
- The root $(a + bi)$ gives a factor $(x - (a + bi))$. What is the factor from the root $2 + 3i$? $(x - (2 + 3i))$
- The root $(a - bi)$ gives a factor $(x - (a - bi))$. What is the factor from the root $2 - 3i$? $(x - (2 - 3i))$
- Using these two roots, what is the quadratic factor Q_1 which they produce in the form $x^2 + bx + c$? $x^2 - 4x + 13$
- By the division of the cubic by the quadratic factor Q_1, what is the other factor?

Answer:

If $2 + 3i$ is a root of $x^3 - 7x^2 + 25x - 39 = 0$, then a factor is $x - (2 + 3i) = x - 2 - 3i$

If $2 + 3i$ is a root of $x^3 - 7x^2 + 25x - 39 = 0$, then another root is its complex conjugate $2 - 3i$

If $2 - 3i$ is a root of $x^3 - 7x^2 + 25x - 39 = 0$, then a factor is $x - (2 - 3i) = x - 2 + 3i$

The quadratic factor of the cubic is therefore given by:
$$(x - 2 - 3i)(x - 2 + 3i) = x^2 - 2x + 3ix - 2x + 4 - 6i - 3ix + 6i - 9i^2$$
$$= x^2 - 4x + 13$$

Using long division:

$$\begin{array}{r} x - 3 \\ x^2 - 4x + 13 \overline{) x^3 - 7x^2 + 25x - 39} \\ -x^3 + 4x^2 - 13x \\ \overline{-3x^2 + 12x - 39} \\ 3x^2 - 12x + 39 \\ \overline{0} \end{array}$$

The cubic roots of $x^3 - 7x^2 + 25x - 39 = 0$ are $2 + 3i$, $2 - 3i$ and $x - 3$.

Q206:
Hints:
- If the complex number $a + bi$ is a root, then its conjugate $a - bi$ is also a root.
- The quadratic is found by multiplying the two factors which are in the form $(x - (a + bi))$ and $(x - (a - bi))$.

Steps:
- The quartic has a root of $-1 - i$. What is another? $-1 + i$
- The root $(a - bi)$ gives a factor $(z - (a - bi))$. What is the factor from the root $-1 - i$? $(z - (-1 - i))$
- The root $(a + bi)$ gives a factor $(z - (a + bi))$. What is the factor from the root $-1 + i$? $(z - (-1 + i))$

© HERIOT-WATT UNIVERSITY

- Using these two roots, what is the quadratic factor Q_1 which they produce in the form $z^2 + bz + c$? $z^2 + 2z + 2$
- By the division of the quartic by the quadratic factor Q_1, what is the other quadratic factor Q_2? $z^2 - 4z + 5$
- For this second quadratic Q_2, what are the complex factors and give the one which is in the form $a + bi$?

Answer:

If $-1 - i$ is a root of $z^4 - 2z^3 - z^2 + 2z + 10 = 0$, then another root is its conjugate $-1 + i$

This gives the factors:

$(z - (-1 - i))$ and $(z - (-1 + i))$

Multiplying these will give another factor:

$$(z - (-1 - i))(z - (-1 + i)) = (z + 1 + i)(z + 1 - i)$$
$$= z^2 + z - iz + z + 1 + iz + i + 1$$
$$= z^2 + 2z + 2$$

Using long division, we can find the remaining factors:

$$\begin{array}{r}
z^2 - 4z + 5 \\
z^2 + 2z + 2 \overline{\smash{\big)}\ z^4 - 2z^3 - z^2 + 2z + 10} \\
\underline{z^4 + 2z^3 + 2z^2} \\
-4z^3 - 2z^2 + 2z \\
\underline{-4z^3 - 8z^2 - 8z} \\
5z^2 + 10z + 10 \\
\underline{5z^2 + 10z + 10} \\
0
\end{array}$$

$z^4 - 2z^3 - z^2 + 2z + 10 = (z^2 + 2z + 2)(z^2 - 4z + 5)$

The other quadratic factor is: $z^2 - 4z + 5$

Using the quadratic formula:

$$z = \frac{-b \pm \sqrt{b^2 - 4ac}}{2a}$$
$$= \frac{4 \pm \sqrt{(-4)^2 - 4(1)(5)}}{2}$$
$$= \frac{4 \pm \sqrt{-4}}{2}$$
$$= \frac{4 \pm 2i}{2}$$
$$= 2 \pm i$$

The quartic roots of $z^4 - 2z^3 - z^2 + 2z + 10 = 0$ are:

- $z = -1 - i$
- $z = -1 + i$
- $z = 2 + i$
- $z = 2 - i$

Q207:

Steps:

- What is the modulus of $\sqrt{3} - i$? $|z| = 2$
- Which quadrant is the argument in? Fourth quadrant
- Using de Moivre's theorem, express $\sqrt{3} - i$ in polar form using exact values. Answer in radians.
  $2\left(\cos\left(-\frac{\pi}{6}\right) + i\sin\left(-\frac{\pi}{6}\right)\right)$
- What is the correct formula for z^n? $r^n(cos(nt) + i\sin(nt))$
- Using de Moivre's theorem, express $(\sqrt{3} - i)^5$ in polar form using exact values.
  $32\left(\cos\left(-\frac{5\pi}{6}\right) + i\sin\left(-\frac{5\pi}{6}\right)\right)$
- What is the modulus of $1 + i$? $|z| = \sqrt{2}$
- Which quadrant is the argument in? First quadrant
- Using de Moivre's theorem, express $(1 + i)^3$ in polar form. Answer in radians.
  $2\sqrt{2}\left(\cos\left(\frac{3\pi}{4}\right) + i\sin\left(\frac{3\pi}{4}\right)\right)$
- What is correct form when dividing in polar form? $\frac{|z_1|}{|z_2|}$, $\arg(\theta_1 - \theta_2)$

Answer:

Convert $\sqrt{3} - i$ into polar form.

Modulus:

$|z| = \sqrt{\left(\sqrt{3}\right)^2 + (-1)^2}$
$= 2$

Argument:

$\tan\alpha = \left|\frac{-1}{\sqrt{3}}\right|$

$\alpha = \frac{\pi}{6}$

$\sqrt{3} - i$ is in the fourth quadrant so $\arg(z) = -\alpha$

$\arg(z) = -\frac{\pi}{6}$

Polar form:

$z = 2\left(\cos\left(-\frac{\pi}{6}\right) + i\sin\left(-\frac{\pi}{6}\right)\right)$

Using de Moivre's theorem:

$$\left(\sqrt{3} - i\right)^5 = 2^5\left(\cos\left(-\frac{\pi}{6}\right) + i\sin\left(-\frac{\pi}{6}\right)\right)^5$$

$$= 32\left(\cos\left(-\frac{5\pi}{6}\right) + i\sin\left(-\frac{5\pi}{6}\right)\right)$$

Convert $1 + i$ into polar form.

Modulus:

$|z| = \sqrt{1^2 + 1^2}$
$= \sqrt{2}$

Argument:

$\tan\alpha = \left|\frac{1}{1}\right|$

$\alpha = \frac{\pi}{4}$

$1 + i$ is in the first quadrant so $\arg(z) = \alpha$

$\arg(z) = \frac{\pi}{4}$

Polar form:

$z = \sqrt{2}\left(\cos\left(\frac{\pi}{4}\right) + i\sin\left(\frac{\pi}{4}\right)\right)$

Using de Moivre's theorem:

$$(1+i)^3 = \sqrt{2}^3 \left(\cos\left(\frac{\pi}{4}\right) + i\sin\left(\frac{\pi}{4}\right)\right)^3$$
$$= 2\sqrt{2}\left(\cos\left(\frac{3\pi}{4}\right) + i\sin\left(\frac{3\pi}{4}\right)\right)$$

So:

$$\frac{(\sqrt{3}-i)^5}{(1+i)^2} = \frac{32\left(\cos\left(-\frac{5\pi}{6}\right)+i\sin\left(-\frac{5\pi}{6}\right)\right)}{2\sqrt{2}\left(\cos\left(\frac{3\pi}{4}\right)+i\sin\left(\frac{3\pi}{4}\right)\right)}$$

Modulus:

$$\frac{32}{2\sqrt{2}} = \frac{16}{\sqrt{2}} \times \frac{\sqrt{2}}{\sqrt{2}}$$
$$= 8\sqrt{2}$$

Argument:

$$-\frac{5\pi}{6} - \frac{3\pi}{4} = -\frac{38\pi}{24}$$
$$= -\frac{19\pi}{12}$$
$$= \frac{5\pi}{12}$$

Solutions: $8\sqrt{2}\left(\cos\left(\frac{5\pi}{12}\right) + i\sin\left(\frac{5\pi}{12}\right)\right)$

Q208:

Steps:

- Expand $(\cos(3\theta) + i\sin(3\theta))^5$ using de Moivre's theorem. $\cos(15\theta) + i\sin(15\theta)$
- What is the correct form of the modulus and argument when dividing in polar form? $\frac{|z_1|}{|z_2|}$, $\arg(\theta_1 - \theta_2)$

Answer:

$$\frac{(\cos(3\theta) + i\sin(3\theta))^5}{(\cos(\theta) + i\sin(\theta))^2} = \frac{\cos(15\theta) + i\sin(15\theta)}{\cos(2\theta) + i\sin(2\theta)}$$
$$= \cos(13\theta) + i\sin(13\theta)$$

Q209:

Steps:

- What is $\cos\left(\frac{\pi}{3}\right) - i\sin\left(\frac{\pi}{3}\right)$ in the form $\cos\theta + i\sin\theta$? $\cos\left(-\frac{\pi}{3}\right) + i\sin\left(-\frac{\pi}{3}\right)$
- What is an expression for $\left(\cos\left(\frac{\pi}{3}\right) - i\sin\left(\frac{\pi}{3}\right)\right)^3$ (using your answer above)? $\cos(-\pi) + i\sin(-\pi)$
- What is the correct form for the modulus and argument when dividing in polar form? $\frac{|z_1|}{|z_2|}$, $\arg(\theta_1 - \theta_2)$

Answer:

$$\frac{\left(\cos\left(\frac{\pi}{3}\right) - i\sin\left(\frac{\pi}{3}\right)\right)^3}{\left(\cos\left(\frac{\pi}{5}\right) + i\sin\left(\frac{\pi}{5}\right)\right)^4} = \frac{\left(\cos\left(-\frac{\pi}{3}\right) + i\sin\left(-\frac{\pi}{3}\right)\right)^3}{\left(\cos\left(\frac{\pi}{5}\right) + i\sin\left(\frac{\pi}{5}\right)\right)^4}$$
$$= \frac{\cos(-\pi) + i\sin(-\pi)}{\cos\left(\frac{4\pi}{5}\right) + i\sin\left(\frac{4\pi}{5}\right)}$$
$$= \cos\left(-\frac{9\pi}{5}\right) + i\sin\left(-\frac{9\pi}{5}\right)$$
$$= \cos\left(\frac{\pi}{5}\right) + i\sin\left(\frac{\pi}{5}\right)$$

ANSWERS: UNIT 1 TOPIC 6   517

Q210:

Steps:

- Using de Moivre's theorem, if $z = \cos\theta + i\sin\theta$, what is the correct formula for z^n? $z^n = \cos(n\theta) + i\sin(n\theta)$
- Using de Moivre's theorem, give an expression for z^6. $z^6 = \cos(6\theta) + i\sin(6\theta)$
- Use Pascal's triangle to evaluate coefficients. $1,\ 6i,\ -15,\ -20i,\ 15,\ 6i,\ -1$
- Using the binomial expansion to express z^6 in the form $u + iv$, where u and v are expressions involving $\sin\theta$ and $\cos\theta$, give the coefficient of each term starting with $\cos^6\theta$.
$(\cos\theta + i\sin\theta)^6 = \cos^6\theta + i6\cos^5\theta\sin\theta - 15\cos^4\theta\sin^2\theta - i20\cos^3\theta\sin^3\theta + 15\cos^2\theta\sin^4\theta + i6\cos\theta\sin^5\theta - \sin^6\theta$
- State $\cos^2\theta$ as an expression of $\sin^2\theta$. $1 - \sin^2\theta$

Answer:

If $z = \cos\theta + i\sin\theta$, then $z^n = \cos(n\theta) + i\sin(n\theta)$
Using de Moivre's theorem: $z^6 = \cos(6\theta) + i\sin(6\theta)$
Using the binomial expansion:
$(\cos\theta + i\sin\theta)^6 = \cos^6\theta + i6\cos^5\theta\sin\theta - 15\cos^4\theta\sin^2\theta - i20\cos^3\theta\sin^3\theta + 15\cos^2\theta\sin^4\theta + i6\cos\theta\sin^5\theta - \sin^6\theta$
Equating the real parts: $\cos(6\theta) = \cos^6\theta - 15\cos^4\theta\sin^2\theta + 15\cos^2\theta\sin^4\theta - \sin^6\theta$
Substituting for $\cos^2\theta = 1 - \sin^2\theta$:
$$\cos(6\theta) = \left(1 - \sin^2\theta\right)^3 - 15\left(1 - \sin^2\theta\right)^2\sin^2\theta + 15\left(1 - \sin^2\theta\right)\sin^4\theta - \sin^6\theta$$
$$= \left(1 - 3\sin^2\theta + 3\sin^4\theta - \sin^6\theta\right) - 15\left(1 - 2\sin^2\theta + \sin^4\theta\right)\sin^2\theta + 15\sin^4\theta - 15\sin^6\theta - \sin^6\theta$$
$$= 1 - 3\sin^2\theta + 3\sin^4\theta - \sin^6\theta - \left(15\sin^2\theta - 30\sin^4\theta + 15\sin^6\theta\right) + 15\sin^4\theta - 15\sin^6\theta - \sin^6\theta$$
$$= 1 - 18\sin^2\theta + 48\sin^4\theta - 32\sin^6\theta$$

Q211:

Steps:

- Using de Moivre's theorem, if $z = \cos\theta + i\sin\theta$, what is the correct formula for z^n? $z^n = \cos(n\theta) + i\sin(n\theta)$
- Using de Moivre's theorem, what is an expression for z^3? $z^3 = \cos(3\theta) + i\sin(3\theta)$
- Using Binomial expansion to express z^3 in the form $u + iv$ where u and v are expressions involving $\sin\theta$ and $\cos\theta$, give the coefficient of each term starting with $\cos^3\theta$. $1,\ 3i,\ -3,\ -i$
- Use Pascal's triangle to evaluate coefficients.
$(cos(3\theta) + i\sin(3\theta)) = \cos^3\theta + 3i\cos^2\theta\sin\theta - 3\cos\theta\sin^2\theta - i\sin^3\theta$
- Evaluate $(i\sin\theta)^3$ remembering that $i^2 = -1$. $\sin(3\theta) = 3\cos^2\theta\sin\theta - \sin^3\theta$

Answer:

If $z = \cos\theta + i\sin\theta$, then $z^n = \cos(n\theta) + i\sin(n\theta)$
Using de Moivre's theorem: $z^3 = \cos(3\theta) + i\sin(3\theta)$
Using the binomial expansion:
$\cos(3\theta) + i\sin(3\theta) = \cos^3\theta + i3\cos^2\theta\sin\theta - 3\cos\theta\sin^2\theta - i\sin^3\theta$

Equating the imaginary parts: $\sin(3\theta) = 3\cos^2\theta \sin\theta - \sin^3\theta$

Substituting for $\cos^2\theta = 1 - \sin^2\theta$:

$$\sin(3\theta) = 3\left(1 - \sin^2\theta\right)\sin\theta - \sin^3\theta$$
$$= 3\sin\theta - 3\sin^3\theta - \sin^3\theta$$
$$= 3\sin\theta - 4\sin^3\theta$$

Q212:

Steps:

- What is the correct expression which is equivalent to $\tan\theta$? $\frac{\sin\theta}{\cos\theta}$
- What is an expression for $\tan(4\theta)$ in terms of $\sin(n\theta)$ and $\cos(n\theta)$? $\frac{\sin(4\theta)}{\cos(4\theta)}$
- Using de Moivre's theorem, if $z = \cos\theta + i\sin\theta$, what is the correct formula for z^n? $z^n = \cos(n\theta) + i\sin(n\theta)$
- Using de Moivre's theorem, what is an expression for z^4? $z^4 = \cos(4\theta) + i\sin(4\theta)$
- Use Pascal's triangle to evaluate coefficients. $1, 4i, -6, -4i, 1$
- Using Binomial expansion to express z^4 in the form $u + iv$ where u and v are expressions involving $\sin\theta$ and $\cos\theta$, give the coefficient of each term starting with $\cos^4\theta$.
  $\cos(4\theta) + i\sin(4\theta) = \cos^4\theta + i4\cos^3\theta\sin\theta - 6\cos^2\theta\sin^2\theta - 4i\cos\theta\sin^3\theta + \sin^4\theta$
- Evaluate $(i\sin\theta)^4$, remembering $i^2 = -1$.
  $\cos(4\theta) = \cos^4\theta - 6\cos^2\theta\sin^2\theta + \sin^4\theta$
  $= 4\cos^3\theta\sin\theta - 4\cos\theta\sin^3\theta$

Answer:

$\tan\theta$ is equivalent to $\frac{\sin\theta}{\cos\theta}$

$\tan(4\theta)$ is therefore equivalent to $\frac{\sin(4\theta)}{\cos(4\theta)}$

If $z = \cos\theta + i\sin\theta$, then $z^n = \cos(n\theta) + i\sin(n\theta)$

Using de Moivre's theorem: $z^4 = \cos(4\theta) + i\sin(4\theta)$

Using the binomial expansion:

$\cos(4\theta) + i\sin(4\theta) = \cos^4\theta + i4\cos^3\theta\sin\theta - 6\cos^2\theta\sin^2\theta - 4i\cos\theta\sin^3\theta + \sin^4\theta$

Equating the real parts: $\cos(4\theta) = \cos^4\theta - 6\cos^2\theta\sin^2\theta + \sin^4\theta$

Equating the imaginary parts: $\sin(4\theta) = 4\cos^3\theta\sin\theta - 4\cos\theta\sin^3\theta$

$$\tan(4\theta) = \frac{\sin(4\theta)}{\cos(4\theta)}$$
$$= \frac{4\cos^3\theta\sin\theta - 4\cos\theta\sin^3\theta}{\cos^4\theta - 6\cos^2\theta\sin^2\theta + \sin^4\theta}$$

Now divide by $\cos^4\theta$:

$\tan(4\theta) = \frac{4\tan\theta - 4\tan^3\theta}{1 - 6\tan^2\theta + \tan^4\theta}$

Q213:

Hints:

- The expression given in the question is raised to a power. $2\cos(n\theta)$ and its equivalent expression must also be raised to this power.

Steps:

- What expression is equivalent to $2\cos(n\theta)$? $z^n + \frac{1}{z^n}$

Answer:

$$(2\cos\theta)^4 = \left(z + \frac{1}{z}\right)^4$$

$$16\cos^4\theta = z^4 + 4z^3\frac{1}{z} + 6z^2\frac{1}{z^2} + 4z\frac{1}{z^3} + \frac{1}{z^4}$$

$$= z^4 + 4z^2 + 6 + 4\frac{1}{z^2} + \frac{1}{z^4}$$

$$= z^4 + \frac{1}{z^4} + 4\left(z^2 + \frac{1}{z^2}\right) + 6$$

$$= 2\cos(4\theta) + 4(2\cos(2\theta)) + 6$$

$$= 2\cos(4\theta) + 8\cos(2\theta) + 6$$

$$\cos^4\theta = \frac{1}{8}(\cos(4\theta) + 4\cos(2\theta) + 3)$$

Q214:

Hints:

- The expression given in the question is raised to a power. $2\cos(n\theta)$ and its equivalent expression must also be raised to this power.

Steps:

- What expression is equivalent to $2\cos(n\theta)$? $z^n - \frac{1}{z^n}$

Answer:

$$(2i\sin\theta)^6 = \left(z - \frac{1}{z}\right)^6$$

$$-64\sin^6\theta = z^6 - 6z^5\frac{1}{z} + 15z^4\frac{1}{z^2} - 20z^3\frac{1}{z^3} + 15z^2\frac{1}{z^4} - 6z\frac{1}{z^5} + \frac{1}{z^6}$$

$$= z^6 - 6z^4 + 15z^2 - 20 + 15\frac{1}{z^2} - 6\frac{1}{z^4} + \frac{1}{z^6}$$

$$= z^6 + \frac{1}{z^6} - 6\left(z^4 + \frac{1}{z^4}\right) + 15\left(z^2 + \frac{1}{z^2}\right) - 20$$

$$= (2\cos(6\theta)) - 6(2\cos(4\theta)) + 15(2\cos(2\theta)) - 20$$

$$= 2\cos(6\theta) - 12\cos(4\theta) + 30\cos(2\theta) - 20$$

$$\sin^6\theta = -\frac{1}{32}(\cos(6\theta) - 6\cos(4\theta) + 15\cos(2\theta) - 10)$$

Q215:

Hints:

- The expression given in the question is raised to a power. $2i\sin(n\theta)$ and its equivalent expression must also be raised to this power.

Steps:

- What expression is equivalent to $2i\sin(n\theta)$? $z^n - \frac{1}{z^n}$

Answer:

$$(2i\sin\theta)^3 = \left(z - \frac{1}{z}\right)^3$$

$$-8i\sin^3\theta = z^3 - 3z^2\frac{1}{z} + 3z\frac{1}{z^2} - \frac{1}{z^3}$$

$$= z^3 - 3z + 3\frac{1}{z} - \frac{1}{z^3}$$

$$= z^3 - \frac{1}{z^3} - 3\left(z - \frac{1}{z}\right)$$

$$= 2i\sin(3\theta) - 3(2i\sin\theta)$$

$$\sin^3\theta = -\frac{1}{4}(\sin(3\theta) - 3\sin\theta)$$

Q216:

Steps:

- What is the modulus and argument of $5 - 2i$?

  Modulus: Argument:

  $|z| = \sqrt{5^2 + (-2)^2}$ $\tan\alpha = \left|\frac{-2}{5}\right|$

  $= \sqrt{29}$ $\alpha = 0.381$

  Since $5 - 2i$ is in the fourth quadrant, $\arg(z) = -\alpha$

- The sine and cosine function repeat themselves every $360°$ or 2π rad. Knowing this and using de Moivre's theorem, what is the correct formula for $z^{\frac{1}{q}}$?

  $z^{\frac{1}{q}} = r^{\frac{1}{q}}\left(\cos\left(\frac{\theta + 2k\pi}{q}\right) + i\sin\left(\frac{\theta + 2k\pi}{q}\right)\right)$

- What is the formula for the nth root?

  $z^{\frac{1}{n}} = \sqrt{29}^{\frac{1}{n}}\left(\cos\left(\frac{-0.381 + 2k\pi}{n}\right) + i\sin\left(\frac{-0.381 + 2k\pi}{n}\right)\right)$

  Use this formula with values of $k = 0, 1, 2$ to find the roots.

Answer:

Modulus: Argument:

$|z| = \sqrt{5^2 + (-2)^2}$ $\tan\alpha = \left|\frac{-2}{5}\right|$

$= \sqrt{29}$ $\alpha = 0.381$

Since $5 - 2i$ is in the fourth quadrant, $\arg(z) = -\alpha$

$\arg(z) = -0 \cdot 381$

$z^3 = \sqrt{29}\left(\cos(-0.381 + 2k\pi) + i\sin(-0.381 + 2k\pi)\right)$

ANSWERS: UNIT 1 TOPIC 6

Using de Moivre's theorem:
$$z = \sqrt[6]{29}\left(\cos\left(\tfrac{-0.381+2k\pi}{3}\right) + i\sin\left(\tfrac{-0.381+2k\pi}{3}\right)\right)$$
The three roots are:

$k = 0$: $\quad z = \sqrt[6]{29}\left(\cos\left(\dfrac{-0\cdot 381}{3}\right) + i\sin\left(\dfrac{-0\cdot 381}{3}\right)\right)$
$\quad\quad = 1\cdot 739 - 0\cdot 222i$

$k = 1$: $\quad z = \sqrt[6]{29}\left(\cos\left(\dfrac{-0\cdot 381 + 2\pi}{3}\right) + i\sin\left(\dfrac{-0\cdot 381 + 2\pi}{3}\right)\right)$
$\quad\quad = -0\cdot 677 + 1\cdot 617i$

$k = 2$: $\quad z = \sqrt[6]{29}\left(\cos\left(\dfrac{-0\cdot 381 + 4\pi}{3}\right) + i\sin\left(\dfrac{-0\cdot 381 + 4\pi}{3}\right)\right)$
$\quad\quad = -1\cdot 062 - 1\cdot 395i$

Q217:

Steps:

- The sine and cosine function repeat themselves every 2π rad. Knowing this and using de Moivre's theorem, what is the correct formula for $z^{\frac{1}{q}}$?
$z^{\frac{1}{q}} = r^{\frac{1}{q}}\left(\cos\left(\dfrac{\theta + 2k\pi}{q}\right) + i\sin\left(\dfrac{\theta + 2k\pi}{q}\right)\right)$

- What is the formula for the nth root?
$z^{\frac{1}{n}} = (27)^{\frac{1}{n}}\left(\cos\left(\dfrac{\tfrac{3\pi}{4} + 2k\pi}{n}\right) + i\sin\left(\dfrac{\tfrac{3\pi}{4} + 2k\pi}{n}\right)\right)$
$\quad = (27)^{\frac{1}{n}}\left(\cos\left(\dfrac{3\pi + 8k\pi}{4n}\right) + i\sin\left(\dfrac{3\pi + 8k\pi}{4n}\right)\right)$
Use this formula with values of $k = 0, 1, 2, 3$ to find the roots.

Answer:

$z^4 = 27\left(\cos\left(\tfrac{3\pi}{4}\right) + i\sin\left(\tfrac{3\pi}{4}\right)\right)$
Using de Moivre's theorem:
$z = 27^{\frac{1}{4}}\left(\cos\left(\dfrac{\tfrac{3\pi}{4} + 2k\pi}{4}\right) + i\sin\left(\dfrac{\tfrac{3\pi}{4} + 2k\pi}{4}\right)\right)$
$= 27^{\frac{1}{4}}\left(\cos\left(\dfrac{3\pi + 8k\pi}{16}\right) + i\sin\left(\dfrac{3\pi + 8k\pi}{16}\right)\right)$
The four roots are:

$k = 0: z = 27^{\frac{1}{4}}\left(\cos\left(\dfrac{3\pi}{16}\right) + i\sin\left(\dfrac{3\pi}{16}\right)\right)$
$\quad\quad z = 1\cdot 895 + 1\cdot 266i$

$k = 1: z = 27^{\frac{1}{4}}\left(\cos\left(\dfrac{11\pi}{16}\right) + i\sin\left(\dfrac{11\pi}{16}\right)\right)$
$\quad\quad z = -1\cdot 266 + 1\cdot 895i$

$k = 2: z = 27^{\frac{1}{4}}\left(\cos\left(\dfrac{19\pi}{16}\right) + i\sin\left(\dfrac{19\pi}{16}\right)\right)$
$\quad\quad z = -1\cdot 895 - 1\cdot 266i$

© HERIOT-WATT UNIVERSITY

$$k = 3 : z = 27^{\frac{1}{4}} \left(\cos \left(\frac{27\pi}{16} \right) + i \sin \left(\frac{27\pi}{16} \right) \right)$$
$$z = 1 \cdot 266 - 1 \cdot 895i$$

Q218:

Steps:

- The sine and cosine function repeat themselves every $360°$. Knowing this and using de Moivre's theorem what is the correct formula for $z^{\frac{1}{q}}$?
$z^{\frac{1}{q}} = r^{\frac{1}{q}} \left(\cos \left(\frac{\theta + 360°k}{q} \right) + i \sin \left(\frac{\theta + 360°k}{q} \right) \right)$
- What is the formula for the nth root?
$z^{\frac{1}{n}} = 32^{\frac{1}{n}} \left(\cos \left(\frac{-135° + 360°k}{n} \right) + i \sin \left(\frac{-135° + 360°k}{n} \right) \right)$
Use this formula with values of $k = 0, 1, 2, 3, 4$ to find the roots.

Answer:

$z^5 = 32(\cos(135°) + i \sin(135°))$

Using de Moivre's theorem:

$z = 32^{\frac{1}{5}} \left(\cos \left(\frac{135 + 360k}{5} \right) + i \sin \left(\frac{135 + 360k}{5} \right) \right)$

The five roots are:

$$k = 0 : z = (32)^{\frac{1}{5}} \left(\cos \left(\frac{135°}{5} \right) + i \sin \left(\frac{135°}{5} \right) \right)$$
$$z = 1 \cdot 782 + 0 \cdot 908i$$

$$k = 1 : z = (32)^{\frac{1}{5}} \left(\cos \left(\frac{495°}{5} \right) + i \sin \left(\frac{495°}{5} \right) \right)$$
$$z = -0 \cdot 313 + 1 \cdot 975i$$

$$k = 2 : z = (32)^{\frac{1}{5}} \left(\cos \left(\frac{855°}{5} \right) + i \sin \left(\frac{855°}{5} \right) \right)$$
$$z = -1 \cdot 975 + 1 \cdot 313i$$

$$k = 3 : z = (32)^{\frac{1}{5}} \left(\cos \left(\frac{1215°}{5} \right) + i \sin \left(\frac{1215°}{5} \right) \right)$$
$$z = -0 \cdot 908 - 1 \cdot 782i$$

$$k = 4 : z = (32)^{\frac{1}{5}} \left(\cos \left(\frac{1575°}{5} \right) + i \sin \left(\frac{1575°}{5} \right) \right)$$
$$z = 1 \cdot 414 - 1 \cdot 414i$$

Q219:

Steps:

- What is 1 in polar form? $\cos(0) + i \sin(0)$
- The sine and cosine function repeat themselves every $360°$ or 2π rad. Knowing this and using de Moivre's theorem, what is the correct formula for $z^{\frac{1}{q}}$?
$z^{\frac{1}{q}} = r^{\frac{1}{q}} \left(\cos \left(\frac{\theta + 2k\pi}{q} \right) + i \sin \left(\frac{\theta + 2k\pi}{q} \right) \right)$
- What is the formula for the nth root?
$z^{\frac{1}{n}} = \left(\cos \left(\frac{2k\pi}{n} \right) + i \sin \left(\frac{2k\pi}{n} \right) \right)$
Use this formula with values of $k = 0, 1, 2$ to find the roots.

Answer:

$z^3 - 1 = 0$

In to polar form: $1 = (cos(0) + i \sin(0))$

Substituting and rearranging: $z^3 = 1(cos(0) + i \sin(0))$

Using de Moivre's theorem:

$z = 1^{\frac{1}{3}} \left(\cos \left(\frac{2k\pi}{3} \right) + i \sin \left(\frac{2k\pi}{3} \right) \right)$

The three roots are:

$k = 0 : z = \left(\cos \left(\frac{0}{3} \right) + i \sin \left(\frac{0}{3} \right) \right)$

$z = 1$

$k = 1 : z = \left(\cos \left(\frac{2\pi}{3} \right) + i \sin \left(\frac{2\pi}{3} \right) \right)$

$z = -0 \cdot 5 + 0 \cdot 866i$

$k = 2 : z = \left(\cos \left(\frac{4\pi}{3} \right) + i \sin \left(\frac{4\pi}{3} \right) \right)$

$z = -0 \cdot 5 - 0 \cdot 866i$

Q220:

Steps:

- If $z = a + ib$, what is the conjugate of this? $\bar{z} = a - ib$
- Give an equation for the real part, and the imaginary part. $3(a - ib) + 2i(a + ib) = 22 + 23i$
- Equate the real and imaginary parts:

$3a - 2b = 22$

$-3b + 2a = 23$

Answer:

The conjugate of $z = a + ib$ is $\bar{z} = a - ib$

$3\bar{z} + 2iz = 22 + 23i$

$3(a - ib) + 2i(a + ib) = 22 + 23i$

$3a - 3ib + 2ia - 2b = 22 + 23i$

Equate the real and imaginary parts:

$3a - 2b = 22$ (1) $\times 2 \Rightarrow 6a - 4b = 44$ (3)

$-3b + 2a = 23$ (2) $\times 3 \Rightarrow 6a - 9b = 69$ (4)

$(3) - (4) \quad 5b = -25$

$b = -5$

Substitute into $(2): \quad 15 + 2a = 23$

$a = 4$

Therefore:

$z = 4 - 5i$

Topic 7: Sequences and series

Simple recurrence relations exercise (page 180)

Q1:
Each term in the sequence is half of the previous term.

$u_{n+1} = 1/2 u_n, u_1 = 100$

Q2:
Each day the driver travels 80 miles further than the previous day so:

$M_{n+1} = M_n + 80, M_0 = 14200$

Q3:
Hints:
$u_1 = 3$
$u_2 = 2 \times 3 + 1 = 7$
$u_3 = 2 \times 7 + 1 = 15$
$u_4 = 2 \times 15 + 1 = 31$
Answer: $u_5 = 63$

Finding a limit exercise (page 185)

Q4:
Hints:

- 2·5 does not lie between -1 and 1.

Answer: no

Q5:
Hints:

- -1 does not lie between -1 and 1.

Answer: no

Q6:
Hints:

- -1 < 0·9 < 1.

Answer: yes

Solving recurrence relations exercise (page 187)

Q7: $-2a + b$

Q8: $-17a + b$

ANSWERS: UNIT 1 TOPIC 7

Q9:
$a = 5$
$b = -7$

Q10:

Steps:

- Let $n = 1$ then $u_{n+1} = au_n + b$ becomes $u_2 = au_1 + b$ hence $7400 = ?\ 9200a + b$
- Let $n = 2$ then $u_{n+1} = au_n + b$ becomes $u_3 = au_2 + b$ hence $6050 = ?\ 7400a + b$
- Solve the above simultaneous equations to find a and b.

Answer: $a = 0 \cdot 75$ and $b = 500$

Q11:

Steps:

- Let $n = 0$ then $u_1 = au_0 + b$. Substitute in the values for u_1, a and b.

Answer: $u_0 = 11600$

Q12:

Steps:

- When $-1 < a < 1$ the recurrence relation $u_1 = au_0 + b$ converges to the limit L as n tends to infinity.
- What is the formula for L? $L = \frac{b}{1-a}$

Answer: $L = 2000$

Q13:

Steps:

- What is u_1? Give your answer in terms of b. $2 + b$
- When $n = 1$ then $u_2 = 2u_1 + b$. Hence $58 = ?$ Give your answer in terms of b. $4 + 3b$

Answer: $b = 18$

Q14:

Steps:

- Let $n = 1$ then $u_{n+1} = au_n + b$ becomes $u_2 = au_1 + b$ hence $1600 = ?$ Give your answer in terms of a and b. $6000a + b$
- Let $n = 2$ then $u_{n+1} = au_n + b$ becomes $u_3 = au_2 + b$ hence $720 = ?$ Give your answer in terms of a and b. $1600a + b$

Answer: $a = 0 \cdot 2$ and $b = 400$

Q15: 80

Q16: 400

Q17: $2x$

Q18: $\frac{10y}{9}$

Q19: $x = \frac{5}{9}y$

© HERIOT-WATT UNIVERSITY

Answers from page 189.

Q20: 10

What is a sequence exercise (page 191)

Q21:
Since $n \in \mathbb{N}$ then the first value of n is 1.

$n = 1: \quad (1-2)^3 + 2 = \quad 1$
$n = 2: \quad (2-2)^3 + 2 = \quad 2$
$n = 3: \quad (3-2)^3 + 2 = \quad 3$
$n = 4: \quad (4-2)^3 + 2 = \quad 10$
$n = 5: \quad (5-2)^3 + 2 = \quad 29$

First five terms: 1, 2, 3, 10, 29

Q22:
Since $n \in \mathbb{W}$, then the first value of n is 0.

$n = 0: \quad \dfrac{2(0)}{0+1} = 0$
$n = 1: \quad \dfrac{2(1)}{1+1} = 1$
$n = 2: \quad \dfrac{2(2)}{2+1} = \dfrac{4}{3}$
$n = 3: \quad \dfrac{2(3)}{3+1} = \dfrac{3}{2}$

Q23:
Since $n \in \mathbb{W}$, then the first value of n is 0.

$n = 0: \quad \dfrac{(0)x^0}{1-x} = 0$
$n = 1: \quad \dfrac{(1)x^1}{1-x} = \dfrac{x}{1-x}$
$n = 2: \quad \dfrac{(2)x^2}{1-x} = \dfrac{2x^2}{1-x}$
$n = 3: \quad \dfrac{(3)x^3}{1-x} = \dfrac{3x^3}{1-x}$
$n = 4: \quad \dfrac{(4)x^4}{1-x} = \dfrac{4x^4}{1-x}$

Q24:

$n = 1: \quad 11 - 3(1) = \quad 8$
$n = 2: \quad 11 - 3(2) = \quad 5$
$n = 3: \quad 11 - 3(3) = \quad 2$
$n = 4: \quad 11 - 3(4) = \quad -1$

Sequences and recurrence relations exercise (page 193)

Q25:

$n = 0: \quad \dfrac{0^2}{0+1} = 0$

$n = 1: \quad \dfrac{1^2}{1+1} = \dfrac{1}{2}$

$n = 2: \quad \dfrac{2^2}{2+1} = \dfrac{4}{3}$

First three terms are: $0, \dfrac{1}{2}, \dfrac{4}{3}$

Q26:

Since it is a recurrence relation we need to find all terms up to an u_3 before we can find u_4.

$u_1 = 15$

$u_2 = \dfrac{1}{3}(15) + 1 \quad \Rightarrow \quad u_2 = 6$

$u_3 = \dfrac{1}{3}(6) + 1 \quad \Rightarrow \quad u_3 = 3$

$u_4 = \dfrac{1}{3}(3) + 1 \quad \Rightarrow \quad u_4 = 2$

First four terms are: 15, 6, 3, 2

Q27:

Since $n \in \mathbb{N}$, then when $n = 5$ we have the fifth term.

$n = 5: \quad 3^{5-1} + 1 = 82$

Fifth term is 82.

Arithmetic sequence exercise (page 197)

Q28:

The general term for an arithmetic sequence is: $u_n = a + (n-1)d$

5th term is given by: $8 = a + 4d \quad (1)$

14th term is given by: $35 = a + 13d \quad (2)$

Using simultaneous equations:

$(2) - (1): \quad 27 = 9d \quad \Rightarrow \quad d = 3$

Substituting this into (1): $8 = a + 4(3) \quad \Rightarrow \quad a = -4$

So the sequence is defined by $\{-4 + 3(n-1)\} = \{3n - 7\}$

Q29:
The general term for an arithmetic sequence is: $u_n = a + (n-1)d$
3rd term is given by: $-7 = a + 2d$ (1)
10th term is given by: $-49 = a + 9d$ (2)
Using simultaneous equations:
$(2) - (1):\quad -42 = 7d \quad \Rightarrow \quad d = -6$
Substituting this into (1): $-7 = a + 2(-6) \quad \Rightarrow \quad a = 5$
So the sequence is defined by $\{5 + (-6)(n-1)\} = \{-6n + 11\}$.

Q30:
The general term for an arithmetic sequence is: $u_n = a + (n-1)d$
Substituting $a = -2$, $d = 5$ and $u_n = 23$ and then simplifying:
$23 = -2 + (n-1)5$
$23 = -2 + 5n - 5$
$5n = 30$
$n = 6$

Q31:
The recurrence relation for an arithmetic sequence is: $u_{n+1} = u_n + d$
So
$u_5 = u_4 + d \quad \Rightarrow \quad x + 8 = 2x + d$ (1)
$u_6 = u_5 + d \quad \Rightarrow \quad 3x + 1 = x + 8 + d$ (2)
Simplifying these:
$-x + 8 = d \quad - (1)$
$2x - 7 = d \quad - (2)$
Subtracting: (2) - (1)
$3x - 15 = 0$
$x = 5$
So
4th term: $2x = 10$
5th term: $x + 8 = 13$
6th term: $3x + 1 = 16$

ANSWERS: UNIT 1 TOPIC 7

Geometric sequence exercise (page 202)

Q32:
The general term for a geometric sequence is: $u_n = ar^{(n-1)}$
4th term is given by: $192 = ar^3$ (1)
7th term is given by: $12288 = ar^6$ (2)
Dividing (2) by (1):
$$\frac{ar^6}{ar^3} = \frac{12288}{192}$$
$$r^3 = 64$$
$$r = 4$$
Substituting this into (1): $192 = a(64) \Rightarrow a = 3$
So the sequence is defined by $\{3 \times 4^{(n-1)}\}$.

Q33:
The general term for an geometric sequence is: $u_n = ar^{(n-1)}$
3rd term is given by: $-0 \cdot 02 = ar^2$ (1)
6th term is given by: $-0 \cdot 00002 = ar^5$ (2)
Dividing (2) by (1):
$$\frac{ar^5}{ar^2} = \frac{-0 \cdot 00002}{-0 \cdot 02}$$
$$r^3 = 0 \cdot 001$$
$$r = 0 \cdot 1$$
Substituting this into (1): $-0 \cdot 02 = a(0 \cdot 01) \Rightarrow a = -2$
So the sequence is defined by $\{-2 \times 0 \cdot 1^{(n-1)}\}$.

Q34:

a) The general term for a geometric sequence is:
$u_n = ar^{(n-1)}$
The population is increasing by the common ratio 1·02.
We have:
$u_n = 1000 \times 1 \cdot 02^{(n-1)}$

b) We need to solve:
$$1000 \times 1 \cdot 02^{(n-1)} > 4000$$
$$1 \cdot 02^{(n-1)} > 4$$
$$\ln 1 \cdot 02^{(n-1)} > \ln 4$$
$$n - 1 > \frac{\ln 4}{\ln 1 \cdot 02}$$
$$n > \frac{\ln 4}{\ln 1 \cdot 02} + 1$$
$$n > 71$$
In 72 years the population will be out of danger.

© HERIOT-WATT UNIVERSITY

Sequence activities (page 206)

Expected answer

2.

Square numbers

$n =$ 1   2   3   4

The area of the square is $n \times n = n^2$

So the general term for the square numbers is $u_n = n^2$.

Pentagonal numbers

$n =$ 1   2   3   4

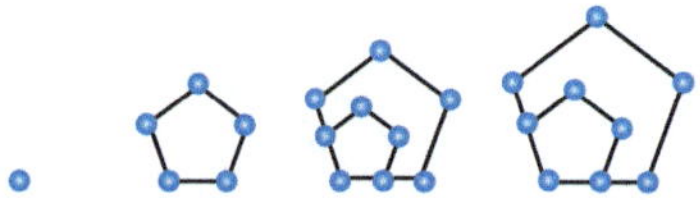

So the general term for the pentagonal numbers is $u_n = \frac{3n^2 - n}{2}$.

Hexagonal numbers

$n =$ 1   2   3   4

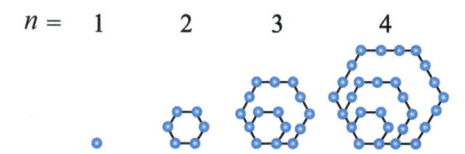

So the general term for the hexagonal numbers is $u_n = \frac{2n(2n-1)}{2}$.

3.

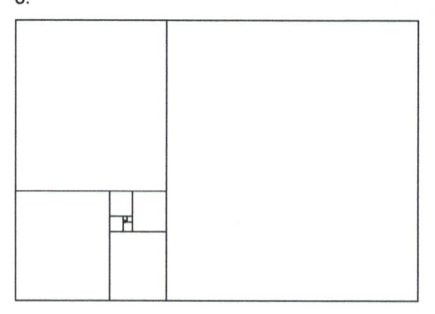

Fibonacci and other sequences exercise (page 207)

Q35:
It is an alternating sequence between 2 and -2.

2, -2, 2, -2, 2, -2, 2

Q36:
It is an alternating sequence where successive terms have a difference of itself plus 2.

$-2, 4, -6, 8, -10, 12, -14, 16$

Q37:
Add the two previous numbers to get the next.

$3, 4, 7, 11, 18, 30, 48, 78$

Q38:
Add the two previous numbers to get the next.

$-2, -5, -7, -12, -19, -31, -50$

Convergence and limits exercise (page 212)

Q39:
This sequence is a null sequence and so has a limit of zero.

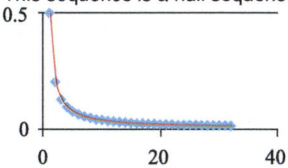

Q40:
This sequence tends to infinity.

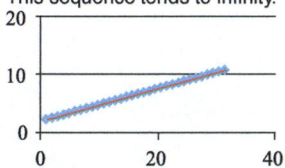

Q41:
This sequence has a limit of 10.

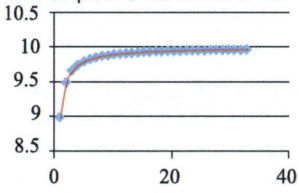

Q42:

This sequence is a null sequence. It is also an alternating sequence.

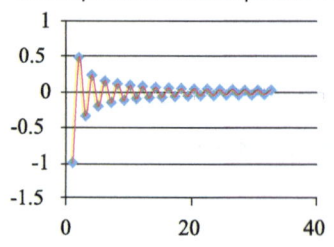

The definition of e as a limit of a sequence (page 213)

Q43:

n	$\{(1+\frac{1}{n})^n\}$	n	$\{(1+\frac{1}{n})^n\}$
1	2	11	2·6042
2	2·25	12	2·6130
3	2·3704	13	2·6206
4	2·4414	14	2·6272
5	2·4883	15	2·6329
6	2·5216	16	2·6379
7	2·5465	17	2·6424
8	2·5658	18	2·6464
9	2·5812	19	2·6500
10	2·5937	20	2·6533

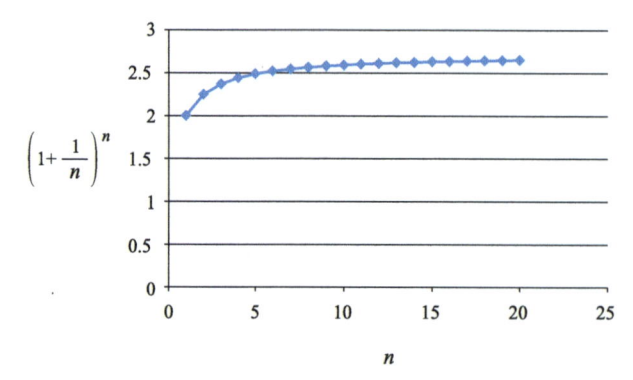

Graph of $\{(1+\frac{1}{n})^n\}$

The calculator gives the value of the twentieth term as 2·6533 (to 4 decimal places). The graph should indicate that the points are levelling out and a limit of under 3 is reasonable to suggest from the graph.

ANSWERS: UNIT 1 TOPIC 7

Q44:

n	1000	2000	3000	4000	5000	6000	7000	8000	9000	10000
$\left\{\left(1+\frac{1}{n}\right)^n\right\}$	2·7169	2·7176	2·7178	2·7179	2·7180	2·7181	2·7181	2·7181	2·7181	2·7181

The term numbers 1000, 2000, 3000 can be examined. At term number 6000 and thereafter the values of the terms of the sequence are equal to 2·7181 (to 4 decimal places).

It is reasonable at this stage to suggest that the sequence has a limit close to 2·7181.

Q45: 2·718281828...

Q46:

In symbols this is $\lim_{n\to\infty} \left(1+\frac{1}{n}\right)^n = e$.

This is read as 'the limit of the sequence one plus one over n all to the power n as n tends to infinity is e.'

The definition of π as a limit of a sequence exercise (page 215)

Q47:

A sensible limit is 6·2832. Each term after number 566 gives this value (to four decimal places).

n	$2n \tan\left(\frac{180}{n}\right)$
3	10·3923
4	8·0000
5	7·2654
6	6·9282
7	6·7420
8	6·6274
9	6·5515
10	6·4984
30	6·3063
50	6·2915
70	6·2874
91	6·2857
110	6·2849
130	6·2844
150	6·2841
170	6·2839
190	6·2838
210	6·2837
211	6·2836
212	6·2836

Q48:

The circumference of the circle is $6 \cdot 2832 = 2\pi \times 1$.

Thus π has a value of $3 \cdot 1416$ using this method.

© HERIOT-WATT UNIVERSITY

Q49:
The suggested formula is $\pi = \frac{1}{2} \lim_{n \to \infty} \{a_n\}$, where $a_n = 2n \tan\left(\frac{180}{n}\right)$ (a_n is the perimeter of a circumscribed polygon with n sides).

Series and sums exercise (page 218)

Q50:
$$\sum_{n=1}^{4} \left(\frac{1}{2}n + 3n^2\right) = \frac{1}{2}\sum_{n=1}^{4} n + 3\sum_{n=1}^{4} n^2$$
$$= \frac{1}{2}(1+2+3+4) + 3(1+4+9+16)$$
$$= 5 + 90 = 95$$

Q51:
This is more difficult since this is a sum to infinity. One way to tackle this is to write it as partial fractions and then see if any terms cancel each other out.

Using partial fractions:
$$\frac{-2}{n^2 + 8n + 15} = \frac{-2}{(n+5)(n+3)}$$
$$= \frac{A}{(n+5)} + \frac{B}{(n+3)}$$
$$-2 = A(n+3) + B(n+5)$$

Substituting $n = -3$: $B = -1$
Substituting $n = -5$: $A = 1$
$$\frac{-2}{n^2 + 8n + 15} = \frac{1}{(n+5)} - \frac{1}{(n+3)}$$
$$\sum_{n=1}^{\infty} \left(\frac{-2}{n^2 + 8n + 15}\right) = \sum_{n=1}^{\infty} \left(\frac{1}{(n+5)} - \frac{1}{(n+3)}\right)$$

Now expand out:
$$= \left(\frac{1}{6} + \frac{1}{7} + \frac{1}{8} + \cdots\right) - \left(\frac{1}{4} + \frac{1}{5} + \frac{1}{6} + \frac{1}{7} + \cdots\right)$$
$$= -\left(\frac{1}{4} + \frac{1}{5}\right)$$
$$= \frac{9}{20}$$

Q52:
$$\sum_{n=1}^{5} (3 - n) = \sum_{n=1}^{5} 3 - \sum_{n=1}^{5} n$$
$$= 3 \times 5 - (1+2+3+4+5)$$
$$= 15 - 15$$
$$= 0$$

ANSWERS: UNIT 1 TOPIC 7

Q53:

$$\sum_{n=1}^{4} (2n^2 - 4n + 1) = \sum_{n=1}^{4} (2n^2) - \sum_{n=1}^{4} (4n) + \sum_{n=1}^{4} (1)$$
$$= 2 \times (1 + 4 + 9 + 16) - 4(1 + 2 + 3 + 4) + 4 \times 1$$
$$= 60 - 20 + 4$$
$$= 24$$

Q54:

First use partial fractions to re-write this and see if any terms cancel each other out.

$$\frac{-2}{n^2 + 2n} = \frac{-2}{n(n+2)}$$
$$= \frac{A}{n} + \frac{B}{n+2}$$
$$-2 = A(n+2) + Bn$$

Substituting $n = 0$: $A = -1$
Substituting $n = -2$: $B = 1$

$$\frac{-2}{n^2 + 2n} = -\frac{1}{n} + \frac{1}{n+2}$$
$$\frac{-2}{n^2 + 2n} = \frac{1}{n+2} - \frac{1}{n}$$
$$\sum_{n=1}^{\infty} \left(\frac{-2}{n^2+2n}\right) = \sum_{n=1}^{\infty} \left(\frac{1}{n+2}\right) - \sum_{n=1}^{\infty} \left(\frac{1}{n}\right)$$

Now expand out:

$$\sum_{n=1}^{\infty} \left(\frac{-2}{n^2+2n}\right) = \sum_{n=1}^{\infty} \left(\frac{1}{n+2}\right) - \sum_{n=1}^{\infty} \left(\frac{1}{n}\right)$$
$$= \left(\frac{1}{3} + \frac{1}{4} + \frac{1}{5} + \frac{1}{6} + \ldots\right) - \left(1 + \frac{1}{2} + \frac{1}{3} + \frac{1}{4} + \ldots\right)$$
$$= -\left(1 + \frac{1}{2}\right)$$
$$= -\frac{3}{2}$$

Q55:

First use partial fractions to re-write this and see if any terms cancel each other out.

$$\frac{-3}{n^2 + 5n + 4} = \frac{-3}{(n+1)(n+4)}$$
$$= \frac{A}{n+1} + \frac{B}{n+4}$$
$$-3 = A(n+4) + B(n+1)$$

Substituting $n = -1$: $A = -1$
Substituting $n = -4$: $B = 1$

© HERIOT-WATT UNIVERSITY

$$\frac{-3}{n^2+5n+4} = -\frac{1}{n+1} + \frac{1}{n+4}$$
$$\frac{-3}{n^2+5n+4} = \frac{1}{n+4} - \frac{1}{n+1}$$
$$\sum_{n=1}^{\infty}\left(\frac{-3}{n^2+5n+4}\right) = \sum_{n=1}^{\infty}\left(\frac{1}{n+4}\right) - \sum_{n=1}^{\infty}\left(\frac{1}{n+1}\right)$$

Now expand out:

$$\sum_{n=1}^{\infty}\left(\frac{-3}{n^2+5n+4}\right) = \sum_{n=1}^{\infty}\left(\frac{1}{n+4}\right) - \sum_{n=1}^{\infty}\left(\frac{1}{n+1}\right)$$
$$= \left(\frac{1}{5} + \frac{1}{6} + \frac{1}{7} + \frac{1}{8}\ldots\right) - \left(\frac{1}{2} + \frac{1}{3} + \frac{1}{4} + \ldots\right)$$
$$= -\left(\frac{1}{2} + \frac{1}{3} + \frac{1}{4}\right)$$
$$= -\frac{13}{12}$$

The sum of the first n integers (page 220)

Expected answer

$S_n = \frac{1}{2}n(n+1)$

Why?

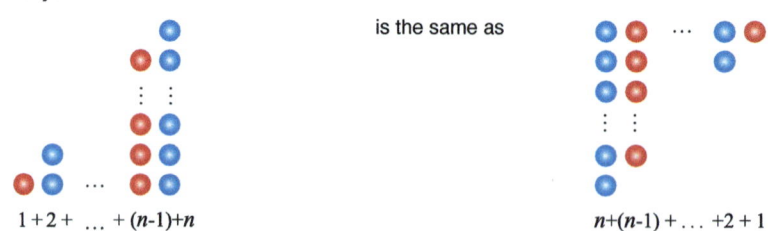

is the same as

1 + 2 + ... + (*n*-1) + *n*

n + (*n*-1) + ... + 2 + 1

Adding the two together forms a rectangle, with $area = n(n+1)$.

It is n and $n+1$ since we are taking two triangles which are exactly the same size and stacking on top of each other. Note in the highlighted box, that this creates a rectangle and not a square. So the lengths are n and $n+1$.

ANSWERS: UNIT 1 TOPIC 7

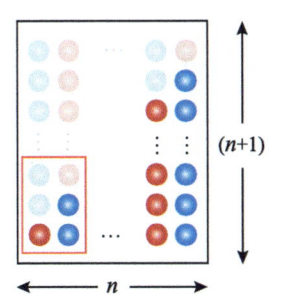

Since the series $1 + 2 + \ldots + (n-1) + n$ is half size.

$S_n = \frac{1}{2}n(n+1)$

Alternatively the formula for the sum of a series could be used to demonstrate this as well:

$S_n = \frac{n}{2}(2a + (n-1)d)$

In this case $a = 1$, $d = 1$

$S_n = \frac{n}{2}(2(1) + (n-1)(1))$
$S_n = \frac{n}{2}(n+1)$

Arithmetic series exercise (page 221)

Q56:

The sum to the nth term of an arithmetic series is given by: $S_n = \frac{n}{2}(2a + (n-1)d)$

We know:

$a = 12$
$d = 7$
$u_n = a + (n-1)d$
$355 = 12 + (n-1)7 \Rightarrow n = 50$

Now substitute into the formula S_n:

$S_{50} = \frac{50}{2}(2(12) + (49-1)7)$
$S_{50} = 9000$

Q57:
The sum to the nth term of an arithmetic series is given by: $S_n = \frac{n}{2}(2a + (n-1)d)$
We know:
$a = 0 \cdot 01$
$d = 0 \cdot 31$
$u_n = a + (n-1)d$
$2 \cdot 8 = 0 \cdot 01 + (n-1)0 \cdot 31 \Rightarrow n = 10$
Now substitute into the formula S_n:
$S_{10} = \frac{10}{2}(2(0 \cdot 01) + (10-1)0 \cdot 31)$
$S_{50} = 14 \cdot 05$

Q58:
The sum to the nth term of an arithmetic series is given by: $S_n = \frac{n}{2}(2a + (n-1)d)$
We know:
$a = 1$
$d = 1$
n in this case tends to infinity and will be denoted by n
Now substitute into the formula S_n:
$S_n = \frac{n}{2}(2(1) + (n-1)1)$
$S_n = \frac{1}{2}n(n+1)$
This is a very useful formula to remember where **sum of the natural numbers** is given by:
$\sum_{r=1}^{n} r = \frac{1}{2}n(n+1)$

Q59:
The sum to the nth term of an arithmetic series is given by:
$S_n = \frac{n}{2}(2a + (n-1)d)$
We know:
$S_{60} = 5190$
$d = 3$
Now substitute into the formula S_n:
$5190 = \frac{60}{2}(2a + (59)3)$
$a = -2$

Q60:
For an arithmetic sequence we have: $u_n = a + (n-1)d$.
so
$45 = 5 + (n-1)d$
$40 = (n-1)d \quad (1)$

For an arithmetic series we have:
$$S_n = \frac{n}{2}(2a + (n-1)d)$$
$$525 = \frac{n}{2}(10 + (n-1)d)$$
Substituting from (1):
$$525 = \frac{n}{2}(10 + 40)$$
$$1050 = 50n$$
$$n = 21$$
There are 21 terms.

Q61:

For an arithmetic sequence: $u_n = a + (n-1)d$
We know:
$u_{17} = 20$
$n = 17$
Now substitute into the formula u_n:
$20 = a + 16d$ (1)
$19 = 2a + 11d$ (2)

The sum to the nth term of an arithmetic series is given by: $S_n = \frac{n}{2}(2a + (n-1)d)$
We know:
$S_{12} = 114$
$n = 12$
Now substitute into the formula S_n:
$$114 = \frac{12}{2}(2a + 11d)$$
$19 = 2a + 11d$ (2)

Solve simultaneously:
(1) × 2 : $40 = 2a + 32d$ − (3)
(3) − (2) : $21 = 21d$ ⇒ $d = 1$
Substituting into (1): $20 = a + 16$ ⇒ $a = 4$
To find S_{18} we know:
$a = 4$
$d = 1$
$n = 18$
Substitute into the arithmetic progression formula: $S_n = \frac{n}{2}(2a + (n-1)d)$
$$S_{18} = \frac{18}{2}(2 \times 4 + 17 \times 1)$$
$$S_{18} = 225$$

Infinite series exercise (page 224)

Q62:

$$\sum_{n=1}^{6}(2n-4) = 2\sum_{n=1}^{6} n - 4\sum_{n=1}^{6}(1)$$
$$= \frac{2 \times (6 \times 7)}{2} - (4 \times 6)$$
$$= 42 - 24$$
$$= 18$$

Q63:

$$\sum_{n=1}^{7}\left(3n - \frac{1}{2}\right) = 3\sum_{n=1}^{7} n - \frac{1}{2}\sum_{n=1}^{7} 1$$
$$= 3 \times \frac{1}{2}(7)(7+1) - \frac{1}{2} \times 7$$
$$= 84 - 3 \cdot 5$$
$$= 80 \cdot 5$$

Using $\sum_{r=1}^{n} r = \frac{1}{2}n(n+1)$.

Q64:

$$\sum_{n=1}^{5}(5n^2 - 6n) = 5\sum_{n=1}^{5} n^2 - 6\sum_{n=1}^{5} n$$
$$= 5 \times \frac{1}{6}(5)(5+1)(10+1) - 6 \times \frac{1}{2} \times 5(5+1)$$
$$= 275 - 90$$
$$= 185$$

Using $\sum_{r=1}^{n} r = \frac{1}{2}n(n+1)$ and $\sum_{r=1}^{n} r^2 = \frac{n(n+1)(2n+1)}{6}$.

Q65:

Notice that we have to separate this into two sums which start from 1. Note also in the second sum that it is from 1 to **3** not 4.

$$\sum_{n=4}^{8}(3-2n) = \sum_{n=1}^{8}(3-2n) - \sum_{n=1}^{3}(3-2n)$$

Separately:

$$\sum_{n=1}^{8}(3-2n) = 3\sum_{n=1}^{8} 1 - 2\sum_{n=1}^{8} n \qquad \sum_{n=1}^{3}(3-2n) = 3\sum_{n=1}^{3} 1 - 2\sum_{n=1}^{3} n$$
$$= 3 \times 8 - 2 \times \frac{1}{2}(8)(8+1) \qquad\qquad = 3 \times 3 - 2 \times \frac{1}{2}(3)(3+1)$$
$$= -48 \qquad\qquad\qquad\qquad\qquad = -3$$

$$\sum_{n=4}^{8}(3-2n) = \sum_{n=1}^{8}(3-2n) - \sum_{n=1}^{3}(3-2n)$$
$$= -48 - (-3)$$
$$= -45$$

Q66:

Notice that we have to separate this into two sums which start from 1. Note also in the second sum that it is from 1 to **2** not 3.

$$\sum_{n=3}^{5}(4n-1) = \sum_{n=1}^{5}(4n-1) - \sum_{n=1}^{2}(4n-1)$$

Separately:

$$\sum_{n=1}^{5}(4n-1) = 4\sum_{n=1}^{5} n - \sum_{n=1}^{5} 1 \qquad\qquad \sum_{n=1}^{2}(4n-1) = 4\sum_{n=1}^{2} n - \sum_{n=1}^{2} 1$$
$$\qquad\qquad\qquad = 4 \times \frac{1}{2}(5)(5+1) - 5 \qquad\qquad\qquad\qquad = 4 \times \frac{1}{2}(2)(2+1) - 2$$
$$\qquad\qquad\qquad = 55 \qquad\qquad\qquad\qquad\qquad\qquad\qquad = 10$$

$$\sum_{n=3}^{5}(4n-1) = \sum_{n=1}^{5}(4n-1) - \sum_{n=1}^{2}(4n-1)$$
$$= 55 - 10$$
$$= 45$$

Q67:

For an arithmetic sequence: $u_n = a + (n-1)d$

We know:

$a = 14$

$d = 4$

$u_n = 90$ — last term

Now substitute into the formula u_n:

$90 = 14 + (n-1)4$

$n = 20$

So in sigma notation the arithmetic progression becomes:

$$\sum_{n=1}^{20}(14 + (n-1)4) = \sum_{n=1}^{20}(10 + 4n)$$

Q68:

For an arithmetic sequence: $u_n = a + (n-1)d$

We know:

$a = 30$

$d = -6$

$u_n = -78$ — last term

Now substitute into the formula u_n:

$-78 = 30 + (n-1)(-6)$

$n = 19$

© HERIOT-WATT UNIVERSITY

So in sigma notation the arithmetic progression becomes:
$$\sum_{n=1}^{19}(30+(n-1)(-6)) = \sum_{n=1}^{19}(36-6n)$$

Q69:
$$\sum_{r=1}^{n}(3n+2) = 3\sum_{r=1}^{n}n + 2\sum_{r=1}^{n}1$$
$$= 3 \times \frac{1}{2}(n)(n+1) + 2n$$
$$= \frac{3}{2}(n)(n+1+2)$$
$$= \frac{3}{2}(n)(n+3)$$

Using $\sum_{r=1}^{n} r = \frac{1}{2}n(n+1)$.

Q70:
$$\sum_{r=4}^{n}(6-8r) = \sum_{r=1}^{n}(6-8r) - \sum_{r=1}^{3}(6-8r)$$

Separately:
$$\sum_{r=1}^{n}(6-8r) = 6\sum_{r=1}^{n}1 - 8\sum_{r=1}^{n}r$$
$$= 6n - 8 \times \frac{1}{2}(n)(n+1)$$
$$= 2n(3-2(n+1))$$
$$= 2n - 4n^2$$

$$\sum_{r=1}^{3}(6-8r) = 6\sum_{r=1}^{3}1 - 8\sum_{r=1}^{3}r$$
$$= 6 \times 3 - 8 \times \frac{1}{2}(3)(3+1)$$
$$= -30$$

So
$$\sum_{r=4}^{n}(6-8r) = \sum_{r=1}^{n}(6-8r) - \sum_{r=1}^{3}(6-8r)$$
$$= 2n - 4n^2 - 30$$
$$= -4n^2 + 2n - 30$$

Geometric series exercise (page 227)

Q71:

The sum to the nth term of a geometric series is given by:
$S_n = \frac{a(1-r^n)}{1-r}$

We know:
$a = 4$
$r = -2$
$n = 9$

To find n:
For the nth term of a geometric sequence: $u_n = ar^{n-1}$

ANSWERS: UNIT 1 TOPIC 7

So $1024 = 4 \times (-2)^{n-1}$

Rearranging and taking ln of both sides: $\ln 256 = \ln (-2)^{n-1}$

Now solving:

$n = \dfrac{\ln 256}{\ln |-2|} + 1$

$n = 9$

Substituting into the formula $S_n = \frac{a(1-r^n)}{1-r}$:

$S_9 = \dfrac{4\left(1 - (-2)^9\right)}{1 - (-2)}$

$S_9 = 684$

Q72:

The sum to the nth term of a geometric series is given by:

$S_n = \frac{a(1-r^n)}{1-r}$

We know:

$a = 4$

$r = -\dfrac{1}{12}$

$n = 12$

Substituting into the formula $S_n = \frac{a(1-r^n)}{1-r}$:

$S_{12} = \dfrac{4\left(1 - \left(-\frac{1}{12}\right)^{12}\right)}{1 - \left(-\frac{1}{12}\right)}$

$S_{12} = 3 \cdot 69$

Q73:

The sum to the nth term of a geometric series is given by:

$S_n = \frac{a(1-r^n)}{1-r}$

We know:

$a = 3$

$r = 3$

$S_n = 9840$

Substituting into the formula $S_n = \frac{a(1-r^n)}{1-r}$:

$9840 = \dfrac{3\left(1 - (3)^n\right)}{1 - (3)}$

$-19680 = 3 - 3^{n+1}$

$19683 = 3^{n+1}$

Taking the ln of both sides and solving:

$\ln 3^{n+1} = \ln 19683$

$n = \dfrac{\ln 19683}{\ln 3} - 1$

$n = 8$

© HERIOT-WATT UNIVERSITY

Q74:
The sum to the nth term of a geometric series is given by:
$S_n = \frac{a(1-r^n)}{1-r}$
We know:
$r = 4$
$S_4 = -850$
Substituting into the formula $S_n = \frac{a(1-r^n)}{1-r}$:

$-850 = \dfrac{a\left(1-(4)^4\right)}{1-(4)}$

$-850 = \dfrac{a \times (-255)}{-3}$

$a = -10$

Q75:
The nth term of a geometric series is given by: $u_n = ar^{n-1}$
We know that:
$u_3 = ar^2 \Rightarrow \dfrac{3}{2}$ — (1)

$u_5 = ar^4 \Rightarrow \dfrac{3}{8}$ — (2)

Dividing (2) by (1):

$\dfrac{ar^4}{ar^2} = \dfrac{\frac{3}{8}}{\frac{3}{2}}$

$r^2 = \dfrac{1}{4}$

$r = \dfrac{1}{2}$

Substituting: $r = \frac{1}{2}$ into (1):

$a\left(\dfrac{1}{2}\right)^2 = \dfrac{3}{2}$

$a = 6$

The sum to the nth term of a geometric series is given by: $S_n = \frac{a(1-r^n)}{1-r}$

$S_4 = \dfrac{6\left(1-\left(\frac{1}{2}\right)^4\right)}{1-\left(\frac{1}{2}\right)}$

$S_4 = \dfrac{45}{4}$

Q76:
Using the definition $S_n = a + ar + ar^2 + \ldots + ar^{n-2} + ar^{n-1}$:
We have: $S_2 = a + ar$
So $4 = a + ar$ — (1)
Starting with: $108 = ar^3 + ar^4$
Factorising: $108 = r^3(a + ar)$
$108 = r^3(4)$
$r = 3$

ANSWERS: UNIT 1 TOPIC 7

So the first term is found by substituting into (1):
$4 = a(1+r)$
$4 = a(1+3)$
$a = 1$

General term is given by: $\sum_{r=1}^{n} ar^{n-1}$

So the series is given by: $S_n = \sum_{k=1}^{n} 3^{k-1}$

Convergent geometric series exercise (page 229)

Q77:

a) This is not convergent since $r = 2$ i.e. r is outwith the range $-1 < r < 1$.
b) This is convergent with $a = 32$ and $r = \frac{1}{2}$ giving $S_\infty = 64$.
c) This is convergent with $a = 32$ and $r = \frac{-1}{2}$ giving $S_\infty = \frac{64}{3}$.
d) This is convergent with $a = \frac{1}{3}$ and $r = \frac{1}{2}$ giving S_∞ as $\frac{2}{3}$.
e) This is not convergent since $r = -3$ i.e. r is outwith the range $-1 < r < 1$.

Q78:

We know that the sum to infinity of a convergent geometric series is given by:
$S_\infty = \frac{a}{1-r}$

We know:
$a = 1$
$r = \frac{1/2}{1} = \frac{1}{2}$

Applying the S_∞:
$S_\infty = \frac{1}{1-\frac{1}{2}}$
$= 2$

Q79:

$9 = \frac{a}{1-r} = \frac{a}{1-\frac{2}{3}} = 3a$

So
$a = 3$

Q80:

We know that the sum to infinity of a convergent geometric series is given by:
$S_\infty = \frac{a}{1-r}$

We know:
$S_\infty = 72$
$r = \frac{5}{6}$

© HERIOT-WATT UNIVERSITY

Applying the S_∞:

$72 = \dfrac{a}{1 - \frac{5}{6}}$

$a = 12$

We know that the nth term of a geometric series is given by: $u_n = ar^{n-1}$

Applying this:

$u_6 = 12 \times \left(\dfrac{5}{6}\right)^{6-1}$

$= \dfrac{3125}{648}$

Q81:

We know that the nth term of a geometric series is given by: $u_n = ar^{n-1}$

Applying this:

We know:　　　　　　　　　　　　　　We know:

$u_2 = ar \Rightarrow 4 = ar \quad -(1)$　　　　$u_4 = ar^3 \Rightarrow 1 = ar^3 \quad -(2)$

Dividing (2) by (1):

$\dfrac{ar^3}{ar} = \dfrac{1}{4} \Rightarrow r^2 = \dfrac{1}{4}$

$r = \dfrac{1}{2}$

Since all terms are positive the common ratio difference must be positive.

Now substitute $r = \frac{1}{2}$ into (1) to find a:

$4 = a\frac{1}{2} \Rightarrow a = 8$

We know that the sum to infinity of a convergent geometric series is given by: $S_\infty = \dfrac{a}{1-r}$

We know:

$a = 8$

$r = \dfrac{1}{2}$

Applying the S_∞:

$S_\infty = \dfrac{8}{1 - \frac{1}{2}}$

$S_\infty = 16$

Q82:

We can express 0·370370... as 0·370 + 0·000370 + 0·000000370 + ...

This is a geometric series where:

$a = 0 \cdot 370$
$r = 0 \cdot 001$, since $|r| < 1$ then S_∞ will exist.

Substituting into: $S_\infty = \dfrac{a}{1-r}$

$S_\infty = \dfrac{0 \cdot 370}{1 - 0 \cdot 001}$

$= \dfrac{10}{27}$

ANSWERS: UNIT 1 TOPIC 7   547

Binomial theorem and the geometric series exercise (page 233)

Q83:
$$(a+x)^{-1} = a^{-1}\left(1+\frac{x}{a}\right)^{-1}$$
$$= \frac{1}{a}\left[1+\left(\frac{-1}{1!}\right)\left(\frac{x}{a}\right)+\left(\frac{-1\times-2}{2!}\right)\left(\frac{x}{a}\right)^2+\left(\frac{-1\times-2\times-3}{3!}\right)\left(\frac{x}{a}\right)^3+\ldots\right]$$
$$= \frac{1}{a}-\frac{x}{a^2}+\frac{x^2}{a^3}-\frac{x^3}{a^4}$$

Q84:
If $|x| < \frac{2}{3}$, then rearranging $\left|\frac{3x}{2}\right| < 1$.
This means we take out a factor of 2^{-1}:
$$2^{-1}\left(1+\frac{3x}{2}\right)^{-1} = \frac{1}{2}\left(1+(-1)\frac{3x}{2}+\frac{(-1)(-2)}{2!}\left(\frac{3x}{2}\right)^2+\frac{(-1)(-2)(-3)}{3!}\left(\frac{3x}{2}\right)^3+\ldots\right)$$
$$= \frac{1}{2}-\frac{3x}{4}+\frac{9x^2}{8}-\frac{27x^3}{16}+\ldots$$

Q85: $\frac{1}{x-5} = (x-5)^{-1}$

Solution 1: Take out a factor of $(-5)^{-1}$
Note: take out $(-5)^{-1}$ so that the variable is the 2nd term in the bracket
For the expansion to be valid $\left|\frac{x}{5}\right| < 1$ i.e. $|x| < 5$.
$$(-5)^{-1}\left(1+\left(-\frac{x}{5}\right)\right)^{-1}$$
$$= -\frac{1}{5}\left(1+(-1)\left(-\frac{x}{5}\right)+\frac{(-1)(-2)}{2!}\left(-\frac{x}{5}\right)^2+\frac{(-1)(-2)(-3)}{3!}\left(-\frac{x}{5}\right)^3+\ldots\right)$$
$$= -\frac{1}{5}-\frac{x}{25}-\frac{x^2}{125}-\frac{x^3}{625}-\ldots$$
or

Solution 2: Take out a factor of $(x)^{-1}$
For the expansion to be valid $\left|\frac{5}{x}\right| < 1$ i.e. $x > 5$ or $x < -5$.
$$(x)^{-1}\left(1+\left(-\frac{5}{x}\right)\right)^{-1}$$
$$= \frac{1}{x}\left(1+(-1)\left(-\frac{5}{x}\right)+\frac{(-1)(-2)}{2!}\left(-\frac{5}{x}\right)^2+\frac{(-1)(-2)(-3)}{3!}\left(-\frac{5}{x}\right)^3+\ldots\right)$$
$$= \frac{1}{x}+\frac{5}{x^2}+\frac{25}{x^3}+\frac{125}{x^4}+\ldots$$

© HERIOT-WATT UNIVERSITY

Q86: $\frac{1}{4x-3} = (4x-3)^{-1}$

Solution 1: Take out a factor of (-3)$^{-1}$

For the expansion to be valid $\left|\frac{4x}{3}\right| < 1$ i.e. $|x| < \frac{3}{4}$.

$(-3)^{-1}\left(1 + \left(-\frac{4x}{3}\right)\right)^{-1}$

$= -\frac{1}{3}\left(1 + (-1)\left(-\frac{4x}{3}\right) + \frac{(-1)(-2)}{2!}\left(-\frac{4x}{3}\right)^2 + \frac{(-1)(-2)(-3)}{3!}\left(-\frac{4x}{3}\right)^3 + \ldots\right)$

$= -\frac{1}{3} - \frac{4x}{9} - \frac{16x^2}{27} - \frac{64x^3}{81} - \ldots$

or

Solution 2: Take out a factor of (4x)$^{-1}$

For the expansion to be valid $\left|\frac{3}{4x}\right| < 1$ i.e. $x > \frac{3}{4}$ or $x < -\frac{3}{4}$.

$(4x)^{-1}\left(1 + \left(-\frac{3}{4x}\right)\right)^{-1}$

$= \frac{1}{4x}\left(1 + (-1)\left(-\frac{3}{4x}\right) + \frac{(-1)(-2)}{2!}\left(-\frac{3}{4x}\right)^2 + \frac{(-1)(-2)(-3)}{3!}\left(-\frac{3}{4x}\right)^3 + \ldots\right)$

$= \frac{1}{4x} + \frac{3}{16x^2} + \frac{9}{64x^3} + \frac{27}{256x^4} + \ldots$

Numeric expansion using the power of -1 exercise (page 234)

Q87:

$(1 + 0 \cdot 05)^{-1} = 1 - (0 \cdot 05) + (0 \cdot 05)^2 - (0 \cdot 05)^3 + \ldots$
$= 1 - 0 \cdot 05 + 0 \cdot 0025 - 0 \cdot 000125 + \ldots$
$= 0 \cdot 9524$

Q88:

$(1 \cdot 21)^{-1} = (1 + 0 \cdot 21)^{-1}$
Using the expansion $(1 + r)^{-1} = 1 - r + r^2 - r^3 + \ldots$
We have:
$(1 + 0 \cdot 21)^{-1} = 1 - 0 \cdot 21 + 0 \cdot 21^2 - 0 \cdot 21^3 + \ldots$
$= 1 - 0 \cdot 21 + 0 \cdot 0441 - 0 \cdot 009261 + \ldots$
$= 0 \cdot 8248$

Q89:

$(0 \cdot 81)^{-1} = (1 - 0 \cdot 19)^{-1}$
Using the expansion $(1 + r)^{-1} = 1 - r + r^2 - r^3 + \ldots$
We have:
$(1 + (-0 \cdot 19))^{-1} = 1 - (-0 \cdot 19) + (-0 \cdot 19)^2 - (-0 \cdot 19)^3 + \ldots$
$= 1 + 0 \cdot 19 + 0 \cdot 0361 + 0 \cdot 0006859 +$
$= 1 \cdot 2268$ **to 4 d.p.**

ANSWERS: UNIT 1 TOPIC 7

Q90:

$(0 \cdot 65)^{-1} = (1 - 0 \cdot 35)^{-1}$

Using the expansion $(1 + r)^{-1} = 1 - r + r^2 - r^3 + \ldots$

We have:

$(1 + (-0 \cdot 35))^{-1} = 1 - (-0 \cdot 35) + (-0 \cdot 35)^2 - (-0 \cdot 35)^3 + \ldots$
$= 1 + 0 \cdot 35 + 0 \cdot 1225 + 0 \cdot 042875 +$
$= 1 \cdot 5154$ to 4 d.p.

Q91:

$(1 \cdot 99)^{-1} = (1 + 0 \cdot 99)^{-1}$

Using the expansion $(1 + r)^{-1} = 1 - r + r^2 - r^3 + \ldots$

We have:

$(1 + 0 \cdot 99)^{-1} = 1 - (0 \cdot 99) + (0 \cdot 99)^2 - (0 \cdot 99)^3 + \ldots$
$= 1 - 0 \cdot 99 + 0 \cdot 9801 - 0 \cdot 970299 + \ldots$
$= 0 \cdot 0198$ to 4 d.p.

If we calculate on a calculator $\frac{1}{1 \cdot 99} = 0 \cdot 5025$. This is very different from the answer above. We would have to expand this series for too many terms to get an accurate answer.

Alternatively, we could have evaluated the number in a slightly different way.

$(1 \cdot 99)^{-1} = (2 - 0 \cdot 01)^{-1} = \frac{1}{2}(1 - 0 \cdot 005)^{-1}$

We have to take out a factor of 2 to put in into the standard form $(1 + r)^{-1}$, remembering that the factor is chosen so that $|r| < 1$.

Using the expansion $(1 + r)^{-1} = 1 - r + r^2 - r^3 + \ldots$

We have:

$\frac{1}{2}(1 - 0 \cdot 005)^{-1} = \frac{1}{2}\left(1 - (-0 \cdot 005) + (-0 \cdot 005)^2 - (-0 \cdot 005)^3 + \ldots\right)$
$= \frac{1}{2}(1 + 0 \cdot 005 + 0 \cdot 000025 + 0 \cdot 00000025 + \cdots)$
$= 0 \cdot 5025$ to 4 d.p.

This is a much more accurate approximation than the one given above.

Q92:

$(2 \cdot 1)^{-1} = (2 + 0 \cdot 1)^{-1}$

Using the expansion $(1 + r)^{-1} = 1 - r + r^2 - r^3 + \ldots$

$(2 \cdot 1)^{-1} = (2 + 0 \cdot 1)^{-1} = \frac{1}{2}(1 + 0 \cdot 05)^{-1}$

We have to take out a factor of 2 to put in into the standard form $(1 + r)^{-1}$, remembering that the factor is chosen so that $|r| < 1$.

Using the expansion $(1 + r)^{-1} = 1 - r + r^2 - r^3 + \ldots$

We have:

$\frac{1}{2}(1 + 0 \cdot 05)^{-1} = \frac{1}{2}\left(1 - (0 \cdot 05) + (0 \cdot 05)^2 - (0 \cdot 05)^3 + \ldots\right)$
$= \frac{1}{2}(1 - 0 \cdot 05 + 0 \cdot 0025 - 0 \cdot 000025 + \cdots)$
$= 0 \cdot 4761$ to 4 d.p.

© HERIOT-WATT UNIVERSITY

Q93:

$(3 \cdot 02)^{-1} = (3 + 0 \cdot 02)^{-1}$

Using the expansion $(1+r)^{-1} = 1 - r + r^2 - r^3 + \ldots$

$(3 \cdot 02)^{-1} = (3 + 0 \cdot 02)^{-1} = \frac{1}{3}(1 + 0 \cdot 01)^{-1}$

We have to take out a factor of 3 to put in into the standard form $(1+r)^{-1}$, remembering that the factor is chosen so that $|r| < 1$.

Using the expansion $(1+r)^{-1} = 1 - r + r^2 - r^3 + \ldots$

We have:

$$\frac{1}{3}(1 + 0 \cdot 01)^{-1} = \frac{1}{3}\left(1 - (0 \cdot 01) + (0 \cdot 01)^2 - (0 \cdot 01)^3 + \ldots\right)$$

$$= \frac{1}{3}(1 - 0 \cdot 01 + 0 \cdot 0001 - 0 \cdot 000001 + \cdots)$$

$$= 0 \cdot 3300 \text{ to 4 d.p.}$$

Partial sums on two common series (page 234)

Expected answer

$\sum\limits_{n=1}^{\infty} \frac{1}{n} = 1 + \frac{1}{2} + \left(\frac{1}{3} + \frac{1}{4}\right) + \left(\frac{1}{5} + \frac{1}{6} + \frac{1}{7} + \frac{1}{8}\right) + \left(\frac{1}{9} + \cdots + \frac{1}{16}\right) + \cdots$

To find the sum of this harmonic series by just adding all the fractions together is a very time consuming task and still leaves doubt as to whether the series converges or not. So we are going to be clever about it.

Notice how the terms have been grouped together into brackets. These have been collected in this way because instead of adding the fractions together to get an exact answer we are going to approximate the answers by underestimating.

Take $\left(\frac{1}{3} + \frac{1}{4}\right)$.

Instead of calculating $\frac{1}{3} + \frac{1}{4}$ we are going to approximate this to $\frac{1}{4} + \frac{1}{4} = \frac{2}{4}$.

We knew that the original addition will be more than this since $\frac{1}{3} > \frac{1}{4}$.

So if we did $1 + \frac{1}{2} + \left(\frac{1}{3} + \frac{1}{4}\right)$, the answer we would get would be less than the real sum.

We then do this for every set of brackets:

$\frac{1}{5} + \frac{1}{6} + \frac{1}{7} + \frac{1}{8}$ becomes $\frac{1}{8} + \frac{1}{8} + \frac{1}{8} + \frac{1}{8} = \frac{4}{8}$, again an underestimation.

What we have at the end then is the sum:

$\sum\limits_{n=1}^{\infty} \frac{1}{n} > 1 + \frac{1}{2} + \frac{2}{4} + \frac{4}{8} + \frac{8}{16} + \cdots$

This is obviously $1 + \frac{1}{2} + \frac{1}{2} + \frac{1}{2} + \frac{1}{2} + \cdots$ which does not have a limit.

We have then shown that $\sum\limits_{n=1}^{\infty} \frac{1}{n} \to \infty$.

Sums to infinity exercise (page 235)

Q94:

a) This is not convergent since $r = 2$ i.e. r is outwith the range $-1 < r < 1$
b) This is convergent with $a = 32$ and $r = 1/2$ giving $S_\infty = 64$
c) This is convergent with $a = 32$ and $r = -1/2$ giving $S_\infty = 64/3$
d) This is convergent with $a = 1/3$ and $r = 1/2$ giving S_∞ as $2/3$
e) This is not convergent since $r = -3$ i.e. r is outwith the range $-1 < r < 1$

Q95: $a = 1$ and $d = 1/2$ which gives $S_\infty = 2$

Q96:
$9 = \frac{a}{1-r} = \frac{a}{1 - 2/3} = 3a$ So $a = 3$

Q97: $\frac{3125}{648}$

Q98: $4 = ar, 1 = ar^3$ so $r^2 = 1/4$. Since $r > 0$ then $r = 1/2$, $a = 8$ and sum to infinity $= 16$

Q99:

$(a+x)^{-1} = a^{-1}\left(1 + \frac{x}{a}\right)^{-1}$

$= a^{-1}\left[1 + \left(\frac{-1}{1!}\right)\left(\frac{x}{a}\right) + \left(\frac{-1 \times -2}{2!}\right)\left(\frac{x}{a}\right)^2 + \left(\frac{-1 \times -2 \times -3}{3!}\right)\left(\frac{x}{a}\right)^3 + \ldots\right]$

$= \frac{1}{a} - \frac{x}{a^2} + \frac{x^2}{a^3} - \frac{x^3}{a^4}$

Q100: The series is $\frac{1}{3x} - \frac{2}{9x^2} + \frac{4}{27x^3} - \frac{8}{81x^4} + \ldots$

Q101: The series is $\frac{1}{3x} - \frac{2}{9x^2} + \frac{4}{27x^3} - \frac{8}{81x^4} + \ldots$

Power series exercise (page 236)

Q102:

$S_{10} = \frac{1}{0!} + \frac{1}{1!} + \frac{1^2}{2!} + \frac{1^3}{3!} + \frac{1^4}{4!} + \frac{1^5}{5!} + \frac{1^6}{6!} + \frac{1^7}{7!} + \frac{1^8}{8!} + \frac{1^9}{9!}$
$= 2 \cdot 718282 \approx e$

Q103:

Power series is of the form:
$\sum_{n=0}^{\infty} a_n x^n = a_0 + a_1 x + a_2 x^2 + a_3 x^3 + \ldots + a_n x^n + \ldots$
$a_0 = 3, a_1 = 6, a_2 = 12, a_3 = 24$
Power series: $3 + 6x + 12x^2 + 24x^3$

Q104:

Power series is of the form:
$\sum_{n=0}^{\infty} a_n x^n = a_0 + a_1 x + a_2 x^2 + a_3 x^3 + \ldots + a_n x^n + \ldots$

© HERIOT-WATT UNIVERSITY

$a_0 = 1$, $a_1 = 1$, $a_2 = \frac{1}{2}$, $a_3 = \frac{1}{3}$, $a_4 = \frac{1}{4}$, $a_5 = \frac{1}{5}$, $a_6 = \frac{1}{6}$

Power series: $1 + x + \frac{1}{2}x^2 + \frac{1}{3}x^3 + \frac{1}{4}x^4 + \frac{1}{5}x^5 + \frac{1}{6}x^6$

Maclaurin series for simple functions exercise (page 238)

Q105:

When $f(x) = \cos(x)$ is repeatedly differentiated we obtain:

$f(x) = \cos(x)$ $\qquad f(0) = 1$
$f^{(1)}(x) = -\sin(x)$ $\qquad f^{(1)}(0) = 0$
$f^{(2)}(x) = -\cos(x)$ $\qquad f^{(2)}(0) = -1$
$f^{(3)}(x) = \sin(x)$ $\qquad f^{(3)}(0) = 0$
$f^{(4)}(x) = \cos(x)$ $\qquad f^{(4)}(0) = 1$
$f^{(5)}(x) = -\sin(x)$ $\qquad f^{(5)}(0) = 0$

Therefore the Maclaurin series generated by $f(x) = \cos(x)$ becomes:

$$\sum_{r=0}^{\infty} f^{(r)}(0) \frac{x^r}{r!} = f(0) + f^{(1)}(0)\frac{x}{1!} + f^{(2)}(0)\frac{x^2}{2!} + f^{(3)}(0)\frac{x^3}{3!} + f^{(4)}(0)\frac{x^4}{4!} + f^{(5)}(0)\frac{x^5}{5!} + \ldots$$

$$= 1 + (0)\frac{x}{1!} + (-1)\frac{x^2}{2!} + (0)\frac{x^3}{3!} + (1)\frac{x^4}{4!} + (0)\frac{x^5}{5!} + \ldots$$

$$= 1 - \frac{x^2}{2!} + \frac{x^4}{4!} - \ldots$$

Q106:

When $f(x) = \ln(1+x)$ is repeatedly differentiated we obtain:
(Write derivatives of $f(x)$ in the form with negative powers)

$f(x) = \ln(1+x)$ $\qquad f(0) = 0$
$f^{(1)}(x) = (1+x)^{-1}$ $\qquad f^{(1)}(0) = 1$
$f^{(2)}(x) = -(1+x)^{-2}$ $\qquad f^{(2)}(0) = -1$
$f^{(3)}(x) = 2(1+x)^{-3}$ $\qquad f^{(3)}(0) = 2$
$f^{(4)}(x) = -6(1+x)^{-4}$ $\qquad f^{(4)}(0) = -6$
$f^{(5)}(x) = 24(1+x)^{-5}$ $\qquad f^{(5)}(0) = 24$

Therefore the Maclaurin series generated by $f(x) = \ln(1+x)$ becomes:

$$\sum_{r=0}^{\infty} f^{(r)}(0) \frac{x^r}{r!} = f(0) + f^{(1)}(0)\frac{x}{1!} + f^{(2)}(0)\frac{x^2}{2!} + f^{(3)}(0)\frac{x^3}{3!} + f^{(4)}(0)\frac{x^4}{4!} + f^{(5)}(0)\frac{x^5}{5!} + \ldots$$

$$= 0 + (1)\frac{x}{1!} + (-1)\frac{x^2}{2!} + (2)\frac{x^3}{3!} + (-6)\frac{x^4}{4!} + (24)\frac{x^5}{5!} + \ldots$$

$$= 0 + (1)\frac{x}{1} + (-1)\frac{x^2}{2} + (2)\frac{x^3}{6} + (-6)\frac{x^4}{24} + (24)\frac{x^5}{120} + \ldots$$

$$= x - \frac{x^2}{2} + \frac{x^3}{3} - \frac{x^4}{4} + \frac{x^5}{5} - \ldots$$

Q107:
Hints:

- Note that
$$\binom{n}{r} = \frac{n!}{r!(n-r)!}$$
$$= \frac{n(n-1)(n-2)(n-3)\ldots(n-(r+1))(n-r)!}{r!(n-r)!}$$
$$= \frac{n(n-1)(n-2)(n-3)\ldots(n-(r+1))}{r!}$$

Answer:

When $f(x) = (1+x)^n$ is repeatedly differentiated we obtain:

$f(x) = (1+x)^n$
$f^{(1)}(x) = n(1+x)^{n-1}$
$f^{(2)}(x) = n(n-1)(1+x)^{n-2}$
$f^{(3)}(x) = n(n-1)(n-2)(1+x)^{n-3}$
$f^{(4)}(x) = n(n-1)(n-2)(n-3)(1+x)^{n-4}$
$f^{(5)}(x) = n(n-1)(n-2)(n-3)(n-4)(1+x)^{n-5}$

and

$f(0) = 1$
$f^{(1)}(0) = n$
$f^{(2)}(0) = n(n-1)$
$f^{(3)}(0) = n(n-1)(n-2)$
$f^{(4)}(0) = n(n-1)(n-2)(n-3)$
$f^{(5)}(0) = n(n-1)(n-2)(n-3)(n-4)$

Therefore the Maclaurin series generated by $f(x) = (1+x)^n$ becomes:

$$\sum_{r=0}^{\infty} f^{(r)}(0)\frac{x^r}{r!} = f(0) + f^{(1)}(0)\frac{x}{1!} + f^{(2)}(0)\frac{x^2}{2!} + f^{(3)}(0)\frac{x^3}{3!} + f^{(4)}(0)\frac{x^4}{4!} + f^{(5)}(0)\frac{x^{52}}{5!} + \ldots$$

$$(1+x)^n = 1 + nx + \frac{n(n-1)}{2!}x^2 + \frac{n(n-1)(n-2)}{3!}x^3 + \frac{n(n-1)(n-2)(n-3)}{4!}x^4 + \ldots$$

$$= 1 + \frac{n!}{1!(n-1)!}x + \frac{n!}{2!(n-2)!}x^2 + \frac{n!}{3!(n-3)!}x^3 + \frac{n!}{4!(n-4)!}x^4 + \frac{n!}{5!(n-5)!}x^5 + \ldots$$

$$= \binom{n}{0} + \binom{n}{1}x + \binom{n}{2}x^2 + \binom{n}{3}x^3 + \binom{n}{4}x^4 + \ldots$$

This is the **binomial expansion**.

Maclaurin's theorem exercise (page 243)

Q108:

$\sin(x) = x - \frac{x^3}{3!} + \frac{x^5}{5!} - \frac{x^7}{7!} + \ldots$

Q109:

$\sin(1) = 1 - \frac{1^3}{3!} + \frac{1^5}{5!} - \frac{1^7}{7!}$

$\sin(1) = 1 - \frac{1}{3!} + \frac{1}{5!} - \frac{1}{7!}$

$\sin(1) = 0 \cdot 841468\ldots$

Q110:

$\cos(x) = 1 - \frac{x^2}{2!} + \frac{x^4}{4!} - \frac{x^6}{6!} + \ldots$

Q111:

$\cos(0 \cdot 5) = 1 - \frac{0 \cdot 5^2}{2!} + \frac{0 \cdot 5^4}{4!} - \frac{0 \cdot 5^6}{6!} + \frac{0 \cdot 5^8}{8!}$

$\cos(0 \cdot 5) = 0 \cdot 874978\ldots$

Q112:

$\ln|1+x| = x - \frac{x^2}{2!} + \frac{x^3}{3!} - \frac{x^4}{4!} + \frac{x^5}{5!} - \frac{x^6}{6!} + \ldots$ for $-1 < x \leqslant 1$

Q113:

$\ln|1 + 0 \cdot 3| = 0 \cdot 3 - \frac{0 \cdot 3^2}{2!} + \frac{0 \cdot 3^3}{3!}$

$\ln|1 + 0 \cdot 3| = 0 \cdot 00045\ldots$

The Maclaurin series for tan⁻¹(x) exercise (page 245)

Q114: The results are:

$f(x) = \ln x$ \qquad $f(0) = \ln 0$

$f^{(1)}(x) = \frac{1}{x}$ \qquad $f^{(1)}(0) = \frac{1}{0}$

$f^{(2)}(x) = \frac{-1}{x^2}$ \qquad $f^{(2)}(0) = \frac{-1}{0}$

$f(x) = \ln x$ and its derivatives are undefined at $x = 0$, therefore a Maclaurin series expansion for $\ln x$ is not possible.

Q115: The results are:

$f(x) = \sqrt{x} = x^{\frac{1}{2}}$ \qquad $f(0) = \sqrt{0}$

$f^{(1)}(x) = \frac{1}{2x^{\frac{1}{2}}}$ \qquad $f^{(1)}(0) = \frac{1}{0}$

$f^{(2)}(x) = -\frac{1}{4x^{\frac{3}{2}}}$ \qquad $f^{(2)}(0) = -\frac{1}{0}$

The derivatives of $f(x) = \sqrt{x}$ are undefined at $x = 0$, therefore we cannot find a Maclaurin series expansion for $\sqrt{x}$.

© HERIOT-WATT UNIVERSITY

Q116: The results are:

$$f(x) = \cot x = \frac{1}{\tan x} \qquad f(0) = \frac{1}{0}$$
$$f^{(1)}(x) = -\cos ec^2 x = -\frac{1}{\sin^2 x} \qquad f^{(1)}(0) = -\frac{1}{0}$$
$$f^{(2)}(x) = 2\cot x \cos ec^2 x = 2\frac{\cos x}{\sin^3 x} \qquad f^{(2)}(0) = \frac{2}{0}$$

$f(x) = \cot x$ and its derivatives are undefined at $x = 0$, therefore a Maclaurin series expansion for $\cot x$ is not possible.

Maclaurin's series expansion to a given number of terms exercise (page 247)

Q117:

$f(x) = e^{x^2}$
$f'(x) = 2xe^{x^2}$
$f''(x) = 2e^{x^2} + 4x^2 e^{x^2}$
$f^{(3)}(x) = 12xe^{x^2} + 8x^3 e^{x^2}$
$f^{(4)}(x) = 12e^{x^2} + 48x^2 e^{x^2} + 16x^4 e^{x^2}$
$f^{(5)}(x) = 120xe^{x^2} + 160x^3 e^{x^2} + 32x^5 e^{x^2}$
$f^{(6)}(x) = 120e^{x^2} + 720x^2 e^{x^2} + 480x^4 e^{x^2} + 64x^6 e^{x^2}$

and

$f(0) = e^0 = 1$
$f'(0) = 2(0)e^0 = 0$
$f''(0) = 2e^0 + 4(0)^2 e^0 = 2$
$f^{(3)}(0) = 12(0)e^0 + 8(0)^2 e^0 = 0$
$f^{(4)}(0) = 12e^0 + 48(0)^2 e^0 + 16(0)^4 e^0 = 12$
$f^{(5)}(0) = 120(0)e^0 + 160(0)^3 e^0 + 32(0)^5 e^0 = 0$
$f^{(6)}(0) = 120e^0 + 720(0)^2 e^{x^2} + 480(0)^4 e^0 + 64(0)^6 e^0 = 120$

Therefore the Maclaurin series generated by $f(x) = e^{x^2}$ becomes:

$$e^{x^2} = 1 + \frac{2}{2!}x^2 + \frac{12}{4!}x^4 + \frac{120}{6!}x^6 + \ldots$$
$$= 1 + x^2 + \frac{1}{2}x^4 + \frac{1}{6}x^6 + \ldots$$

Q118:

$$f(x) = \frac{1}{2}\sin(2x)$$
$$f'(x) = \cos(2x)$$
$$f''(x) = -2\sin(2x)$$
$$f^{(3)}(x) = -4\cos(2x)$$
$$f^{(4)}(x) = 8\sin(2x)$$
$$f^{(5)}(x) = 16\cos(2x)$$
$$f^{(6)}(x) = -32\sin(2x)$$
$$f^{(7)}(x) = -64\cos(2x)$$

$$f(0) = \frac{1}{2}\sin(2(0)) = 0$$
$$f'(0) = \cos(2(0)) = 1$$
$$f''(0) = -2\sin(2(0)) = 0$$
$$f^{(3)}(0) = -4\cos(2(0)) = -4$$
$$f^{(4)}(0) = 8\sin(2(0)) = 0$$
$$f^{(5)}(0) = 16\cos(2(0)) = 16$$
$$f^{(6)}(0) = -32\sin(2(0)) = 0$$
$$f^{(7)}(0) = -64\cos(2(0)) = -64$$

Therefore the Maclaurin series generated by $f(x) = \frac{1}{2}\sin(2x)$ becomes:

$$\frac{1}{2}\sin(2x) = x - \frac{4}{3!}x^3 + \frac{16}{5!}x^5 - \frac{64}{7!}x^7 + \ldots$$
$$= x - \frac{2}{3}x^3 + \frac{2}{15}x^5 - \frac{4}{315}x^7 + \ldots$$

Q119:

Recall from the standard results that $\frac{d}{dx}\tan^{-1}(x) = \frac{1}{1+x^2}$. Using this result and applying the chain rule to $f(x) = \frac{2}{3}\tan^{-1}(3x)$ gives the first derivative. i.e.

$$f'(x) = \frac{2}{3} \times \frac{1}{1+(3x)^2} \times \frac{d}{dx}(3x)$$
$$= \frac{2}{3} \times \frac{1}{1+(3x)^2} \times 3$$
$$= \frac{2}{1+9x^2}$$

$$f(x) = \frac{2}{3}\tan^{-1}(3x)$$
$$f'(x) = \frac{2}{1+9x^2}$$
$$f''(x) = \frac{-36x}{(1+9x^2)^2}$$
$$f^{(3)}(x) = \frac{-36}{(1+9x^2)^2} + \frac{1296x^2}{(1+9x^2)^3}$$
$$f^{(4)}(x) = \frac{3888x}{(1+9x^2)^3} - \frac{69984x^3}{(1+9x^2)^4}$$
$$f^{(5)}(x) = \frac{3888}{(1+9x^2)^3} - \frac{419904x^2}{(1+9x^2)^4} - \frac{5038848x^4}{(1+9x^2)^5}$$

And

ANSWERS: UNIT 1 TOPIC 7

$$f(0) = \frac{2}{3}\tan^{-1}(3(0)) = 0$$

$$f'(0) = \frac{2}{1+9(0)^2} = 2$$

$$f''(0) = \frac{-36(0)}{\left(1+9(0)^2\right)^2} = 0$$

$$f^{(3)}(0) = \frac{-36}{\left(1+9(0)^2\right)^2} + \frac{1296(0)^2}{\left(1+9(0)^2\right)^3} = -36$$

$$f^{(4)}(0) = \frac{3888(0)}{\left(1+9(0)^2\right)^3} - \frac{69984(0)^3}{\left(1+9(0)^2\right)^4} = 0$$

$$f^{(5)}(0) = \frac{3888}{\left(1+9(0)^2\right)^3} - \frac{419904(0)^2}{\left(1+9(0)^2\right)^4} - \frac{5038848(0)^4}{\left(1+9(0)^2\right)^5} = 3888$$

Therefore the Maclaurin series generated by $f(x) = \frac{2}{3}\tan^{-1}(3x)$ becomes:

$$\frac{2}{3}\tan^{-1}(3x) = 2x - \frac{36}{3!}x^3 + \frac{3888}{5!}x^5 - \ldots$$

$$= 2x - 6x^3 + \frac{162}{5}x^5 - \ldots$$

Q120:

$$f(x) = \ln(1-3x) \qquad\qquad f(0) = \ln(1-3(0)) = 0$$

$$f'(x) = \frac{-3}{1-3x} \qquad\qquad f'(0) = \frac{-1}{1-3(0)} = -3$$

$$f''(x) = \frac{-9}{(1-3x)^2} \qquad\qquad f''(0) = \frac{-9}{(1-3(0))^2} = -9$$

$$f^{(3)}(x) = \frac{-54}{(1-3x)^3} \qquad\qquad f^{(3)}(0) = \frac{-54}{(1-3(0))^3} = -54$$

$$f^{(4)}(x) = \frac{-486}{(1-3x)^4} \qquad\qquad f^{(4)}(0) = \frac{-486}{(1-3(0))^4} = -486$$

Therefore the Maclaurin series generated by $f(x) = \ln(1-3x)$ becomes:

$$\ln(1-3x) = -3x - \frac{9}{2!}x^2 - \frac{54}{3!}x^3 - \frac{486}{4!}x^4 - \ldots$$

$$= -3x - \frac{9}{2}x^2 - 9x^3 - \frac{81}{4}x^4 - \ldots$$

© HERIOT-WATT UNIVERSITY

Q121:

$f(x) = \ln(\cos(x))$

Recall from the standard results that $\frac{d}{dx} \ln |x| = \frac{1}{x}$.
Using this result and applying the chain rule to $f(x) = \ln|\cos(x)|$ gives the first derivative, i.e.

$$f'(x) = \frac{1}{\cos(x)} \times \frac{d}{dx}(\cos(x))$$
$$= \frac{-\sin(x)}{\cos(x)}$$
$$= -\tan(x)$$

Using the standard result for $\frac{d}{dx} \tan(x) = \sec^2(x)$ gives the second derivative.
Alternatively take $\frac{-\sin(x)}{\cos(x)}$ and apply the quotient rule to differentiate.

$f''(x) = -\sec^2(x)$

Again using the standard result for $\frac{d}{dx} \sec(x) = \sec(x) \tan(x)$ and applying the chain rule we obtain the third derivative.

$$f^{(3)}(x) = -2\sec(x) \times \frac{d}{dx} \sec(x)$$
$$= -2\sec(x) \sec(x) \tan(x)$$
$$= -2\sec^2(x) \tan(x)$$

To obtain the fourth derivatives we use the standard results $\frac{d}{dx} \sec(x) = \sec(x) \tan(x)$ and $\frac{d}{dx} \tan(x) = \sec^2(x)$ and apply the product and chain rules.
Let $u = -2\sec^2(x)$ and $v = \tan(x)$.
Then $u' = -4\sec^2(x) \tan(x)$ and $v' = \sec^2(x)$.
Note that the chain rule was used to differentiate $-2\sec^2(x)$. This was demonstrated for the third derivative.
Applying the product rule gives the fourth derivative:

$$f^4(x) = -4\sec^2(x) \tan(x) \tan(x) + \left(-2\sec^2(x)\right) \sec^2(x)$$
$$= -4\sec^2(x) \tan^2(x) - 2\sec^4(x)$$

and

$f(0) = \ln(\cos(0)) = 0$
$f'(0) = -\tan(0) = 0$
$f''(0) = -\sec^2(0) = -1$
$f^{(3)}(0) = -2\sec^2(0) \tan(0) = 0$
$f^{(4)}(0) = -4\sec^2(0) \tan^2(0) - 2\sec^4(0) = -2$

Therefore the Maclaurin series generated by $f(x) = \ln(\cos(x))$ becomes:

$$\ln(\cos(x)) = -\frac{1}{2!}x^2 - \frac{2}{4!}x^4 - \ldots$$
$$= -\frac{x^2}{2} - \frac{x^4}{12} - \ldots$$

ANSWERS: UNIT 1 TOPIC 7   559

Composite Maclaurin series expansion exercise (page 248)

Q122:
When $f(x) = \sin(3x)$ is repeatedly differentiated we obtain:

$f(x) = \sin(3x)$ $\qquad f(0) = 0$
$f^{(1)}(x) = 3\cos(3x)$ $\qquad f^{(1)}(0) = 3$
$f^{(2)}(x) = -9\sin(3x)$ $\qquad f^{(2)}(0) = 0$
$f^{(3)}(x) = -27\cos(3x)$ $\qquad f^{(3)}(0) = -27$
$f^{(4)}(x) = 81\sin(3x)$ $\qquad f^{(4)}(0) = 0$
$f^{(5)}(x) = 243\cos(3x)$ $\qquad f^{(5)}(0) = 243$

Therefore the Maclaurin series generated by $f(x) = \sin(3x)$ becomes:

$$\sum_{r=0}^{\infty} f^{(r)}(0) \frac{x^r}{r!} = f(0) + f^{(1)}(0) \frac{x}{1!} + f^{(2)}(0) \frac{x^2}{2!} + f^{(3)}(0) \frac{x^3}{3!} + f^{(4)}(0) \frac{x^4}{4!} + f^{(5)}(0) \frac{x^5}{5!} + \ldots$$

$$= 3x - \frac{27x^3}{3!} + \frac{243x^5}{5!} - \ldots$$

$$= 3x - \frac{9x^3}{2} + \frac{81x^5}{40} - \ldots$$

Q123:
When $f(x) = \cos(2x)$ is repeatedly differentiated we obtain:

$f(x) = \cos(2x)$ $\qquad f(0) = 1$
$f^{(1)}(x) = -2\sin(2x)$ $\qquad f^{(1)}(0) = 0$
$f^{(2)}(x) = -4\cos(2x)$ $\qquad f^{(2)}(0) = -4$
$f^{(3)}(x) = 8\sin(2x)$ $\qquad f^{(3)}(0) = 0$
$f^{(4)}(x) = 16\cos(2x)$ $\qquad f^{(4)}(0) = 16$
$f^{(5)}(x) = -32\sin(2x)$ $\qquad f^{(5)}(0) = 0$

Therefore the Maclaurin series generated by $f(x) = \cos(2x)$ becomes:

$$\sum_{r=0}^{\infty} f^{(r)}(0) \frac{x^r}{r!} = f(0) + f^{(1)}(0) \frac{x}{1!} + f^{(2)}(0) \frac{x^2}{2!} + f^{(3)}(0) \frac{x^3}{3!} + f^{(4)}(0) \frac{x^4}{4!} + f^{(5)}(0) \frac{x^5}{5!} + \ldots$$

$$= 1 - \frac{4x^2}{2!} + \frac{16x^4}{4!} - \ldots$$

$$= 1 - 2x^2 + \frac{2x^4}{3} - \ldots$$

© HERIOT-WATT UNIVERSITY

Q124:

When $f(x) = e^{-x}$ is repeatedly differentiated we obtain:

$f(x) = e^{-x}$	$f(0) = 1$
$f^{(1)}(x) = -e^{-x}$	$f^{(1)}(0) = -1$
$f^{(2)}(x) = e^{-x}$	$f^{(2)}(0) = 1$
$f^{(3)}(x) = -e^{-x}$	$f^{(3)}(0) = -1$
$f^{(4)}(x) = e^{-x}$	$f^{(4)}(0) = 1$
$f^{(5)}(x) = -e^{-x}$	$f^{(5)}(0) = -1$

Therefore the Maclaurin series generated by $f(x) = e^{-x}$ becomes:

$$\sum_{r=0}^{\infty} f^{(r)}(0) \frac{x^r}{r!} = f(0) + f^{(1)}(0)\frac{x}{1!} + f^{(2)}(0)\frac{x^2}{2!} + f^{(3)}(0)\frac{x^3}{3!} + f^{(4)}(0)\frac{x^4}{4!} + f^{(5)}(0)\frac{x^5}{5!} + \ldots$$

$$= 1 - x + \frac{x^2}{2!} - \frac{x^3}{3!} + \frac{x^4}{4!} - \frac{x^5}{5!} + \ldots$$

$$= 1 - x + \frac{x^2}{2} - \frac{x^3}{6} + \frac{x^4}{24} - \frac{x^5}{120} + \ldots$$

Q125:

When $f(x) = \ln(1 - 2x)$ is repeatedly differentiated we obtain:
(Write derivatives of $f(x)$ in the form with negative powers)

$f(x) = \ln(1 - 2x)$	$f(0) = 0$
$f^{(1)}(x) = -2(1-2x)^{-1}$	$f^{(1)}(0) = -2$
$f^{(2)}(x) = -4(1-2x)^{-2}$	$f^{(2)}(0) = -4$
$f^{(3)}(x) = -16(1-2x)^{-3}$	$f^{(3)}(0) = -16$
$f^{(4)}(x) = -96(1-2x)^{-4}$	$f^{(4)}(0) = -96$
$f^{(5)}(x) = -768(1-2x)^{-5}$	$f^{(5)}(0) = -768$

Therefore the Maclaurin series generated by $f(x) = \ln(1 - 2x)$ becomes:

$$\sum_{r=0}^{\infty} f^{(r)}(0) \frac{x^r}{r!} = f(0) + f^{(1)}(0)\frac{x}{1!} + f^{(2)}(0)\frac{x^2}{2!} + f^{(3)}(0)\frac{x^3}{3!} + f^{(4)}(0)\frac{x^4}{4!} + f^{(5)}(0)\frac{x^5}{5!} + \ldots$$

$$= -2x - \frac{4x^2}{2!} - \frac{16x^3}{3!} - \frac{96x^4}{4!} - \frac{768x^5}{5!} - \ldots$$

$$= -2x - 2x^2 - \frac{8x^3}{3} - 4x^4 - \frac{32x^5}{5} - \ldots$$

Q126:

$f(x) = \ln(1 + e^x)$

$f'(x) = \dfrac{e^x}{1 + e^x}$

To obtain the second derivative we must use the quotient rule $\dfrac{dy}{dx} = \dfrac{u'v - uv'}{[v]^2}$

Let $u = e^x$ and $v = 1 + e^x$.
Then $u' = e^x$ and $v' = e^x$.

The second derivative becomes:
$$\frac{dy}{dx} = \frac{e^x(1+e^x) - e^x(e^x)}{(1+e^x)^2}$$
$$= \frac{e^x + e^{2x} - e^{2x}}{(1+e^x)^2}$$
$$= \frac{e^x}{(1+e^x)^2}$$

Therefore $f^{(2)}(x) = \frac{e^x}{(1+e^x)^2}$

To obtain the third derivative we must use the quotient rule again.

Let $u = e^x$ and $v = (1+e^x)^2$.

Then $u' = e^x$ and $v' = 2e^x(1+e^x)$.

Note that we differentiate v using the chain rule.

The third derivative becomes:
$$\frac{dy}{dx} = \frac{e^x(1+e^x)^2 - e^x \times 2e^x(1+e^x)}{\left((1+e^x)^2\right)^2}$$
$$= \frac{e^x(1+e^x)^2 - 2e^{2x}(1+e^x)}{(1+e^x)^4}$$

Take out a common factor of $(1+e^x)$
$$= \frac{(1+e^x)\left(e^x(1+e^x) - 2e^{2x}\right)}{(1+e^x)^4}$$
$$= \frac{e^x(1+e^x) - 2e^{2x}}{(1+e^x)^3}$$
$$= \frac{e^x + e^{2x} - 2e^{2x}}{(1+e^x)^3}$$
$$= \frac{e^x - e^{2x}}{(1+e^x)^3}$$
$$= \frac{e^x(1-e^x)}{(1+e^x)^3}$$

Therefore the third derivative is:

$f^{(3)}(x) = \frac{e^x(1-e^x)}{(1+e^x)^3}$

Now we differentiate again using the quotient rule to obtain the fourth derivative.

Simplify the numerator and

let $u = e^x - e^{2x}$ and $v = (1+e^x)^3$.

Then $\begin{array}{l} u' = e^x - 2e^{2x} \\ = e^x(1-2e^x) \end{array}$ and $v' = 3e^x(1+e^x)^2$

Note that we differentiate v using the chain rule.

$$\frac{dy}{dx} = \frac{e^x(1-2e^x)(1+e^x)^3 - e^x \times 3e^x(1+e^x)^2}{\left((1+e^x)^3\right)^2}$$
$$= \frac{e^x(1-2e^x)(1+e^x)^3 - 3e^{2x}(1+e^x)^2}{(1+e^x)^6}$$

© HERIOT-WATT UNIVERSITY

Take out a common factor of $(1+e^x)^2$

$$=\frac{(1+e^x)^2\left(e^x(1-2e^x)(1+e^x)-3e^{2x}\right)}{(1+e^x)^6}$$

$$=\frac{e^x(1-2e^x)(1+e^x)-3e^{2x}}{(1+e^x)^4}$$

$$=\frac{e^x-4e^{2x}-3e^{3x}}{(1+e^x)^4}$$

$$=\frac{e^x\left(1-4e^x-3e^{2x}\right)}{(1+e^x)^4}$$

Therefore the fourth derivative is:

$$f^{(4)}(x)=\frac{e^x\left(1-4e^x-3e^{2x}\right)}{(1+e^x)^4}$$

and

$f(0)=\ln(1+e^0)=\ln(2)$

$f'(0)=\dfrac{e^0}{1+e^0}=\dfrac{1}{2}$

$f''(0)=\dfrac{e^0}{(1+e^0)^2}=\dfrac{1}{4}$

$f^{(3)}(0)=\dfrac{e^0(1-e^0)}{(1+e^0)^3}=0$

$f^{(4)}(0)=\dfrac{e^0\left(1-4e^0+e^{2(0)}\right)}{(1+e^0)^4}=-\dfrac{1}{8}$

Therefore the Maclaurin series generated by $f(x)=\ln(1+e^x)$ becomes:

$$\ln(1+e^x)=\ln(2)+\frac{1}{2}x+\frac{1}{4}\times\frac{x^2}{2!}-\frac{1}{8}\times\frac{x^4}{4!}+\ldots$$

$$=\ln(2)+\frac{1}{2}x+\frac{1}{8}x^2-\frac{1}{192}x^4$$

Standard Maclaurin series expansions (page 250)

Q127:

Function	Maclaurin expansion	Interval of convergence
e^x	$=1+x+\frac{x^2}{2!}+\frac{x^3}{3!}+\frac{x^4}{4!}+\frac{x^5}{5!}+\ldots$	Converges for all $x\in\mathbb{R}$
$\sin(x)$	$=x-\frac{x^3}{3!}+\frac{x^5}{5!}-\frac{x^7}{7!}+\frac{x^9}{9!}-\frac{x^{11}}{11!}+\ldots$	Converges for all $x\in\mathbb{R}$
$\cos(x)$	$=1-\frac{x^2}{2!}+\frac{x^4}{4!}-\frac{x^6}{6!}+\frac{x^8}{8!}-\frac{x^{10}}{10!}+\ldots$	Converges for all $x\in\mathbb{R}$
$\tan^{-1}(x)$	$=x-\frac{x^3}{3}+\frac{x^5}{5}-\frac{x^7}{7}+\frac{x^9}{9}-\frac{x^{11}}{11}+\ldots$	$|x|\leqslant 1$
$\ln(1+x)$	$=x-\frac{x^2}{2}+\frac{x^3}{3}-\frac{x^4}{4}+\frac{x^5}{5}-\frac{x^6}{6}+\ldots$	$-1<x\leqslant 1$
$(1-x)^{-1}$	$=1+x+x^2+x^3+x^4+x^5\ldots$	$|x|<1$

ANSWERS: UNIT 1 TOPIC 7

These expansions are not on the formula sheet and it would be beneficial to remember them along with the ranges of convergence, but it is not essential as long as you are able to derive them using the Maclaurin's Theorem.

Standard Maclaurin series expansions exercise (page 250)

Q128:
The Maclaurin series expansion for $f(x) = e^x$ is:
$$e^x = 1 + x + \frac{x^2}{2!} + \frac{x^3}{3!} + \frac{x^4}{4!} + \frac{x^5}{5!} + \ldots, x \in \mathbb{R}$$
Let $x = -4x$,
$$e^{-4x} = 1 + (-4x) + \frac{(-4x)^2}{2!} + \frac{(-4x)^3}{3!} + \frac{(-4x)^4}{4!} + \frac{(-4x)^5}{5!} + \ldots \qquad x \in \mathbb{R}$$
$$= 1 - 4x + 8x^2 - \frac{32x^3}{3} + \frac{32x^4}{3} - \frac{128x^5}{15} + \ldots$$

Q129:
The Maclaurin series expansion for $f(x) = \sin(x)$ is: $\sin(x) = x - \frac{x^3}{3!} + \frac{x^5}{5!} - \frac{x^7}{7!} + \ldots, x \in \mathbb{R}$
Let $x = 5x$,
$$\sin(5x) = (5x) - \frac{(5x)^3}{3!} + \frac{(5x)^5}{5!} - \frac{(5x)^7}{7!} + \ldots \qquad x \in \mathbb{R}$$
$$= 5x - \frac{125}{6}x^3 + \frac{625}{24}x^5 - \frac{15625}{1008}x^7 + \ldots$$

Q130:
The Maclaurin series expansion for $f(x) = \cos(x)$ is: $\cos(x) = 1 - \frac{x^2}{2!} + \frac{x^4}{4!} - \frac{x^6}{6!} + \ldots, x \in \mathbb{R}$
Let $x = 3x$,
$$\cos(3x) = 1 - \frac{(3x)^2}{2!} + \frac{(3x)^4}{4!} - \frac{(3x)^6}{6!} + \ldots \qquad x \in \mathbb{R}$$
$$= 1 - \frac{9x^2}{2} + \frac{27x^4}{8} - \frac{81x^6}{80} + \ldots$$

Q131:
The Maclaurin series expansion for $f(x) = \tan^{-1}(x)$ is: $\tan^{-1}(x) = x - \frac{x^3}{3} + \frac{x^5}{5} - \frac{x^7}{7} + \ldots, |x| \leqslant 1$
Let $x = 2x$,
$$\tan^{-1}(2x) = (2x) - \frac{(2x)^3}{3} + \frac{(2x)^5}{5} - \frac{(2x)^7}{7} + \ldots \qquad |x| \leqslant \frac{1}{2}$$
$$= 2x - \frac{8x^3}{3} + \frac{32x^5}{5} - \frac{128x^7}{7} + \ldots$$

© HERIOT-WATT UNIVERSITY

Q132:

The Maclaurin series expansion for $f(x) = \ln(1 + 2x)$ is:
$\ln(1+x) = x - \frac{x^2}{2} + \frac{x^3}{3} - \frac{x^4}{4} + \frac{x^5}{5} - \frac{x^6}{6} + \ldots,\ -1 < x \leqslant 1$
Let $x = 2x$,
$$\ln(1+2x) = (2x) - \frac{(2x)^2}{2} + \frac{(2x)^3}{3} - \frac{(2x)^4}{4} + \frac{(2x)^5}{5} - \frac{(2x)^6}{6} + \ldots \quad -\tfrac{1}{2} < x \leqslant \tfrac{1}{2}$$
$$= 2x - 2x^2 + \frac{8}{3}x^3 - 4x^4 + \frac{32}{5}x^5 - \frac{32}{3}x^6 + \ldots$$

Q133:

Hints:

- If you cannot recall the standard result for $(1-x)^{-1}$ derive it using Maclaurin's expansion, then substitute for $3x$.

Answer: The Maclaurin series expansion for $f(x) = (1-3x)^{-1}$ is:
$(1-x)^{-1} = 1 + x + x^2 + x^3 + x^4 + x^5 + \ldots,\ -1 < x < 1$
Let $x = 3x$,
$$(1-3x)^{-1} = 1 + (3x) + (3x)^2 + (3x)^3 + (3x)^4 + (3x)^5 + \ldots \quad |x| < \tfrac{1}{3}$$
$$= 1 + 3x + 9x^2 + 27x^3 + 81x^4 + 243x^5 + \ldots$$

Q134:

The Maclaurin series expansion for $f(x) = e^x$ is:
$e^x = 1 + x + \frac{x^2}{2!} + \frac{x^3}{3!} + \frac{x^4}{4!} + \frac{x^5}{5!} + \frac{x^6}{6!} + \frac{x^7}{7!} + \ldots,\ x \in \mathbb{R}$
Let $x = ix$,
$$e^{ix} = 1 + (ix) + \frac{(ix)^2}{2!} + \frac{(ix)^3}{3!} + \frac{(ix)^4}{4!} + \frac{(ix)^5}{5!} + \frac{(ix)^6}{6!} + \frac{(ix)^7}{7!} + \ldots$$
$$= 1 + ix - \frac{x^2}{2!} - i\frac{x^3}{3!} + \frac{x^4}{4!} + i\frac{x^5}{5!} - \frac{x^6}{6!} - i\frac{x^7}{7!} + \ldots$$

Q135:

Separating the real and imaginary parts:
$$e^{ix} = \left(1 - \frac{x^2}{2!} + \frac{x^4}{4!} - \frac{x^6}{6!} + ..\right) + i\left(x - \frac{x^3}{3!} + \frac{x^5}{5!} - \frac{x^7}{7!} + \ldots\right)$$
$$e^{ix} = \cos(x) + i\sin(x)$$

Composite function examples using standard results exercise (page 255)

Q136:
Hints:

- $e^{x-x^2} = e^x e^{-x^2}$

Answer:

a)
The Maclaurin series expansion for e^x is given by: $e^x = 1 + x + \frac{x^2}{2!} + \frac{x^3}{3!} + \frac{x^4}{4!} + \ldots, x \in \mathbb{R}$

b)
Let $x = -x^2$,

$$e^{-x^2} = 1 + (-x^2) + \frac{(-x^2)^2}{2!} + \frac{(-x^2)^3}{3!} + \frac{(-x^2)^4}{4!} + \ldots$$

$$= 1 - x^2 + \frac{x^4}{2!} - \ldots$$

$$= 1 - x^2 + \frac{x^4}{2!}$$

c)
Using indice rules:

$$e^{x-x^2} = e^x e^{-x^2}$$

$$= \left(1 + x + \frac{x^2}{2!} + \frac{x^3}{3!} + \frac{x^4}{4!} + \ldots\right)\left(1 - x^2 + \frac{x^4}{2!} - \ldots\right)$$

$$= 1\left(1 - x^2 + \frac{x^4}{2!}\right) + x\left(1 - x^2 + \frac{x^4}{2!}\right) + \frac{x^2}{2!}\left(1 - x^2 + \frac{x^4}{2!}\right) + \frac{x^3}{3!}\left(1 - x^2 + \frac{x^4}{2!}\right) + \frac{x^4}{4!}\left(1 - x^2 + \frac{x^4}{2!}\right)$$

$$= 1 - x^2 + \frac{x^4}{2!} + x - x^3 + \ldots + \frac{x^2}{2!} - \frac{x^4}{2!} + \ldots + \frac{x^3}{3!} - \ldots + \frac{x^4}{4!} - \ldots$$

$$= 1 + x - x^2 + \frac{x^2}{2!} - x^3 + \frac{x^3}{3!} + \frac{x^4}{2!} - \frac{x^4}{2!} + \frac{x^4}{4!}$$

$$= 1 + x - \frac{x^2}{2} - \frac{5}{6}x^3 + \frac{x^4}{24}$$

Q137:

Hints:

- $\cos^2(x) = \cos(x)\cos(x)$

Answer: The Maclaurin series expansion for $\cos(x)$ is given by:
$\cos(x) = 1 - \frac{x^2}{2!} + \frac{x^4}{4!} - \ldots, x \in \mathbb{R}$
Therefore,
$\cos^2(x) = \cos(x)\cos(x)$

$$= \left(1 - \frac{x^2}{2!} + \frac{x^4}{4!} - \ldots\right)\left(1 - \frac{x^2}{2!} + \frac{x^4}{4!} - \ldots\right)$$

$$= 1\left(1 - \frac{x^2}{2!} + \frac{x^4}{4!}\right) - \frac{x^2}{2!}\left(1 - \frac{x^2}{2!} + \frac{x^4}{4!} - \ldots\right) + \frac{x^4}{4!}\left(1 - \frac{x^2}{2!} + \frac{x^4}{4!} - \ldots\right)$$

$$= 1 - \frac{x^2}{2!} + \frac{x^4}{4!} - \frac{x^2}{2!} + \frac{x^4}{4} - \ldots + \frac{x^4}{4!} - \ldots$$

$$= 1 - x^2 + \frac{1}{3}x^4$$

Q138:

Hints:

- Replace $\sin(x)$ with it's Maclaurin expansion up to terms in x^3. Substitute this into the Maclaurin expansion for $\tan^{-1}(x)$.

Answer: The Maclaurin series expansion for $\sin(x)$ and $\tan^{-1}(x)$ are given by:
$\sin(x) = x - \frac{x^3}{3!} + \frac{x^5}{5!} - \ldots, x \in \mathbb{R}$
$\tan^{-1}(x) = x - \frac{x^3}{3} + \frac{x^5}{5} - \ldots, |x| \leqslant 1$
Therefore,

$$\tan^{-1}(\sin(x)) = \left(x - \frac{x^3}{3!}\right) - \frac{1}{3}\left(x - \frac{x^3}{3!}\right)^3 + \ldots$$

$$= x - \frac{1}{3!}x^3 - \frac{1}{3}\left(x^3 - \frac{1}{4}x^4 - \ldots\right) + \ldots$$

$$= x - \frac{1}{2}x^3$$

End of topic 7 test (page 262)

Q139:
$$u_n = a + (n-1)d$$
$$= 17 + (25-1)9$$
$$= 233$$

Q140:
The summation of an arithmetic series is given by: $S_n = \frac{n}{2}(2a + (n-1)d)$
$a = 13$
$d = -5$
$n = 28$
So
$$S_{28} = \frac{28}{2}(2(13) + (27)(-5))$$
$$S_{28} = -1526$$

Q141:
a)
$$u_n = a + (n-1)d$$
$$= -11 + (14-1)4$$
$$= 41$$
b)
$$S_n = \frac{n}{2}(2a + (n-1)d)$$
$$= \frac{14}{2}(2 \times -11 + (14-1)4)$$
$$= 210$$

Q142:
The general term in a geometric sequence is: $u_n = ar^{n-1}$
$a = 2$
$r = 4$
$n = 8$
So
$u_8 = 2 \times 4^7$
$u_8 = 32768$

Q143:
$S_n = \frac{162\left(1 - \left(\frac{1}{3}\right)^n\right)}{1 - \frac{1}{3}}$

© HERIOT-WATT UNIVERSITY

Q144:

a) The general term in a geometric sequence is: $u_n = ar^{n-1}$
$a = 625$
$r = \dfrac{1}{5}$
$n = 9$
So
$u_9 = 625 \times \dfrac{1}{5}^8$
$u_9 = \dfrac{1}{625}$

b) The summation of a geometric series is given by: $S_n = \dfrac{a(1-r^n)}{1-r}$
$a = 625$
$r = \dfrac{1}{5}$
$n = 6$
So
$S_6 = \dfrac{625\left(1 - \dfrac{1}{5}^6\right)}{1 - \dfrac{1}{5}}$
$S_6 = 781.2$

Q145:

a)
$u_n = a + (n-1)\,d$
$= 5 + (120 - 1)\,2$
$= 243$

b)
$S_n = \dfrac{n}{2}(2a + (n-1)\,d)$
$= \dfrac{200}{2}(2 \times 5 + (200 - 1)\,2)$
$= 40800$

Q146:

a)
$u_n = ar^{n-1}$
$r = \dfrac{3645}{1215} \Rightarrow r = 3$
$98415 = a \times 3^9 \Rightarrow a = 5$
$u_n = 5 \times 3^{n-1}$

b)
$S_n = \dfrac{a(1-r^n)}{1-r}$
$S_{12} = \dfrac{5(1 - 3^{12})}{1-3}$
$S_{12} = 1328600$

Q147:
a)
$u_n = a + (n-1)d$
$= 8 + (40-1)22$
$= 866$

b)
$S_n = \dfrac{n}{2}(2a + (n-1)d)$
$= \dfrac{50}{2}(2 \times 8 + (50-1)22)$
$= 27350$

Q148:
a)
$u_n = ar^{n-1}$
$r = \dfrac{14}{7} \Rightarrow r = 2$
$7 = a \times 2^0 \Rightarrow a = 7$
$u_n = 7 \times 2^{n-1}$
So
$u_{11} = 7 \times 2^{10}$
$u_{11} = 7168$

b)
$S_n = \dfrac{a(1-r^n)}{1-r}$
$S_9 = \dfrac{7(1-2^9)}{1-2}$
$S_9 = 3577$

Q149:
$S_\infty = \dfrac{a}{1-r}$
$S_\infty = \dfrac{2}{1-\frac{3}{5}}$
$S_\infty = 5$

Q150:
$S_\infty = \dfrac{a}{1-r}$
$\dfrac{18}{5} = \dfrac{a}{1-\frac{1}{6}}$
$a = \dfrac{18}{5} \times \dfrac{5}{6} \Rightarrow a = 3$
$u_3 = 3 \times \left(\dfrac{1}{6}\right)^2$
$u_3 = \dfrac{1}{12}$

Q151:

Sequence

$u_n = a + (n-1)d$
$9 = a + 3d$
$a = 9 - 3d$

Series

$S_n = \dfrac{n}{2}(2a + (n-1)d)$
$36 = \dfrac{8}{2}(2a + 7d)$
$9 = 2a + 7d$

Substituting in for $a = 9 - 3d$:
$9 = 2(9 - 3d) + 7d$
$d = -9$

Substituting to find a:
$a = 9 - 3(-9)$
$a = 36$

Finding the sum of the first 21 terms:
$S_{21} = \dfrac{21}{2}(2(36) + (20)(-9))$
$= -1134$

Q152:

a)
$\dfrac{n^3 + 2n - 3}{n(n^4 + 5)} = \dfrac{n^3 + 2n - 3}{n^5 + 5n}$

Dividing by the highest power of n:
$= \dfrac{\frac{1}{n^2} + \frac{2}{n^4} - \frac{3}{n^5}}{1 + \frac{5}{n^4}}$

Now as $n \to \infty$ then:
$\dfrac{\frac{1}{n^2} + \frac{2}{n^4} - \frac{3}{n^5}}{1 + \frac{5}{n^4}} \to 0$

b)
$\dfrac{(3n+1)(2n-3)}{5n(n+2)} = \dfrac{6n^2 - 7n - 3}{5n^2 + 10n}$

Dividing by the highest powers of n:
$= \dfrac{6 - \frac{7}{n} - \frac{3}{n^2}}{5 + \frac{10}{n}}$

Now as $n \to \infty$ then:
$\dfrac{6 - \frac{7}{n} - \frac{3}{n^2}}{5 + \frac{10}{n}} \to \dfrac{6}{5}$

c)
$7 - \dfrac{3(n^2 - n + 1)}{n^2} = 7 - \dfrac{3n^2 - 3n + 3}{n^2}$

Dividing by the highest powers of n:
$= 7 - \dfrac{3 - \frac{3}{n} + \frac{3}{n^2}}{1}$

Now as $n \to \infty$ then:
$7 - \dfrac{3 - \frac{3}{n} + \frac{3}{n^2}}{1} \to 7 - 3 = 4$

ANSWERS: UNIT 1 TOPIC 7

Q153:

$3 = ar^2$ and $\frac{1}{3} = ar^4$

Dividing: $\frac{ar^4}{ar^2} = \frac{\frac{1}{3}}{3} \Rightarrow r^2 = \frac{1}{9} \Rightarrow r = \frac{1}{3}$

Substituting back in: $3 = a\left(\frac{1}{3}\right)^2 \Rightarrow a = 27$

So

$S_\infty = \frac{a}{1-r}$

$S_\infty = \frac{27}{1-\frac{1}{3}} \Rightarrow S_\infty = \frac{81}{2}$

Q154:

$$(1 - 0 \cdot 19)^{-1} = 1 + 0 \cdot 19 + 0 \cdot 19^2 + 0 \cdot 19^3 + \cdots$$
$$= 1 + 0 \cdot 19 + 0 \cdot 0361 + 0 \cdot 006859 + \cdots$$
$$= 1 \cdot 2330 \text{ to 4 d.p.}$$

Q155:

$$(1 + 0 \cdot 15)^{-1} = 1 - 0 \cdot 15 + 0 \cdot 15^2 - 0 \cdot 15^3 + \cdots$$
$$= 1 - 0 \cdot 15 + 0 \cdot 0225 - 0 \cdot 003375 + \cdots$$
$$= 0 \cdot 8691 \text{ to 4 d.p.}$$

Q156:

$$(3 \cdot 23)^{-1} = \frac{1}{3}\left(1 + \frac{0 \cdot 23}{3}\right)^{-1}$$
$$= \frac{1}{3}\left(1 - \frac{0 \cdot 23}{3} + \left(\frac{0 \cdot 23}{3}\right)^2 - \left(\frac{0 \cdot 23}{3}\right)^3 + \cdots\right)$$
$$= 0 \cdot 3096 \text{ to 4 d.p.}$$

Q157:

The geometric series is

$$(3 - 0 \cdot 05)^{-1} = 3^{-1}\left(1 + \frac{0 \cdot 05}{3} + \frac{0 \cdot 0025}{9} + \frac{0 \cdot 000125}{27} + \cdots\right)$$
$$= \frac{1}{3} + \frac{0 \cdot 05}{9} + \frac{0 \cdot 0025}{9} + \frac{0 \cdot 000125}{27} + \cdots$$

and this approximates to 0·338983

Q158:

$$(4 + 7x)^{-1} = \frac{1}{7x}\left(1 + \frac{4}{7x}\right)^{-1}$$
$$= \frac{1}{7x}\left(1 - \frac{4}{7x} + \left(\frac{4}{7x}\right)^2 - \left(\frac{4}{7x}\right)^3 + \cdots\right)$$
$$= \frac{1}{7x} - \frac{4}{49x^2} + \frac{16}{343x^3} - \frac{64}{2401x^4} + \cdots$$

© HERIOT-WATT UNIVERSITY

Q159:

$$(3x-4)^{-1} = -\frac{1}{4}\left(1-\frac{3x}{4}\right)^{-1}$$

$$= -\frac{1}{4}\left(1+\frac{3x}{4}+\left(\frac{3x}{4}\right)^2+\left(\frac{3x}{4}\right)^3+\cdots\right)$$

$$= -\frac{1}{4}-\frac{3x}{16}-\frac{9x^2}{64}-\frac{27x^3}{256}-\cdots$$

Q160:

For the condition $\left|\frac{-4x}{3}\right| < 1$ the series is $\frac{1}{3}+\frac{4x}{9}+\frac{16x^2}{27}+\frac{64x^3}{81}+\cdots$.

For the condition $\left|\frac{-3}{4x}\right| < 1$ the series is $-\frac{1}{4x}-\frac{3}{16x^2}-\frac{9}{64x^3}-\frac{27}{256x^4}-\cdots$.

Q161:

- For years 1 to 5 inclusive the geometric sequence is $\{500 \times 1 \cdot 075^n\}$ and for years 6 to 16 the geometric sequence is $\{1217 \cdot 81 \times 1 \cdot 05^n\}$.
- The first sequence has $a = 500$ and $r = 1 \cdot 075$.
  The second sequence has $a = 1217 \cdot 81$ and $r = 1 \cdot 05$.
- The amount at the end of the 16th year (the start of the 17th year) is \$1983·68 (or \$1983·69 if no rounding takes place at the end of year 5).

Q162:

$$\sum_{n=1}^{4}(7-2n) = \sum_{n=1}^{4}7 - 2\sum_{n=1}^{4}n$$

$$= 4 \times 7 - 2 \times \frac{1}{2} \times 4 \times (4+1)$$

$$= 8$$

Q163:

$$\sum_{n=1}^{9}(2n^2+3n^3) = 2\sum_{n=1}^{9}n^2 + 3\sum_{n=1}^{9}n^3$$

$$= 2 \times \frac{9(9+1)(2(9)+1)}{6} + 3 \times \frac{9^2(9+1)^2}{4}$$

$$= 6645$$

Q164:

$$\sum_{n=1}^{10}(4+3n^3-9n) - \sum_{n=1}^{3}(4+3n^3-9n)$$

$$= \sum_{n=1}^{10}4 + 3\sum_{n=1}^{10}n^3 - 9\sum_{n=1}^{10}n - \left(\sum_{n=1}^{3}4 + 3\sum_{n=1}^{3}n^3 - 9\sum_{n=1}^{3}n\right)$$

$$= 4 \times 10 + 3 \times \frac{10^2 \times 11^2}{4} - 9 \times \frac{10 \times 11}{2} - \left(4 \times 3 + 3 \times \frac{3^2 \times 4^2}{4} - 9 \times \frac{3 \times 4}{2}\right)$$

$$= 40 + 9075 - 495 - (12 + 108 - 54)$$

$$= 8554$$

Q165:

Hints:

- Use partial fractions.

Answer: $\frac{1}{3}$

The trick is to find the partial fractions, use the combination rules, write out the first few terms of each series and subtract.

$\frac{1}{n^2+5n+6} = \frac{1}{(n+3)(n+2)}$

So

$$\frac{1}{(n+3)(n+2)} = \frac{A}{(n+3)} + \frac{B}{(n+2)}$$

$$1 = A(n+2) + B(n+3)$$

Sub $n = -2$: $B = 1$

Sub $n = -3$: $A = -1$

So

$\frac{1}{(n+3)(n+2)} = \frac{1}{n+2} - \frac{1}{n+3}$

$$\sum_{n=1}^{\infty}\left(\frac{1}{n^2+5n+6}\right) = \sum_{n=1}^{\infty}\left(\frac{1}{n+2} - \frac{1}{n+3}\right)$$

$$= \sum_{n=1}^{\infty}\left(\frac{1}{n+2}\right) - \sum_{n=1}^{\infty}\left(\frac{1}{n+3}\right)$$

$$= \left(\frac{1}{3} + \frac{1}{4} + \frac{1}{5} + \frac{1}{6} + \frac{1}{7} + \cdots\right) - \left(\frac{1}{4} + \frac{1}{5} + \frac{1}{6} + \frac{1}{7} + \cdots\right)$$

$$= \frac{1}{3}$$

Q166:

$$\sum_{r=1}^{n} r(r-1) = \sum_{r=1}^{n} r^2 - r$$

$$= \sum_{r=1}^{n} r^2 - \sum_{r=1}^{n} r$$

$$= \frac{n(n+1)(2n+1)}{6} - \frac{n(n+1)}{2}$$

$$= \frac{n(n+1)}{6}[(2n+1) - 3]$$

$$= \frac{n(n+1)}{6}[2n - 2]$$

$$= \frac{2n(n+1)(n-1)}{6}$$

$$= \frac{1}{3}n(n+1)(n-1)$$

Q167:

$$\sum_{r=1}^{n} r(r+1)(r+2) = \sum_{r=1}^{n} r^3 + 3\sum_{r=1}^{n} r^2 + 2\sum_{r=1}^{n} r$$

$$= \frac{n^2(n+1)^2}{4} + 3 \times \frac{n(n+1)(2n+1)}{6} + 2 \times \frac{n(n+1)}{2}$$

Take out a factor of $\frac{n(n+1)}{4}$:

$$= \frac{n(n+1)}{4}\left\{n(n+1) + \frac{3 \times 4 \times (2n+1)}{6} + \frac{2 \times 4}{2}\right\}$$

$$= \frac{n(n+1)}{4}[n(n+1) + 2(2n+1) + 4]$$

$$= \frac{1}{4}n(n+1)(n+2)(n+3)$$

Q168:

$$\sum_{r=1}^{n} r(r^2 + 2) = \sum_{r=1}^{n} r^3 + 2r$$

$$= \sum_{r=1}^{n} r^3 + 2\sum_{r=1}^{n} r$$

$$= \frac{n^2(n+1)^2}{4} + 2 \times \frac{n(n+1)}{2}$$

Take out a factor of $\frac{n(n+1)}{4}$:

$$= \frac{n(n+1)}{4}\left\{n(n+1) + \frac{2 \times 4}{2}\right\}$$

$$= \frac{n(n+1)}{4}[n(n+1) + 4]$$

$$= \frac{1}{4}n(n+1)(n^2 + n + 4)$$

Q169:

The Maclaurin series expansion for $\sin(x)$ is given by: $\sin(x) = x - \frac{x^3}{3!} + \frac{x^5}{5!} - \frac{x^7}{7!} + \ldots \quad x \in \mathbb{R}$

For the expansion of $2\sin(2x)$ we need to replace x with $2x$ in the expansion for $\sin(x)$. Note that we only want the first three non-zero terms so we can ignore any subsequent terms.

$$2\sin(2x) = 2\left((2x) - \frac{(2x)^3}{3!} + \frac{(2x)^5}{5!} - \ldots\right)$$

$$= 2\left(2x - \frac{4x^3}{3} + \frac{4x^5}{15} - \ldots\right)$$

$$= 4x - \frac{8x^3}{3} + \frac{8x^5}{15}$$

ANSWERS: UNIT 1 TOPIC 7

Q170:

The Maclaurin series expansion for $\cos(x)$ is given by: $\cos(x) = 1 - \frac{x^2}{2!} + \frac{x^4}{4!} - \frac{x^6}{6!} + \ldots \quad x \in \mathbb{R}$

For the expansion of $2\cos(2x)$ we need to replace x with $2x$ in the expansion for $\cos(x)$. Note that we only want the first three non-zero terms so we can ignore any subsequent terms.

$$2\cos(2x) = 2\left(1 - \frac{(2x)^2}{2!} + \frac{(2x)^4}{4!} - \ldots\right)$$
$$= 2\left(1 - 2x^2 + \frac{2x^4}{3} - \ldots\right)$$
$$= 2x - 4x^2 + \frac{4x^4}{3}$$

Q171:

The Maclaurin series expansion for $\tan^{-1}(x)$ is given by: $\tan(x) = x - \frac{x^3}{3} + \frac{x^5}{5} - \frac{x^7}{7} + \ldots \quad |x| \leqslant 1$

For the expansion of $2\tan^{-1}(2x)$ we need to replace x with $2x$ in the expansion for $\tan^{-1}(x)$. Note that we only want the first three non-zero terms so we can ignore any subsequent terms.

$$2\tan^{-1}(2x) = 2\left((2x) - \frac{(2x)^3}{3} + \frac{(2x)^5}{5} - \ldots\right)$$
$$= 2\left(2x - \frac{8x^3}{3} + \frac{32x^5}{5} - \ldots\right)$$
$$= 4x - \frac{16x^3}{3} + \frac{64x^5}{5}$$

Q172:

The Maclaurin series expansion for $\ln(1+x)$ is given by: $\ln(1+x) = x - \frac{x^2}{2} + \frac{x^3}{3} - \frac{x^4}{4} + \ldots \quad -1 < x \leqslant 1$

For the expansion of $2\ln(1+2x)$ we need to replace x with $2x$ in the expansion for $\ln(1+x)$. Note that we only want terms to x^3 so we can ignore any subsequent terms.

$$2\ln(1+2x) = 2\left((2x) - \frac{(2x)^2}{2} + \frac{(2x)^3}{3} - \ldots\right)$$
$$= 2\left(2x - 2x^2 + \frac{8x^3}{3} - \ldots\right)$$
$$= 4x - 4x^2 + \frac{16x^3}{3}$$

Q173:

The Maclaurin series expansion for $(1-x)^{-1}$ is given by: $(1-x)^{-1} = 1 + x + x^2 + x^3 + x^4 + \ldots \quad |x| < 1$

For the expansion of $2(1+2x)^{-1}$ we need to replace x with $-2x$ in the expansion for $(1-x)^{-1}$. Note that we only want terms to x^3 so we can ignore any subsequent terms.

$$2(1+2x)^{-1} = 2\left(1 + (-2x) + (-2x)^2 + (-2x)^3 + \ldots\right)$$
$$= 2\left(1 - 2x + 4x^2 - 8x^3 + \ldots\right)$$
$$= 2 - 4x + 8x^2 - 16x^3$$

© HERIOT-WATT UNIVERSITY

Q174:

The Maclaurin series expansion for e^x is given by: $e^x = 1 + x + \frac{x^2}{2!} + \frac{x^3}{3!} + \frac{x^4}{4!} + \ldots \quad x \in \mathbb{R}$

For the expansion of $2e^{2x}$ we need to replace x with $2x$ in the expansion for e^x. Note that we only want terms in x^2 so we can ignore any order of x greater than 2.

$$2e^{2x} = 2\left(1 + (2x) + \frac{(2x)^2}{2!} + \ldots\right)$$
$$= 2\left(1 + 2x + 2x^2 + \ldots\right)$$
$$= 2 + 4x + 4x^2$$

Q175:

Using correct formula for the Maclaurin series:

$f(x) = f(0) + f^{(1)}(0)x + f^{(2)}(0)\frac{x^2}{2!} + f^{(3)}(0)\frac{x^3}{3!} + f^{(4)}(0)\frac{x^4}{4!} + \ldots$

Continue with the correct Maclaurin series expansion for $\sqrt{(1+2x)}$

$f(x) = \sqrt{(1+2x)}$ $\qquad f(0) = 1$
$f^{(1)}(x) = (1+2x)^{-\frac{1}{2}}$ $\qquad f^{(1)}(0) = 1$
$f^{(2)}(x) = -(1+2x)^{-\frac{3}{2}}$ $\qquad f^{(2)}(0) = -1$
$f^{(3)}(x) = 3(1+2x)^{-\frac{5}{2}}$ $\qquad f^{(3)}(0) = 3$

Correct answer:

$1 + x - \frac{x^2}{2} + \frac{x^3}{2} - \ldots$

Q176:

Using correct formula for the Maclaurin series:

$f(x) = f(0) + f^{(1)}(0)x + f^{(2)}(0)\frac{x^2}{2!} + f^{(3)}(0)\frac{x^3}{3!} + f^{(4)}(0)\frac{x^4}{4!} + \ldots$

Continue with the correct Maclaurin series expansion for $(1+2x)^{-3}$

$f(x) = (1+2x)^{-3}$ $\qquad f(0) = 1$
$f^{(1)}(x) = -6(1+2x)^{-4}$ $\qquad f^{(1)}(0) = -6$
$f^{(2)}(x) = 48(1+2x)^{-5}$ $\qquad f^{(2)}(0) = 48$
$f^{(3)}(x) = -480(1+2x)^{-6}$ $\qquad f^{(3)}(0) = -480$

Correct answer:

$1 - 6x + 24x^2 - 80x^3 + \ldots$

Q177:

Correct formula for $\sin(x)$:

$x - \frac{x^3}{3!} + \frac{x^5}{5!} - \frac{x^7}{7!} + \ldots$

Continue with correct expansion for $\sin(0 \cdot 5)$:

$(0 \cdot 5) - \frac{(0 \cdot 5)^3}{3!} + \frac{(0 \cdot 5)^5}{5!} - \frac{(0 \cdot 5)^7}{7!} + \ldots$

Correct answer:

0·4794 (to 4 d.p.)

ANSWERS: UNIT 1 TOPIC 7

Q178:

Correct formula for e^x:

$1 + x + \frac{x^2}{2!} + \frac{x^3}{3!} + \frac{x^4}{4!} + \ldots$

Continue with correct expansion for $e^{0.5}$:

$1 + (0 \cdot 5) + \frac{(0 \cdot 5)^2}{2!} + \frac{(0 \cdot 5)^3}{3!} + \frac{(0 \cdot 5)^4}{4!} + \ldots$

Correct answer:

1·649 (to 3 d.p.)

Q179:

Standard results:

$$e^x = 1 + x + \frac{x^2}{2!} + \frac{x^3}{3!} + \frac{x^4}{4!} + \ldots$$

$$\sin(x) = x - \frac{x^3}{3!} + \frac{x^5}{5!} - \ldots$$

Substituting $3x$ into $\sin(x)$ gives:

$$\sin(3x) = 3x - \frac{(3x)^3}{3!} + \frac{(3x)^5}{5!} - \ldots$$

$$= 3x - \frac{9x^3}{2} + \frac{81x^5}{40} - \ldots$$

Multiplying the expansions for e^x and $\sin(3x)$ as far as the term x^4 gives:

$$f(x) = e^x \sin(3x)$$

$$= \left(1 + x + \frac{x^2}{2} + \frac{x^3}{6} + \frac{x^4}{24} + \ldots\right)\left(3x - \frac{9x^3}{2} + \frac{81x^5}{40} - \ldots\right)$$

$$= 1\left(3x - \frac{9x^3}{2}\right) + x\left(3x - \frac{9x^3}{2}\right) + \frac{x^2}{2}(3x) + \frac{x^3}{6}(3x)$$

$$= 3x - \frac{9x^3}{2} + 3x^2 - \frac{9x^4}{2} + \frac{3x^3}{2} + \frac{x^4}{2}$$

$$= 3x + 3x^2 - 3x^3 - 4x^4$$

Q180:

Standard results:

$\sin(x) = x - \frac{x^3}{3!} + \frac{x^5}{5!} - \ldots$

$\tan^{-1}(x) = x - \frac{x^3}{3} + \frac{x^5}{5} - \ldots, \; |x| \leqslant 1$

Therefore, as far as the term in x^3 we have:

$$\tan^{-1}(\sin(x)) = \left(x - \frac{x^3}{3!}\right) - \frac{1}{3}\left(x - \frac{x^3}{3!}\right)^3 + \ldots$$

$$= x - \frac{1}{3!}x^3 - \frac{1}{3}\left(x^3 - \frac{1}{4}x^4 - \ldots\right) + \ldots$$

$$= x - \frac{1}{2}x^3$$

Topic 8: Curve sketching
Determining stationary points exercise (page 276)

Q1:

$y = 3x^3 + 18x^2 + 27x + 1$

$\frac{dy}{dx} = 9x^2 + 36x + 27$

Stationary points occur when $\frac{dy}{dx} = 0$.

$9x^2 + 36x + 27 = 0$

$9(x^2 + 4x + 3) = 0$

$9(x+1)(x+3) = 0$

$x + 1 = 0 \quad \text{or} \quad x + 3 = 0$

$x = -1 \quad \text{or} \quad x = -3$

When $x = -3, y = 3(-3)^3 + 18(-3)^2 + 27(-3) + 1 = 1$
When $x = -1, y = 3(-1)^3 + 18(-1)^2 + 27(-1) + 1 = -11$

Thus the stationary points occur at (-3,1) and (-1,-11).
To determine their nature we draw a nature table.

x	-4 →	-3	-2 →	-1	0 →
$(x+1)$	-	-	-	0	+
$(x+3)$	-	0	+	+	+
$9(x+1)(x+3)$	+	0	-	0	+
shape	/	—	\	—	/

This gives us a maximum turning point at (-3,1) and a minimum turning point at (-1,-11).

Q2:

$y = x^4 - 4x^3 + 1$

$\frac{dy}{dx} = 4x^3 - 12x^2 = 4x^2(x-3)$

Stationary points occur when $\frac{dy}{dx} = 0$.

$4x^2(x-3) = 0$

$4x^2 = 0 \quad \text{or} \quad x - 3 = 0$

$x = 0 \quad \text{or} \quad x = 3$

When $x = 0$ then $y = 1$
When $x = 3$ then $y = -26$

Thus the coordinates of the stationary points are (0,1) and (3,-26).
To determine their nature we draw a nature table.

ANSWERS: UNIT 1 TOPIC 8

x	-1 $\rightarrow$	0	1 $\rightarrow$	3	4 $\rightarrow$
$4x^2$	+	0	+	+	+
$(x-3)$	-	-	-	0	+
$4x^2(x-3)$	-	0	-	0	+
shape	\	—	\	—	/

This gives us a falling point of inflection at (0,1) and a maximum turning point at (3,-26).

Closed intervals exercise (page 278)

Q3:
It is much easier to spot the maximum and minimum values if we make a sketch.
Step 1: Find the stationary points and their nature.
$\frac{dy}{dx} = 3x^2 - 12$
Stationary points occur when $\frac{dy}{dx} = 0$.

$3x^2 - 12 = 0$
$3(x^2 - 4) = 0$
$3(x - 2)(x + 2) = 0$
$x - 2 = 0$ or $x + 2 = 0$
$x = 2$ or $x = -2$

When $x = 2$ then $y = 2^3 - 12 \times 2 = -16$.
When $x = -2$ then $y = (-2)^3 - 12 \times (-2) = 16$.
Thus the stationary points are (2, -16) and (-2, 16)

x	-3 $\rightarrow$	-2	0 $\rightarrow$	2	3 $\rightarrow$
$(x-2)$	-	-	-	0	+
$(x+2)$	-	0	+	+	+
$3(x-2)(x+2)$	+	0	-	0	+
shape	/	—	\	—	/

This gives us a maximum turning point at (-2,16) and a minimum turning point at (2, -16).
Step 2: Find the end points
When $x = -5$ then $y = -65$
When $x = 3$ then $y = -9$

Step 3: Make a sketch

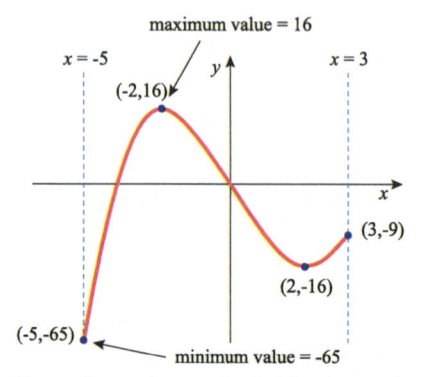

Hence the maximum value is 16 at the maximum turning point (-2,16) and the minimum value is -65 at the end point (-5,-65).

Transformations of graphs exercise (page 292)

Q4: a)

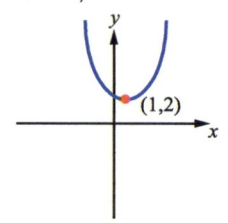

Q5: (-2,-5)

Q6:
The transformation $g(x + 1) - 3$ tells us that the graph of the function $g(x)$ should be moved left horizontally by 1 unit and down vertically by 3 units.
To do this we subtract 1 from the x-coordinates and subtract 3 from the y-coordinates.
The identifiable points are:

$g(x)$		$g(x + 1) - 3$
(-3,0)	⇒	(-4,-3)
(-2,7)	⇒	(-3,4)
(0,3)	⇒	(-1,0)
(1,5)	⇒	(0,2)
(6,0)	⇒	(5,-3)

ANSWERS: UNIT 1 TOPIC 8

Giving,

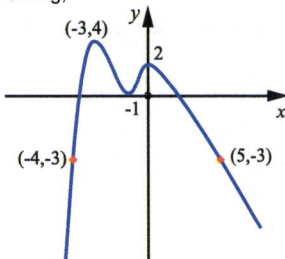

Q7: (-2,-3)

Q8: b)

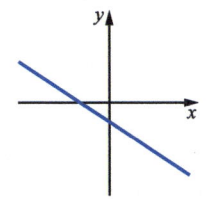

Q9: b)

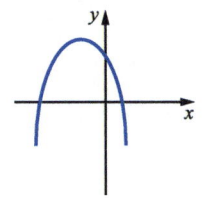

Q10: (2,2)

Q11: a)

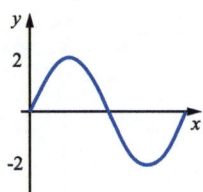

Q12:

The transformation $1/2 f(x + 3)$ moves the graph of $f(x)$ horizontally left by 3 units and squashes it vertically by a factor of a half.
To do this we subtract 3 from the x-coordinates and half the y-coordinates.
The identifiable points are:

f(x)		1/2f(x + 3)
(-4,0)	⇒	(-7,0)
(0,6)	⇒	(-3,3)
(4,0)	⇒	(1,0)

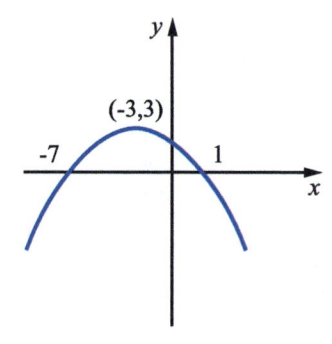

Q13: d)

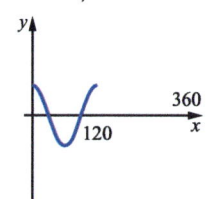

ANSWERS: UNIT 1 TOPIC 8 583

Domain and codomain range (page 304)

Q14:

Graph of $y = \sqrt{5x-3}$	Graph of $y = \frac{1}{x-3}$
Domain: $x \geqslant \frac{3}{5}$	**Domain:** $x \neq 3$
Range: $y \geqslant 0$	**Range:** $y \in \mathbb{R}$

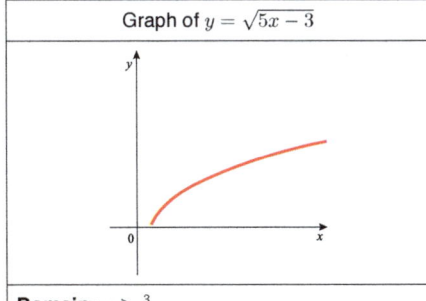

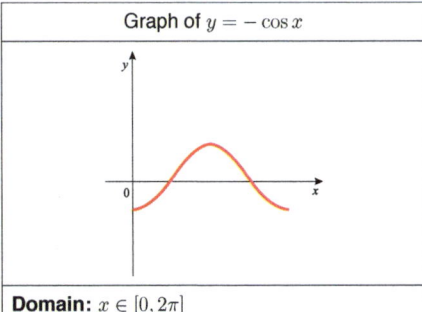

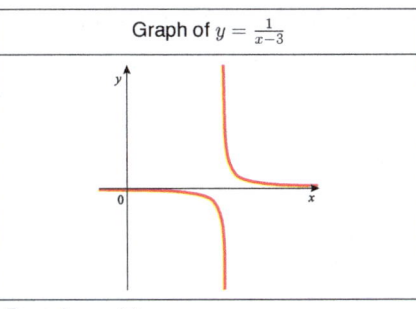

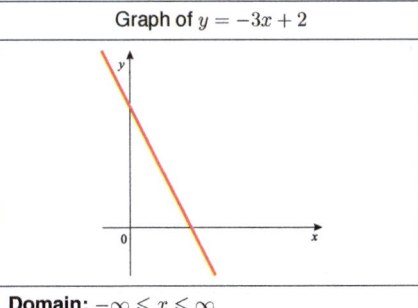

Graph of $y = -\cos x$	Graph of $y = -3x + 2$
Domain: $x \in [0, 2\pi]$	**Domain:** $-\infty \leqslant x \leqslant \infty$
Range: $y \in [-1, 1]$	**Range:** $-\infty \leqslant y \leqslant \infty$

Note that $y \in \mathbb{R}$ can be interchanged for $-\infty \leqslant y \leqslant \infty$.

1. **Graph of** $y = \sqrt{5x-3}$
   Using real numbers, $y = \sqrt{5x-3}$ is not defined for negative numbers so $5x - 3 \geqslant 0$. This means $x \geqslant \frac{3}{5}$. If $x = \frac{3}{5}$ then $y = 0$, therefore $y \geqslant 0$.
2. **Graph of** $y = \frac{1}{x-3}$
   $\frac{1}{x-3}$ is undefined when $x - 3 = 0$, therefore $x \neq 3$. So the domain is $\{x \in \mathbb{R} : x \neq 3\}$. That is all the real numbers except 3. The range is then all the real numbers: $y \in \mathbb{R}$.
3. **Graph of** $y = -\cos x$
   The domain of the function is $x \in [0, 2\pi]$. It can be seen from the graph that within this range the minimum value is given when $x = 0$ and the maximum when $x = \pi$. So the range is $y \in [-1, 1]$.
4. **Graph of** $y = -3x + 2$
   This line is infinitely long so has a domain of all the real numbers: $-\infty \leqslant x \leqslant \infty$
   Corresponding the function is clearly defined for all values in the y-axis: $y \in \mathbb{R}$.

© HERIOT-WATT UNIVERSITY

Function definition exercise (page 305)

Q15:
$\frac{1}{x+5}$ is undefined for $x \neq -5$, so the domain is $\{x \in \mathbb{R} : x \neq -5\}$.
The range is $\{y \in \mathbb{R}\}$.

Q16:
$\frac{1}{x^2+3x+2}$ is undefined when $x^2 + 3x + 2 = 0$.
Factorising:
$(x+2)(x+1) = 0$
$x = -2$ and $x = -1$
So the domain is $\{x \in \mathbb{R} : x \neq -2, x \neq -1\}$.
The range $\{f(x) \in \mathbb{R}\}$.

Q17:
There are no values that this function is undefined for so the domain is $\{x \in \mathbb{R} : -\infty \leqslant x \leqslant \infty\}$.
The range is also unrestricted and is $\{y \in \mathbb{R} : -\infty \leqslant y \leqslant \infty\}$.

Q18:
The graph of $y = \tan x$ should be familiar.

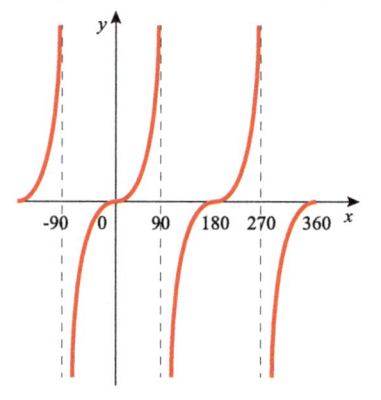

It is clear that the smallest value is when $x = -45°$.
$\tan(-45°) = -\frac{1}{\sqrt{2}}$
It is clear that the largest value is when $x = 45°$.
$\tan(45°) = \frac{1}{\sqrt{2}}$
The range is:
$-\frac{1}{\sqrt{2}} \leqslant y \leqslant \frac{1}{\sqrt{2}}$

ANSWERS: UNIT 1 TOPIC 8

Q19:
Evaluating $f(2) = 0$. This means the smallest number that $f(x)$ can be is > 0. Since we are only allowed integers ($\mathbb{Z}$) then essentially we are left with the natural numbers ($\mathbb{N}$). This can be written in several ways:

$\{f(x) \in \mathbb{N}\}$, or $\{f(x) \in \mathbb{Z} : x > 0\}$, or $\{f(x) \in \mathbb{Z}^+\}$

Q20: We can only consider integer input values. When evaluating x^2 with integers we have the all the positive integers. Since we are given $y = x^2 + 4$, then it is all the positive integers greater or equal to 4.

$\{y \in \mathbb{N} : y \geqslant 4\}$, or $\{y \in \mathbb{Z} : y \geqslant 4\}$, or $\{y \in \mathbb{Z}^+ : y \geqslant 4\}$

Q21:
When:

$x = 4$	then	$y = \frac{1}{5}$
$x = 5$	then	$y = \frac{1}{6}$
$x = 6$	then	$y = \frac{1}{7}$
...		

$y \to 0$ and the biggest value that y can be is $\frac{1}{5}$.
So the range is $\{y \in \mathbb{Q} : 0 < y < \frac{1}{5}\}$, (where $y = \frac{1}{x+1}$).

Q22:
$\sqrt{x^2 - 1}$ is undefined when $x^2 - 1$ is negative, so $x^2 - 1$ must be bigger than or equal to zero i.e. $x^2 - 1 \geqslant 0$.

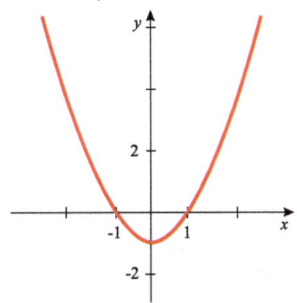

From the sketch we can see that for $x^2 - 1 \geqslant 0$, this is the part of the graph above the x-axis, then $x \leqslant -1$ and $x \geqslant 1$.
So the domain is $\{x \in \mathbb{R} : x \leqslant -1 \text{ and } x \geqslant 1\}$.
Now since we have defined $x^2 - 1 \geqslant 0$ then $\sqrt{x^2 - 1} \geqslant 0$.
The range is $\{y \in \mathbb{R} : y \geqslant 0\}$.

Types of functions (page 308)

Q23: a) Many-to-one

Q24: b) One-to-one

Q25: a) Many-to-one

Q26: a) Many-to-one

Q27: b) One-to-one

Q28: a) Many-to-one

One-to-one and onto functions exercise (page 310)

Q29:

$y = 4x^2 - 1$ is a many-to-one mapping. If we take $x = 1$ and $x = -1$, then $y = 3$ in both cases.

It is not an onto mapping because the range is $y > 1$. The codomain is $y \in \mathbb{R}$.

Q30:

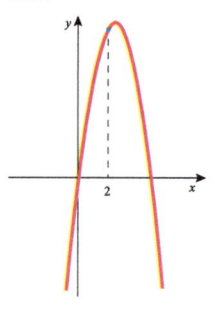

[Graph of a downward parabola with vertex above the x-axis, crossing x-axis near 2]

From the sketch of the graph when $x \leqslant 2$ there is only one x for one y, and one y for one x. It is a one-to-one mapping.

Q31:

This is a one-to-one mapping. There is only one x for one y, and one y for one x.

Q32:

This is a one-to-one mapping. There is only one x for one y, and one y for one x.

Q33:

We can define any interval of x values that would define one x to one y and one y to one x.

e.g. the interval : $\left[-\frac{3\pi}{2}, -\frac{\pi}{2}\right] \left[-\frac{\pi}{2}, \frac{\pi}{2}\right]$ (the most commonly used), $\left[\frac{\pi}{2}, \frac{3\pi}{2}\right], \ldots$.

Q34:

The domain of this function is $x \in \mathbb{R}$ and the range is $y \in \mathbb{R}$.

The codomain is also $y \in \mathbb{R}$.

This range is the same as the codomain. Therefore this is an onto mapping.

Q35:
Evaluating the domain:

$x = -1$	then	$y = 4$
$x = 0$	then	$y = 0$
$x = 1$	then	$y = -2$
$x = 2$	then	$y = 0$

The range $\{-2, 0, 4\}$. The codomain is $\{-2, -1, 0, 1, 2, 3, 4, 5, 6\}$. These are not the same so it is not an onto mapping.

Curve sketching proofs exercise (page 313)

Q36:
$f(s) = 2s - 4$
$f(t) = 2t - 4$
Let $f(s) = f(t)$
Then $2s - 4 = 2t - 4$
So $s = t$

The function is one-to-one.

Q37:
$g(s) = 2s^5$
$g(t) = 2t^5$
Let $g(s) = g(t)$
Then $2s^5 = 2t^5$
So $s = t$

The function is one-to-one.

Q38:
$h(s) = \dfrac{1}{s+2}$
$h(t) = \dfrac{1}{t+2}$
Let $h(s) = h(t)$
Then, $\dfrac{1}{s+2} = \dfrac{1}{t+2}$
So $s = t$

The function is one-to-one.

Q39:
Let $t \in \mathbb{R}$
Let $t = 3x$ giving $s = \frac{t}{3}$
$$\begin{aligned}f(s) &= f\left(\frac{t}{3}\right)\\ &= 3 \times \frac{t}{3}\\ &= t\end{aligned}$$

The function is onto.

Q40:
Let $t \in \mathbb{R}^+$
Let $t = s^3$ giving $s = \sqrt[3]{t}$
$$\begin{aligned}g(s) &= g\left(\sqrt[3]{t}\right)\\ &= \left(\sqrt[3]{t}\right)^3\\ &= t\end{aligned}$$

The function is onto.

Q41:
Let $t \in \{z \in \mathbb{R}; z \neq 1\}$
Let $t = 1 + \frac{1}{s} \;\Rightarrow\; t - 1 = \frac{1}{s}$ giving $s = \frac{1}{t-1}$.
$$\begin{aligned}h(s) &= h\left(\frac{1}{t-1}\right)\\ &= 1 + \frac{1}{\frac{1}{t-1}} = 1 + \frac{t-1}{1}\\ &= t\end{aligned}$$

The function is onto.

Q42:
Take $s = -1$ and $t = 1$ then $f(s) = \frac{1}{2}$ and $f(t) = \frac{1}{2}$. But $s \neq t$ and so the function is not one-to-one.

Q43:
Let $f(s) = -4$. There is no element $s \in \mathbb{R}^+$, the domain of f such that $f(s) = -4$ as $f(s) = \sqrt{s} \geqslant 0$ for all s. Hence the function is not onto.

Inverse functions (page 318)

Q44:

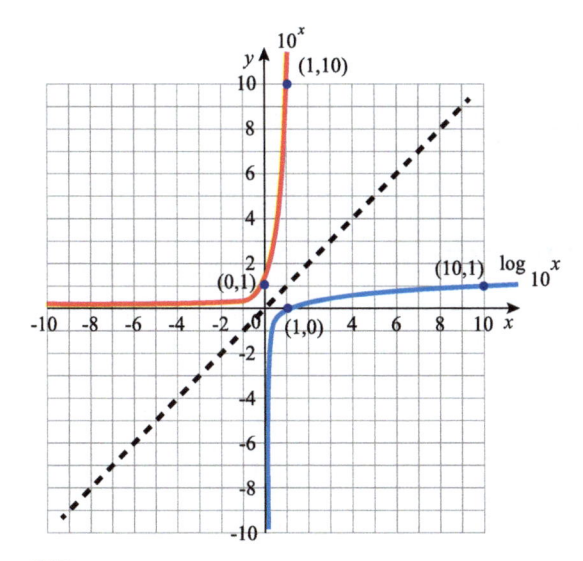

Q45:

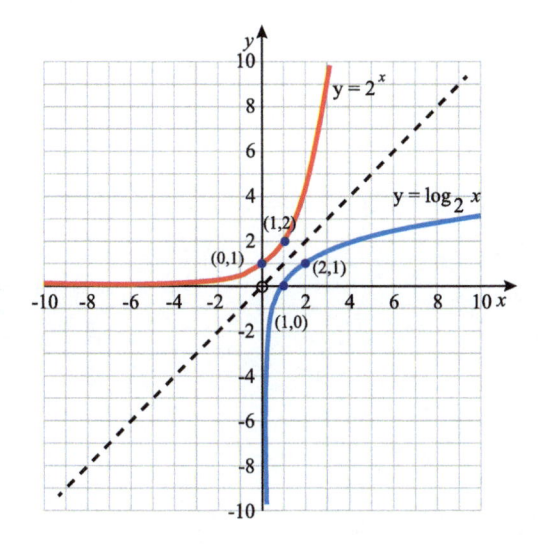

Q46:

$f(x)^{-1} = \frac{8-x}{2}$, where $x \in \mathbb{R}$.

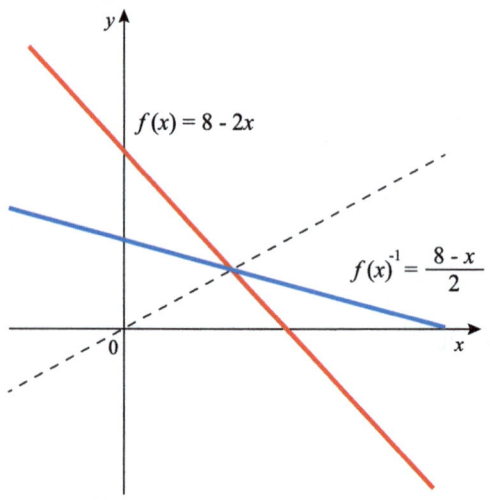

Q47:

$g(x)^{-1} = 2 - \frac{1}{x}$ where $x \in \mathbb{R}$, $x > 0$.

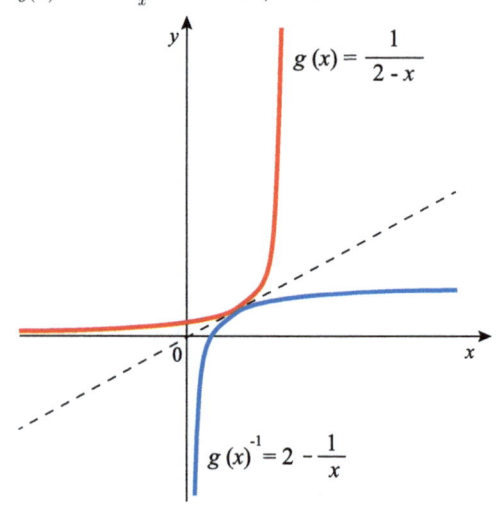

Q48:

$h^{-1}(x) = \frac{x+6}{4}$ where $x \in \mathbb{R}$.

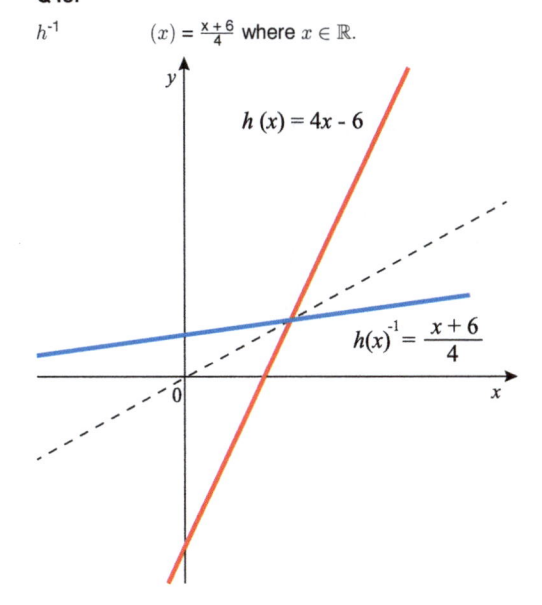

$h(x) = 4x - 6$

$h(x)^{-1} = \frac{x+6}{4}$

Exploring inverse trigonometric functions (page 322)

Q49:

Using your ruler place it horizontally over the graph of $\sin(x)$. Move the ruler up and down the y-axis. The ruler intersects the curve at one point at any given moment. Now take the ruler and place it vertically parallel to the y-axis. Move the ruler back and forward along the x-axis. At any point there is only one point of intersection with the ruler and the curve. Since vertical and horizontal movement gives only one point of contact, $\sin(x)$ where $\left[\frac{-\pi}{2}, \frac{\pi}{2}\right]$ is a one-to-one function.

a) The function is one-to-one.

  The domain generates a range of [-1,1]
  This is equal to the codomain shown.

  The function is onto.

b) A possible restricted domain for $\cos x$ is $[0, \pi]$. This is the most common restricted domain given. Other possibilities are: ... $[-3\pi, -2\pi]$, $[-2\pi, -\pi]$, $[-\pi, 0]$, $[0, \pi]$, $[\pi, 2\pi]$, $[2\pi, 3\pi]$, ...

c) A possible restricted domain for $\tan x$ is $\left[\frac{-\pi}{2}, \frac{\pi}{2}\right]$. This is the most common domain. Other possible domains are: ... $\left[\frac{-3\pi}{2}, \frac{-\pi}{2}\right]$, $\left[\frac{-\pi}{2}, \frac{\pi}{2}\right]$, $\left[\frac{\pi}{2}, \frac{3\pi}{2}\right]$ ....

d) The domain of $\sin^{-1} x$ is [-1, 1]. This is the range of $\sin(x)$.

e) $y = \tan^{-1} x$

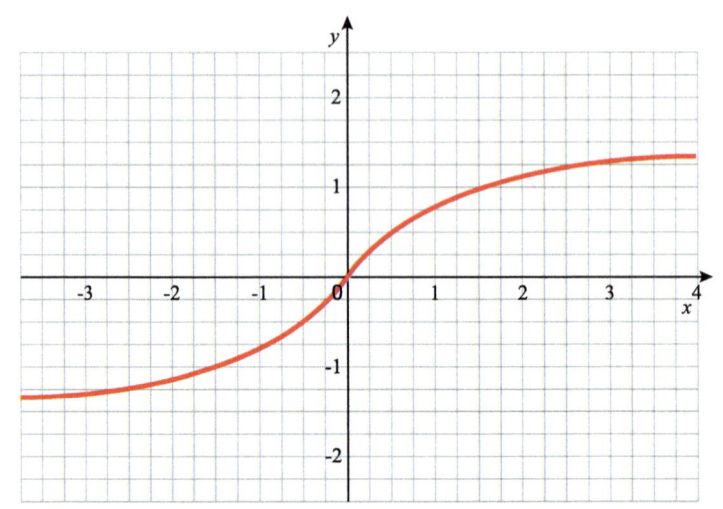

This inverse would be obtained from $y = \tan x$ with the domain of $\left[\frac{-\pi}{2}, \frac{\pi}{2}\right]$.
Note the horizontal asymptotes at $y = \frac{\pi}{2}$ and $y = \frac{-\pi}{2}$.

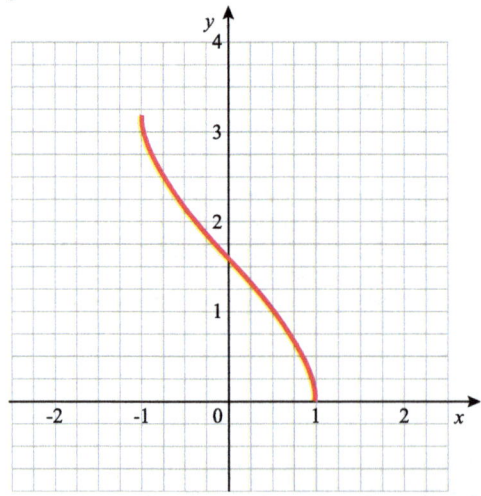

f) $y = \cos^{-1} x$

This inverse was produced using a domain for $y = \cos x$ of $[0, \pi]$.
Note that it touches the x-axis at $x = 1$.

ANSWERS: UNIT 1 TOPIC 8

Q50:

a) **Inverse of** $y = k \sin(x)$
For the function $y = k \sin x$ to be a one-to-one function the domain has to be $\left[\frac{-\pi}{2}, \frac{\pi}{2}\right]$ and the range is $[-k, k]$.
For an inverse the range must be the same as the codomain.
For the inverse function $y = \sin^{-1}\left(\frac{x}{k}\right)$ the domain is therefore $[-k, k]$ and the range is $\left[\frac{-\pi}{2}, \frac{\pi}{2}\right]$.
The inverse function is found by,
$$\frac{y}{k} = \sin x$$
$$\sin^{-1}\left(\frac{y}{k}\right) = x$$
$$y = \sin^{-1}\left(\frac{y}{k}\right)$$
This range is not unique. In general the range is $\left[\frac{n\pi}{2}, \frac{n\pi}{2} + \pi\right]$, where n is an odd integer.

b) **Inverse of** $y = \cos(bx)$
For the function $y = \cos(bx)$ to be a one-to-one function the domain has to be $\left[0, \frac{\pi}{b}\right]$ and the range is $[-1, 1]$.
For an inverse the range must be the same as the codomain.
For the inverse function $y = \frac{1}{b}\cos^{-1}(x)$ the domain is therefore $[-1, 1]$ and the range is $\left[0, \frac{\pi}{b}\right]$.
The inverse function is found by,
$$\cos^{-1}(y) = bx$$
$$\frac{1}{b}\cos^{-1}(y) = x$$
$$y = \frac{1}{b}\cos^{-1}(x)$$
This range is not unique. In general the range is $\left[\frac{n\pi}{2}, \frac{(n+1)\pi}{2}\right]$, where n is an integer.

c) **Inverse of** $y = \tan(cx)$
For the function $y = \tan(cx)$ to be a one-to-one function the domain has to be $\left[-\frac{\pi}{2c}, \frac{\pi}{2c}\right]$ and the range is $[-\infty, \infty]$.
For an inverse the range must be the same as the codomain.
For the inverse function $y = \frac{1}{c}\tan^{-1}(x)$ the domain is therefore $[-\infty, \infty]$ and the range is $\left[-\frac{\pi}{2c}, \frac{\pi}{2c}\right]$.
This range is not unique. In general the range is $\left[\frac{n\pi}{2c}, \frac{n\pi}{2c} + \frac{\pi}{2}\right]$, where n is an odd integer.

Inverse functions exercise (page 324)

Q51:

x	$\sin^{-1} x$	$\tan^{-1} x$	$\cos^{-1} x$
$\frac{\sqrt{3}}{2}$	60°	********	30°
1	90°	45°	0°
$\frac{1}{2}$	30°	********	60°
$\frac{1}{\sqrt{2}}$	45°	********	45°
$\sqrt{3}$	********	60°	********

© HERIOT-WATT UNIVERSITY

Q52:

x	$\sin^{-1} x$	$\tan^{-1} x$	$\cos^{-1} x$
$\frac{\sqrt{3}}{2}$	$\frac{\pi}{3}$	********	$\frac{\pi}{6}$
1	$\frac{\pi}{2}$	$\frac{\pi}{4}$	0
$\frac{1}{\sqrt{2}}$	$\frac{\pi}{4}$	********	$\frac{\pi}{4}$
$\sqrt{3}$	********	$\frac{\pi}{3}$	********
$\frac{1}{2}$	$\frac{\pi}{6}$	********	$\frac{\pi}{3}$

Q53:

If the original function has a coordinate (x, y) then the inverse swaps the x and y around to give (y, x).

Therefore, the point (1,0) becomes (0,1)

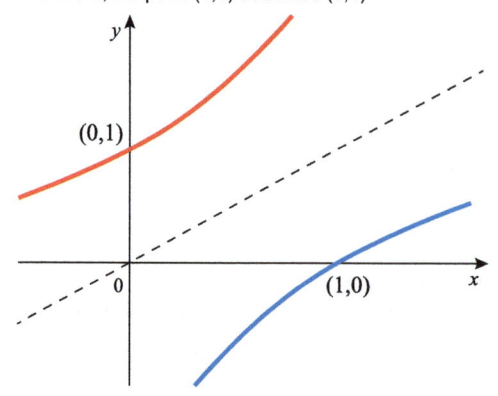

Q54:

To find the inverse of the function $y = \frac{1}{a}\ln|x|$, $x > 0$ make x the subject of the formula first then swap x and y.

$x = \frac{1}{a}\ln|y|$

$ax = \ln|y|$

$e^{ax} = y$

$y = e^{ax}$

Odd, even or neither function exercise (page 331)

Q55:

a) $f(x) = \cos x$
$f(-x) = \cos(-x) = \cos x$
Since $f(x) = f(-x)$ the function is even.

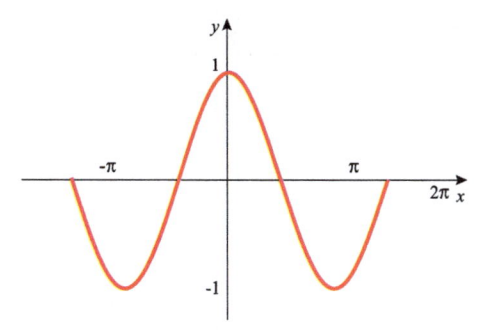

Note the reflection in the y-axis.

b) $g(x) = \frac{1}{2x-1}$
$g(-x) = \frac{1}{-2x-1} \neq g(x)$ or $-g(x)$
The function is neither odd or even.

c) $k(x) = \frac{1}{x}$
$k(-x) = \frac{1}{-x} = -\frac{1}{x}$
$k(x) = -k(x)$ so the function is odd.

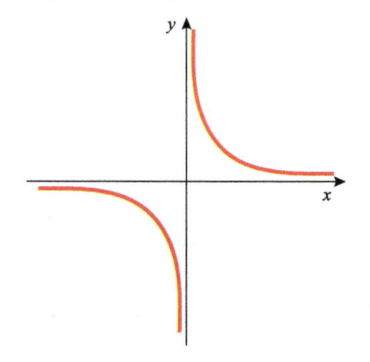

Q56:

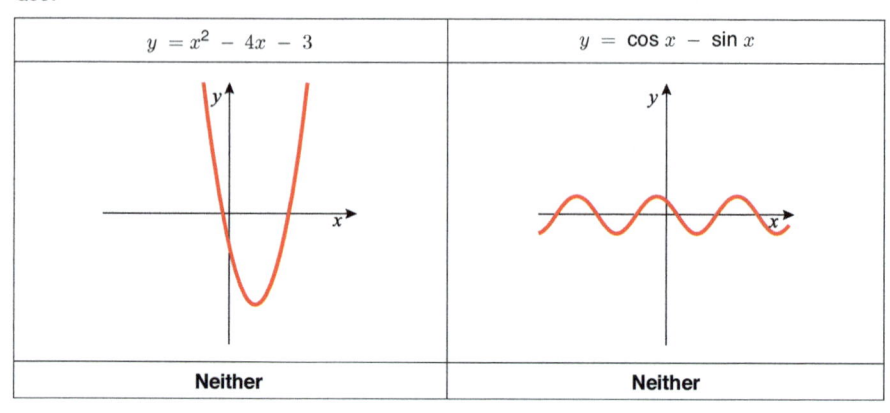

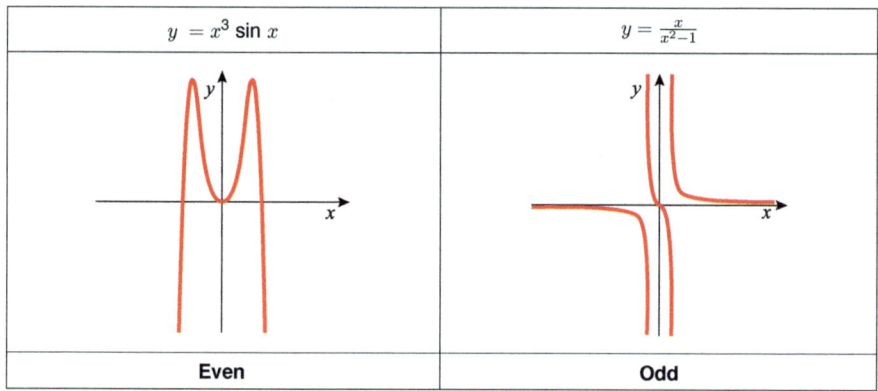

$y = x^2 - 4x - 3$	$y = \cos x - \sin x$
(graph)	*(graph)*
Neither	**Neither**

$y = x^3 \sin x$	$y = \frac{x}{x^2 - 1}$
(graph)	*(graph)*
Even	**Odd**

- Graph $y = x^2 - 4x - 3$
  $f(x) = x^2 - 4x - 3$
  $f(-x) = x^2 + 4x - 3$
  $-f(x) = -x^2 + 4x + 3$
  $f(-x) \neq f(x)$ not even and $f(-x) \neq -f(x)$. It is neither.
- Graph $y = \cos x - \sin x$
  $f(x) = \cos x - \sin x$
  $f(-x) = \cos x + \sin x$
  $-f(x) = -\cos x + \sin x$
  $f(-x) \neq f(x)$ not even and $f(-x) \neq -f(x)$. It is neither.
- Graph $y = x^3 \sin x$
  $f(x) = x^3 \sin x$
  $f(-x) = x^3 \sin x$
  $-f(x) = -x^3 \sin x$

ANSWERS: UNIT 1 TOPIC 8

$f(-x) \neq f(x)$ is even.
- Graph $y = \frac{x}{x^2-1}$

$f(x) = \frac{x}{x^2 - 1}$

$f(-x) = \frac{-x}{x^2 - 1}$

$-f(x) = \frac{-x}{x^2 - 1}$

$f(-x) = -f(x)$ it is odd.

Q57:

$f(-x) = x^2 e^x$

$-f(x) = -x^2 e^{-x}$

$f(-x) \neq f(x)$ not even and $f(-x) \neq -f(x)$. It is neither.

Q58:

$f(-x) = \frac{-4x}{(-x-1)^2} = \frac{-4x}{(x+1)^2}$

$-f(x) = \frac{-4x}{(x-1)^2}$

$f(-x) \neq f(x)$ not even and $f(-x) \neq -f(x)$. It is neither.

Q59:

$f(-\theta) = \frac{\sin(-\theta)}{(-\theta)^2 + 3}$

$f(-\theta) = \frac{-\sin\theta}{\theta^2 + 3}$

$-f(\theta) = \frac{-\sin\theta}{\theta^2 + 3}$

$f(-\theta) \neq -f(\theta)$ it is odd.

Q60:

$f(-x) = \sqrt{5x^2 - 2}$

$-f(x) = -\sqrt{5x^2 - 2}$

$f(-x) = f(x)$ it is even.

Q61: a) Odd

© HERIOT-WATT UNIVERSITY

Critical and stationary points exercise (page 335)

Q62:

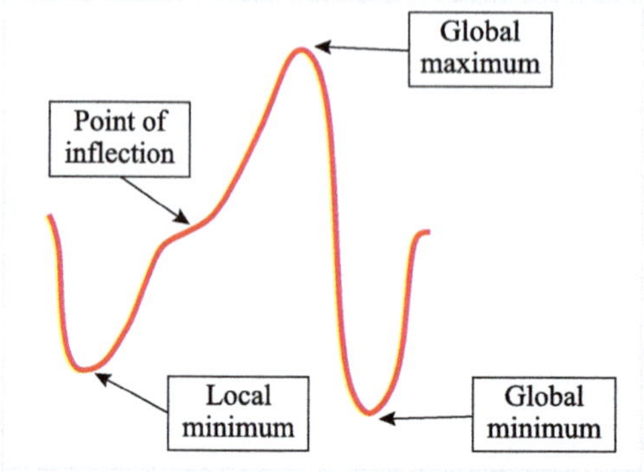
Global
maximum
Point of
inflection
Local
minimum
Global
minimum

Q63:

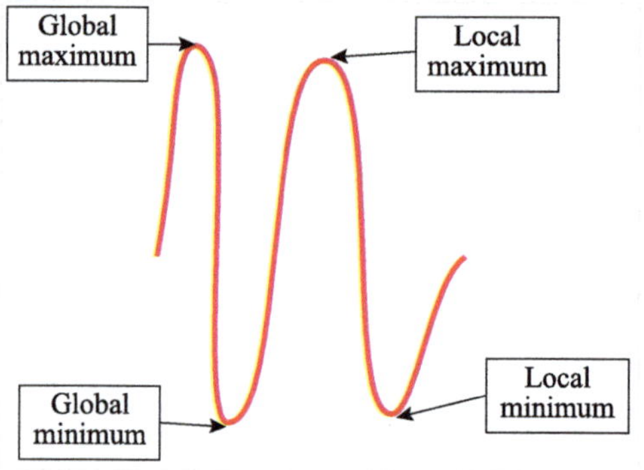
Global
maximum
Local
maximum
Global
minimum
Local
minimum

Q64:

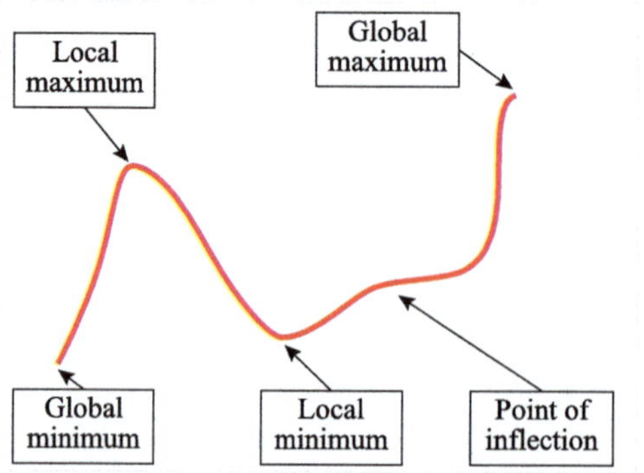
Local
maximum
Global
maximum
Global
minimum
Local
minimum
Point of
inflection

Q65:

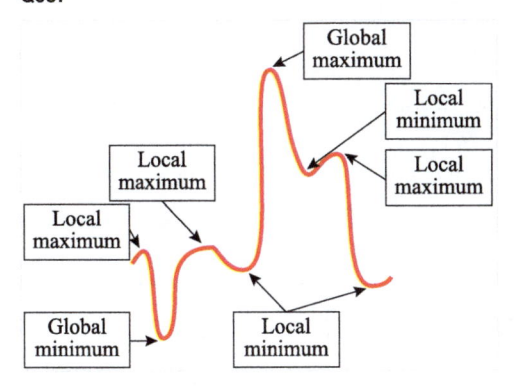

- Global maximum
- Local minimum
- Local maximum
- Local maximum
- Local maximum
- Global minimum
- Local minimum

The first derivative test exercise (page 339)

Q66:

$y = -5x^2 + 4x - 6$

First derivative: $\frac{dy}{dx} = -10x + 4$

For stationary points $\frac{dy}{dx} = 0$: $-10x + 4 = 0 \Rightarrow x = \frac{2}{5}$

Corresponding y-coordinate is: $y = -5\left(\frac{2}{5}\right)^2 + 4\left(\frac{2}{5}\right) - 6 \Rightarrow y = -\frac{26}{5}$

Nature tables:

x	$\left(\frac{2}{5}\right)^-$	$\frac{2}{5}$	$\left(\frac{2}{5}\right)^+$
$\frac{dy}{dx}$	+	0	-
slope	↗	→	↘

There is a minimum turning point at $\left(\frac{2}{5}, -\frac{26}{5}\right)$.

Q67:

$y = 2x^3 + 6$

First derivative: $\frac{dy}{dx} = 6x^2$

For stationary points $\frac{dy}{dx} = 0$: $6x^2 = 0 \Rightarrow x = 0$

Corresponding y-coordinate is: $y = 2(0)^3 + 6 \Rightarrow y = 6$

Nature tables:

x	0^-	0	0^+
$\frac{dy}{dx}$	+	0	+
slope	↗	→	↗

There is a rising point of inflection (0,6).

Q68:

$y = \frac{1}{3}x^3 - 4x$

First derivative: $\frac{dy}{dx} = x^2 - 4$

For stationary points $\frac{dy}{dx} = 0$: $x^2 - 4 = 0 \Rightarrow x = 2, x = -2$

Corresponding y-coordinate is:

$$y = \frac{1}{3}(2)^3 - 4(2) \Rightarrow y = -\frac{16}{3}$$

$$y = \frac{1}{3}(-2)^3 - 4(-2) \Rightarrow y = \frac{16}{3}$$

Nature table:

x	-2^-	-2	→	2	2^+
$\frac{dy}{dx}$	+	0	-	0	+
slope	↗	→	↘	→	↗

There is a maximum turning point at $(-2, \frac{16}{3})$ and a minimum turning point at $(2, -\frac{16}{3})$.

Q69:

$y = 4x^2 + 8x - 3$

First derivative: $\frac{dy}{dx} = 8x + 8$

For stationary points $\frac{dy}{dx} = 0$: $8x + 8 = 0 \Rightarrow x = -1$

Corresponding y-coordinate is: $y = 4(-1)^2 + 8(-1) - 3 \Rightarrow y = -7$

Nature tables:

x	-1^-	-1	-1^+
$\frac{dy}{dx}$	-	0	+
slope	↘	→	↗

There is a minimum turning point at (-1,-7).

ANSWERS: UNIT 1 TOPIC 8

The second derivative test exercise (page 342)

Q70:
$y = 4x^3 + 5x^2$
First derivative:
$y = 4x^3 + 5x^2$
$\frac{dy}{dx} = 12x^2 + 10x$
For stationary points $\frac{dy}{dx} = 0$:
$12x^2 + 10x = 0$
$2x(6x + 5) = 0$
$x = 0$ or $x = -\frac{5}{6}$
Second derivative: $\frac{d^2y}{dx^2} = 24x + 10$
When $x = 0$ then $\frac{d^2y}{dx^2} = 10$ so $\frac{d^2y}{dx^2} > 0$ it is a minimum turning point at $x = 0$.
When $x = -\frac{5}{6}$ then $\frac{d^2y}{dx^2} = -20$ so $\frac{d^2y}{dx^2} < 0$ it is a maximum turning point at $x = -\frac{5}{6}$.

Q71:
$y = 2x + \frac{3}{x+2}$
First derivative:
$y = 2x + 3(x+2)^{-1}$
$\frac{dy}{dx} = 2 - 3(x+2)^{-2}$
For stationary points $\frac{dy}{dx} = 0$:
$2 - \frac{3}{(x+2)^2} = 0$
$2 = \frac{3}{(x+2)^2}$
$(x+2)^2 = \frac{3}{2}$
$x = -2 \pm \sqrt{\frac{3}{2}}$
$x = -0 \cdot 78$ or $x = -3 \cdot 22$
Second derivative:
$\frac{d^2y}{dx^2} = 6(x+2)^{-3}$
$\frac{d^2y}{dx^2} = \frac{6}{(x+2)^3}$
When $x = -0 \cdot 78$ then $\frac{d^2y}{dx^2} = 3 \cdot 3$, so $\frac{d^2y}{dx^2} > 0$ it is a minimum turning point at $x = -0 \cdot 78$.
When $x = -3 \cdot 22$ then $\frac{d^2y}{dx^2} = -3 \cdot 3$, so $\frac{d^2y}{dx^2} < 0$ it is a maximum turning point at $x = -3 \cdot 22$.

© HERIOT-WATT UNIVERSITY

Q72:

$y = \frac{x-1}{x+2}$

First derivative:

By using the quotient rule: $\left(\frac{f}{g}\right)' = \frac{f'g-fg'}{[g]^2}$

$\frac{dy}{dx} = \frac{(x+2)-(x-1)}{(x+2)^2} \Rightarrow \frac{dy}{dx} = \frac{3}{(x+2)^2}$

For stationary points $\frac{dy}{dx} = 0$: $\frac{3}{(x+2)^2} = 0 \Rightarrow 3 = 0$

Which is not true. There are no turning points.

Q73:

$y = \frac{x^3}{x-1}$

First derivative:

$y = \frac{x^3}{x-1}$

$\frac{dy}{dx} = \frac{3x^2(x-1) - x^3}{(x-1)^2}$

$\frac{dy}{dx} = \frac{2x^3 - 3x^2}{(x-1)^2}$

For stationary points $\frac{dy}{dx} = 0$:

$\frac{2x^3 - 3x^2}{(x-1)^2} = 0$

$\frac{x^2(2x-3)}{(x-1)^2} = 0$

$x = 0$ or $x = \frac{3}{2}$

Second derivative:

By using the quotient rule: $\left(\frac{f}{g}\right)' = \frac{f'g-fg'}{[g]^2}$

$\frac{d^2y}{dx^2} = \frac{(6x^2 - 6x)(x-1)^2 - (2x^3 - 3x^2) 2(x-1)}{(x-1)^4}$

$\frac{d^2y}{dx^2} = \frac{6x(x-1)(x-1)^3 - x^2(2x-3) 2(x-1)}{(x-1)^4}$

Take out a common factor of $2x(x-1)$

$\frac{d^2y}{dx^2} = \frac{2x(x-1)(x^2 - 3x + 3)}{(x-1)^4}$

$\frac{d^2y}{dx^2} = \frac{2x(x^2 - 3x + 3)}{(x-1)^3}$

When $x = \frac{3}{2}$ then $\frac{d^2y}{dx^2} = 18$, so $\frac{d^2y}{dx^2} > 0$ it is a minimum turning point at $x = \frac{3}{2}$.

When $x = 0$ then $\frac{d^2y}{dx^2} = 0$.

Since $\frac{d^2y}{dx^2} = 0$ when $x = 0$ then we need to use a nature table for this turning point.

ANSWERS: UNIT 1 TOPIC 8

x	0⁻	0	0⁺
$\frac{dy}{dx}$	-	0	-
slope	↘	→	↘

There is a falling point of inflection at $x = 0$.

Q74:

$y = \frac{x^2+1}{x-2}$

First derivative:
$$\frac{dy}{dx} = \frac{2x(x-2) - (x^2+1)}{(x-2)^2}$$
$$\frac{dy}{dx} = \frac{x^2 - 4x - 1}{(x-2)^2}$$

For stationary points $\frac{dy}{dx} = 0$:
$$\frac{x^2 - 4x - 1}{(x-2)^2} = 0$$
$$x^2 - 4x - 1 = 0$$
$$x = 2 \pm \sqrt{5}$$
$x = -0 \cdot 24$ or $x = 4 \cdot 24$

Second derivative:
$$\frac{d^2y}{dx^2} = \frac{(2x-4)(x-2)^2 - (x^2 - 4x - 1) \, 2(x-2)}{(x-2)^4}$$
$$\frac{d^2y}{dx^2} = \frac{2(x-2)(x-2)^2 - (x^2 - 4x - 1) \, 2(x-2)}{(x-2)^4}$$

Take out a comman factor of $2(x-2)$
$$\frac{d^2y}{dx^2} = \frac{2(x-2)\left[(x-2)^2 - (x^2 - 4x - 1)\right]}{(x-2)^4}$$
$$\frac{d^2y}{dx^2} = \frac{2(x-2)\left[(x^2 - 4x + 4) - (x^2 - 4x - 1)\right]}{(x-2)^4}$$
$$\frac{d^2y}{dx^2} = \frac{2(x-2)[5]}{(x-2)^4}$$
$$\frac{d^2y}{dx^2} = \frac{10}{(x-2)^3}$$

When $x = -0 \cdot 24$ then $\frac{d^2y}{dx^2} = -5 \cdot 24$, so $\frac{d^2y}{dx^2} < 0$ it is a maximum turning point at $x = -0 \cdot 24$.

When $x = 4 \cdot 24$ then $\frac{d^2y}{dx^2} = 0 \cdot 29$, so $\frac{d^2y}{dx^2} > 0$ it is a minimum turning point at $x = 4 \cdot 24$.

© HERIOT-WATT UNIVERSITY

Q75:
First derivative:

$y = \frac{2x^3}{x+5}$

$\frac{dy}{dx} = \frac{6x^2(x+5) - 2x^3}{(x+5)^2}$

$\frac{dy}{dx} = \frac{4x^3 + 30x^2}{(x+5)^2}$

For stationary points $\frac{dy}{dx} = 0$:

$\frac{4x^3 + 30x^2}{(x+5)^2} = 0$

$\frac{2x^2(2x+15)}{(x+5)^2} = 0$

$x = 0$ or $x = -\frac{15}{2}$

Second derivative:

$\frac{d^2y}{dx^2} = \frac{(12x^2 + 60x)(x+5)^2 - (4x^3 + 30x^2) 2(x+5)}{(x+5)^4}$

$\frac{d^2y}{dx^2} = \frac{12x(x+5)(x+5)^2 - 2x^2(2x+15) 2(x+5)}{(x+5)^4}$

Take out a common factor of $4(x+5)$

$\frac{d^2y}{dx^2} = \frac{4x(x+5)\left[3(x+5)^2 - x(2x+15)\right]}{(x+5)^4}$

$\frac{d^2y}{dx^2} = \frac{4x(x+5)\left[3(x^2 + 10x + 25) - (2x^2 + 15x)\right]}{(x+5)^4}$

$\frac{d^2y}{dx^2} = \frac{4x(x+5)(x^2 + 15x + 75)}{(x+5)^4}$

$\frac{d^2y}{dx^2} = \frac{4x(x^2 + 15x + 75)}{(x+5)^3}$

When $x = -\frac{15}{2}$ then $\frac{d^2y}{dx^2} = 36$, so $\frac{d^2y}{dx^2} > 0$ it is a maximum turning point at $x = -\frac{15}{2}$.

When $x = 0$ then $\frac{d^2y}{dx^2} = 0$.

Since $\frac{d^2y}{dx^2} = 0$ when $x = 0$ then we need to use a nature table for this turning point.

x	0⁻	0	0⁺
$\frac{dy}{dx}$	+	0	+
slope	↗	→	↗

There is a rising point of inflection at $x = 0$.

Q76:

$y = \frac{3}{5}x^5 - 4x^3 + 8x$

First derivative: $\frac{dy}{dx} = 3x^4 - 12x^2 + 8$

For stationary points $\frac{dy}{dx} = 0$:

$3x^4 - 12x^2 + 8 = 0$

$$x = \frac{12 \pm \sqrt{144 - 4(3)(8)}}{2(3)}$$

$$x = 2 \pm \frac{2}{3}\sqrt{3}$$

$x = 0 \cdot 85$ and $x = 3 \cdot 15$

Second derivative: $\frac{d^2y}{dx^2} = 12x^3 - 24x$

When $x = 0 \cdot 85$ then $\frac{d^2y}{dx^2} = -13 \cdot 0$, so $\frac{d^2y}{dx^2} < 0$ it is a maximum turning point at $x = 0 \cdot 85$.

When $x = 3 \cdot 15$ then $\frac{d^2y}{dx^2} = 299 \cdot 5$, so $\frac{d^2y}{dx^2} > 0$ it is a minimum turning point at $x = 3 \cdot 15$.

Concavity exercise (page 349)

Q77:

$$y = x^3 - 4x^2$$
$$\frac{dy}{dx} = 3x^2 - 8x$$
$$\frac{d^2y}{dx^2} = 6x - 8$$

When $\frac{d^2y}{dx^2} = 0 \Rightarrow x = \frac{4}{3}$

When $x = \frac{4}{3}$, then $\frac{dy}{dx} = -\frac{16}{3} \neq 0$

At $x = \frac{4}{3}$, $\frac{dy}{dx} \neq 0$ but $\frac{d^2y}{dx^2} = 0$, possible non-horizontal point of inflection.

When $x = \frac{4}{3}^-$, then $\frac{d^2y}{dx^2} < 0 \Rightarrow concave\ down$

When $x = \frac{4}{3}^+$, then $\frac{d^2y}{dx^2} > 0 \Rightarrow concave\ up$

There is a change in concavity at $x = \frac{4}{3}$. It is a point of inflection.

There is a non-horizontal point of inflection at $\left(\frac{4}{3}, -\frac{128}{27}\right)$.

Q78:

$$y = 3x^4 - 2x^2 + 1$$
$$\frac{dy}{dx} = 12x^3 - 4x$$
$$\frac{d^2y}{dx^2} = 36x^2 - 4$$

When $\frac{d^2y}{dx^2} = 0 \Rightarrow x = \pm\frac{1}{3}$

When $x = \frac{1}{3}$, then $\frac{dy}{dx} = -\frac{8}{9} \neq 0$

When $x = -\frac{1}{3}$, then $\frac{dy}{dx} = \frac{8}{9} \neq 0$

At $x = \pm\frac{1}{3}$, $\frac{dy}{dx} \neq 0$ but $\frac{d^2y}{dx^2} = 0$, possible non-horizontal point of inflection.

When $x = \frac{1}{3}^-$, then $\frac{d^2y}{dx^2} < 0 \Rightarrow$ concave down, when $x = \frac{1}{3}^+$, then $\frac{d^2y}{dx^2} > 0 \Rightarrow$ concave up

When $x = -\frac{1}{3}^-$, then $\frac{d^2y}{dx^2} > 0 \Rightarrow$ concave up, when $x = -\frac{1}{3}^+$, then $\frac{d^2y}{dx^2} < 0 \Rightarrow$ concave down

There is a change in concavity at $x = \frac{1}{3}$ and $x = -\frac{1}{3}$. They are both points of inflection.

There is a non-horizontal point of inflection at $\left(\frac{1}{3}, \frac{22}{27}\right)$ and $\left(-\frac{1}{3}, \frac{22}{27}\right)$.

Q79:

$$y = 5x^2 + 3x - 2$$
$$\frac{dy}{dx} = 10x + 3$$
$$\frac{d^2y}{dx^2} = 10, \frac{d^2y}{dx^2} \neq 0$$

There is no non-horizontal point of inflection.

Q80:

$$y = -3x + 7$$
$$\frac{dy}{dx} = -3$$
$$\frac{d^2y}{dx^2} = 0$$

Even though $\frac{d^2y}{dx^2} = 0$ and $\frac{dy}{dx} \neq 0$, there is no non-horizontal point of inflection since this is a straight line.

Q81:

$$y = x^5 + 3x^4 - 2x$$
$$\frac{dy}{dx} = 5x^4 + 12x^3 - 2$$
$$\frac{d^2y}{dx^2} = 20x^3 + 36x^2$$

When $\frac{d^2y}{dx^2} = 0 \Rightarrow x = 0, x = -\frac{9}{5}$

When $x = 0$, then $\frac{dy}{dx} = -2 \neq 0$

When $x = -\frac{9}{5}$, then $\frac{dy}{dx} = -\frac{2437}{125} \neq 0$, possible non-horizontal point of inflection.

At $x = 0$, $x = -\frac{9}{5}$, $\frac{dy}{dx} \neq 0$ but $\frac{d^2y}{dx^2} = 0$, possible non-horizontal point of inflection.

When $x = 0^-$, then $\frac{d^2y}{dx^2} > 0$ ⇒ concave up, when $x = 0^+$, then $\frac{d^2y}{dx^2} > 0$ ⇒ concave up
When $x = -\frac{9}{5}^-$, then $\frac{d^2y}{dx^2} < 0$ ⇒ concave down, when $x = -\frac{9}{5}^+$, then $\frac{d^2y}{dx^2} > 0$ ⇒ concave up
There is a change in concavity at $x = -\frac{9}{5}$. It is a point of inflection.
There is a non-horizontal point of inflection at $\left(-\frac{9}{5}, 16.2\right)$ only.

Q82:

$$y = \frac{x^3}{6} - x^2 + 2x$$
$$\frac{dy}{dx} = \frac{x^2}{2} - 2x + 2$$
$$\frac{d^2y}{dx^2} = x - 2 \;\Rightarrow\; x = 2$$

When $x = 2$, then $\frac{dy}{dx} = 0$

Since the first and second derivatives are zero we need to use a nature table.

x	2⁻	2	2⁺
$\frac{dy}{dx}$	+	0	+
slope	↗	→	↗

There is a non-horizontal point of inflection at $\left(2, \frac{4}{3}\right)$.

Q83:

$$y = \frac{5}{x^3} + x$$
$$\frac{dy}{dx} = -\frac{15}{x^4} + 1$$
$$\frac{d^2y}{dx^2} = \frac{60}{x^5} \text{ then } \frac{d^2y}{dx^2} \neq 0 \text{ and } \frac{d^2y}{dx^2} \text{ is undefined when } x = 0$$

$\frac{d^2y}{dx^2}$ is undefined when $x = 0$. There is a possibility of a point of inflection. However, looking at the original function $x \neq 0$.
There is no non-horizontal point of inflection.

Q84:

$$y = 4x^{\frac{1}{2}} + 2x$$
$$\frac{dy}{dx} = \frac{2}{x^{\frac{1}{2}}} + 2$$
$$\frac{d^2y}{dx^2} = -\frac{1}{x^{\frac{3}{2}}} \quad \text{then} \quad \frac{d^2y}{dx^2} \neq 0$$

$\frac{d^2y}{dx^2}$ is undefined when $x = 0$ and $\frac{dy}{dx}$ is also undefined when $x = 0$. In fact $\frac{dy}{dx}$ cannot equal zero.

Since $\frac{dy}{dx}$ cannot equal zero and $\frac{d^2y}{dx^2}$ is undefined at zero there is a possible non-horizontal point of inflection at $x = 0$.

When $x = 0^-$, then $\frac{d^2y}{dx^2}$ is undefined.

When $x = 0^+$, then $\frac{d^2y}{dx^2} < 0 \quad \Rightarrow \quad$ concave down

There is no point of inflection as the concavity does not change. (It does not exist below 0)

Q85:

$$y = 2x^{\frac{3}{5}} + x^2$$
$$\frac{dy}{dx} = \frac{6}{5x^{\frac{2}{5}}} + 2x$$
$$\frac{d^2y}{dx^2} = -\frac{12}{25x^{\frac{7}{5}}} + 2$$

So if $\frac{d^2y}{dx^2} = 0$

$$-\frac{12}{25x^{\frac{7}{5}}} + 2 = 0$$
$$\frac{12}{25x^{\frac{7}{5}}} = 2$$
$$\frac{12}{50} = x^{\frac{7}{5}}$$
$$x = \sqrt[7]{\left(\frac{12}{50}\right)^5}$$
$$x = 0 \cdot 36$$

When $x = 0 \cdot 36$, $\frac{d^2y}{dx^2} = 0$ and $\frac{dy}{dx} \neq 0$. There is a possibility of a non-horizontal point of inflection.

When $x = 0 \cdot 36^-$, then $\frac{d^2y}{dx^2} < 0 \quad \Rightarrow \quad$ concave down

When $x = 0 \cdot 36^-$, then $\frac{d^2y}{dx^2} > 0 \quad \Rightarrow \quad$ concave up

The concavity changes. There is a point of inflection at (0·36, 1·21).

ANSWERS: UNIT 1 TOPIC 8

Answers from page 352.

Q86:

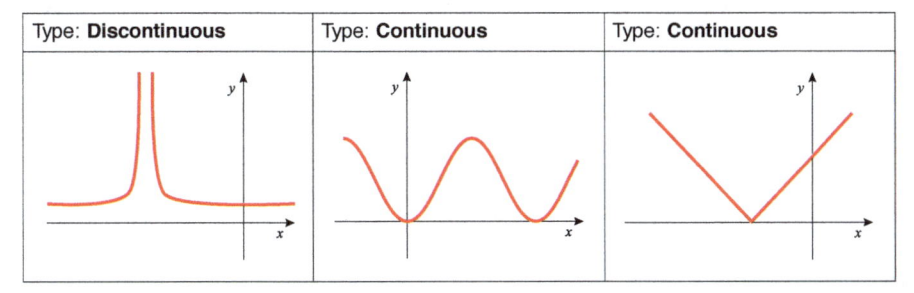

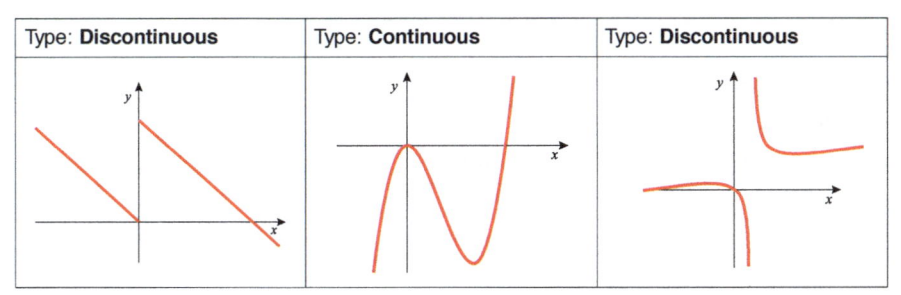

Type: **Discontinuous**	Type: **Continuous**	Type: **Continuous**

Type: **Discontinuous**	Type: **Continuous**	Type: **Discontinuous**

Identifying asymptotes exercise (page 361)

Q87:

Function	$y = \frac{5}{x-3}$
Vertical asymptote	$x = 3$
Horizontal asymptote	$y = 0$
Slant asymptote	none

$y = \frac{5}{x-3}$

Vertical asymptote: When denominator is zero.

$x - 3 = 0 \Rightarrow x = 3$

Horizontal Asymptote: Divide all terms by the highest power of x and let $x \to \pm\infty$.

$y = \frac{\frac{5}{x}}{1 - \frac{3}{x}}$

As $x \to +\infty$, then $y \to 0^+$

As $x \to -\infty$, then $y \to 0^-$

Slant Asymptote: Exists when the degree of the numerator is bigger than the denominator. Then divide numerator by denominator and let $x \to \pm\infty$.

© HERIOT-WATT UNIVERSITY

Degree of numerator is not bigger than the degree of the denominator so there is no slant asymptote.

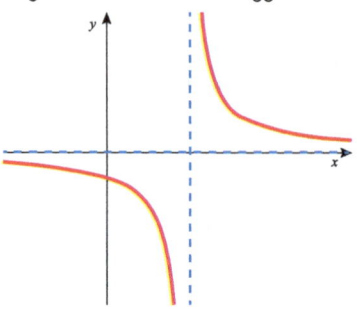

Q88:

Function	$y = \frac{3x+1}{x-2}$
Vertical asymptote	$x = 2$
Horizontal asymptote	$y = 3$
Slant asymptote	none

$y = \frac{3x+1}{x-2}$

Vertical asymptote: When denominator is zero.

$x - 2 = 0 \Rightarrow x = 2$

Horizontal Asymptote: Divide all terms by the highest power of x and let $x \to \pm\infty$.

$y = \frac{3 + \frac{1}{x}}{1 - \frac{2}{x}}$

As $x \to +\infty$, then $y \to 3^+$

As $x \to -\infty$, then $y \to 3^-$

Slant Asymptote: Exists when the degree of the numerator is bigger than the denominator. Then divide numerator by denominator and let $x \to \pm\infty$.

Degree of numerator is not bigger than the degree of the denominator so there is no slant asymptote.

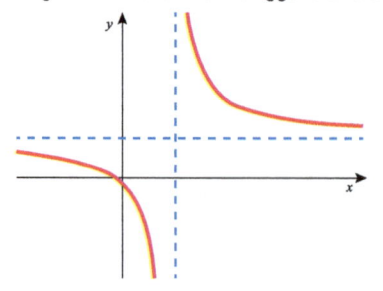

ANSWERS: UNIT 1 TOPIC 8

Q89:

Function	$y = \frac{2x+3}{x^2-3x+2}$
Vertical asymptote	$x = 1,\ x = 2$
Horizontal asymptote	$y = 0$
Slant asymptote	none

$y = \frac{2x+3}{x^2-3x+2}$

Vertical asymptote: When denominator is zero.

$x^2 - 3x + 2 = 0 \quad \Rightarrow \quad (x-2)(x-1) = 0 \quad \Rightarrow \quad x = 1, x = 2$

As $x \to 1^-$, then $y \to +\infty$ and as $x \to 1^+$, then $y \to -\infty$

As $x \to 2^-$, then $y \to -\infty$ and as $x \to 2^+$, then $y \to +\infty$

Horizontal Asymptote: Divide all terms by the highest power of x and let $x \to \pm\infty$.

$y = \frac{\frac{2}{x} + \frac{3}{x^2}}{1 - \frac{3}{x} - \frac{2}{x^2}}$

As $x \to +\infty$, then $y \to 0^+$

As $x \to -\infty$, then $y \to 0^-$

Slant Asymptote: Exists when the degree of the numerator is bigger than the denominator. Then divide numerator by denominator and let $x \to \pm\infty$.

Degree of numerator is not bigger than the degree of the denominator so there is no slant asymptote.

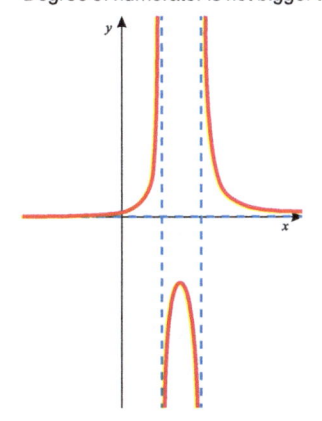

Q90:

Function	$y = x + 1 - \frac{1}{x}$
Vertical asymptote	$x = 0$
Horizontal asymptote	none
Slant asymptote	$y = x + 1$

$y = x + 1 - \frac{1}{x}$

Vertical asymptote: When denominator is zero.

$x = 0$

Horizontal Asymptote: Divide all terms by the highest power of x and let $x \to \pm\infty$.

$y = x + 1 - \frac{1/x}{1}$

We only divide the fraction part through by x because we know how $x + 1$ behaves as $x \to \pm\infty$:
$y \pm \infty$

As $x \to +\infty$, then $y \to \infty$

As $x \to -\infty$, then $y \to -\infty$

No horizontal asymptote.

Slant Asymptote: Exists when the degree of the numerator is bigger than the denominator. Then divide numerator by denominator and let $x \to \pm\infty$.

$y = x + 1 - \frac{1}{x}$

As $x \to +\infty$, then $y \to x + 1$ from below

As $x \to -\infty$, then $y \to x + 1$ from above

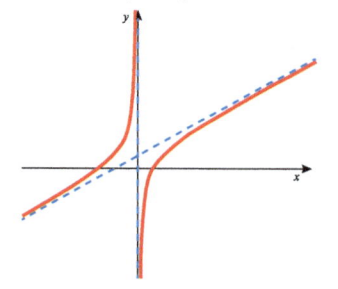

Q91:

Function	$y = \frac{x^2 + 5}{x + 1}$
Vertical asymptote	$x = -1$
Horizontal asymptote	none
Slant asymptote	$y = x - 1$

$y = \frac{x^2 + 5}{x + 1}$

ANSWERS: UNIT 1 TOPIC 8

Vertical asymptote: When denominator is zero.

$x + 1 = 0 \Rightarrow x = -1$

Horizontal Asymptote: Divide all terms by the highest power of x and let $x \to \pm\infty$.

$y = \dfrac{1 + \frac{5}{x^2}}{\frac{1}{x} + \frac{1}{x^2}}$

As $x \to \pm\infty$, then y is undefined.

No horizontal asymptote.

Slant Asymptote: Exists when the degree of the numerator is bigger than the denominator. Then divide numerator by denominator and let $x \to \pm\infty$.

$y = x - 1 + \dfrac{6}{x+1}$

As $x \to +\infty$, then $y \to x - 1$ from above

As $x \to -\infty$, then $y \to x - 1$ from below

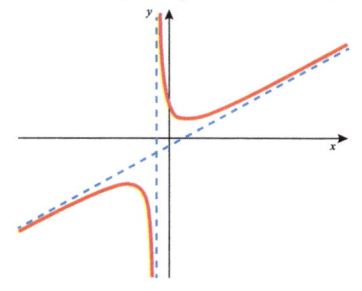

Q92:

Function	$y = \dfrac{x+2}{x^2+x-2}$
Vertical asymptote	$x = 1$
Horizontal asymptote	$y = 0$
Slant asymptote	none

$y = \dfrac{x+2}{x^2+x-2} \Rightarrow y = \dfrac{x+2}{(x+2)(x-1)} \Rightarrow y = \dfrac{1}{x-1}$

Vertical asymptote: When denominator is zero.

$x - 1 = 0 \Rightarrow x = 1$

Horizontal Asymptote: Divide all terms by the highest power of x and let $x \to \pm\infty$.

$y = \dfrac{\frac{1}{x}}{1 - \frac{1}{x}}$

As $x \to +\infty$, then $y \to 0^+$.

As $x \to -\infty$, then $y \to 0^-$.

Slant Asymptote: Exists when the degree of the numerator is bigger than the denominator. Then divide numerator by denominator and let $x \to \pm\infty$.

Degree of numerator is not greater than the degree of the denominator. There is no slant asymptote.

© HERIOT-WATT UNIVERSITY

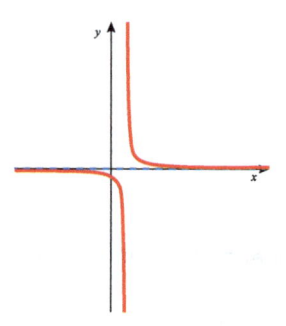

[Graph showing a hyperbola-like curve with vertical asymptote near y-axis and horizontal asymptote at x-axis]

Type 1 rational function exercise (page 371)

Q93:

a)
Symmetry

$$f(x) = \frac{2}{3x+2}$$

$$f(-x) = \frac{2}{-3x+2}$$

$$-f(x) = -\frac{2}{3x+2}$$

$f(-x) \neq f(x)$ and $f(-x) \neq -f(x)$, it is neither odd nor even.

b)
Crosses the x-axis

When $y = 0$: $\frac{2}{3x+2} = 0 \;\Rightarrow\; 2 = 0$

Which is impossible so does not cross the x-axis.

Crosses the y-axis

When $x = 0$:

$$y = \frac{2}{0+2}$$

$$y = 1$$

Crosses the y axis at (0,1).

c)
Turning points

Let $y = f(x)$ and $y = \frac{2}{3x+2}$

First derivative

$\frac{dy}{dx} = -\frac{6}{(3x+2)^2}$

For $\frac{dy}{dx} = 0$: $-\frac{6}{(3x+2)^2} = 0 \;\Rightarrow\; -6 = 0$

Therefore there are no turning points.

d)
Second derivative

$\frac{d^2y}{dx^2} = \frac{36}{(3x+2)^3}$

For possible points of inflection $\frac{d^2y}{dx^2} = 0$ or is undefined.

$\frac{36}{(3x+2)^3} = 0 \Rightarrow 36 = 0$

$\frac{d^2y}{dx^2}$ is undefined at $x = -\frac{2}{3}$. There is a possible point of inflection since $\frac{dy}{dx} \neq 0$.

However, the original function is undefined for $x = -\frac{2}{3}$. Therefore there are no non-horizontal points of inflection.

e)
Asymptotes

Vertical asymptote

When the denominator is zero

$3x + 2 = 0$

$\Rightarrow x = -\frac{2}{3}$ is a vertical asymptote.

As $x \to -\frac{2}{3}^-$, $y \to -\infty$

As $x \to -\frac{2}{3}^+$, $y \to +\infty$

Horizontal asymptote

Divide by the highest power and take the limit as $x \to \infty$.

$\lim\limits_{x \to \infty} \left(\frac{\frac{2}{x}}{3 + \frac{2}{x}} \right) = 0$

There is a horizontal asymptote at $y = 0$.

As $x \to -\infty$, $y \to 0^-$

As $x \to +\infty$, $y \to 0^+$

Slant asymptote

When the degree of the numerator is bigger than the denominator.

The degree of the numerator is less than the denominator so there is no slant asymptote.

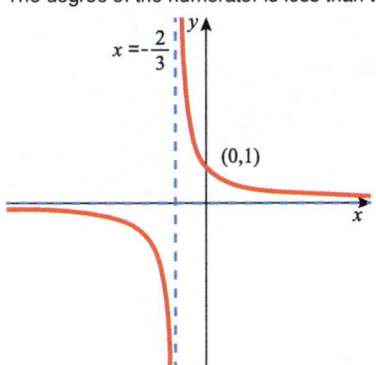

Q94:

a)
Asymptotes
Vertical asymptote
When the denominator is zero

$x = 0 \Rightarrow \quad x = 0$ is a vertical asymptote.

As $x \to 0^-$, $y \to +\infty$

As $x \to 0^+$, $y \to -\infty$

Horizontal asymptote
Divide by the highest power and take the limit as $x \to \infty$.

$$\lim_{x \to \infty} \left(\frac{-\frac{4}{x}}{1}\right) = 0$$

There is a horizontal asymptote at $y = 0$.

As $x \to -\infty$, $y \to 0^+$

As $x \to +\infty$, $y \to 0^-$

Slant asymptote
When the degree of the numerator is bigger than the denominator.

The degree of the numerator is less than the denominator so there is no slant asymptote.

b)
First derivative
$\frac{dy}{dx} = \frac{4}{x^2}$

For $\frac{dy}{dx} = 0$: $\frac{4}{x^2} = 0 \quad \Rightarrow \quad 4 = 0$

Therefore there are no turning points.

c)
Points of Inflexion
Second derivative
$\frac{d^2y}{dx^2} = -\frac{8}{x^3}$

For possible points of inflection $\frac{d^2y}{dx^2} = 0$ or is undefined.

$-\frac{8}{x^3} = 0 \quad \Rightarrow \quad -8 = 0$

$\frac{d^2y}{dx^2}$ is undefined when $x = 0$. There is a possible point of inflection since $\frac{dy}{dx} \neq 0$.
However, the original function is undefined for $x = 0$.

Therefore there are no non-horizontal points of inflection.

d)
$f(x) = -\frac{4}{x}$

Symmetry

$f(x) = -\frac{4}{x}$

$f(-x) = \frac{4}{x}$

$-f(x) = \frac{4}{x}$

$f(-x) = -f(x)$ it is odd and has rotational symmetry of order 2.

Q95:
Symmetry

$$f(x) = \frac{5}{4x-1}$$

$$f(-x) = \frac{5}{-4x-1}$$

$$-f(x) = -\frac{5}{4x-1}$$

$f(-x) \neq f(x)$ and $f(-x) \neq -f(x)$, it is neither odd nor even.

Crosses the x-axis
When $y = 0$: $\frac{5}{4x-1} = 0 \Rightarrow 5 = 0$
Which is impossible so does not cross the x-axis.

Crosses the y-axis
When $x = 0$
$$y = \frac{5}{4(0)-1}$$
$$y = -5$$
Crosses the y axis at (0,-5).

Turning points and points of inflection
First derivative
$\frac{dy}{dx} = -\frac{20}{(4x-1)^2}$
For $\frac{dy}{dx} = 0$: $-\frac{20}{(4x-1)^2} = 0 \Rightarrow -20 = 0$
Therefore there are no turning points.

Second derivative
$\frac{d^2y}{dx^2} = \frac{160}{(4x-1)^3}$
For possible points of inflection $\frac{d^2y}{dx^2} = 0$ or is undefined.
$\frac{160}{(4x-1)^3} = 0 \Rightarrow 160 = 0$
$\frac{d^2y}{dx^2}$ is undefined at $x = \frac{1}{4}$. There is a possible point of inflection since $\frac{dy}{dx} \neq 0$.
However, the original function is undefined for $x = \frac{1}{4}$.
Therefore there are no non-horizontal points of inflection.

Asymptotes
Vertical asymptote
When the denominator is zero:
$4x - 1 = 0$
$\Rightarrow x = \frac{1}{4}$ is a vertical asymptote.
As $x \to \frac{1}{4}^-$, $y \to -\infty$
As $x \to \frac{1}{4}^+$, $y \to +\infty$

Horizontal asymptote
Divide by the highest power and take the limit as $x \to \infty$.

$\lim_{x \to \infty} \left(\frac{\frac{5}{x}}{4 - \frac{1}{x}} \right) = 0$

There is a horizontal asymptote at $y = 0$.

As $x \to -\infty$, $y \to 0^-$

As $x \to +\infty$, $y \to 0^+$

Slant asymptote

When the degree of the numerator is bigger than the denominator.

The degree of the numerator is less than the denominator so there is no slant asymptote.

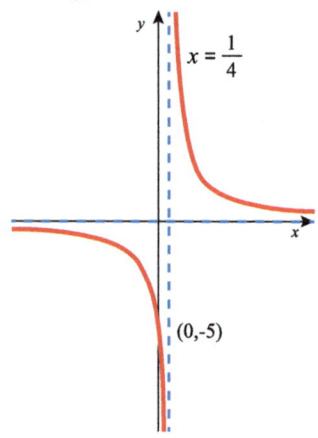

Type 2 rational function exercise (page 375)

Q96:

a)
Asymptotes

Vertical asymptote

When the denominator is zero $2x + 1 = 0$

$\Rightarrow \quad x = -\frac{1}{2}$ is a vertical asymptote.

As $x \to -\frac{1}{2}^-$, $y \to +\infty$

As $x \to -\frac{1}{2}^+$, $y \to -\infty$

Horizontal asymptote

Divide by the highest power and take the limit as $x \to \infty$.

$\lim_{x \to \infty} \left(\frac{3 - \frac{2}{x}}{2 + \frac{1}{x}} \right) = 0$

There is a horizontal asymptote at $y = \frac{3}{2}$.

As $x \to -\infty$, $y \to \frac{3}{2}^+$

As $x \to +\infty$, $y \to \frac{3}{2}^-$

ANSWERS: UNIT 1 TOPIC 8

Slant asymptote

When the degree of the numerator is bigger than the denominator.
The degree of the numerator is the same as the denominator so there is no slant asymptote.

b)
Turning points

$\frac{dy}{dx} = \frac{7}{(2x+1)^2}$

For $\frac{dy}{dx} = 0$: $\frac{7}{(2x+1)^2} = 0 \Rightarrow 7 = 0$

Therefore there are no turning points.

c)
Points of Inflexion

Second derivative

$\frac{d^2y}{dx^2} = \frac{-28}{(2x+1)^3}$

For possible points of inflection $\frac{d^2y}{dx^2} = 0$ or is undefined.

$\frac{-28}{(2x+1)^3} = 0 \Rightarrow -28 = 0$

$\frac{d^2y}{dx^2}$ is undefined when $x = -\frac{1}{2}$. There is a possible point of inflection here since $\frac{dy}{dx} \neq 0$.
However, the original function is undefined for $x = -\frac{1}{2}$.
Therefore there are no non-horizontal points of inflection.

d)
Crosses the x-axis

When $y = 0$:

$\frac{3x-2}{2x+1} = 0$

$\Rightarrow 3x - 2 = 0$

$\Rightarrow x = \frac{2}{3}$

Crosses at $\left(\frac{2}{3}, 0\right)$.

Crosses the y-axis

When $x = 0$:

$y = \frac{0-2}{0+1}$

$y = -2$

It crosses the y-axis at (0,-2).

e)
Let $y = f(x)$ and $y = \frac{3x-2}{2x+1}$

Symmetry

$f(-x) = \frac{-3x-2}{-2x+1}$

$-f(x) = -\frac{3x-2}{2x+1}$

$f(x) = \frac{3x-2}{2x+1}$

$f(-x) \neq -f(x)$ and $f(-x) \neq f(x)$, therefore the function is neither even nor odd.

© HERIOT-WATT UNIVERSITY

Q97:

a)
Asymptotes

Vertical asymptote

When the denominator is zero $2x + 3 = 0$

$\Rightarrow \quad x = -\frac{3}{2}$ is a vertical asymptote.

As $x \to \frac{3}{2}^-$, $y \to +\infty$

As $x \to \frac{3}{2}^+$, $y \to -\infty$

Horizontal asymptote

Divide by the highest power and take the limit as $x \to \infty$.

$\lim\limits_{x \to \infty} \left(\frac{2 + \frac{1}{x}}{2 + \frac{3}{x}} \right) = 1$

There is a horizontal asymptote at $y = 1$.

As $x \to -\infty$, $y \to 1^+$

As $x \to +\infty$, $y \to 1^-$

Slant asymptote

When the degree of the numerator is bigger than the denominator.

The degree of the numerator is the same as the denominator so there is no slant asymptote.

b)
Stationary points

First derivative

$\frac{dy}{dx} = \frac{4}{(2x+3)^2}$

For $\frac{dy}{dx} = 0$: $\frac{4}{(2x+3)^2} = 0 \quad \Rightarrow \quad 4 = 0$

Therefore there are no turning points.

c)

Let $y = f(x)$ and $y = \frac{2x+1}{2x+3}$

Symmetry

$f(-x) = \frac{-2x+1}{-2x+3}$

$-f(x) = -\frac{2x+1}{2x+3}$

$f(x) = \frac{2x+1}{2x+3}$

$f(-x) \neq -f(x)$ and $f(-x) \neq f(x)$, therefore the function is neither even nor odd.

Crosses the x-axis

When $y = 0$:

$\frac{2x+1}{2x+3} = 0$

$\Rightarrow 2x + 1 = 0$

$\Rightarrow x = -\frac{1}{2}$

Crosses at $\left(-\frac{1}{2}, 0\right)$.

Crosses the y-axis

When $x = 0$:
$$y = \frac{0+1}{0+3}$$
$$y = \frac{1}{3}$$

It crosses the y-axis at $\left(0, \frac{1}{3}\right)$.

Turning points

Second derivative

$\frac{d^2y}{dx^2} = \frac{-16}{(2x+3)^3}$

For possible points of inflection $\frac{d^2y}{dx^2} = 0$ or is undefined.

$\frac{-16}{(2x+3)^3} = 0 \quad \Rightarrow \quad -16 = 0$

$\frac{d^2y}{dx^2}$ is undefined when $x = -\frac{3}{2}$. There is a possible point of inflection here since $\frac{dy}{dx} \neq 0$.
However, the original function is undefined for $x = -\frac{3}{2}$.
Therefore there are no non-horizontal points of inflection.

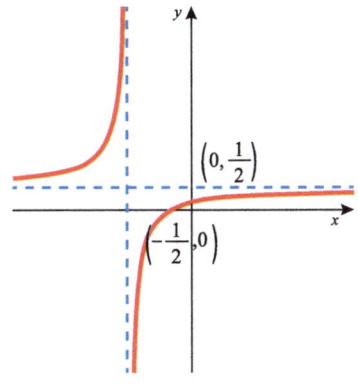

Q98: Let $y = f(x)$ and $y = \frac{5x+1}{4x-3}$.

Symmetry

$f(-x) = \frac{-5x+1}{-4x-3}$

$-f(x) = -\frac{5x+1}{4x-3}$

$f(x) = \frac{5x+1}{4x-3}$

$f(-x) \neq -f(x)$ and $f(-x) \neq f(x)$, therefore the function is neither even nor odd.

Crosses the x-axis
When $y = 0$:
$\frac{5x+1}{4x-3} = 0$
$\Rightarrow 5x + 1 = 0$
$\Rightarrow x = -\frac{1}{5}$
Crosses at $\left(-\frac{1}{5}, 0\right)$.

Crosses the y-axis
When $x = 0$:
$y = \frac{0+1}{0-3}$
$y = -\frac{1}{3}$
It crosses the y-axis at $\left(0, -\frac{1}{3}\right)$.

Turning points
First derivative
$\frac{dy}{dx} = \frac{-19}{(4x-3)^2}$
For $\frac{dy}{dx} = 0$: $\frac{-19}{(4x-3)^2} = 0 \Rightarrow -19 = 0$
Therefore there are no turning points.

Second derivative
$\frac{d^2y}{dx^2} = \frac{152}{(4x-3)^3}$
For possible points of inflection $\frac{d^2y}{dx^2} = 0$ or is undefined.
$\frac{152}{(4x-3)^3} = 0 \Rightarrow 152 = 0$
$\frac{d^2y}{dx^2}$ is undefined when $x = \frac{3}{4}$. There is a possible point of inflection here since $\frac{dy}{dx} \neq 0$.
However, the original function is undefined for $x = \frac{3}{4}$.
Therefore there are no non-horizontal points of inflection.

Asymptotes
Vertical asymptote
When the denominator is zero $4x - 3 = 0$
$\Rightarrow x = \frac{3}{4}$ is a vertical asymptote.
As $x \to \frac{3}{4}^-, y \to -\infty$
As $x \to \frac{3}{4}^+, y \to +\infty$

Horizontal asymptote

Divide by the highest power and take the limit as $x \to \infty$.

$\lim_{x \to \infty} \left(\frac{5 + \frac{1}{x}}{4 - \frac{3}{x}} \right) = \frac{5}{4}$

There is a horizontal asymptote at $y = \frac{5}{4}$.

As $x \to -\infty$, $y \to \frac{5}{4}^-$

As $x \to +\infty$, $y \to \frac{5}{4}^+$

Slant asymptote

When the degree of the numerator is bigger than the denominator.

The degree of the numerator is the same as the denominator so there is no slant asymptote.

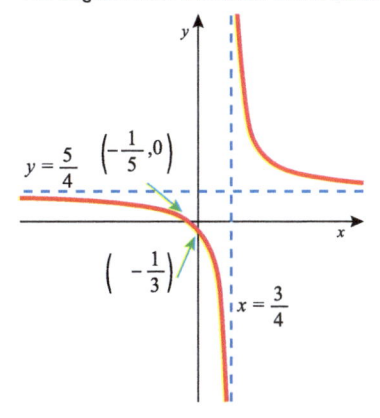

Type 3 rational function exercise (page 382)

Q99:

a)
Stationary points
First derivative

$\frac{dy}{dx} = \frac{2 - 4x}{(x^2 - x - 6)^2}$

For $\frac{dy}{dx} = 0$: $\frac{2 - 4x}{(x^2 - x - 6)^2} = 0 \Rightarrow x = \frac{1}{2}$

If $x = \frac{1}{2}$, then

$y = \frac{2}{\left(\frac{1}{2}\right)^2 - \left(\frac{1}{2}\right) - 6}$

Turning point at $\left(\frac{1}{2}, -0 \cdot 32\right)$.

Second derivative

$\frac{d^2y}{dx^2} = \frac{4(3x^2 - 3x + 7)}{(x^2 - x - 6)^3}$

At $x = \frac{1}{2}$: $\frac{d^2y}{dx^2} < 0$

Which is a maximum turning point.

b)
Second derivative
$$\frac{d^2y}{dx^2} = \frac{4(3x^2-3x+7)}{(x^2-x-6)^3}$$
Non-horizontal point of inflection
$\frac{d^2y}{dx^2} = 0 \Rightarrow 3x^2 - 3x + 7 = 0$
$b^2 - 4ac < 0$ so there are no real roots.
$\frac{d^2y}{dx^2} \neq 0$
$\frac{d^2y}{dx^2}$ and $\frac{dy}{dx}$ are undefined when $x = 3$ and $x = -2$. However, so is y.
Therefore, there are no non-horizontal points of inflection.

c)
Asymptotes

Vertical asymptote

When the denominator is zero
$x^2 - x - 6 = 0$
$(x+2)(x-3) = 0$
$\Rightarrow x = -2, x = 3$ is a vertical asymptote.
As $x \to -2^-$, $y \to +\infty$
As $x \to -2^+$, $y \to -\infty$
As $x \to 3^-$, $y \to -\infty$
As $x \to 3^+$, $y \to +\infty$

Horizontal asymptote

Divide by the highest power and take the limit as $x \to \infty$.
$$\lim_{x \to \infty} \left(\frac{\frac{2}{x^2}}{1-\frac{1}{x}-\frac{6}{x^2}} \right) = 0$$
There is a horizontal asymptote at $y = 0$.
As $x \to -\infty$, $y \to 0^+$
As $x \to +\infty$, $y \to 0^+$

Slant asymptote

When the degree of the numerator is bigger than the denominator.
The degree of the numerator is the same as the denominator so there is no slant asymptote.

d)
Crosses the y-axis

When $x = 0$:
$y = \frac{2}{0-0-6}$
$y = -\frac{1}{3}$
Crosses the y-axis at $\left(0, -\frac{1}{3}\right)$.

… ANSWERS: UNIT 1 TOPIC 8

e)
$y = f(x)$ and $y = \frac{2}{x^2-x-6}$

Symmetry

$f(-x) = \frac{2}{x^2+x-6}$

$-f(x) = -\frac{2}{x^2-x-6}$

$f(x) = \frac{2}{x^2-x-6}$

$f(-x) \neq -f(x)$ and $f(-x) \neq f(x)$ therefore the function is neither even nor odd.

Q100:

a)
Asymptotes

Vertical asymptote

When the denominator is zero

$x^2 - 1 = 0$

$\Rightarrow \quad x = -1, x = 1$ is a vertical asymptote.

As $x \to -1^-$, $y \to -\infty$
As $x \to -1^+$, $y \to +\infty$
As $x \to 1^-$, $y \to -\infty$
As $x \to 1^+$, $y \to +\infty$

Horizontal asymptote

Divide by the highest power and take the limit as $x \to \infty$.

$\lim_{x \to \infty} \left(\frac{\frac{2}{x}}{1-\frac{1}{x^2}} \right) = 0$

There is a horizontal asymptote at $y = 0$.

As $x \to -\infty$, $y \to 0^-$
As $x \to +\infty$, $y \to 0^+$

Slant asymptote

When the degree of the numerator is bigger than the denominator.

The degree of the numerator is the same as the denominator so there is no slant asymptote.

b)
Stationary points

First derivative

$\frac{dy}{dx} = \frac{-2(x^2+1)}{(x^2-1)^2}$

For $\frac{dy}{dx} = 0$: $\frac{-2(x^2+1)}{(x^2-1)^2} = 0 \Rightarrow x^2 + 1 = 0$

There are no stationary points.

c)
Non-horizontal points of inflection
Second derivative

$$\frac{d^2y}{dx^2} = 0$$

$$\frac{d^2y}{dx^2} = -\frac{-4x(x^2-1)^2 - (-2x^2-2)\, 2(x^2-1)\, 2x}{(x^2-1)^4}$$

$$\frac{d^2y}{dx^2} = -\frac{-4x(x^2-1)^2 + 8x(x^2+1)(x^2-1)}{(x^2-1)^4}$$

Taking out a common factor of $4x(x^2-1)$

$$\frac{d^2y}{dx^2} = -\frac{-4x(x^2-1)\left[(x^2-1) - 2(x^2+1)\right]}{(x^2-1)^4}$$

$$\frac{d^2y}{dx^2} = -\frac{-4x(x^2-1)\left[-x^2-3\right]}{(x^2-1)^4}$$

$$\frac{d^2y}{dx^2} = \frac{4x(x^2+3)}{(x^2-1)^3}$$

$$\frac{d^2y}{dx^2} = 4x(x^2+3) \;\Rightarrow\; x = 0$$

$\frac{dy}{dx} \neq 0$ at $x = 0$. There is a possible point of inflection at $x = 0$.

When $x = 0^-$, then $\frac{d^2y}{dx^2} > 0 \;\Rightarrow\;$ *concave up*

When $x = 0^+$, then $\frac{d^2y}{dx^2} < 0 \;\Rightarrow\;$ *concave down*

Concavity changes at $x = 0$.

There is a non-horizontal point of inflection at (0,0).

d)
Sketch the curve

Let $y = f(x)$ and $y = \frac{2x}{x^2-1}$

Symmetry

$$f(-x) = \frac{-2x}{x^2-1}$$

$$-f(x) = -\frac{2x}{x^2-1}$$

$$f(x) = \frac{2x}{x^2-1}$$

$f(-x) = -f(x)$, therefore the function is odd and has rotational symmetry of order 2.

Crosses the x-axis

When $y = 0$: $\frac{2x}{x^2-1} = 0 \;\Rightarrow\; 2x = 0 \;\Rightarrow\; x = 0$

Crosses the x-axis at (0,0).

Crosses the y-axis

When $x = 0$:

$y = \frac{2x}{x^2-1}$

$y = 0$

ANSWERS: UNIT 1 TOPIC 8

Crosses the y-axis at (0,0).

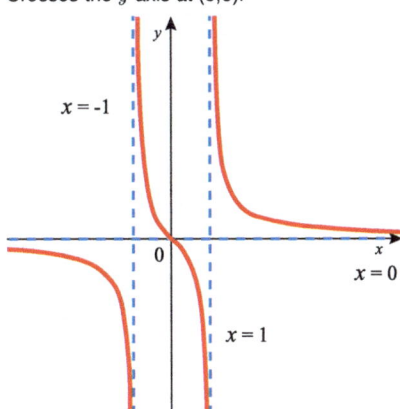

Q101:
$y = f(x)$ and $y = \frac{3}{1+2x^2}$

Symmetry

$f(-x) = \dfrac{3}{1+2x^2}$

$-f(x) = -\dfrac{3}{1+2x^2}$

$f(x) = \dfrac{3}{1+2x^2}$

$f(-x) = f(x)$ therefore the function is even and symmetrical about the y-axis.

Crosses the x-axis

When $y = 0$:

$\frac{3}{1+2x^2} = 0$

$\Rightarrow \quad 3 = 0$

Does not cross the x-axis.

Crosses the y-axis

When $x = 0$: $y = \frac{3}{1+2(0)^2} \quad \Rightarrow \quad y = 3$

Crosses the y-axis at (0,3).

Turning points and points of inflection

First derivative

$\frac{dy}{dx} = \frac{-12x}{(1+2x^2)^2}$

For $\frac{dy}{dx} = 0$: $\frac{-12x}{(1+2x^2)^2} = 0 \Rightarrow \quad -12x = 0 \quad \Rightarrow \quad x = 0$

There is a turning point.

If $x = 0$ then $y = \frac{3}{1+2(0)^2} \quad \Rightarrow \quad y = 3$

Turning point at (0,3).

© HERIOT-WATT UNIVERSITY

Second derivative

$$\frac{d^2y}{dx^2} = \frac{12(6x^2-1)}{(1+2x^2)^3}$$

At $x = 0$: $\frac{d^2y}{dx^2} < 0$

Which is a maximum turning point.

Non-horizontal point of inflection

$$\frac{d^2y}{dx^2} = \frac{-12(1+2x^2)^2 + 12x \times 2(1+2x^2) 4x}{(1+2x^2)^4}$$

$$\frac{d^2y}{dx^2} = \frac{-12(1+2x^2)^2 + 96x^2(1+2x^2)}{(1+2x^2)^4}$$

Taking out a common factor of $12(1+2x^2)$

$$\frac{d^2y}{dx^2} = \frac{12(1+2x^2)\left[-(1+2x^2)+8x^2\right]}{(1+2x^2)^4}$$

$$\frac{d^2y}{dx^2} = \frac{12\left[6x^2-1\right]}{(1+2x^2)^3}$$

$6x^2 - 1 = 0 \Rightarrow x = \pm\frac{1}{\sqrt{6}}$

$\frac{dy}{dx} \neq 0$ at $x = \pm\frac{1}{\sqrt{6}}$. These are possible points of inflection.

When $x = \frac{1}{\sqrt{6}}^-$, then $\frac{d^2y}{dx^2} < 0 \Rightarrow$ concave down

When $x = \frac{1}{\sqrt{6}}^+$, then $\frac{d^2y}{dx^2} > 0 \Rightarrow$ concave up

When $x = -\frac{1}{\sqrt{6}}^-$, then $\frac{d^2y}{dx^2} > 0 \Rightarrow$ concave up

When $x = -\frac{1}{\sqrt{6}}^+$, then $\frac{d^2y}{dx^2} < 0 \Rightarrow$ concave down

Change in concavity at $x = -\frac{1}{\sqrt{6}}$ and $x = \frac{1}{\sqrt{6}}$

There are non-horizontal points of inflection at $\left(\frac{1}{\sqrt{6}}, \frac{9}{4}\right)$ and $\left(-\frac{1}{\sqrt{6}}, \frac{9}{4}\right)$.

Asymptotes

Vertical asymptote

When the denominator is zero

$1 + 2x^2 = 0$

There are no vertical asymptotes.

Horizontal asymptote

Divide by the highest power and take the limit as $x \to \infty$.

$$\lim_{x \to \infty} \left(\frac{\frac{3}{x^2}}{\frac{1}{x^2}+2}\right) = 0$$

There is a horizontal asymptote at $y = 0$

As $x \to -\infty$, $y \to 0^+$

As $x \to +\infty$, $y \to 0^+$

ANSWERS: UNIT 1 TOPIC 8

Slant asymptote
When the degree of the numerator is bigger than the denominator.
The degree of the numerator is the same as the denominator so there is no slant asymptote.

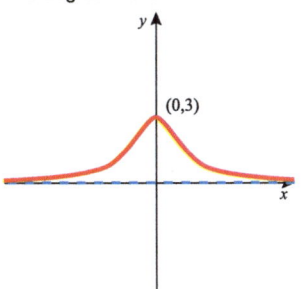

[Graph showing a bell-shaped curve peaking at (0,3) with horizontal asymptote along the x-axis]

Type 4 rational function exercise (page 386)

Q102:
a)
Asymptotes
Vertical asymptote
When the denominator is zero
$$x^2 - 1 = 0$$
$(x-1)(x+1) = 0$
$\Rightarrow$ $x = 1$ and $x = -1$ are vertical asymptotes.
As $x \to -1^-$, $y \to +\infty$
As $x \to -1^+$, $y \to -\infty$
As $x \to 1^-$, $y \to -\infty$
As $x \to 1^+$, $y \to +\infty$

Horizontal asymptote
Divide by the highest power and take the limit as $x \to \infty$.
$$\lim_{x \to \infty} \left(\frac{1}{1 - \frac{1}{x^2}} \right) = 1$$
There is a horizontal asymptote at $y = 1$.
As $x \to -\infty$, $y \to 1^+$
As $x \to +\infty$, $y \to 1^+$

Slant asymptote
When the degree of the numerator is bigger than the denominator.
The degree of the numerator is the same as the denominator so there is no slant asymptote.

b)
Stationary points
First derivative
$$\frac{dy}{dx} = \frac{-2x}{(x^2-1)^2}$$

For $\begin{array}{l}\frac{dy}{dx} = 0 \\ -2x = 0 \\ x = 0\end{array}$

If $x = 0$, then $f(0) = \frac{0^2}{0^2-1}$ $\Rightarrow$ $f(0) = 0$
There is a turning point at (0,0).

Second derivative
Turning point
$$\frac{d^2y}{dx^2} = \frac{6x^2+2}{(x^2-1)^3}$$
At $x = 0$: $\frac{d^2y}{dx^2} < 0$ which is a maximum turning point.

c)
Non-horizontal point of inflection
$$\frac{d^2y}{dx^2} = \frac{6x^2+2}{(x^2-1)^3}$$
For non-horizontal points of inflection $\frac{d^2y}{dx^2} \neq 0$ or is undefined.
$\frac{d^2y}{dx^2} = 0$ $\Rightarrow$ $x^2 = -\frac{1}{3}$, which is not possible for $x \in \mathbb{R}$.
$\frac{d^2y}{dx^2}$ is undefined when $x = 1$ and $\frac{dy}{dx} \neq 0$. Possible point of inflection. However, $x \neq 1$, so there is not point of inflection.

d)
Crosses the x-axis
When $y = 0$: $\frac{x^2}{x^2-1} = 0 \Rightarrow$ $x^2 = 0$ $\Rightarrow$ $x = 0$
Crosses the x-axis at (0,0).

Crosses the y-axis
When $x = 0$:
$$y = \frac{0^2}{0^2-1}$$
$y = 0$
Crosses the y-axis at (0,0).

e)
Symmetry
$$f(-x) = \frac{x^2}{x^2-1}$$
$$-f(x) = -\frac{x^2}{x^2-1}$$
$$f(x) = \frac{x^2}{x^2-1}$$
$f(-x) = f(x)$, therefore the function is even.

ANSWERS: UNIT 1 TOPIC 8

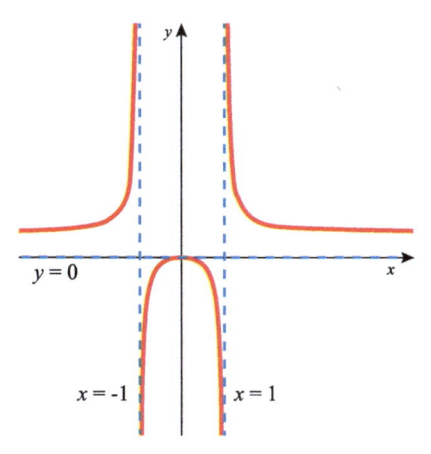

[Graph showing curve with vertical asymptotes at $x = -1$ and $x = 1$, and horizontal asymptote $y = 0$.]

Q103:

a)
Asymptotes

Vertical asymptote

When the denominator is zero
$$x^2 - 4x + 4 = 0$$
$$(x-2)(x-2) = 0$$
$$\Rightarrow \quad x = 2 \text{ is a vertical asymptote.}$$
As $x \to 2^-$, $y \to +\infty$
As $x \to 2^+$, $y \to +\infty$

Horizontal asymptote

Divide by the highest power and take the limit as $x \to \infty$.
$$\lim_{x \to \infty} \left(\frac{1 + \frac{3}{x} - \frac{3}{x^2}}{1 - \frac{4}{x} + \frac{4}{x^2}} \right) = 1$$
There is a horizontal asymptote at $y = 1$.
As $x \to -\infty$, $y \to 1^-$
As $x \to +\infty$, $y \to 1^+$

Slant asymptote

When the degree of the numerator is bigger than the denominator.
The degree of the numerator is the same as the denominator so there is no slant asymptote.

b)
Stationary points

First derivative

$\frac{dy}{dx} = \frac{-7x}{(x-2)^3}$

For $\frac{dy}{dx} = 0$
$-7x = 0$
$x = 0$

© HERIOT-WATT UNIVERSITY

If $x = 0$, then
$$y = \frac{0+0-3}{0-0+4}$$
$$y = -\frac{3}{4}$$
Turning point at $\left(0, -\frac{3}{4}\right)$.

Second derivative

Turning point

$\frac{d^2y}{dx^2} = \frac{14(x+1)}{(x-2)^4}$

At $x = 0$: $\frac{d^2y}{dx^2} > 0$ which is a minimum turning point.

c)
Non-horizontal point of inflection

$$\frac{d^2y}{dx^2} = \frac{14(x+1)}{(x-2)^4}$$

$\frac{d^2y}{dx^2} = 0 \quad \Rightarrow \quad x+1 = 0$

$\Rightarrow \quad x = -1$

When $x = -1$, $\frac{d^2y}{dx^2} = 0$ and $\frac{dy}{dx} \neq 0$. Possible point of inflection.

When $x = -1^-$, then $\frac{d^2y}{dx^2} < 0 \quad \Rightarrow concave\ down$

When $x = -1^+$, then $\frac{d^2y}{dx^2} > 0 \quad \Rightarrow concave\ up$

There is a change in concavity at $x = -1$. It is a point of inflection.

There is a non-horizontal point of inflection at $\left(-1, -\frac{5}{9}\right)$.

d)
Let $y = f(x)$ and $y = \frac{x^2+3x-3}{x^2-4x+4}$

Symmetry

$$f(-x) = \frac{x^2-3x-3}{x^2+4x+4}$$

$$-f(x) = -\frac{x^2+3x-3}{x^2-4x+4}$$

$$f(x) = \frac{x^2+3x-3}{x^2-4x+4}$$

$f(-x) \neq -f(x)$ and $f(-x) \neq f(x)$ therefore the function is neither even nor odd.

Crosses the x-axis

When $y = 0$:

$\frac{x^2+3x-3}{x^2-4x+4} = 0 \Rightarrow \quad x^2 + 3x - 3 = 0$

Using the quadratic formula:

$x = \frac{-3 \pm \sqrt{21}}{2}$

Crosses the x-axis at $\left(\frac{-3-\sqrt{21}}{2}, 0\right)$ and $\left(\frac{-3+\sqrt{21}}{2}, 0\right)$.

# ANSWERS: UNIT 1 TOPIC 8

Crosses the y-axis
When $x = 0$:
$$y = \frac{0 + 0 - 3}{0 - 0 + 4}$$
$$y = -\frac{3}{4}$$
Crosses the y-axis at $\left(0, -\frac{3}{4}\right)$.

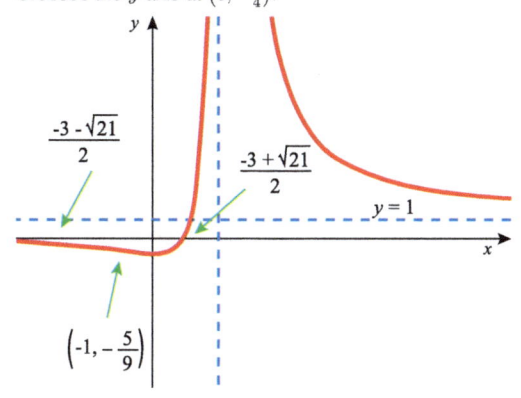

[Graph showing curve with asymptote $y = 1$, x-intercepts at $\frac{-3-\sqrt{21}}{2}$ and $\frac{-3+\sqrt{21}}{2}$, and minimum point $\left(-1, -\frac{5}{9}\right)$.]

Q104:

Let $y = f(x)$ and $y = \frac{3x(x+1)}{(x+3)^2}$

Symmetry

$$f(-x) = -\frac{3x(-x+1)}{(-x+3)^2}$$

$$-f(x) = -\frac{3x(x+1)}{(x+3)^2}$$

$$f(x) = \frac{3x(x+1)}{(x+3)^2}$$

$f(-x) \neq -f(x)$ and $f(-x) \neq f(x)$, therefore the function is neither odd nor even.

Crosses the x-axis
When $y = 0$: $\frac{3x(x+1)}{(x+3)^2} = 0 \Rightarrow \quad 3x(x+1) = 0$
$x = 0$ or $x = -1$
Crosses the x-axis at (0,0), (-1,0).

Crosses the y-axis
When $x = 0$: $y = \frac{3x(x+1)}{(x+3)^2} \quad \Rightarrow \quad y = 0$
Crosses the y-axis at (0,0).

© HERIOT-WATT UNIVERSITY

Turning points
First derivative
$\frac{dy}{dx} = \frac{3(5x+3)}{(x+3)^3}$

For $\frac{dy}{dx} = 0$: $5x + 3 = 0 \Rightarrow x = -\frac{3}{5}$

There is a turning point.

If $x = -\frac{3}{5}$, then

$y = \frac{3\left(-\frac{3}{5}\right)\left(\left(-\frac{3}{5}\right)+1\right)}{\left(\left(-\frac{3}{5}\right)+3\right)^2} \Rightarrow y = -\frac{1}{8}$

Turning point at $\left(-\frac{3}{5}, -\frac{1}{8}\right)$.

Second derivative
Turning point
$\frac{d^2y}{dx^2} = -\frac{6(5x-3)}{(x+3)^4}$

At $x = -\frac{3}{5}$: $\frac{d^2y}{dx^2} > 0$

Which is a minimum turning point at $\left(-\frac{3}{5}, -\frac{1}{8}\right)$.

Non-horizontal point of inflection
$\frac{d^2y}{dx^2} = 0 \Rightarrow -\frac{6(5x-3)}{(x+3)^4} = 0$

$5x - 3 = 0 \Rightarrow x = \frac{3}{5}$

When $x = \frac{3}{5}$, $\frac{d^2y}{dx^2} = 0$ and $\frac{dy}{dx} \neq 0$. There is a possible non-horizontal point of inflection if there is a change in concavity.

When $x = \frac{3}{5}^-$, then $\frac{d^2y}{dx^2} > 0 \Rightarrow$ concave up

When $x = \frac{3}{5}^+$, then $\frac{d^2y}{dx^2} < 0 \Rightarrow$ concave down

There is a change in concavity at $x = \frac{3}{5}$. It is a point of inflection.

There is a non-horizontal point of inflection at $\left(\frac{3}{5}, \frac{2}{9}\right)$.

When $x = -3$, $\frac{d^2y}{dx^2}$ and $\frac{dy}{dx}$ are undefined.

However, $f(x)$ is also undefined for these values.

Asymptotes
Vertical asymptote
When the denominator is zero

$x + 3 = 0$

$x = -3$

Vertical asymptotate at $x = -3$

As $x \to -3^-$, $y \to \infty^+$

As $x \to -3^+$, $y \to \infty^+$

Horizontal asymptote
Divide by the highest power and take the limit as $x \to \infty$.

ANSWERS: UNIT 1 TOPIC 8

$\lim\limits_{x \to \infty} \left(\frac{3 + \frac{3}{x}}{1 + \frac{6}{x} + \frac{9}{x^2}} \right) = 3$

There is a horizontal asymptote at $y = 3$.

As $x \to -\infty$, $y \to 3^+$

As $x \to +\infty$, $y \to 3^-$

Slant asymptote

When the degree of the numerator is bigger than the denominator.

The degree of the numerator is the same as the denominator so there is no slant asymptote.

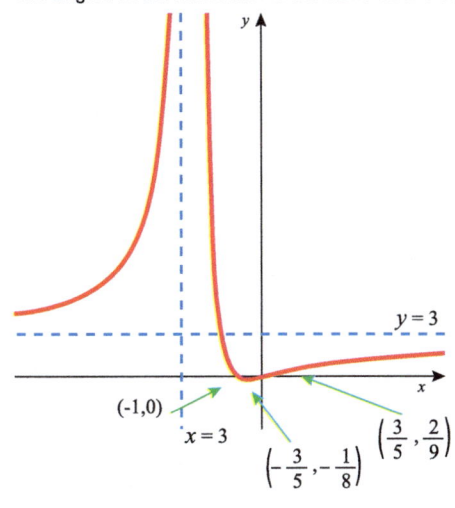

Type 5 rational function exercise (page 390)

Q105:

a)

This function is an improper function as the numerator has a higher degree than the denominator. To make it easier to differentiate we will re-write it.

First divide by the denominator to give: $y = x + 1 + \frac{4}{x+1}$

Let $y = f(x)$ and $y = x + 1 + \frac{4}{x+1}$

Stationary points

First derivative

$\frac{dy}{dx} = 1 - \frac{4}{(x+1)^2}$

For $\frac{dy}{dx} = 0$

$1 - \frac{4}{(x+1)^2} = 0$

$x^2 + 2x - 3 = 0$

$(x+3)(x-1) = 0$

© HERIOT-WATT UNIVERSITY

When $x = 1$ and $x = -3$

If $x = 1$, then

$y = 1 + 1 + \dfrac{4}{1+1}$

$y = 4$

Turning point: (1,4)

If $x = -3$, then

$y = -3 + 1 + \dfrac{4}{-3+1}$

$y = -4$

Turning point: (-3,-4)

Second derivative

Turning point

$\dfrac{d^2y}{dx^2} = \dfrac{8}{(x+1)^3}$

At $x = 1$: $\dfrac{d^2y}{dx^2} > 0$

Which is a minimum turning point at (1,4).

$\dfrac{d^2y}{dx^2} = \dfrac{8}{(x+1)^3}$

At $x = -3$: $\dfrac{d^2y}{dx^2} < 0$

Which is a maximum turning point at (-3,-4).

b)

Asymptotes

Vertical asymptote

When the denominator is zero

$x + 1 = 0 \Rightarrow \quad x = -1$ is a **vertical asymptote**.

As $x \to 1^-$, $y \to -\infty$

As $x \to 1^+$, $y \to +\infty$

Horizontal asymptote

Divide by the highest power and take the limit as $x \to \infty$.

$\lim\limits_{x \to \infty} \left(x + 1 + \dfrac{\frac{4}{x}}{1+\frac{1}{x}} \right) = 0$

As $x \to \infty$, $y \to x + 1$ so there is not a horizontal asymptote, but a slant asymptote.

Slant asymptote

When the degree of the numerator is bigger than the denominator.

$y = x + 1$

As $x \to +\infty$, $y \to (x+1)^+$

As $x \to -\infty$, $y \to (x+1)^-$

c)

Non-horizontal point of inflection

For non-horizontal points of inflection $\dfrac{d^2y}{dx^2} = 0$ or be undefined.

$\frac{d^2y}{dx^2}$ is undefined when $x = -1$. However, the original function is undefined for $x = -1$, meaning there is a vertical asymptote here.
Therefore, there cannot be a point of inflection at $x = -1$.
$$\frac{d^2y}{dx^2} = \frac{8}{(x+1)^3}$$
$$\frac{d^2y}{dx^2} = 0 \Rightarrow 8 = 0$$
There are no non-horizontal points of inflection.

Q106:
a)
This function is an improper function as the numerator has a higher degree than the denominator.
Divide by the denominator to give: $f(x) = x - \frac{2}{x-3}$

b)
Asymptote
Vertical asymptote
When the denominator is zero
$x - 3 = 0 \Rightarrow x = 3$ is a **vertical asymptote**.
As $x \to 3^-$, $y \to +\infty$
As $x \to 3^+$, $y \to -\infty$

Horizontal asymptote
Divide by the highest power and take the limit as $x \to \infty$.
$$\lim_{x \to \infty} \left(\frac{\frac{2}{x}}{1 - \frac{3}{x}} \right) = x$$
As $x \to \infty$, $y \to x$ so there is not a horizontal asymptote, but a slant asymptote.

Slant asymptote
When the degree of the numerator is bigger than the denominator.
$y = x$
As $x \to -\infty$, $y \to x^+$
As $x \to +\infty$, $y \to x^-$

c)
Stationary Points
First derivative
$\frac{dy}{dx} = 1 + \frac{2}{(x-3)^2}$
$$\frac{dy}{dx} = 0$$
For $1 + \frac{2}{(x-3)^2} = 0$
$(x-3)^2 + 2 = 0$
There are no real solutions for this. There are no stationary points.

Non-horizontal point of inflection

$\frac{d^2y}{dx^2} = -\frac{4}{(x-3)^3}$

For non-horizontal points of inflection $\frac{d^2y}{dx^2} = 0$ or is undefined.

$\frac{d^2y}{dx^2} = 0 \Rightarrow -4 = 0$

This is impossible, so there is no non-horizontal point of inflection here.

$\frac{d^2y}{dx^2}$ is undefined when $x = 3$. However, in the original function $x \neq 3$, it is undefined here. There is a vertical asymptote here, which means there cannot be a non-horizontal point of inflection here.

There are no non-horizontal points of inflection.

Symmetry

$f(-x) = -x + \dfrac{2}{x+3}$

$-f(x) = -x + \dfrac{2}{x-3}$

$f(x) = x - \dfrac{2}{x-3}$

$f(-x) \neq -f(x)$ and $f(-x) \neq f(x)$, therefore the function is neither even nor odd.

Crosses the x-axis

When $y = 0$: $x - \frac{2}{x-3} = 0 \Rightarrow x^2 - 3x - 2 = 0$

Using the quadratic formula:

$x = \dfrac{-b \pm \sqrt{b^2 - 4ac}}{2a}$

$x = \dfrac{3 \pm \sqrt{9 - 4(1)(-2)}}{2(1)}$

$x = 3 \cdot 6$ or $x = -0 \cdot 56$ to 2 d.p.

Crosses the x-axis at (3·6 ,0) and (-0·56 ,0).

Crosses the y-axis

When $x = 0$:

$y = 0 - \dfrac{2}{0-3}$

$y = \dfrac{2}{3}$

It crosses the y-axis at $\left(0, \frac{2}{3}\right)$.

ANSWERS: UNIT 1 TOPIC 8

[Graph showing a function with vertical asymptote at x = 3, oblique asymptote y = x, passing through (0, 2/3), (-0.56, 0), and (3.6, 0)]

Q107:
This function is an improper function as the numerator has a higher degree than the denominator. Divide by the denominator to give: $f(x) = 2x - 5 + \frac{4}{x+2}$

Symmetry

$$f(-x) = -2x - 5 + \frac{4}{2-x}$$

$$-f(x) = -2x + 5 - \frac{4}{x+2}$$

$$f(x) = 2x - 5 + \frac{4}{x+2}$$

$f(-x) \neq f(x)$ and $f(-x) \neq -f(x)$, therefore the function is neither odd nor even.

Crosses the x-axis

When $y = 0$: $2x - 5 + \frac{4}{x+2} = 0$

$\Rightarrow \quad 2x^2 - x - 6 = 0$

$\Rightarrow \quad (2x+3)(x-2) = 0$

$\Rightarrow \quad x = -\frac{3}{2}$ or $x = 2$

Crosses the x-axis at $(-\frac{3}{2}, 0)$, $(2, 0)$

Crosses the y-axis

When $x = 0$: $y = 2(0) - 5 + \frac{4}{(0)+2} \quad \Rightarrow \quad y = -3$

It crosses the y-axis at (0, -3).

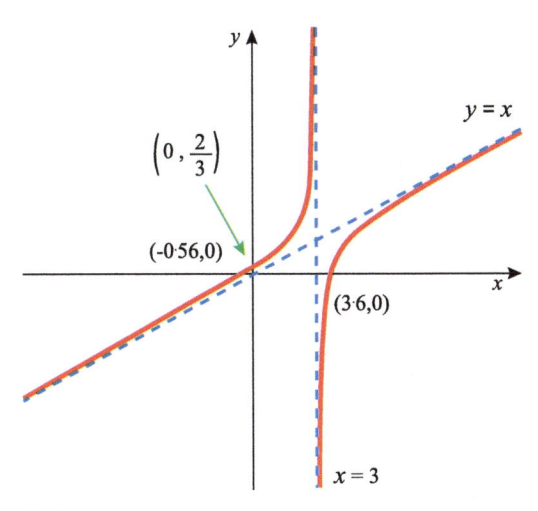

Stationary points

First derivative

$\frac{dy}{dx} = 2 - \frac{4}{(x+2)^2}$

For $\frac{dy}{dx} = 0$: $x^2 + 4x + 2 = 0$

© HERIOT-WATT UNIVERSITY

Using the quadratic formula:

$$x = \frac{-b \pm \sqrt{b^2 - 4ac}}{2a}$$

$$x = \frac{-4 \pm \sqrt{16 - 4(1)(2)}}{2(1)}$$

$x = -2 + \sqrt{2}$ or $x = -2 - \sqrt{2}$.

Second derivative

Turning point

$\frac{d^2y}{dx^2} = \frac{8}{(x+2)^3}$

At $x = -2 + \sqrt{2}$: $\frac{d^2y}{dx^2} > 0$

Which is a minimum turning point at $(-2 + \sqrt{2}, -9 + 4\sqrt{2})$.

$\frac{d^2y}{dx^2} = \frac{8}{(x+2)^3}$

At $x = -2 - \sqrt{2}$: $\frac{d^2y}{dx^2} < 0$

Which is a maximum turning point at $(-2 - \sqrt{2}, -9 - 4\sqrt{2})$.

Non-horizontal point of inflection

This occurs when $\frac{d^2y}{dx^2} = 0$ or is undefined.

Since $x \neq -2$, there is no point of inflection here.

$\frac{d^2y}{dx^2} = 0 \quad \Rightarrow \quad \frac{8}{(x+2)^3} = 0$

$\Rightarrow \quad 8 = 0$ this is impossible. There are no non-horizontal points of inflection.

Asymptotes

Vertical asymptote

When the denominator is zero $x + 2 = 0 \quad \Rightarrow \quad x = -2$

There are vertical asymptotes $x = -2$.

As $x \to -2^-$, $y \to -\infty$

As $x \to -2^+$, $y \to +\infty$

Horizontal asymptote

Divide by the highest power and take the limit as $x \to \infty$.

$\lim_{x \to \infty} \left(2x - 5 + \frac{\frac{4}{x}}{1 + \frac{2}{x}}\right) = 2x - 5$

There is a not a horizontal asymptote but a slant asymptote at $y = 2x - 5$.

Slant asymptote

When the degree of the numerator is bigger than the denominator.

As $x \to -\infty$, $y \to (2x - 5)^-$

As $x \to +\infty$, $y \to (2x - 5)^+$

ANSWERS: UNIT 1 TOPIC 8

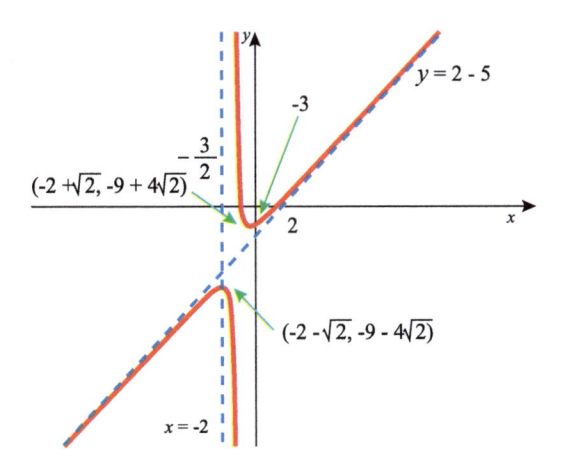

[Graph showing curve with asymptote $x = -2$, line $y = 2-5$, points $(-2+\sqrt{2}, -9+4\sqrt{2})$, $(-2-\sqrt{2}, -9-4\sqrt{2})$, and markings at $-\frac{3}{2}$, -3, and 2]

Modulus function exercise (page 394)

Q108: c) C

Q109:

Hints:

- Original function is $y = x + 4$

Steps:

- First sketch the graph of $f^{-1}(x)$

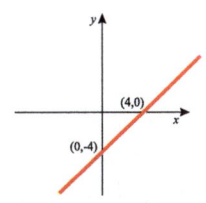

[Graph showing line through $(4,0)$ and $(0,-4)$]

- Then sketch the graph of $f^{-1}(x - 1)$

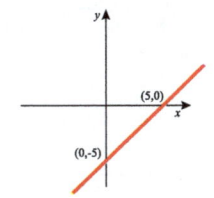

[Graph showing line through $(5,0)$ and $(0,-5)$]

- Finally sketch the graph of $\left|f^{-1}(x-1)\right|$

Answer: B

© HERIOT-WATT UNIVERSITY

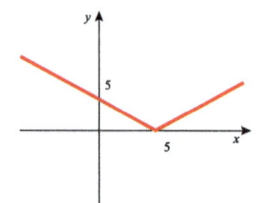

Q110:

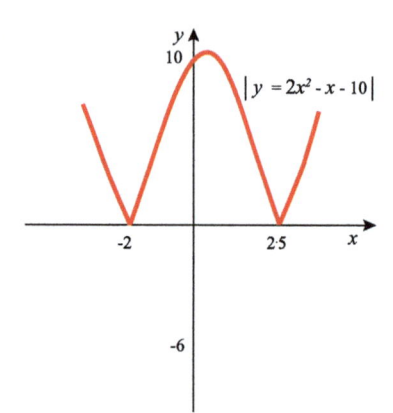

Q111: $y = |\tan x|$

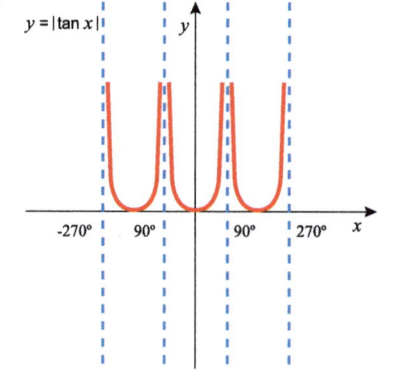

ANSWERS: UNIT 1 TOPIC 8

Q112:

Steps:

- Sketch $f(-x)$

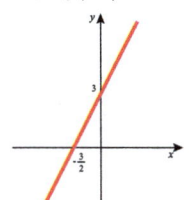

- Sketch $|f(-x)|$

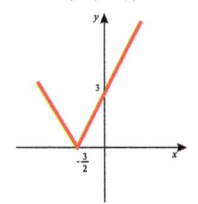

- Sketch $|f(-x)| + 3$

Answer:

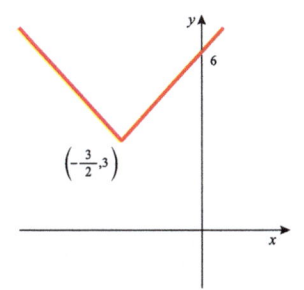

Q113:

Steps:

- Sketch $3f(x)$

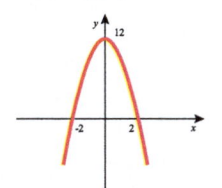

© HERIOT-WATT UNIVERSITY

- Sketch $3f(x+2)$

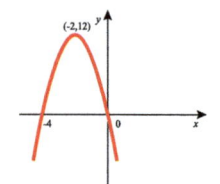

- Sketch $|3f(x+2)|$

Answer:

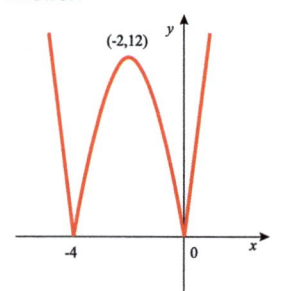

Q114:

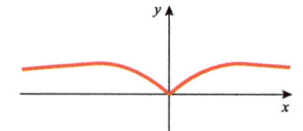

Q115:

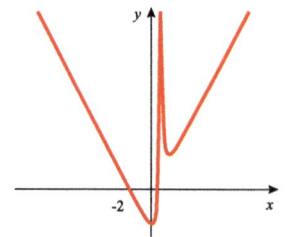

End of topic 8 test (page 403)

Q116:

Since we cannot square root a negative number then

$3x - 2 \geqslant 0$

$\quad x \geqslant \dfrac{2}{3}$

Q117:

Since we cannot square root a negative number then

$2x + 1 \geqslant 0$

$\quad x \geqslant -\dfrac{1}{2}$

Q118:

Since we cannot divide by zero then

$5x - 2 \neq 0$

$\quad x \neq \dfrac{2}{5}$

Q119:

Looking at the graph of $\tan x$, $x \in \left[-\dfrac{\pi}{4}, \dfrac{\pi}{4}\right]$

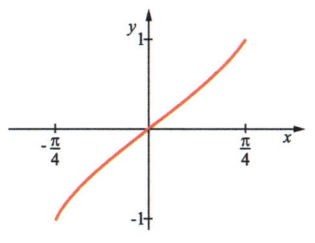

The range is [-1,1]

Q120:

Domain	Range: Evaluate $3x^2 + x - 2$
-3	0
-1	2
1	22

0, 2, 22

Q121:

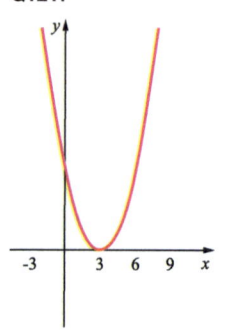

Looking at the graph of $y = (x-3)^2$ the biggest domain for on-to-one is $x \geqslant 3$.

Q122:

Domain	Range: Evaluate $x^2 - 5$
1	-4
3	4
5	20
7	44

To be an onto function the codomain has to be the same as the range.
Codomain [-4,4,20,44]

Q123:

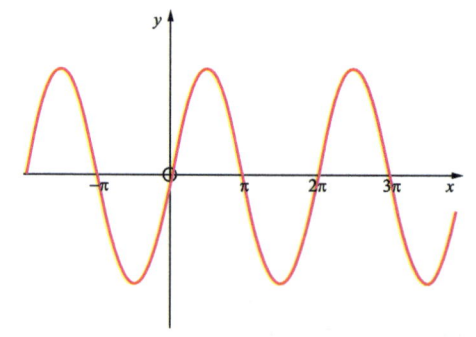

Look at the sin x function. To be one-to-one is $\left[-\frac{\pi}{2}, \frac{\pi}{2}\right]$

ANSWERS: UNIT 1 TOPIC 8

Q124:

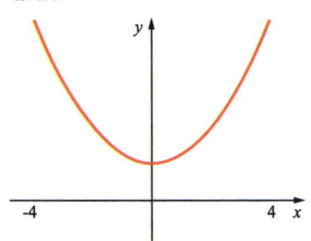

To be one-to-one $x \leqslant 0$

Q125:

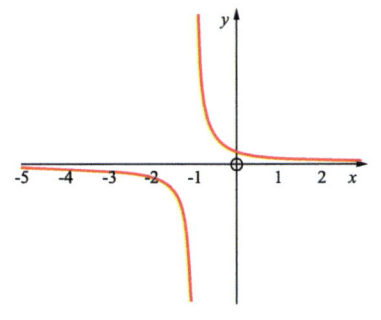

To be one-to-one $x \neq -1$

Q126:

Domain	Range: Evaluate $(x + 1)^2$
-10	81
-8	49
-6	25

Codomain $x \in \{25, 49, 81\}$

Q127:

$$y = 3x + 5$$
$$y - 5 = 3x$$
$$x = \frac{y - 5}{3}$$
$$y = \frac{x - 5}{3}$$

© HERIOT-WATT UNIVERSITY

Q128:

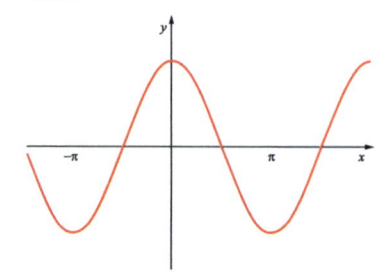

Looking at the cosine graph, to be one-to-one the domain is $[0, \pi]$.

Q129:

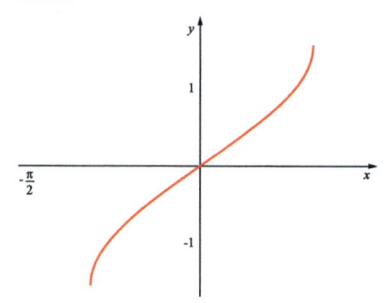

Looking at $y = \sin^{-1}\theta$ graph, to be one-to-one the domain is $\left[-\frac{\pi}{2}, \frac{\pi}{2}\right]$.

Q130:

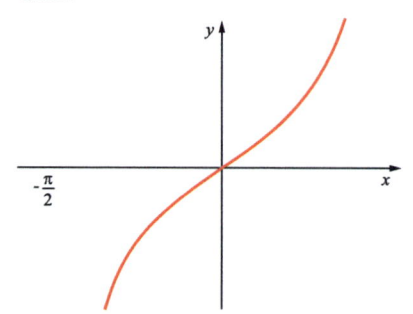

The range of $\tan^{-1}\theta$ is $[-\infty, \infty]$.

Q131: C

Q132: D

ANSWERS: UNIT 1 TOPIC 8

Q133:

$f(x) = x^3 \cos x$
$-f(x) = -x^3 \cos x$
$f(-x) = (-x)^3 \cos(-x)$
$ = -x^3 \cos(x)$
$-f(x) = f(-x)$ therefore it is odd.

Q134:

$f(x) = \dfrac{1}{x} \sin x$
$-f(x) = -\dfrac{1}{x} \sin x$
$f(-x) = \dfrac{1}{(-x)} \sin(-x)$
$ = \dfrac{1}{x} \sin x$
$f(x) = f(-x)$ therefore it is even.

Q135:

$f(x) = x^2 e^x$
$-f(x) = -x^2 e^x$
$f(-x) = (-x)^2 e^{(-x)}$
$ = x^2 e^{-x}$
$f(-x) \neq -f(x)$ and $f(-x) \neq f(-x)$ so this function is neither even nor odd.

Q136:

$f(x) = \dfrac{x^3 + 3}{x}$
$-f(x) = -\dfrac{x^3 + 3}{x}$
$f(-x) = \dfrac{(-x)^3 + 3}{(-x)}$
$ = \dfrac{-x^3 + 3}{-x}$
$ = \dfrac{x^3 - 3}{x}$
$f(-x) \neq -f(x)$ and $f(-x) \neq f(-x)$ so this function is neither even nor odd.

Q137:

$f(x) = \dfrac{x^2 - 2x + 1}{x^2 + 3x + 2}$
$-f(x) = -\dfrac{x^2 - 2x + 1}{x^2 + 3x + 2}$
$f(-x) = \dfrac{(-x)^2 - 2(-x) + 1}{(-x)^2 + 3(-x) + 2}$
$ = \dfrac{x^2 + 2x + 1}{x^2 - 3x + 2}$

© HERIOT-WATT UNIVERSITY

$f(-x) \neq -f(x)$ and $f(-x) \neq f(-x)$ so this function is neither even nor odd.

Q138:

a)

Vertical asymptotes occur when $x - 3 = 0$, so vertical asymptote at $x = 3$.

b)

```
                    3x    +    11
        _____
x - 3 | 3x²   +   2x   +   1
        3x²   -   9x
        _____
                  11x   +   1
                  11x   -   33
                  _____
                            34
```

$\frac{3x^2+2x+11}{x-3} = 3x + 11 + \frac{34}{x-3}$

Slant asymptote is $y = 3x + 11$

Q139:

First derivative

$f'(x) = 6x^2 + 24x + 9$

Second derivative

$f''(x) = 12x + 24$

Non-horizontal point of inflection

Non-horizontal point of inflection $f''(x) = 0$ or is undefined.

Solve
$12x + 24 = 0$
$12(x + 2) = 0$
$x = -2$

Possible non-horizontal point of inflection at $x = -2$.

Check concavity:

x	→	-2	→
$f''(x)$	-	0	+
concavity	down		up

There is a change in concavity, so there is a non-horizontal point of inflection at $x = -2$.

Q140:

For a non-horizontal point of inflection to occur $f'(x) \neq 0$ and $f''(x) = 0$ or is undefined at x.

First derivative

$f'(x) = -2\sin\left(x + \frac{\pi}{2}\right)$

Check for $x = 0$:

$f'(0) = -2\sin\left(0 + \frac{\pi}{2}\right)$
$= 2\sin\left(\frac{\pi}{2}\right)$
$= 2$

$f'(x) \neq 0$ therefore it is not a stationary point.

Second derivative

$f''(x) = -2\cos\left(x + \frac{\pi}{2}\right)$
Check for $x = 0$:
$f''(0) = -2\cos\left(0 + \frac{\pi}{2}\right)$
$= -2\cos\left(\frac{\pi}{2}\right)$
$= 0$

$f''(0) = 0$ therefore there is a possibility of a non-horizontal point of inflection at $x = 0$.
Check concavity:

x	$\rightarrow$	0	$\rightarrow$
$f''(x)$	-	0	+
concavity	down		up

There is a change in concavity, so there is a non-horizontal point of inflection at $x = 0$.

Q141:

For a non-horizontal point of inflection to occur $f'(x) \neq 0$ and $f''(x) = 0$ or is undefined at x.

First derivative

$f'(x) = 3\cos\left(x - \frac{\pi}{4}\right)$
Check for $x = \frac{\pi}{4}$:
$f'(0) = 3\cos\left(\frac{\pi}{4} - \frac{\pi}{4}\right)$
$= 3\cos(0)$
$= 3$

$f'(x) \neq 0$ therefore it is not a stationary point.

Second derivative

$f''(x) = -3\sin\left(x - \frac{\pi}{4}\right)$
Check for $x = \frac{\pi}{4}$:
$f''(0) = -3\sin\left(\frac{\pi}{4} - \frac{\pi}{4}\right)$
$= -3\sin(0)$
$= 0$

$f''(x) = 0$ therefore there is a possibility of a non-horizontal point of inflection at $x = \frac{\pi}{4}$.
Check concavity:

x	$\rightarrow$	$\frac{\pi}{4}$	$\rightarrow$
$f''(x)$	+	0	-
concavity	up		down

© HERIOT-WATT UNIVERSITY

There is a change in concavity, so there is a non-horizontal point of inflection at $x = \frac{\pi}{4}$

Q142: B

Q143: C

Q144: C

Q145: C

Q146: B

ND - #0072 - 260623 - C0 - 210/148/35 - PB - 9781911057772 - Matt Lamination